The Mechanics of Biological Materials

This book introduces the mechanical principles governing the behaviour of a wide range of biological materials, which are materials produced by a biological system. The approach is systematic and based on one-dimensional (fibres), two-dimensional (membrane), and three-dimensional (bulk) biological materials. The essential mathematical tools are developed from first principles and applied to materials as diverse as spider silk, blood vessels, and bone. It offers a progressive introduction of mathematics and mechanics concepts and offers detailed solutions to numerous worked examples.

- Provides an essential gateway to access complex treatises in the important and demanding models governing the observed behaviour of biological materials.
- Teaches readers to exploit the wondrous properties found in nature in the development of biomimetic applications.
- Offers a progressive introduction of mathematics and mechanics concepts.
- Includes detailed solutions to numerous worked examples, case studies, and homework problems.

The text is aimed at students of chemistry, materials engineering, and biology who do not have a background in mechanics but wish to further their knowledge of the mechanics of biological materials.

Manuel Elices is a Foreign Member of the National Academy of Engineering, USA, Member of the Royal Spanish Academy of Sciences, Member of the Royal Spanish Academy of Engineering, and Emeritus Professor of Materials Science at the Polytechnical University of Madrid, Spain.

Gustavo Guinea is a Corresponding Member of the Royal Spanish Academy of Sciences, Professor of Materials Science, former Director of the Centre for Biomedical Technology at the Polytechnic University of Madrid, Spain, and Head of the Laboratory for Biomaterials and Regenerative Engineering at this Centre.

John Morton is a Fellow in Engineering Science at University College Oxford, UK.

The Mechanics of Biological Materials

by

Manuel Elices, Gustavo Guinea, and John Morton

CRC Press
Taylor & Francis Group
Boca Raton London New York

CRC Press is an imprint of the
Taylor & Francis Group, an **informa** business

Designed cover image: Shutterstock

First edition published 2026
by CRC Press
2385 NW Executive Center Drive, Suite 320, Boca Raton FL 33431

and by CRC Press
4 Park Square, Milton Park, Abingdon, Oxon, OX14 4RN

CRC Press is an imprint of Taylor & Francis Group, LLC

ISBN: 978-1-032-89277-1 (hbk)
ISBN: 978-1-032-95228-4 (pbk)
ISBN: 978-1-003-58379-0 (ebk)

DOI: 10.1201/9781003583790

Typeset in FranklinGothic-Book font
by KnowledgeWorks Global Ltd.

Publisher's Note

This book has been prepared from camera-ready copy provided by the authors.

https://routledge.com/9781032892771

CONTENTS

INTRODUCTION

This book is about the mechanical behaviour of biological materials. It is meant to help the reader to understand and apply the concepts to, the design of new materials and devices. To some extent, the book tries to unravel the mysteries of nature and make the millions of years of nature's own R&D a source of inspiration for the optimisation of man-made materials.

The book provides a bridge between the mechanical models and theories developed for conventional (synthetic) materials and structures, and the properties and behaviour of biological materials. As such, it is useful not only for science and engineering students who want to strengthen and broaden their mechanical training, but also for life science students who wish to delve deeper into the mechanisms and material properties of the constituents of living organisms.

It is worth clarifying that this book deals only with the mechanical properties of **Biological Materials**[1] and, therefore, is not a book that deals with **Biomaterials**, of which there are many excellent texts. This book is specifically directed at students of **Materials Science and Engineering**, **Chemistry and Physics of Materials**, **Biology** or **Biomedicine** who are interested in understanding more about the mechanical properties of the biological materials that nature has given us.

The level of the book is *intermediate*; it lies between books that focus on a qualitative description of biological materials and those that specialise in structural and continuum mechanics. This book is intended to be an aid to students with little or no mechanics training so that they can successfully assimilate articles that use the more elaborate time-dependent models[2] of material behaviour or continuum mechanics[3]. The book is constructed of three sections of increasing complexity. The first one addresses the mechanical properties of materials that are for all practical purposes one-dimensional, *fibres*. In the second section the focus is on materials which can be characterised by two dimensions: *membranes* and *laminas*. Finally, consideration is given to materials for which all three dimensions must be taken into account, in order to determine their mechanical behaviour. In each section detailed illustrative examples of biological materials are presented, such as cotton and silk fibres, blood vessels, bone or wood, among others.

Another benefit of this book is that it contains descriptions of the more common test methods for the determination of the mechanical properties of biological materials. These descriptions are not limited to how to measure the material properties[4] but also how to interpret and evaluate the experimental results.

Finally, the book contains 81 exercises which are solved in detail, in order to facilitate understanding of the mechanics of biological materials and to assist the application of the concepts and techniques introduced in the text.

[1] The notion of a biological material has evolved over time. A **Biological Material** is anything which has a biological origin. However, members of the European Society for Biomaterials decided, in September 1991, to reserve the word **Biomaterial** exclusively for those "materials used to correct or replace tissue or functions in the human body". Under this definition, titanium can be a *biomaterial*. However, it is not a *biological material*.

[2] Viscoelastic or viscoplastic models, for example.

[3] Green-Lagrange and Almansi strain tensors, or Cauchy and Piola-Kirchhoff stress tensors, for example.

[4] ASTM or ISO standards.

SECTION 1

BIOLOGICAL FIBRES

1 CHAPTER Biological Fibres. General Concepts.

1.1. INTRODUCTION

The vast majority of biological materials, and in particular all those with a structural (load bearing) function, are formed of fibres because, either on their own or as reinforcement in an organic or inorganic matrix, they constitute unidirectional elements, 1D (tendons, for example); two-dimensional, 2D (membranes, such as the pericardium); or three-dimensional, 3D (such as bone), with a variety of properties not yet achievable in artificial materials.

In this section the mechanical properties of biological materials which are essentially unidirectional (along the length of the fibres) are analysed. This assumption simplifies the mathematical treatment not only for the interpretation of the tests performed on the materials but also the modelling of their behaviour.

Fibres produced by animals or plants are much more complex than artificial fibres. This is hardly surprising since biological fibres are often multifunctional and possess hierachical structures. Plant fibres such as cotton, linen, hemp or wood fibres are **polysaccharides,** basically cellulose[1] materials, while animal-produced fibres, such as wool, silk and tendon are **polypeptides** amino acid polymers.

[1] In 1838, Anselme Payen suggested that the walls of many plant cells were formed of the same substance, which he called cellulose. Cellulose is a linear polymer of linked D-glucose units.

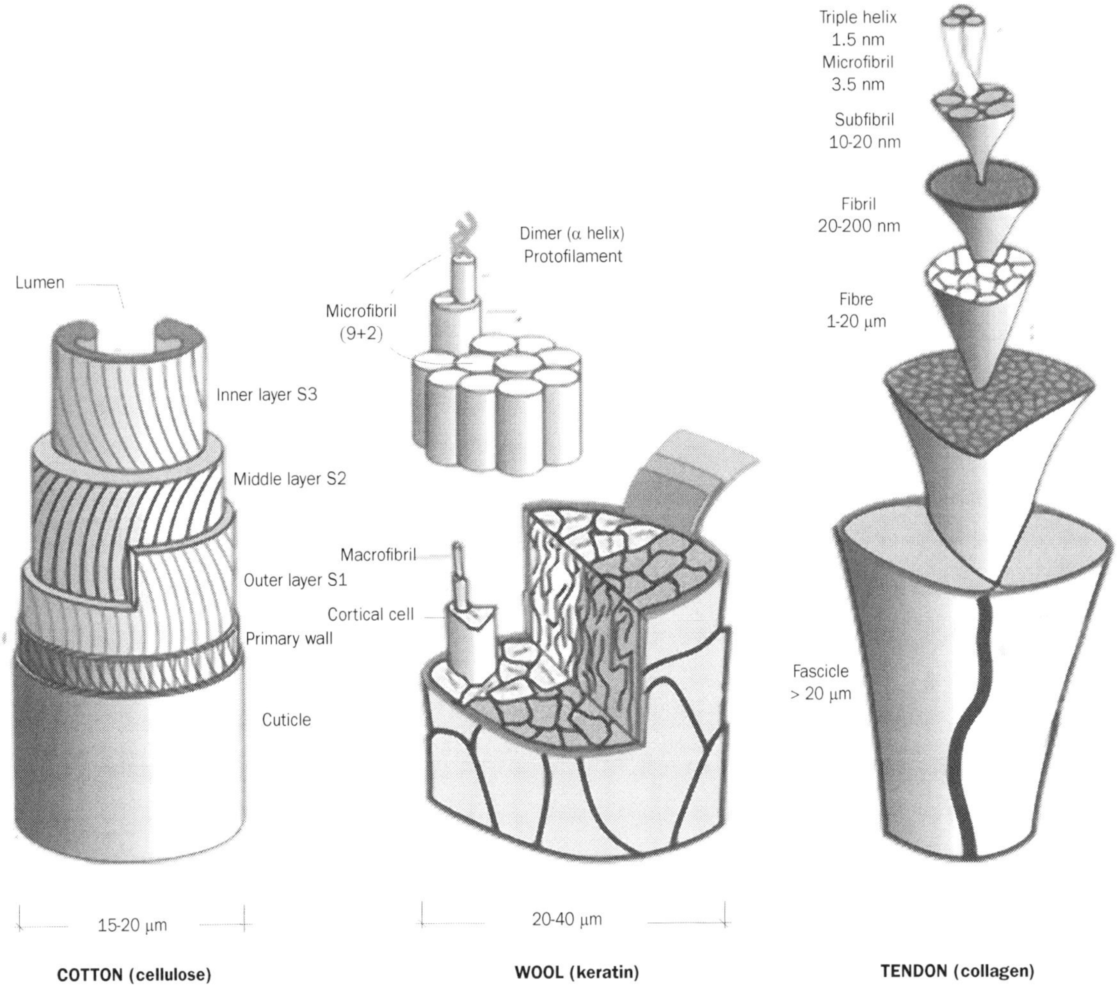

Fig. 1.1. *Hierarchical structure of several biological fibres.*

The hierarchical structure of various fibres: cotton, wool and tendon are shown in figure 1.1. In biological materials, hierachical structures can encompass many orders of magnitude; in a muscle, from the diameter of a tropocollagen fibril (10^{-9} m) to the diameter of a tendon (10^{-2} m) or, in the case of a tree, from the diameter of a cellulose fibre (10^{-9} m) to the diameter of a giant sequoya (10 m). Stiffness, strength and other mechanical properties are controlled and optimised by the interaction between the various hierachical structures, to such an extent that it is very often difficult to distinguish between the material and the structure, which hugely complicates the development of models for the behaviour (constitutive equations) of these fibres.

Fibres from the **animal kingdom** can be considered, in a very simplified form, as being composed of proteins –aligned to the axis of the fibre– embedded in a proteic matrix. *Keratin* fibres are the principal component of skin, wool and bird feathers. Keratin is rich in cysteine –an amino acid containing sulphur– which produces the characteristic smell when these fibres are burned. Structures based on keratin can be very strong, such as rhinoceros horn, and very flexible, such as human hair. Silk fibres produced by silkworms and spiders also contribute structurally; they form the cocoons of the silkworms and the threads of spider webs, whose tenacity has not been surpassed by artificial fibres. *Collagen* fibres are the most abundant structural component of the extracellular matrix of vertebrates and invertebrates and the principal ingredient of tendons and ligaments.

Fibres produced from **plants** have the advantages of low cost, biodegradablity, ready availablity, low density and reasonable mechanical properties. In plants they perform several functions, but principally they play a structural role –providing support to stems and leaves– which allows them to carry out their own functions and to withstand inclement weather. Another characteristic of vegetable fibres is their capacity to absorb moisture, since cellulose –which is hygroscopic– is a major component. This absorption changes their weight, dimensions and mechanical properties, as will be commented upon later.

Biological fibres have many industrial applications, as well as being important in the textile sector. They are environmentally friendly, renewable and have sufficiently good mechanical properties for many applications –resistant to impact and low density– which make them useful, for example, in the automotive sector (*Mercedes-Benz* E Class cars use them in several components) and in geotextiles, to stabilise floors and banks. Silkworm fibres, whose principal destiny is the textile industry, are used in biomedicine for sutures and, recently, to make scaffolds in tissue engineering. Spider silk fibres are opening promising applications due to their extraordinary tenacity. Collagen fibres have numerous uses in medicine, bioengineering and in the cosmetics and food industries. Their excellent mechanical and biological properties place them at the forefront of future biomaterials.

The aim of this book is to describe the mechanical properties of biological fibres, and their possible applications for structural purposes.

In the first chapter, a short summary of the principal natural fibres –polymers, both of vegetable and animal origin– is presented.

In order to characterise the mechanical behaviour of fibres, it is essential to test them, record the results and to know how to interpret them. The second chapter is dedicated to that end.

First, the most common mechanical tests are described in detail: tensile and torsion tests, the former being the more frequently employed.

The interpretation of the results is not without difficulty with these biological materials, because they can undergo large deformations, and the response to a stimulus can depend upon time and the load history. This behaviour falls within the field of visco-elasto-plasticity whose mathematical treatment is quite complex.

In order to facilitate the interpretation of the experimental results, to be able to model the material and make reliable predictions, studies have been centred around several elemental behaviours: **Hookean elastic, elastomeric,** and **linear viscoelastic.** It must be remembered that the conclusions derived from these studies are only valid within the bounds of the assumptions made previously.

In the third chapter, four examples of fibres are presented; one of vegetable origin (cotton) and three of animal origin (wool, silk and collagen). In all of these, a brief description of the structure, mechanical properties and possible applications are given.

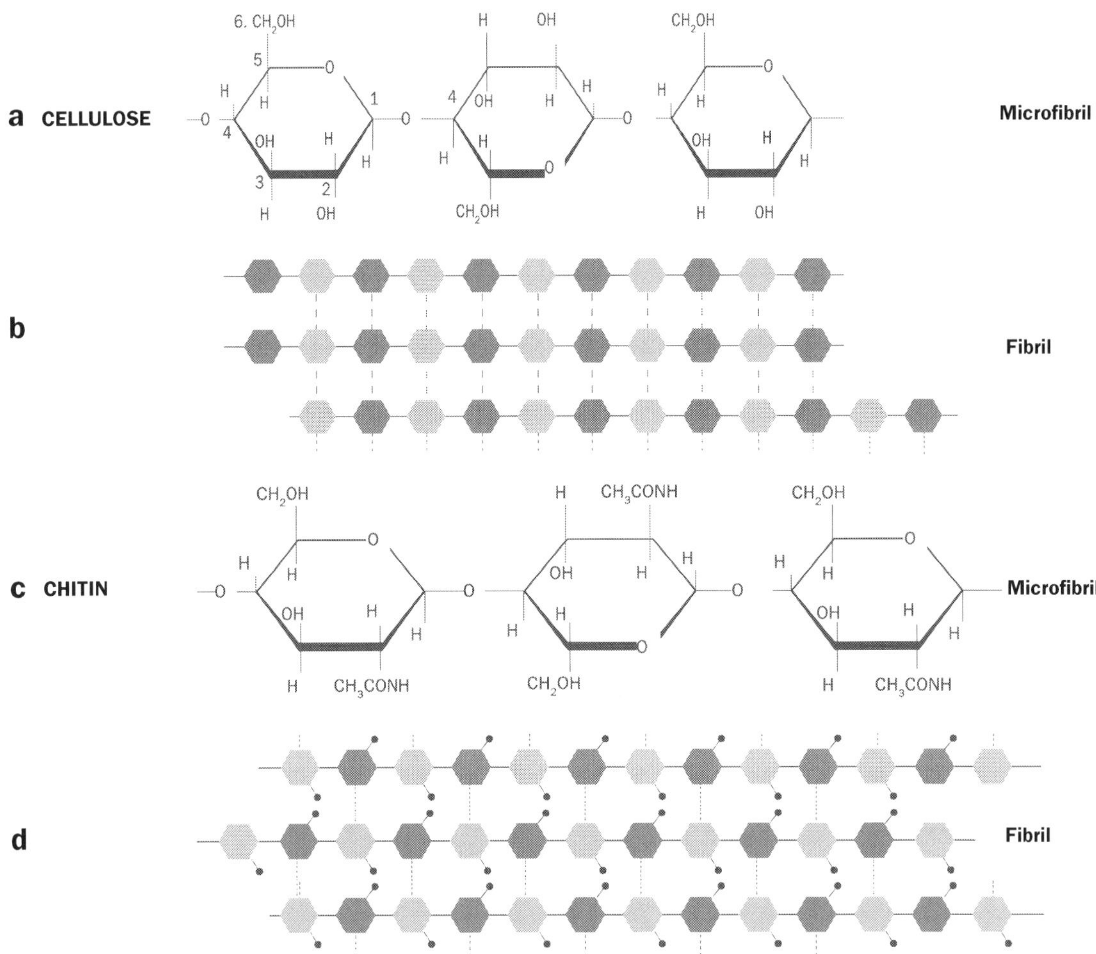

Fig. 1.2. *Structure of cellulose (a, b) and chitin (c, d) fibres.*

1.2. POLYSACCHARIDE FIBRES

Polysaccharides are natural polymers made up of a large number of monosaccharides (carbohydrates) connected by acetal links. The most common fibres are *cellulose* and *chitin,* both of which have similar primary structures, as shown in figure 1.2. Cellulose is formed of D- glucose monomers which condense through β (1→4) glycosidic bonds (the plane of the monomer rotates 180°) between 1 and 4 carbons. Cellulose is present in the cell walls of plants, in most algae and some classes of mushroom. Chitin is formed of N-acetylglucosamine units and is found in the cell walls of fungi, yeast and arthropods, as well as in the exoskeletons of the last mentioned.

1.2.1. Cellulose fibres

Cellulose is the most abundant naturally occurring polymer and is the structural component in plant cell walls.

The most stable form of cellulose is shown in figure 1.2a, a configuration that allows the formation of long straight chains. In this structure, all the hydroxyl groups (-OH) can form hydrogen bonds, both within and between molecules. This promotes the creation of highly stable supramolecular fibres with excellent mechanical properties.

The fibril elements of cellulose have a diameter of 3.5 nm and are about 30 nm long, typically formed of 40 or so molecules [Vincent 1990]. When ordered into groups, they form small monoclinic crystals, with parameters: a = 0.835 nm, b= 1.03 nm, c = 0.79 nm [Bledzki, Gassan 1999].

The fibrils are arranged in different forms around the axis of the fibre creating a hierarchical structure. In a mature cotton fibre, six distinct levels can be observed (figure 1.1a): the cuticle, the primary wall, the outer layer S1, the secondary wall or middle layer S2, the lumen wall or inner layer S3 and the sixth level is the lumen itself, a channel that runs along the centre of the fibre.

Cellulose fibres in general are used as reinforcement for composite materials (Bledzki, Gassan, 1999, for instance).

Cotton fibre contains more than 90% cellulose. Generally, cellulose is associated with other amorphous polymers of lignin (polyphenol) and hemicelluloses (polysaccharides of glucose, galactose and fructose). For example, wood contains between 40-55% cellulose, 15-35% lignin and 25-40% hemicellulose.

As a result of the variation in their specific components, the differences between vegetable fibres lies more in the micro- and macrostructure of the fibres than in their chemical composition.

1.2.2. Chitin fibres

After cellulose, chitin is the most abundant naturally occurring polymer. It forms part of the cell walls of mushrooms and yeast, mollusc shells and the exoskeletons of arthropods and other invertebrates. Chitin takes its name from the Greek *chitón* (meaning tunic or robe).

The microfibril elements of chitin have a diameter of about 3 nm and are typically formed of 20 or so molecules reaching a length of up to 300 nm [Vincent 1990]. The dominant acetic group ($-CO-CH_3$) promotes the formation of a large number of hydrogen bonds between the molecules which is difficult in cellulose, because of its size, and relative mobilty of the fibres. For these reasons, the stiffness of chitin is greater than that of cellulose.

1.3. POLYPEPTIDE FIBRES

Polypeptides are amino acids (figure 1.3a). The joints between the amino acids are achieved through peptide links in which the nitrogen of an amino acid is connected to the carbonyl group (–CO–) of an amino acid and releasing water (figure 1.3b). Most natural polypeptides are a syntesis of the 20 naturally occurring amino acids, and which are given in the Appendix. The difference between them lies in the R (residue) group, linked to the central carbon of the amino acid, or carbon α (C^{α}) whose position is shown in figure 1.3a. A typical protein can contain about 500 amino acids, although some can have as many as 6000.

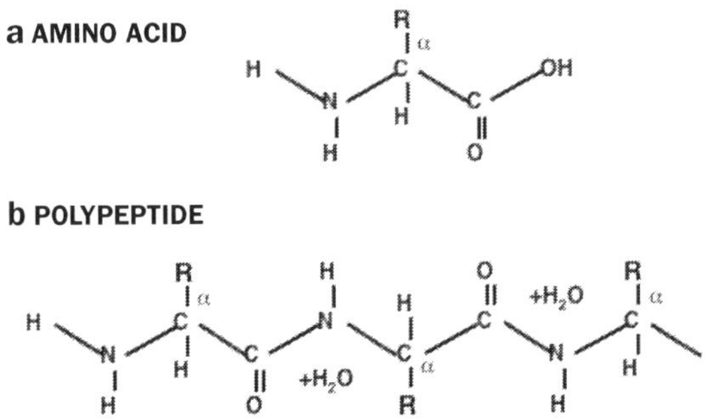

Fig. 1.3. *a. The basic structure of amino acid. R (the residue) represents the side group that characterises each amino acid. b. Polymerisation (by condensation) of three amino acids.*

There are several families of protein fibres of significant structural relevance: *keratins*, proteins that contain sulphur and are the principal component of hair and wool, silks, produced by silkworms and spiders; *collagen* fibres that constitute tissue in most animals; and *actin* and *myosin* fibres that form muscles. Other important protein fibres are: elastin, usually associated with collagen; *abductin* that occurs in the ligament of the hinge of bivalves; and *resilin* that is found in some insects and arthopds.

1.3.1. Keratin fibres

There are many types of keratin fibres, all of which are characterised by their high cysteine content, amino acid containing sulphur (–SH) and capable of forming disulphide bridges (–S– S–) with other molecules. The sulphur present gives a characteristic odour in burnt fibres. One way of classifying them is in terms of the type of the animal in which they are found; keratin in mammals (hair, wool...), in birds (feathers) and in other animals (reptiles, among others). A wool fibre has an almost circular cross-section with a diameter ranging from 20 to 40 microns and length between 5 and 50 cm (figure 1.3b). Human hair has a diameter of about 70 microns.

The structure of wool fibre is very complicated. The smallest unit is the *α-helix* that forms the dimer (figure 1.3b). The second hierarchical level is the *protofibril* or *protofilament*, formed by two double helices. The next level of complexity is the microfibril, formed of various protofibrils. The fibres are formed by the superposition of *microfibrils*, reinforced by hydrogen bonds and sulphur bridges, giving way to *macrofibrils* that, in turn, are grouped into the cortical cell.

The properties of the fibres depend upon the nature and number of crosslinks between fibres and are very sensitive to the ambient humidity. Indeed ancient weather sensors were based upon changes in the flexibilty of hair with humidity.

1.3.2. Silk fibres

Silk fibres are produced in particular glands of some creatures. They are of great interest owing to their biosynthesis, processing and functional properties. Silks have different compositions, structures and properties depending upon the animal and glands that produce them. The silks that have been most studied are those of the silkworm *(Bombyx mori)* and some spiders *(Nephila, Araneus* and *Argiope).*

The silkworm cocoon thread –whose diameters range from 10 to 25 microns– contain structural proteins; the heavy fibroin chain (about 325 kDa) and lighter chain (about 25 kDa), with a sericin protein system (between 20 and 310 kDa) that surrounds the fibroins and behaves like a glue between the different threads. The sericins are more hydrophilic than the fibroins, which is why they are more easily eliminated when the cocoon is boiled in an alkaline solution, a step known as degumming, and which allows later spinning of the fibroin filaments.

Orb-weaver spiders produce silk thread in different glands in the their abdomens, according to its purpose; mooring and security threads, circular web threads, threads to wrap the prey, cocoons, etc. In general, the diameter of these fibres –between 2 and 5 microns– is less than that of silkworms. In particular, the security threads of spiders contain two proteins: MASp1 (for Major Ampullate Silk proteins, spidroin 1) with a molcular mass of 275 kDa, and the other, called MASp2, with a molcular mass and composition similar to the former.

The most common amino acids in webs are alanine (A) and glycine (G) that constitute 70% of the total, followed by tyrosine (Y), leucine (L), glutamine (Q), proline (P) and valine (V). These polypeptides are characterised by highly repetitive sequencing (AAA..., GGXGGX...) that makes them difficult to manipulate biochemically.

Despite being a material of great scientific and technological, interest, there is still little that is known about the structure and organisation of silk fibre.

1.3.3. Collagen fibres

Collagen is the most important protein fibre in the animal kingdom; it is the most common structural component in the extracellular matrix of vertebrates, invertebrates and sponges, and the principal component of tendons and ligaments (tendon connects the muscle to the bone and ligament joins one bone to another). Collagen is a generic term used to describe a variety of polypeptides composed essentially of glycine, proline and hydroxyproline, having the triple helix in common as an elementary component as shown in figure 1.4c). To date, more than 20 distinct families of collagen have been identified (with each type denoted by a Roman numeral). Each family is distributed among particular tissues, having a specific function and characteristic structure with a different size of the triple helix domain.

Types of collagen I, II, III and IV are the most abundant (Kadler 1994). Type I collagen is found in a tendons, ligaments, bones and in the cornea, representing approximately 90% of the collagen in the body. Type II collagen forms the fibrils of cartilage and the notochord. Type III is present in the fibrils of blood vessels and skin. Type IV collagen appears in the basement membrane of the epithelium, forming a sort of molecular filter with protection and absorption functions.

The mechanical properties of collagen depend upon the diameter of the fibrils, the nature and number of cross-links between molecules, the structural hierarchy of the fibrils and the interaction with other components of the matrix in which they are embedded. In tendon, the fibril diameter varies widely –between 50 and 500 nm– and is connected by numerous cross-links and bundled to form fibres of diameter between 1 and 20 microns. The fibres, in turn, are bundled to form fascicles, as shown in figure 1.4c. This highly aligned structure is responsible for the excellent mechanical properties of tendons.

1.3.4. Actine and myosine fibres

Actine and myosine are the principal protein fibres in muscles. These stretch and contract thanks to a gliding and gripping mechanism between the actine and myosine fibres, as shown schematically in figure 1.4. The thicker myosine possesses heads that join and unlock the thinner actine fibres as it moves, providing a relative displacement mechanism –similar to a rowing action– the origin of contraction and distension of muscle fibre.[2]

The structure of muscle is hierarchical, as shown in figure 1.4. Actine and myosine fibres form the basic unit called the *sarcomere*. These, in turn, are organised into microfibrils, and bundles of microfibrils to form muscle fibres (figure 1.4). Examination of the muscle, reveals characteristic striations, called Z-discs or Z-lines, that are the junctions between the sarcomeres. The actine filaments are fixed to the Z-discs, while the myosine can glide parallel to the actine thanks to the mechanism outlined above. The sarcomeres of vertebrates are about 1.5 microns long when fully contracted but can reach double that size when stretched.

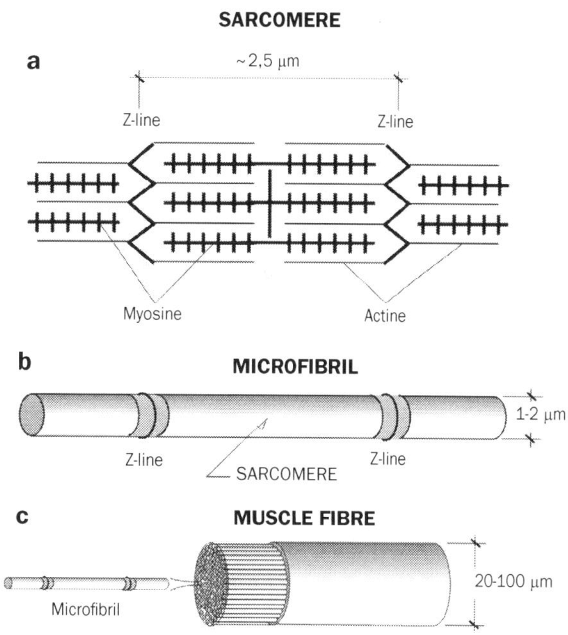

Fig. 1.4. *Structure of a muscle fibre.*
a. Sarcomere.
b. Microfibril.
c. Fibre.

2 https://www.nature.com/articles/nature09450

1.3.5. Elastin, resilin and abductin

These fibres exhibit an elastomeric behaviour, similar to rubber and other elastic materials (the *elastomeric* concept will be expanded upon later in the study of the mechanical properties of materials).

Elastin is a very stable insoluble protein rich in valine, glycine, alanine and proline with VPVGP being the most recurrent motif. It is present only in vertebrates, in elastic organs such as the walls of arteries and veins, in skin, and in the lungs. The ligament in the nape of ungulates is an organ rich in elastin; a kind of rope that runs along the upper part of the neck and which is always in tension, allowing the animals to keep their heads raised without the constant aid of their muscles.

Resilin is found in and is used to store and release elastic energy; insects use it in the elastic joints of their wings, and grasshoppers at the base of their hind legs, as a catapult when jumping. The reason for this is the recurrent GGRPSDSYGAPGGGN sequence.

Abductin occurs in the internal ligaments of the shells of bivalve molluscs, in which they act in opposition to the adductor muscles in the mollusc. The reason for this is the repeated FGGMGGGNAG sequence.

Abductin, like resilin, recovers elastic energy stored during deformation when the fibre is unloaded. Thanks to this phenomenon very little energy is dissipated, principally taking the form of heat in each unloading cycle. This permits not only great efficiency in flight, or in jumping, but it also avoids an overheating that could prove fatal.

1.4. CONCLUDING REMARKS

Following this brief introduction it might be thought that disentangling the relationships between the **microstructure** and **mechanical behaviour** of biological fibres is no easy task. The structure of natural fibres is complex because it is so hierarchical and when attempts are made to relate it to the mechanical response it is not easy to identify the most relevant aspects. On the other hand, the mechanical behaviour is also complex because these materials often sustain large deformations, which means that, in many cases, the theory of linear elasticity cannot be applied. Moreover, the deformations can depend upon time, temperature and humidity, among other parameters, making analysis even more difficult.

However, it is possible to add some order to these ideas and to propose simple, but realistic, models that allow an understanding of these materials and, in some cases, make predictions about their response to external mechanical stimuli. In general, for amorphous polymers, the most important variables that must be considered are the **duration of the stimulus** and **temperature.** When these materials are subjected to very fast actions, or tested at low temperatures, they demonstrate a rigid response, similar to a conventional solid. In this situation the molecular chains move with difficulty and the structure of the material remains essentially unchanged. The material deforms while maintaining the relative positions of its components and, as will be seen later, the change in internal energy due to the deformation governs its response. This situation is termed the glassy state and is usually studied –for small reversible deformations– applying models of **elastic behaviour.**

At high temperatures or when the tests are slow enough to allow molecular movement, the chains reorganise themselves during loading, producing a complex time-dependent response, which results in *hysteresis* phenomena (that is to say, the loading response is different from that which is observed during unloading), *creep* (the deformation increases with temperature although the load is held constant) and *relaxation* (material subjected to constant deformation sustains a gradually reducing force). When the deformations are not too great and can be considered reversible these phenomena can be approximated to **viscoelastic behaviour.**

If due to the temperature or other conditions (such as the presence of solvents that weaken the molecular interactions) the mobility of the chains is greatly facilitated or, also, if the load is applied in experiments over times much longer than that needed for molecular reorganisation, these materials demonstrate elastic behaviour (that is to say, there is always a one-to-one correspondence between the load and applied deformation, and the material recovers its original dimensions on unloading) although this may be nonlinear on account of the large deformations sustained. In these circumstances the material is said to exhibit **elastomeric behaviour,** with a change in entropy (that is, its internal organisation) being principally responsible for this response.

The simplest representation of these concepts is shown in figure 1.5. The mechanical property of the material is represented on the vertical axis (in this case, the elastic modulus), and temperature or the duration of the test is on the horizontal axis. Note that the vertical scale is logarithmic, as is the time scale. T_g is the glass transition temperature and T_m is the melting point.

The three types of behaviour *–elastic, elastomeric* and *viscoelastic–* will be analysed in the next chapter.

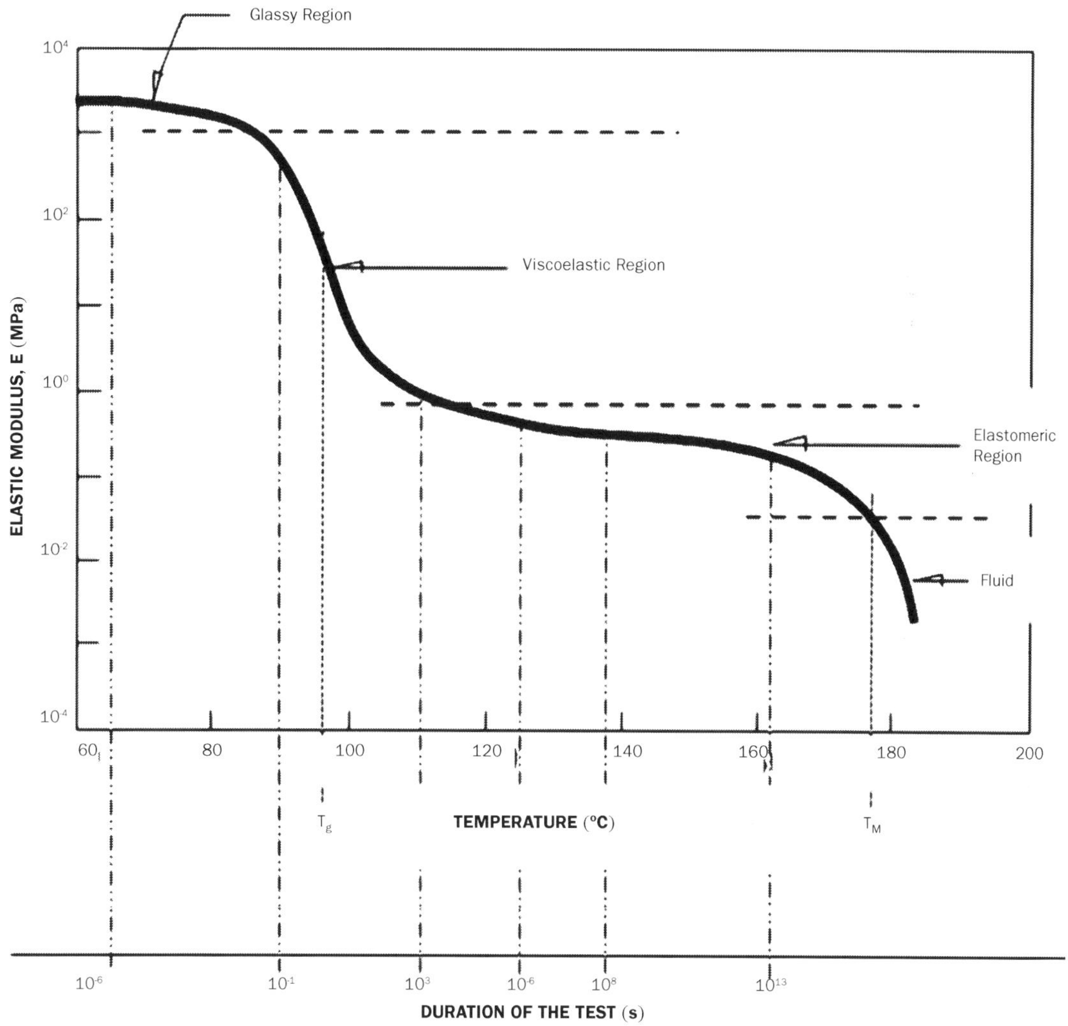

Fig. 1.5. *Schematic representation of the behaviour of the elastic modulus of a fibre as a function of temperature or time.*

As far as the microscruture is concerned, it is also difficult to give general guidelines, although at the most basic level all the materials have some common aspects: they are long polymer chains with sequences that are repeated many times (known as motifs), as shown in Table 1.1. These chains can adopt particular configurations, depending upon the temperature, the external loading and particular composition (not only their monomers and their associated "motifs" –see the table– but, also, their side chains). Thanks to the repetition of the established sequences, part of these long chains can organise themselves into small ordered structures (called crystallites, or microcrystals). In this case it is possible to model the behaviour of these polymers as composites of microcrystals embedded in an amorphous matrix; little crystalline tablets joined by fringes that come from their edges, as shown in figure 1.6. The simple model of a composite material –microcrystals embedded in an amorphous matrix– allows, as will be seen later, prediction of the mechanical behaviour of many biological materials.

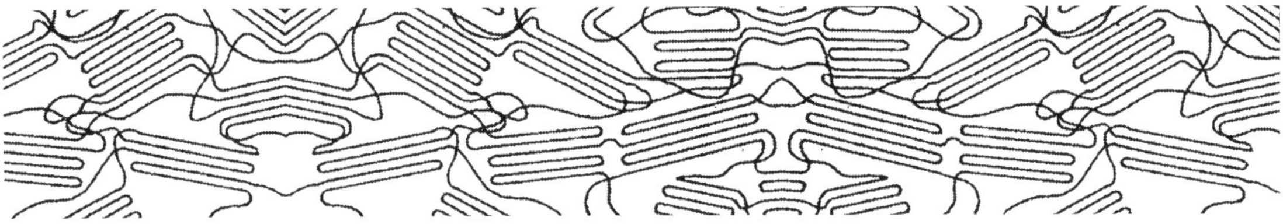

Fig. 1.6. *Model of fibre, idealised as a composited material of micro- (or nano) crystals embedded in an amorphous matrix. Fibre crystalinity is related with the percentage of crystal zones.*

TABLE 1.1 MOTIFS AND SECONDARY STRUCTURES IN SOME POLYPEPTIDES

POLYPEPTIDE	MOTIF	SECONDARY STRUCTURE
Keratin	$(abbabbb)_n$*	β Sheets, Coils
Silkworm silk	$(GAGAGS)_n$	β Sheets
Spider silk	A_n	β Sheets
	$(GXG)_n$	Coils
Collagen	$(GXP)_n$	Coils
Elastin	$(VPGVG)_n$	–
Resilin	GGRPSDSYGAPGGGN	–
Abductin	FGGMGGGNAG	–

* a = nonpolar residue
b = polar residues
A = alanine
D = aspartic acid
G = glycine
M = methionine
N = asparagine
R = arginine
S = serine
P = proline
V = valine
Y = tyrosine

1.5. APPENDIX

Abbreviations for the 20 most common amino acids.

AMINO ACID	ABBREVIATION (3 letters)	ABBREVIATION (1 letter)
Alanine	Ala	A
Arginine	Arg	R
Asparagine	Asn	N
Aspartic acid	Asp	D
Cisteine	Cys	C
Glutamine	Gln	Q
Glutamic acid	Glu	E
Glycine	Gly	G
Histidine	His	H
Isoleucine	Ile	I
Leucine	Leu	L
Lisine	Lys	K
Methionine	Met	M
Phenylalanine	Phe	F
Proline	Pro	P
Serine	Ser	S
Treonine	Thr	T
Tryptofan	Trp	W
Tyrosine	Tyr	Y
Valine	Val	V

2 CHAPTER Mechanical Behaviour of Fibres

2.1. INTRODUCTION

In order to study the mechanical properties of a material experiments must be performed, in which predetermined forces and/or deformations are applied to a body and to measure the resulting deformation and/or forces. In the case of fibres, in which one of the geometrical dimensions –the length– is much greater than the other two –the cross-section–, the most useful experiment is the *tensile* test, so the greater part of the following section will be dedicated to it.

The information obtained from a tensile test allows, in most cases, the determination of the one-dimensional constitutive equation for the material. Constitutive equations relate the applied forces on a material to the resulting deformations. In general, the information derived from a tensile test is macroscopic, that is; it allows for example prediction of the deformations if the forces are known, or vice versa, but it does not provide information about the microscopic menchanisms that give rise to the phenomenon and that are usually multiple and act simultaneously. For example a tensile test on a tendon yields a relationship between the applied force and the extension of the tendon. However, if the intimate deformation mechanisms in the collagen fibres, responsible for the tendon extension, are required, complementary experiments and hypotheses about the composition and structure are needed.

From tensile tests on materials it is apparent that there are three broad classes of behaviour, according to the type of deformation produced.

i) When the deformation produced is immediate, a function only of the applied force and not how it is applied, the material undergoes **elastic behaviour**. The relation between the force (F) and deformation (Δ) is one-to-one (see figure 2.1), and the constitutive equation of the material is represented by a function $F=f(\Delta)$, which can be inverted to yield $\Delta = f^{-1}(F)$.

Hookean[3] *elastic* or *linear elastc behaviour* is simplest type of elastic behaviour. In this case the relation between the force and deformation is linear: $F = k \cdot \Delta$, where k is a constant. The mechanical response is determined, to a great extent, by the internal energy stored or released when the material deforms.

If the relation between the force and deformation is not linear, the behaviour is termed *hyperelastic* or *nolinear elastic*. A case of particular importance is that of *elastomeric behaviour* which is seen in some materias such as rubber. In these materials the mechanical response is principally due to a change in the configurational entropy of the molecules, associated with its conformation in chains, as will be discussed later.

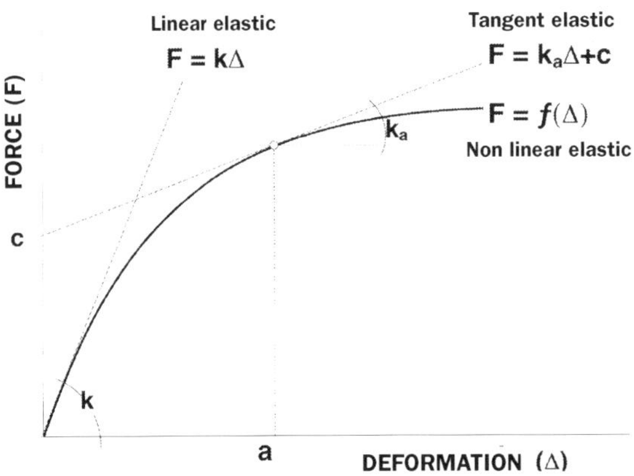

Fig. 2.1. *Nonlinear behaviour and linear approximations (tangents).*

Since any elastic constitutive equation $F=f(\Delta)$ can be approximated at the origin ($\Delta \approx 0$) by its tangent line, the initial behaviour (= for small deformations) of any elastic material can be taken as linear elastic. Equally, for small increments of deformation from a particular point Δ_a, the elastic constitutive equation can be approximated by the tangent at $F=f(\Delta)=-k_a \cdot \Delta + C$ at this point, and once more the behaviour can be considered as locally linear elastic.

ii) When the deformation produced is instantaneous, but the value of which *depends on the loading sequence*, for example, because there are loading and unloading processes the *behaviour* is termed **plastic**. In this case there is no fixed relationship between the force and the deformation, it being necessary to know the complete loading history suffered by the material in order to determine the final deformation. The material generally displays permanent deformation (or plasticity) once the applied force is removed since the mechanical response is governed by internal irreversible dissipative processes.

iii) When the deformation of a material subjected to a constant force varies with time this is **time dependent** or *deferred mechanical* behaviour. This is, without doubt, the most complex situation and, unfortunately, the most frequently encountered in the study of biological materials.

One situation of great interest, and something simpler, is when under a constant applied force the deformation stabilses over a long period of time (theoretically an infinite time), and its final value Δ_∞ depends only upon the applied load, in a manner similar to the elastic case, through biunivocal relationship $F=f(\Delta_\infty)$. In this case the *behaviour* of the material is *viscoelastic* and the mechanical response is associated with changes in both internal energy and entropy.

The behaviours described in ii) and iii) are called *anelastic* or *inelastic*, in contrast to the elastic behaviour shown in i) because there is no one-to-one relationship between forces and deformations. In this section the most important behaviours in the analysis of biological fibres such as **Hookean elastic, elastomeric** and **linear viscoelastic** will be considered.

In some experiments the fibre may fracture. The fracture process is complex and depends as much on the fibre as on the applied loading. It is usual to distinguish between two types of fracture: *brittle*, characterised by the absence of permanent deformation after the specimen has broken, and *ductile*, when the fibre suffers large plastic deformation during the fracture process. However, the boundaries between both types of behaviour are not welldefined.

When the fibres are brittle, or slightly ductile, the fracture behaviour can be approximated by means of a probabilistic fracture model based on Weibull's theory of the weakest link. This model is helpful in accounting for the variation in strength values obtained experimentally.

In this section a brief introduction to Weibull's model is presented, but a more detailed study of the fracture of biological materials will be left until Section 3.

[3] Robert Hooke (1635-1703) was an English experimental scientist who in 1660 first enunciated the law of proportionality between the extension of an elastic body and the force on it (Hooke's Law).

2.2. MECHANICAL TESTS

The simplest and most frequently performed tests on fibres are tensile tests (also called uniaxial tension or simple tension tests) and torsion tests.

2.2.1. The tensile test

This consists of a sample of material in the form of a cylinder or flat strip (in this second case the length is the predominant dimension) which is subjected to an increasing traction, causing force F and displacement, $\Delta L = L - L_0$, with respect to a base measurement L_0, as indicated in figure 2.

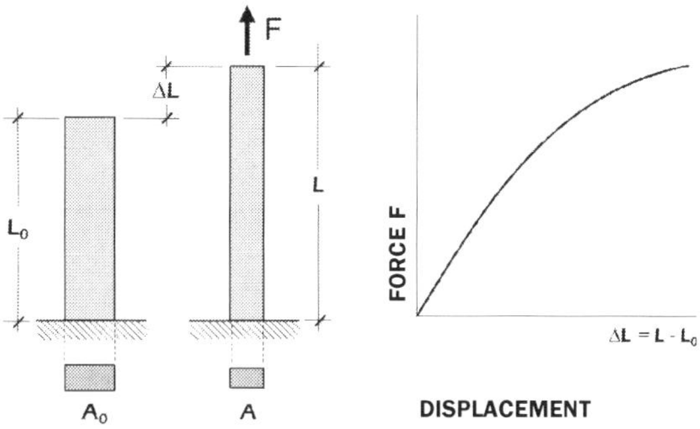

Fig. 2.2. *Specimen concept: L_0 and A_0, initial length and cross-sectional area. Schematic of the result from a tensile test: Force displacement curve.*

In general, the force F is measured as a function of the displacement ΔL, this being the variable that controls the process. Occasionally the test is performed by varying the force and measuring the corresponding displacement. The test may be influenced by the rate of change of the displacement with time (or of the force, if it is that which is controlled) and by external factors such as temperature, pressure, relative humidity or the medium surrounding the sample.

Even though, in the region of the grips, there is a complex stress state, if the specimen is sufficiently long –for practical purposes a length which is greater than five times the diameter is sufficient– the effects of the teeth decay rapidly. As a result it is possible to assume that the loading is uniform across the transverse section and that the deformation is uniform throughout the specimen (excluding the gripping regions). Furthermore, in what follows, it is assumed that the material is isotropic.

In order to study properties of materials which are independent of the geometry of the testpiece stresses and strains are used, which are values of the force and displacement normalised with respect to the specimen dimensions.

- Engineering or nominal stresses and strains

The **engineering stress** s is defined as:

$$s = \frac{F}{A_0} \quad (1)$$

where F is the force on the sample and A_0 is its initial cross-sectional area.

The **engineering strain** e is defined as:

$$e = \frac{\Delta L}{L_0} = \lambda - 1 \quad (2)$$

where $\Delta L = L - L_0$ is the change in length with respect to the base measurement L_0 and λ is the elongation, defined as the ratio of lengths $\lambda = L/L_0$.

As a consequence of these definitions, the force-displacement curves (F–ΔL) and engineering stress-engineering strain curves (s-e) are similar, and differ only in scaling factors, which are the constants A_0 and L_0.

- True stresses and strains

It is common for biological materials to undergo large deformations before fracture, with significant changes in the length and cross-section. At a given time, the initial values A_0 and L_0 can be quite different from the actual cross-sectional area and length of the specimen, so that it is usually preferable to employ the true stress and strain instead of engineering ones.

The **true stress** σ is defined as:

$$\sigma = \frac{F}{A} \quad (3)$$

where A is now the cross-sectional area of the specimen under the applied load F.

For an infinitesimal process in which length increases from L to $L+dL$ the **true strain** ε produced is defined as:

$$d\varepsilon = \frac{dL}{L} \quad (3bis)$$

When the base measurement L_0 has deformed as much as L, the total true strain will be, by integration of $d\varepsilon$:

$$\varepsilon = \int_{L_0}^{L} \frac{dL}{L} = \ln\frac{L}{L_0} = \ln\lambda \quad (4)$$

The relationship between the true and engineering strains is simple given that:

$$\varepsilon = \ln\frac{L}{L_0} = \ln\left(1+\frac{\Delta L}{L_0}\right) = \ln(1+e) \tag{6}$$

The relationship between the stresses can be readily deduced if *conservation of volume* of the specimen is assumed, that is, if:

$$A_0 L_0 = AL = \text{constant} \tag{5}$$

Although this assumption may appear limiting, it can be applied to almost all biological materials, since they may be considered incompressible for the levels of stress usually experienced by these materials given their high water content[4].

In these circumstances:

$$\sigma = \frac{F}{A} = \frac{F}{A_0}\frac{A_0}{A} = \frac{F}{A_0}\frac{L}{L_0} = s\lambda = s(1+e) \tag{7}$$

The sign convention for stresses and strains is the same; tension is considered as positive and compression negative.

The results of tensile tests on two different materials are shown in figure 3. The s-e curve represents engineering values which, as mentioned above, are proportional to the force F and displacement ΔL. The curve in figure 3a has a maximum that coincides with the beginning of an instability in the test (a phenomenon known as *necking* and which will be discussed below). The $\sigma - \varepsilon$ curve corresponds to values that are called true. When volume is conserved these values can be calculated from equations (6) and (7) up to the point of instability (maximum of the s-e curve). Beyond this point the true stress varies throughout the specimen, since the cross-section is not uniform due to necking. The curve in figure 3b does not have the instability mentioned above and true values can be calulated for the entire experimental curve from equations (6) and (7), if volume is conserved during the test.

Interpretation of the curves shown in figure 3 is not trivial, in particular at large strains. For small deformations the behaviour is usually elastic; if the specimen is unloaded, the curve returns to zero and, moreover, it follows the same path as on loading.

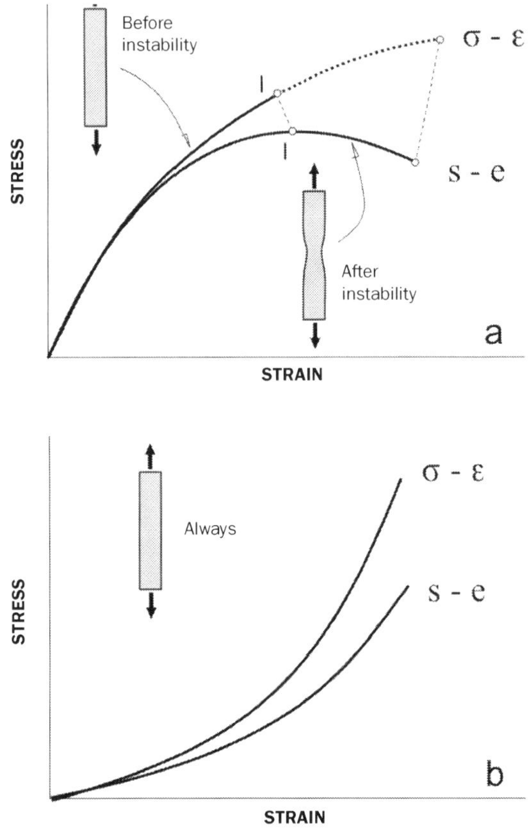

Fig. 2.3. *Stress strain curves (iengineering and true).*
a. Material with a response that exhibits necking (point I).
b. Material with a response that does not show necking.

When the strains are large –a common situation in biological materials– unloading does not usually follow the loading path and a new phenomenon called hysteresis occurs again and will be discussed later. It is also possible that the deformation is not totally recovered after the specimen is completely unloaded –including when the unloading takes place over a long period of time– in these cases the deformation is called plastic. When time is a factor and there is plasticity, interpretation of the tensile test is much more complcated, as will be seen in the examples presented later in this chapter.

[4] The ratio of the pressure Δp applied to a body and the resulting change in volume ΔV/V, with sign reversed, is called the bulk modulus. The bulk modulus of water (2GPa) is similar to that of polymeric materials (~2 to 4GPa), and much less than those of metals and ceramics (~30 to 200GPa). The water contained in soft biological materials such as the fibres –which can be treated as polymers– makes an important contribution to changes in rigidity under hydrostatic stresses, causing the change in deformation. This fact, together with the usually low values of these stresses (≈1/3 of the axial stress), supports the incompressible hypothesis.

EXERCISE 1

Calculate up to what point it is possible to consider the engineering and true strains as measured in a tensile test as equivalent, if the allowable error is 3%.

Solution:
From equation (6) that defines the relationship between the true-strain (ε) and engineering strain (e), we obtain

$$e = \exp(\varepsilon) - 1$$

hence

$$\text{relative error} = \frac{e-\varepsilon}{\varepsilon} = \frac{1}{\varepsilon}\left(\exp(\varepsilon) - 1 - \varepsilon\right) = \frac{1}{\varepsilon}\left(1 + \varepsilon + \frac{\varepsilon^2}{2} + \ldots - 1 - \varepsilon\right) \approx \frac{\varepsilon}{2}$$

We see that the error will be less than 3% provided that $(\varepsilon, e) < 0.06$, or in other words they will be the same so long as the engineering (or true) strain does not exceed 6%.

EXERCISE 2

Demonstrate that when the strains are not small, only true strains are additive, but not the engineering ones. That is, if starting from L_0 the final length is L_2 with an intermediate value L_1 then $e_{02} \neq e_{01} + e_{12}$ and that $\varepsilon_{02} = \varepsilon_{01} + \varepsilon_{12}$

Solution:
From equation (4) we obtain

$$\varepsilon_{01} = \ln\left(\frac{L_1}{L_0}\right), \quad \varepsilon_{12} = \ln\left(\frac{L_2}{L_1}\right), \quad \varepsilon_{02} = \ln\left(\frac{L_2}{L_0}\right)$$

from which

$$\varepsilon_{01} + \varepsilon_{12} = \ln\left(\frac{L_1}{L_0}\right) + \ln\left(\frac{L_2}{L_1}\right) = \ln\left(\frac{L_2}{L_0}\right) = \varepsilon_{02}$$

Similarly, from equation (2)

$$e_{01} = \frac{L_1}{L_0} - 1, \quad e_{12} = \frac{L_2}{L_1} - 1, \quad e_{02} = \frac{L_2}{L_0} - 1$$

from which

$$e_{01} + e_{12} = \frac{L_1}{L_0} + \frac{L_2}{L_1} - 2 \neq \frac{L_2}{L_0} - 1 = e_{02}$$

- *Necking*

Necking is an instability that can occur in a tensile test before rupture. During a tensile test $F = \sigma A$, where F is the applied force, σ the true stress, and A the cross-sectional area at that instant, and which is the same throughout the specimen. As the specimen deforms, σ increases and A decreases. If the specimen does not break, force F can reach a maximum value that can be determined from the condition $dF = 0$, or:

$$dF = \sigma\, dA + A\, d\sigma = 0$$

from which

$$\sigma = -\, A\, d\sigma/dA$$

If the volume of the specimen is constant, AL is constant and

$$dA\, L + dL\, A = 0$$

which leads to

$$-\frac{A}{dA} = \frac{L}{dL} = 1/d\varepsilon$$

then the condition for the maximum can be expressed as

$$\sigma = d\sigma/d\varepsilon \tag{8}$$

Equation (8) indicates that the maximum stress is achieved when the slope of the strain hardening, $d\sigma/d\varepsilon$, is numerically equal to the true stress, σ. Provided that $d\sigma/d\varepsilon > 0$, the strain will be uniform along the length of the specimen, but when $\sigma = d\sigma/d\varepsilon$ ($\equiv dF = 0$) he strain localises. A particular section deforms a little more than the others (due to some geometrical imperfection or some defect in the material) so it will have a lower load-carrying capacity and the load will fall to this level. The other regions will not continue deforming and the defromation will be concentrated around the weakened section, creating a neck by striction (or simply *necking*).

When the constitutive equation of the material is known it is possible to calulate the maximum load that a specimen can support as well as the maximum uniform strain.

EXERCISE 3

In 1885 the French scientist A. Considère proposed a graphical construction for the maximum load that can be supported by the specimen in a tensile test. The maximum value is obtained by drawing the tangent from the origin (0,0) to the stress-elongation (σ - λ) curve.

Illustrate this graphical construction and show that the result coincides with the instability criterion of equation (8)

Solution:

From equation (8) we can write:

$$\sigma = \frac{d\sigma}{d\varepsilon} = \frac{d\sigma}{d\lambda}\frac{d\lambda}{d\varepsilon} = \frac{d\sigma}{d\lambda}\lambda$$

Since equation (4) yields $d\lambda/\lambda = d\varepsilon$.

The equation can be rewritten as

$$\frac{\sigma}{\lambda} = \frac{d\sigma}{d\lambda}$$

which means that in a diagram the tangent at the point of instability coincides with the value of the secant slope σ/λ. Therefore, at this point of instability a straight line from the origin is tangent to the curve, as shown in the figure.

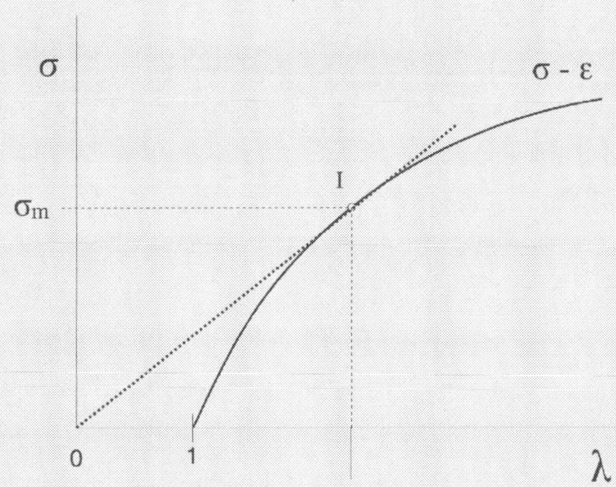

EXERCISE 4

A common model for a pair of blood vessels is based on Deimirai's constitutive equation, which in unidirectional form can be written as $\sigma = C_1(\lambda^2 - \lambda^{-1}) \exp [C_2 (\lambda^2 + 2/\lambda - 3)]$, where λ is the elongation of the sample L/L_0.

Calculate the values of $\sigma-\varepsilon$ at the instigation of necking if C_1= 50kPa and C_2=–0.1

What values must C_1 and C_2 have for necking to occur? Compare the results obtained with typical values for the ascending aorta C_1=64kPa and C_2=2. What conclusions can be drawn?

Solution:

Instability occurs when $\sigma = (d\sigma/d\lambda)\,\lambda$ (see the previous exercise). Therefore

$$\sigma = C_1\left(\lambda^2 - \frac{1}{\lambda}\right)\exp\left[C_2\left(\lambda^2 + \frac{2}{\lambda} - 3\right)\right] = \frac{d\sigma}{d\lambda}\lambda = \lambda C_1 \exp\left[C_2\left(\lambda^2 + \frac{2}{\lambda} - 3\right)\right]\left[2\lambda - \frac{1}{\lambda^2} + C_2\left(\lambda^2 - \frac{1}{\lambda}\right)\left(2\lambda - \frac{2}{\lambda^2}\right)\right]$$

from which

$$\left(\lambda^2 - \frac{1}{\lambda}\right) = \lambda\left[2\lambda - \frac{1}{\lambda^2} + C_2\left(\lambda^2 - \frac{1}{\lambda}\right)\left(2\lambda - \frac{2}{\lambda^2}\right)\right]$$

which reduces to

$$-\lambda - \frac{2}{\lambda^2} = C_2\left(\lambda^2 - \frac{1}{\lambda}\right)\left(2\lambda - \frac{2}{\lambda^2}\right)$$

since the values of the parentheses are always ≥0 for λ≥1, there is only a solution if C_2<0. In the case of the aorta C_2=2 so no instability is observed. This explains the absence of aneurisms.

For C_1= 50kPa and C_2=–0.1 we obtain

$$\lambda^3 - 5\lambda - 2 - \frac{10}{\lambda^2} + \frac{1}{\lambda^3} = 0$$

whose solutions are λ = 0.09975 (<1, compression; so this is ignored) and λ =2.5272 which leads to ε = 0.9271 and σ=197.3kPa. These results are shown in the figure. Note that when C_2=2 it is not possible to draw a tangent from the origin. However, for C_2=–0.1, a tangent can be constructed.

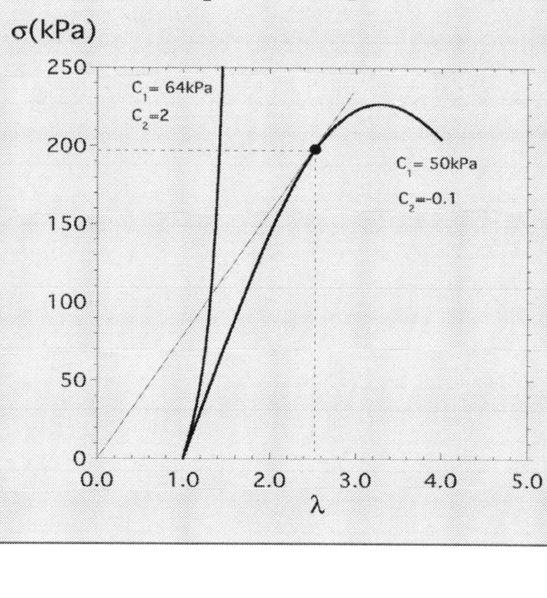

- Units used in the textile industry

Measurements of the cross-sections of fibres employed in the textile industry create problems because of their small dimensions and shapes. It is common, then, to take as a reference some measurements based on the mass of a specific length of fibre. Initially, the *denier* (the mass in grammes of a 9-km length of fibre) was used. The selection of a 9-km length is not arbitrary and is based on the fact that a thread of natural silk of that length usually ways about 1 gramme. Currently the *tex* is used, this is the mass in grammes, corresponding to a length of 1 km. Since this unit results in such a small quantity it is common to use the *decatex* (dtx) which is a value close to the denier.

If it is assumed that a fibre has a constant cross section, the value of the cross-sectional area can be deduced from the tex (T) by means of the simple relationship

$$T = 10^6 \rho A \tag{9}$$

provided that the density ρ and cross-sectional area A are expressed in SI units; that is, in kgm^{-3} and m^2, respectively.

Another unit used in the textile industry is the specific strength (also known as the *tenacity*, although this term can lead to confusion, because it is a term specific to Fracture Mechanics, as will be seen later). The specific strength (SR) is defined as the breaking force divided by the tex and is measured in the SI system in newton/tex ($Ntex^{-1}$). The relationship between the specific strength in $Ntex^{-1}$ (SR) and the engineering fracture stress (ultimate tensile strength) s_R is:

$$\text{SR } (Ntex^{-1}) = 10^{-6}\, s_R\, \rho^{-1} \tag{10}$$

where s_R and ρ must be expressed, again in SI units, that is in Pa and kgm^{-3}, respectively.

The units used in the textile industry (in Europe the **tex** is preferred while in the USA it is the **denier**) are rarely used in other engineering fields, so for this reason the following summary is presented.

Tex = mass in grammes of 1,000 metres of fibre
denier = mass in grammes of 9,000 metres of fibre
specific strength = failure load in Newtons divided by tex
= grammes force/denier

The following equivalences may be useful (*A* cross-sectional area)

1 gf/denier = 0.088 N/tex 1 tex = $\rho(kg/m^3).A(m^2)$ 1 Pa = $1(N/tex).10^6\rho(kg/m^3)$

EXERCISE 5

A polyester fibre has an average strength of 40cN/tex and an engineering failure strain of 20%. Given that the density of polyester is 1380 kg/m^3 caculate the true stress and strain at failure of the fibre.

Solution:
From equation (10) we obtain

$$s_R(\text{Pa}) = 10^6 \rho\, (\text{kgm}^{-3})\, SR\, (\text{Ntex}^{-1})$$

from which

$$s_R(\text{Pa}) = 10^6 \cdot 1380 \cdot 0.01 \cdot 40 = 552\ 10^6\,\text{Pa} = 552\ \text{MPa}$$

Equations (6) and (7) then provide the true strain and stress

$$\varepsilon_R = \ln(1+e_R) = \ln(1+0.2) = 0.182$$
$$\sigma_R = s_R(1+e_R) = 652\ \text{MPa}$$

2.2.2. The torsion test

In the torsion test a *cylindrical* sample of material is subjected to an increasing torque (or torsional moment) T, while recording the angle of twist ϕ, as shown in figure 2.4a.

This test admits the possibility of large strains because it is not limited by the necking instability that occurs in a tensile test. As in the latter, the experiment is influenced, especially in biological fibres, by the loading rate and by external factors such as temperature, pressure, relative humidity or the type of medium surrounding the specimen.

Note that the torsion test presents important differences from the tensile test since the forces are now parallel (tangential) to the cross-section and not perpendicular, as in the tensile test. In the torsion test, moreover, the strain is not uniform (equal at every point, as happens in the tensile test) since it varies as a function of the distance from the axis of rotation (figure 2.4d).

Therefore, when the term $R\phi / L$ is **small,** it is possible to make hyptheses that simplify calculation of the stress state and to establish a linear relationship between the torsional moment T, and the angle of twist ϕ, a result that agrees with experiment when this requirement is satisfied.

In these conditions (small $R\phi / L$) every element of material deforms under a state of pure shear (as shown in figure 2.4b, and which will be detailed later in Section 2.3.2). The Coulomb[5] hypothesis that sections perpendicular to the axis of the fibre remain plane and parallel to each other after deformation, keeping their shape, the motion of the section being a rotation relative to its centre is also satisfied.

The *shear strain* $\gamma(r)$, defined as the angular change of a right angle (see figure 2.4b), is the same at two points located at a distance r from the axis, does not depend upon z, and is given by

$$\gamma(r) = r\phi / L \qquad (11)$$

The corresponding *tangential* or *shear stress* τ, is defined as the tangential (or shear) force divided by the area over which it acts. (Recall that it has been assumed that deformations are small, so that the area is essentially constant, so it is not necessary to distinguish between true and engineering stresses). The value of τ cannot be

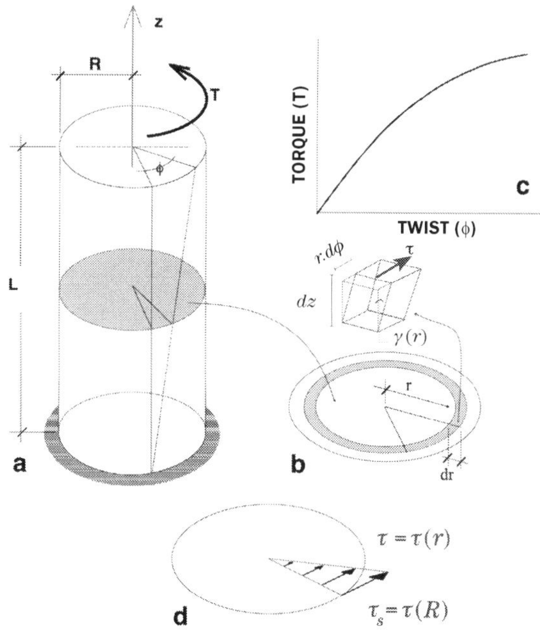

Fig. 2.4. *a. Torsion test. Notation used.*
b. Detail of an element of the specimen subjected to pure shear.
c. Example of the torsional moment curve as a function of the angle of twist.
d. Distribution of the stesses on a plane perpendicular to the axis z.

measured directly, or calculated from the external torsional moment T, because τ depends on $\gamma(r)$, and this relationship is unknown *a priori* because it depends on the material properties.

The relationship between the torsional moment and the angle of twist can be calculated –provided $R\phi / L$ is small– as follows:

The shear stress acting on an annular element of radius r and width dr makes an infinitesimal contribution to the total torsional moment T equal to:

$$dT = \tau \text{ x area x radius} = \tau(2\pi r dr)r = 2\pi\tau r^2 dr$$

and, as a result

$$T = 2\pi \int_0^R \tau r^2\, dr \qquad (12)$$

[5] Charles-Augustin de Coulomb (1736-1806) was a French engineer most famous for his law of attraction between two electric charges (the S.I. unit of charge is named after him). He also made many contributions to physics and the mechanics of solids. He is credited with the theory of simple torsion and the definition of the coefficient of friction.

This equation cannot be integrated directly because τ depends on r through γ and this relationship is unknown.

The way to solve it is to assume *a priori* some relationship between τ and γ, the simplest being, a linear relationship between stress and strain, that is, $\tau = G\gamma$. This approximation is equivalent to assuming that the material is Hookean elastic. The constant G is measured in Pascals (Pa = Nm^{-2}) and is usually called the *shear modulus, stiffness modulus* or *transverse modulus* of elasticity. With this hypothesis:

$$T = \frac{\pi R^4}{2}\frac{G}{L}\phi = \frac{IG}{L}\phi \tag{13}$$

where $I=\pi R^4/2$; the polar moment of area of the circular section of the cylindrical specimen under consideration.

Clearly, the relationship between T and ϕ is linear in these circumstances and the slope of the straight line can be used to estimate the value of G.

The result obtained in equation (13) can also be expressed as a function of the shear stress on the boundary with the lateral surface $\tau_S = \tau(R)$, in the following

$$T = \frac{\pi R^4}{2}\frac{G}{L}\phi = \frac{\pi R^3}{2}\tau_S \tag{14}$$

When the curve departs slightly from a straight line, the value of G determined from the slope at the origin is normally used, obtained from

$$G = (L/I)\,(dT/d\phi)_{\phi=0}$$

As already noted, one of the advantages of this test is the possibility of achieving larger deformations than in the tensile test without instability problems. (In any case, the deformation must be **sufficiently small** so that the shear stress and strain can be defined as indicated and that equations (11) and (12) remain valid). The danger is that for large deformations the material can become nonlinear so that equations (13) and (14) cease to be valid.

EXERCISE 6

Derive a relationship between τ and γ from two experiments with values R and $R + \Delta R$.
From this relationship determine the curve $T = T(\phi)$ for a torsion test.

Solution:

For a particular value of the angle of twist ϕ the torque on a fibre of radius R will be (equation (12))

$$T_R(\phi) = 2\pi \int_0^R \tau r^2\, dr$$

while for a fibre of radius $R+\Delta R$, for the same twist ϕ, the value of the torque will be

$$T_{R+\Delta R}(\phi) = 2\pi \int_0^{R+\Delta R} \tau r^2\, dr$$

splitting the integral in this last equation we can write

$$T_{R+\Delta R}(\phi) = T_R(\phi) + 2\pi \int_R^{R+\Delta R} \tau r^2\, dr \approx T_R(\phi) + 2\pi\ \tau(R)\int_R^{R+\Delta R} r^2\, dr \approx T_R(\phi) + 2\pi\ \tau(R) R^2\,\Delta R$$

in which we have approximated τ in the interval $(R, R+\Delta R)$ by its value at $r=R$, and neglecting second order and higher terms in ΔR. Rearranging, $\tau(R)$ becomes

$$\tau(R) = \frac{T_{R+\Delta R}(\phi) - T_R(\phi)}{2\pi R^2\,\Delta R} \tag{a}$$

this equation, together with (11) when $r=R$

$$\gamma(R) = \frac{R\phi}{L} \tag{b}$$

provides the parametric equation for the curve τ-γ in terms of T and ϕ.

If the function $\tau = \tau(\gamma)$ is known, it is possible to obtain the relation $T = T(\phi)$ without use of equation (12)

$$T(\phi) = 2\pi \int_0^R \tau(\gamma) r^2\, dr = 2\pi \int_0^R \tau\left[\gamma(r\phi/L)\right] r^2\, dr$$

in which we used equation (11)

EXERCISE 7

Deduce a general relationship similar to equation (14) for elements of arbitrary cross-section, assuming that τ and γ are linearly related. Is it necessary that the section has a centre of symmetry in order to satisfy the conditions for the torsion test

Solution:

Figure a) shows an arbitrary section subjected to a torque. If we assume that Coulomb's hypothesis is valid (sections perpendicular to the axis remain plane and parallel to one another after deformation, keeping their shape) the motion of the section will be simply a rotation with respect to some point O in the section.

The shear stress at a point A acts parallel to the section and perpendicular to the segment OA, as shown in the figure. The torsional moment about point O will be,

$$T = \int_{\Omega} \tau r d\Omega$$

where Ω is the area of the section and $d\Omega$ is an infinitesimal element of area centred around point A. Imposing a linear relationship between τ and γ we can write

$$\tau = G\gamma = G\frac{\phi}{L}r$$

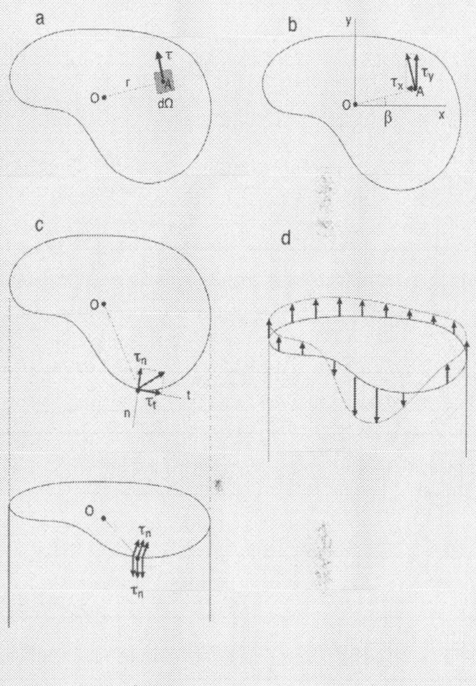

in which we have also used equation (11), substituting this in the previous expression, the integration for T gives us

$$T = G\frac{\phi}{L}\int_{\Omega} r^2 d\Omega = G\frac{\phi}{L}I_O$$

where the integral is none other than the polar moment of area of the section I_O about point O. This is a general form of equation (14).

Point O cannot be arbitrary since equilibrium requires that the resultant force in the plane of the section must be zero (see figure b))

$$0 = \int_{\Omega} \tau_y d\Omega = \int_{\Omega} \tau_x d\Omega$$

and since $\tau_y = \tau \cos\beta = G\frac{\phi}{L}r\cos\beta = G\frac{\phi}{L}x$

then $0 = \int_{\Omega} G\frac{\phi}{L}x d\Omega = G\frac{\phi}{L}\int_{\Omega} x d\Omega$

and similarly for coordinate *y*. Hence point O must be the centroid of the section. If the section has a centre of symmetry then it must coincide with the centroid.

Note that for noncircular sections, Coulomb's hypothesis is never strictly satisfied. The reason is that points on the contour in which the edge is not perpendicular to r (see figure c)) have a normal component of shear stress (τ_n) that requires a shear stress component in the direction parallel to the axis of the fibre, as shown in figure c), which is impossible if the surface is stress-free. The inability to satisfy the equilibrium condition means that the section warps, and there will be nonuniform deformation in the axial direction of the fibre (figure d)).

In analysing torsion tests on biological fibres it is usually assumed that the cross-section of the fibre is circular. A more detailed examination shows that this is not always so; the cross-section of silk from the silkworm is triangular and that of cotton is kidney-shaped, as mentioned previously.

2.3. ELASTIC BEHAVIOUR

2.3.1. Introduction

As mentioned above, the experiment most frequently performed on fibres is the tensile test. In order to interpret its results it is necessary to make certain assumptions about the relationships between true stresses and strains, relationships that are known as *constitutive equations*.

The simplest constitutive equation is that for the behaviour of an elastic material, for which the value of the stress is only a function of the strain at that instant; it does not depend upon the loading history, nor time and other variables. Figure 2.5 shows two types of stress-strain curve for elastic materials following a tensile test. The (approximately) linear curves in figure 5a are characteristic of many metallic and ceramic fibres, as well as some biological fibres such as linen and cotton. Linear-elastic behaviour is only possible if the deformation is small, and in this case it is not usual to distinguish between true and engineering stresses and strains ($e \approx \varepsilon$; $s \approx \sigma$).

The curve in figure 5b is typical of elastomeric fibres, such as rubber, in which the stress is no longer proportional to the strain. However, the strain returns to zero –and along the same load path– when the stress is removed. A common mistake is to assume that elastic behaviour is always linear. The behaviour of rubber fibres shows that ther are notable exceptions.

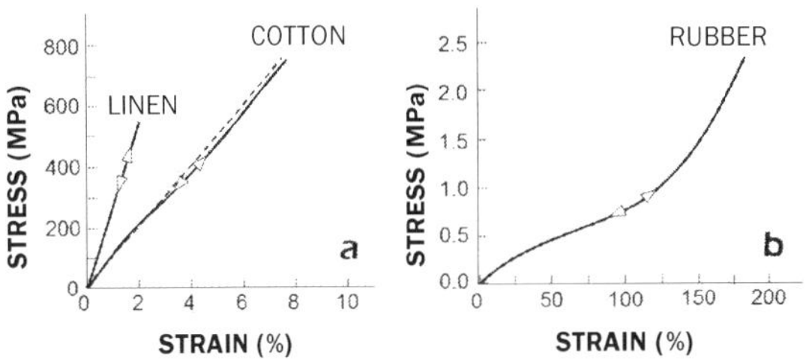

Fig. 2.5. *Exampes of stress-strain curves showing elastic behaviour.*
a. Linear elastic (linen and cotton).
b. Nonlinear elastic (rubber) (Treloar 1975).

2.3.2. Linear elasticity

Linear-elastic behaviour of a fibre and, also, the initial part of nonlinear behaviour (that part of the curve that can be represented by the tangent at the origin, which is usually in the region of small deformations) can be characterised using two parametrs: the *elastic modulus* E or Young's[6] modulus and *Poisson's*[7] ratio υ. Under these conditions it is said that the response of the fibre is **Hookean,** or **linear elastic.**

The *elastic modulus* E is measured in Pascals (Nm^{-2}) and is the slope of the stress-strain line (σ-ε) recorded in a tensile test (figure 2.6a)

$$E=\sigma/\varepsilon \tag{15}$$

The *initial elastic modulus* for a fibre of a nonlinear elastic material, such as that shown in figure 2.6b, is given by

$$E_0 = (d\sigma/d\epsilon)_{\epsilon=0}$$

As noted above, it is always possible to represent the response of a nonlinear elastic material in the region of a point ε_0 by its local tangent. In this way it is possible to talk about a *tangent elastic* modulus (figure 2.6b)

$$E' = (d\sigma/d\epsilon)_{\epsilon_0}$$

The *Poisson ratio* ν, is a nondimensional quantity and is defined as the ratio between the transverse and longitudinal strains in the fibre (see figure 2.7)

$$\nu = - \epsilon_x/\epsilon_z \tag{16}$$

where ε_z is the strain in the direction of the applied load (longitudinal) and ε_x is the strain in the transverse direction ($\varepsilon_x=\varepsilon_y$, since it is assumed that the material is isotropic).

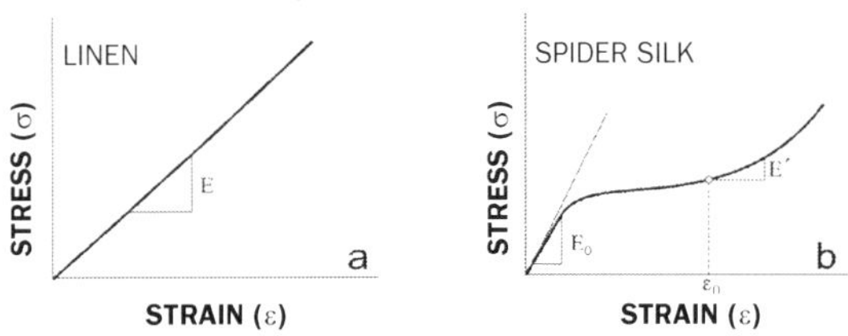

Fig. 2.6. *Concept of the elasic modulus of a fibre.*
a. Linear elastic.
b. Nonlinear elastic E_0 (Initial modulus), E' (Tangent modulus).

6 Although it was the Swiss mathematican Leonhard Euler (1707 - 1783) who introduced this parameter, it was the English scientist Thomas Young (1773-1829) who popularised it in a work published in 1807.

7 Siméon Denis Poisson (1781-1840) was a French physicist and mathematician who obtained the relationship.

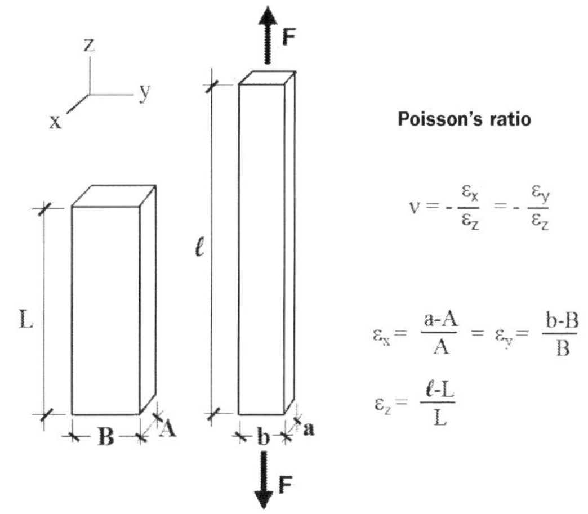

Fig. 2.7. *Definition of Poisson's ratio. Homogeneous isotrpic material and small deformations* $(e \approx \varepsilon; s \approx \sigma)$

A specimen usually contracts transversely under tensile loading. Since ε_x and ε_y will be negative then the Poisson ratio will be positive (although for some very rare materials, called *auxetics*, the opposite is true (Kolken *et al.* 2020)). Possible values of ν lie in the range –1 to +0.5 and are usually about 0.3 for metals, 0.2 for ceramics and approximately 0.5 for polymers and biological materials. When $\nu = 0.5$ the volume of the material remains constant during deformation.

Expressions relating the stresses and strains for a fibre subjected to tension σ in the axial direction, assuming Hookean elastic behaviour, are summarised in figure 2.8. The relationships will be generalised later for materials with two and predominantly three dimensions.

For a cylindrical fibre loaded in torsion (as described in 2.2.2) an elemental volume of material will be in a state of pure shear and, if the behaviour is Hookean elastic, the stresses and strains will be as shown in figure 2.8b. Note that all of the components are zero except for the shear stress $\tau_{z\theta}$, which depends upon the distance, r, from the centre of the specimen. (The subindices "$z\theta$" indicate that shears act on a section perpendicular to the axis "z" pointing in the direction of increasing "θ". Likewise, subindex "zr" means that the forces lie in a plane parallel to "z=constant" acting in the radial direction "r", and in the case of "$r\theta$" they are applied to a surface with constant "r" and direction defined by "θ"). The relationship between stress and strain is given (according to 2.2.2) by

$$\tau_{z\theta} = G\,\gamma_{z\theta} = G\,r\,(d\phi/dz)_{\phi=0} \tag{17}$$

The shear modulus G is a function of E and ν and is given by

$$G = E/2\,(1+\nu) \tag{18}$$

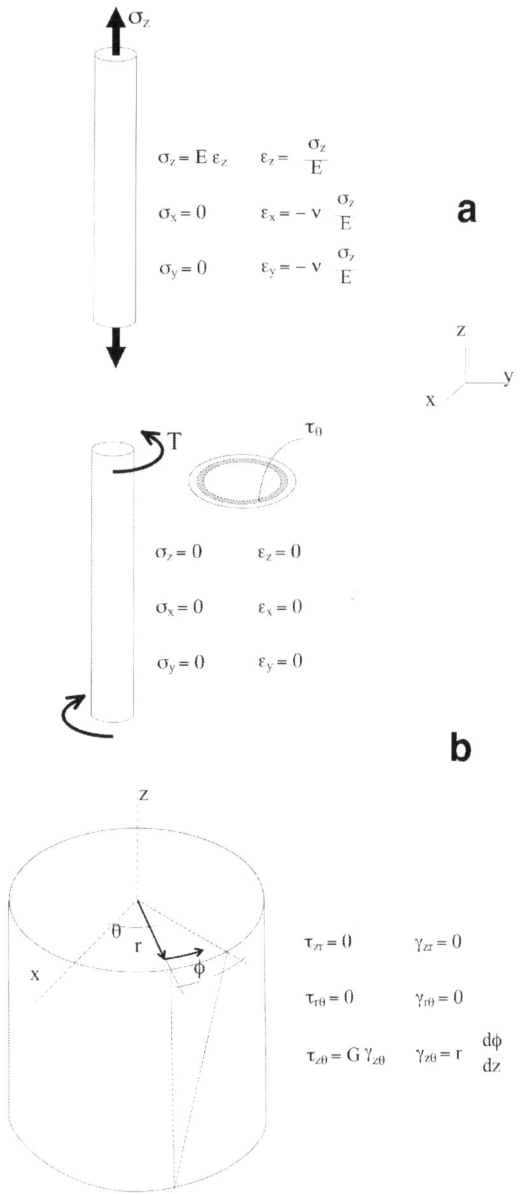

Fig. 2.8. *Summary of the stress state for a fibre in the linear elatic regime.*
a. Tensile test.
b. Torsion test.

EXERCISE 8
Derive equation (18) $G = E/2(1+\nu)$ for a Hookean elastic material.

Solution:
Consider a cubic element of side L subjected to shear stresses $\tau_{zy} = \tau_{yx} = \tau$ as shown in figure a. At all times we consider small deformations and, in consequence, that the geometry remains unaltered because changes are small. However, in the figures the movements are exaggerated to provide greater clarity.

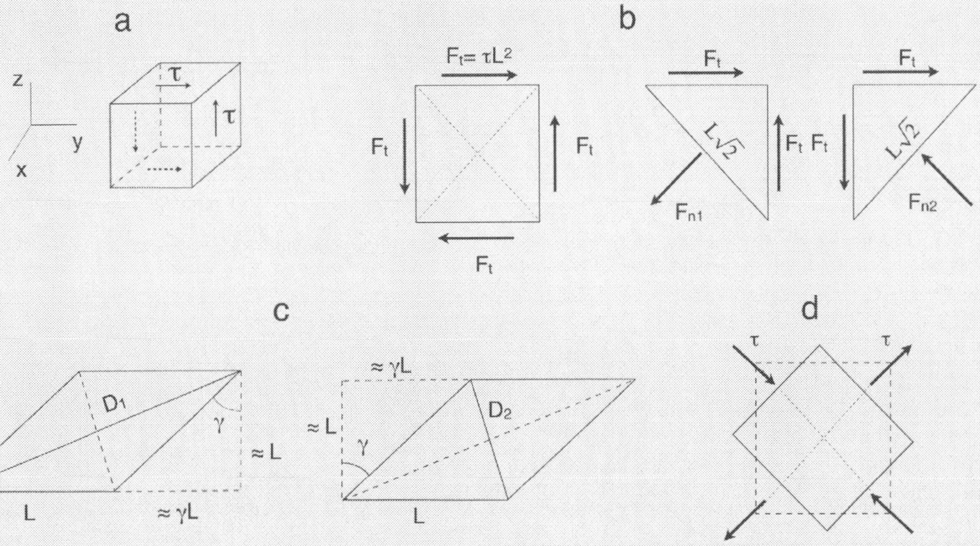

The effect produced by shear forces $F_t = \tau L^2$ is equivalent to two forces in the diagonal directions (figure b),

$$F_{n1} = F_{n2} = \sqrt{2}\, F_t = \sqrt{2}\, \tau L^2$$

that give rise to a tensile stress along the diagonal that is stretched (direction 1), of value

$$\sigma_1 = \frac{F_{n1}}{L^2\sqrt{2}} = \tau$$

and a compressive stress along the diagonal that is shortened (direction 2)

$$\sigma_2 = \frac{-F_{n2}}{L^2\sqrt{2}} = -\tau$$

The length of the stretched diagonal is (figure c)

$$D_1 = \sqrt{L^2 + (L + L\gamma)^2} \approx L\sqrt{2}(1 + \gamma/2)$$

with a corresponding strain $\varepsilon_1 = \dfrac{D_1 - L\sqrt{2}}{L\sqrt{2}} \approx \gamma/2$

and the diagonal that is shortened sees its length reduced to

$$D_2 = \sqrt{L^2 + (L - L\gamma)^2} \approx L\sqrt{2}(1 - \gamma/2)$$

with a strain

$$\varepsilon_2 = \frac{D_2 - L\sqrt{2}}{L\sqrt{2}} \approx -\gamma/2$$

Considering the stress state in the diagonal directions (1) and (2) (figure d), we see that it is possible to represent them as a superposition of two states of simple traction

$$\sigma_1 = \tau \; ; \qquad \sigma_2 = 0 \; ; \qquad \sigma_z = 0$$

and

$$\sigma_1 = 0; \qquad \sigma_2 = -\tau \; ; \qquad \sigma_z = 0$$

whose strains, represented in figure 2.8a,are respectively

$$\varepsilon_1 = \tau/E \; ; \qquad \varepsilon_2 = -\nu\,\tau/E \; ; \qquad \varepsilon_z = -\nu\,\tau/E$$

and

$$\varepsilon_1 = \nu\,\tau/E \; ; \qquad \varepsilon_2 = -\tau/E \; ; \qquad \varepsilon_z = \nu\,\tau/E$$

and therefore the total strain seen by an element will be their sum

$$\varepsilon_1 = \tau(1+\nu)/E \; ; \varepsilon_2 = -\tau(1+\nu)/E. \qquad \varepsilon_z = 0$$

Since $\varepsilon_1 = \gamma/2$ we obtain

$$\varepsilon_1 = \gamma/2 = \tau(1+\nu)/E$$

from which, with the aid of the expression $\gamma = \tau/G$, we determine G

$$G = \frac{E}{2(1+\nu)}$$

EXERCISE 9

Justify that when ν = 0,5 a material deforms while maintaining a constant volume.

Solution:

The change in volume produced during the deformation represented in figure 2.7 is

$$abl - ABL = A(1+\varepsilon_x)\,B(1+\varepsilon_y)\,L(1+\varepsilon_z) - ABL \approx ABL(\varepsilon_x + \varepsilon_y + \varepsilon_z)$$

in which we have neglected products of the deformations, since they are very small.

Substituting the value of the Poisson ratio (following figure 2.7) we obtain

$$abl - ABL \approx ABL(\varepsilon_x + \varepsilon_y + \varepsilon_z) = ABL(-\nu\varepsilon_z - \nu\varepsilon_z + \varepsilon_z) = ABL\;\varepsilon_z(1-2\nu)$$

an expression in which we can verify that when ν=0.5 the change in volume is zero.

EXERCISE 10
Show that $-1 < \nu < +0.5$

Solution:
The shear modulus G is given by equation (18) $G = \dfrac{E}{2(1+\nu)}$

which must be always positive, as must Young's modulus E, so these conditions require that $\nu > -1$.

The bulk moduls K, defined as the relationship between the uniform pressure applied to a body and the resulting change in volume, must also be positive. The value of K can be deduced from an experiment in which stresses $\sigma_x = \sigma_y = \sigma_z = -p$ are applied to a body.

This problem can be solved as a sum of the cases:

$$\sigma_x = -p \; ; \; \sigma_y = \sigma_z = 0$$
$$\sigma_y = -p \; ; \; \sigma_x = \sigma_z = 0$$

and

$$\sigma_z = -p \; ; \; \sigma_x = \sigma_y = 0$$

whose solution can be deduced from figure 2.8a, yielding for each case

$$\varepsilon_x = -p/E \; ; \; \varepsilon_y = \varepsilon_z = -\nu p/E$$
$$\varepsilon_y = -p/E \; ; \; \varepsilon_x = \varepsilon_z = -\nu p/E$$

and

$$\varepsilon_z = -p/E \; ; \; \varepsilon_x = \varepsilon_y = -\nu p/E$$

The total strain is the sum of each case, being equal to

$$\varepsilon_x = \varepsilon_y = \varepsilon_z = -p(1-2\nu)/E$$

The change in volume produced will be (see the previous exercise)

$$\Delta V = V(\varepsilon_x + \varepsilon_y + \varepsilon_z) = -Vp3(1-2\nu)/E$$

where V is the initial volume and ΔV its change, that will be negative. The value of K is, by definition

$$K = -\frac{p}{\Delta V/V} = \frac{E}{3(1-2\nu)}$$

In order to have $K > 0$ we see that $\nu < +0.5$.

2.3.3. Concluding remarks

Hookean, or linear-elastic behaviour is the simplest approximation to the response of a material. Figure 2.8 summarises the relationships between stresses and strains for a fibre that behaves in a Hookean manner.

At the microscopic level, this behaviour can be interpreted as the result of tiny reversible displacements of particles (atoms, molecules, etc.) from their eqilibrium positions. Consequently, the internal structure of the material does not change and the variations in the internal energy can be calulated with respect to the equilibrium positions, through a linear approach in which only the first two terms are considered; the constant equilibrium value and a linear term with the displacements of the particles. This implies that the force –as the energtically conjugate of the displacement– appears as the first derivative of the internal energy with respect to the dispacement.

As will be seen in the next section, when the deformation implies changes in the arrangement of the particles of the material the analysis is complicated, the *linear approximation* to the internal energy is not valid, and quantities such as entropy, related to the degree of disorder, come into play.

In general, in biological materials the deformation tends not to be proportional to the load. Fibres that exhibit *nonlinear elastic* behaviour in a tensile test cannot be characterised in such a simple way nor is it possible to obtain such a simple constitutive equation.

2.4. ELASTOMERIC BEHAVIOUR

2.4.1. Introduction

There is a family of polymers, capable of standing large reversible deformations, called elastomers; these are polymers whose chains are cross-linked. The classical example of an elastomer is vulcanised rubber. Natural rubber (cis-polyisoprene) is not properly an elastomer, because during deformation of the polymer chains slide over each other and the orginal length is not recovered on unloading; the deformation is not reversible and the material cannot be considered as elastic. Only when cross-links between the polymer chains are created –for example, by means of sulphur bonds, a process known as vulcanisation– is a polymer produced that recovers its original shape.

For this type of material it is possible to assume that under the action of a force the arrangement of its molecules change immediately, so that it is possible to apply the laws of thermodynamics and to relate the stress state to changes in the internal energy and entropy. For this approach to provide information about the physical nature of the parameters used it is essential, in a second step, to know how to relate the internal energy and the entropy of the materials to the movement of the molecular chains.

Typical curves obtained from a tensile test on an elastomer are shown in figure 2.9. Note that true strains approaching 170% are achieved, and that this is reversible (the rise of the curve is coincident with the descent), and nonlinear. Only for small strains, of the order of 1%, can it be assumed that the relationship between stress and strain is linear, that is to say Hookean. This behaviour indicates that the fundamental mechanism of the deformation of elastomers must be very different from metallic materials, ceramics and other types of solids. The nonlinearity that is exhibited for large deformations must be, as will be seen later, due more to changes in entropy than to variations in internal energy.

The entropy of an elastomer is reduced during deformation because the material is more ordered –the chains align themselves– and only recover their initial shape if they are unloaded, since the entropy of an isolated system tends to increase. The entropy of a completely extended polymer chain is zero because it is the only conguration possible. On the other hand, in the initial equilibrium position the chain is curled up, there are many possible configuration that produce the same contraction, and the entropy is large. However, the internal energy of a long chain is practically the same whatever the configuration, since its origin is mainly in the stretching of its "links" and not in their relative curling. For this reason, in the absence of external forces, a stretched chain will tend to return to its equilibrium position basically due to the effect of entropy.

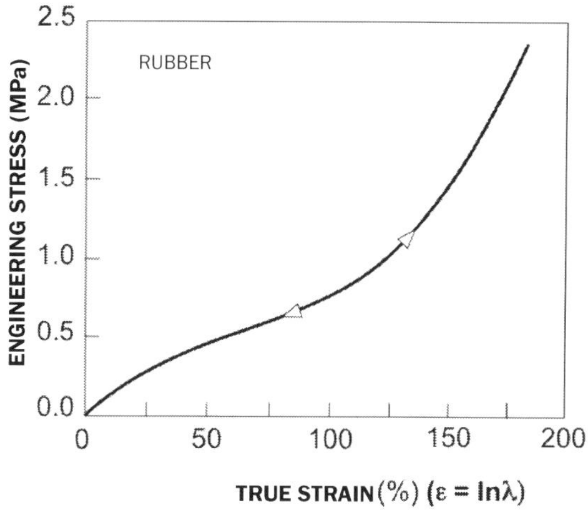

Fig. 2.9. *True stress-true strain curve for natural rubber vulcanised 8% sulphur. Up to strains of 200% the behaviour is elastomeric. (Data from Treloar, 1944)*

2.4.2. Thermodynamics of elastomeric deformation

The objective of this section is to find an expression that relates the force acting on a fibre to the change in entropy during deformation.

The First Law of Thermodynamics –conservation of energy– implies that:

$$dU = dQ + dW \tag{19}$$

where dU is the change in internal energy of the system, dQ is the heat energy contribution which, in a reversible process, is given by TdS (where T is the absolute temperature and S is the entropy), and dW is the contribution from the work done by the external forces.

Considering a tensile test, the work done by the external forces, dW will have two components, one due to the applied force F and the other resulting from the external pressure p (which also exerts forces on the surface of the specimen) - see figure 2.10.

$$dW = F\,dl - p\,dV$$

where dl is the displacement of the applied force F and dV the change in volume during the deformation. Therefore, equation (19), for a tensile test, can be written as:

$$dU = TdS + Fdl - p\,dV \tag{20}$$

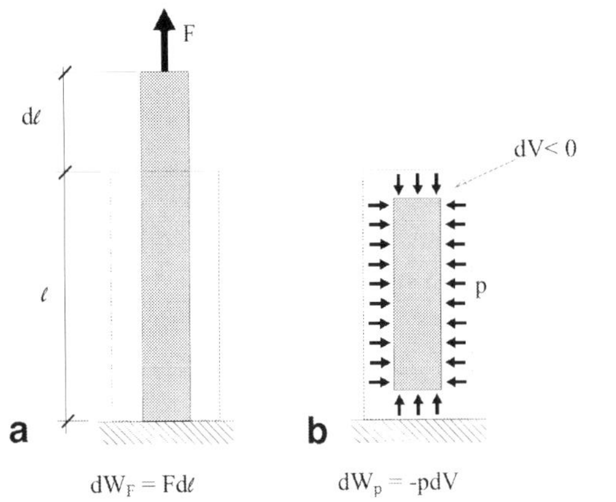

Fig. 2.10. *Work done by the external forces on a fibre (the initial position is shown dotted).*
a. External tensile force.
b. Contraction due to the effects of the external pressure.

As observed above, most biological materials deform under constant volume and dV=0. In the case of elastomers this assumption can be justified for small deformations but, in general, it is not obvious. For small deformations the bulk modulus[8], K, is about 10,000 times greater than the shear modulus, G. The relationship between both moduli and the Poisson ratio ν, is (see, for example, Landau 1986):

$$\nu = \frac{3K-2G}{2\,(3K+G)}$$

Therefore, when $K>>G$ is $\nu \simeq 0.5$, and the deformation occurs essentially at constant volume. In any case, for typical values of pressure (about 1 atm.) and solids or liquids (which unlike gases undergo little change in volume), the term pdV is negligible compared to Fdl.

It is possible then to ignore the effects of a change in volume and to consider only the thermodynamic variables of the problem which are the applied force F, specimen length l, and temperature T.

$$dU = TdS + Fdl \qquad (21)$$

It is interesting to note that, in this thermodynamic description of the deformation process, the equation of state of the thermodynamic system (for the problem considered) coincides with the constitutive relationship that exists between the applied force and the length, and, also, depends upon the temperature

$$F = f(l,T)$$

thus only three variables (F, l, T) are truly independent. Or, in other words, it is not possible to arbitrarily set values to the three variables at the same time, since the third is determined by the other two.

In 1935, Meyer and Ferri showed that, for some *elastomers*, if the specimen length l was kept constant, the force F would be proportional to the temperature T, that is:

$$F = a\,T \qquad (22)$$

where a is a positive constant. Similar results have been obtained for biological materials which behave elastomerically, such as resilin (Weis-Fogh, 1961) and abductin (Alexander, 1966). Note that elastomers tend to shrink when heated, since according to equation (22) they develop tractions that increase with temperature for constant length. For this reason, if the length is held fixed, the specimen will tighten with increased temperature, as can be observed in figure 2.11.

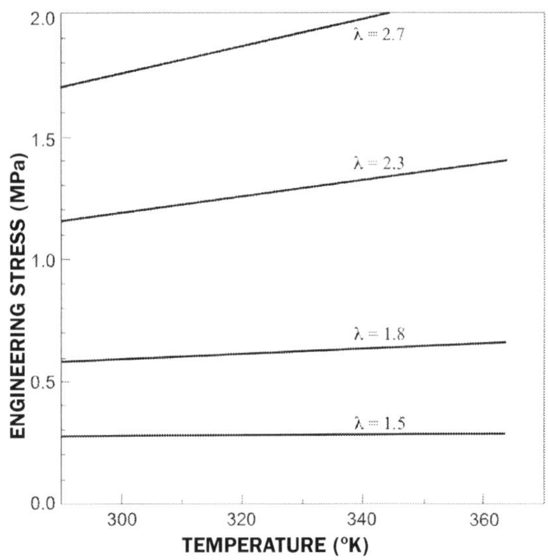

Fig. 2.11. *Variation of the engineering stress as a function of temperature for a resilin fibre in which the length L is held constant (for different elongations λ ($\lambda = L/L_0$), where L_0 is the initial length in each case, of dry fibre. The data are for the resilin in the libelula tendon Aeshna grandis (based on Weis-Fogh, 1961).*

8 The bulk modulus is defined as the ratio of the infinitesimal pressure increase to the resulting relative decrease of the volume (see Section 3, 8.3.2).

This behaviour (where it is assumed that $F = a\,T$, when l is constant) justifies the role of entropy in the calulation of force F.

From equation (20), and bearing in mind what was said above, the force F (see the exercise below) is given by:

$$F = \left(\frac{\partial U}{\partial l}\right)_T - T\left(\frac{\partial S}{\partial l}\right)_T \tag{23}$$

but the first term of equation (23) is zero, since (see the exercise that follows):

$$\left(\frac{\partial U}{\partial l}\right)_T = F - T\left(\frac{\partial F}{\partial T}\right)_l = F - T\frac{F}{T} = 0 \tag{24}$$

because $(\partial F/\partial T)_l = a = F/T$ according to the experimental results expressed by equation (22).

This shows that the sole contribution to the force F (under the assumptions made) is the entropy given, according to equation (23), by:

$$F = -T\left(\frac{\partial S}{\partial l}\right)_T \tag{25}$$

In order to proceed to the determination of the constitutive equation $F = f(l, T)$ it is necessary to obtain information about $S(l)$, the entropy of the fibre.

EXERCISE 11
Derive equations (23) and (24).

Note: the use of the thermodynamic function $A = U - TS$, known as the Helmholtz free energy, helps the derivation.

Solution:
Differentiating A and observing that $dU = TdS + Fdl$

$$dA = dU - TdS - SdT = Fdl - SdT \tag{a}$$

from this result, and noting that A is a function of the variables and T,

$$dA = \left(\frac{\partial A}{\partial l}\right)_T dl + \left(\frac{\partial A}{\partial T}\right)_l dT \tag{b}$$

it is deduced, comparing (a) and (b), that

$$F = \left(\frac{\partial A}{\partial l}\right)_T = \left(\frac{\partial U}{\partial l}\right)_T - T\left(\frac{\partial S}{\partial l}\right)_T \quad \text{and} \quad -S = \left(\frac{\partial A}{\partial T}\right)_l$$

which yields equation (23).

Using Schwartz's theorem it is possible to write

$$\frac{\partial}{\partial T}\left(\frac{\partial A}{\partial l}\right)_T = \frac{\partial}{\partial l}\left(\frac{\partial A}{\partial T}\right)_l$$

in which

$$\left(\frac{\partial F}{\partial T}\right)_l = -\left(\frac{\partial S}{\partial l}\right)_T \tag{c}$$

then, taking into account equations (23) and (c)

$$\left(\frac{\partial U}{\partial l}\right)_T = F + T\left(\frac{\partial S}{\partial l}\right)_T = F - T\left(\frac{\partial F}{\partial T}\right)_l$$

resulting in equation (24).

2.4.3. Molecular theory of elastomeric deformation

The mechanical behaviour of elastomers can be modelled by assuming that they are composed of long molecular chains, capable of adopting a large number of configurations as a result of the thermal vibrations of their atoms. Although the molecular chains are linked together, it is assumed that the number of links is sufficiently small and that they do not significantly impede the movement of the chains so that, in the absence of external forces, the chains adopt positions corresponding to the state of maximum entropy. When a force is applied, the chains will tend to deform in the direction of the force, and entropy will be reduced, causing a state of tension.

Therefore, in order to obtain a relationship between stress and strain from equation (25) it will be necessary to calculate the entropy of a network of chains as a function of the length of the network. This calculation can be performed in two stages; first determining the entropy of an isolated chain and, then, calculating the variation of the entropy of the network as a function of its length.

The purpose of this section is to show the reader the route to an expression for the entropy as a function of the length of the polymer chain. The objective of this book is not to develop the detail of each step, the interested reader will find the fine detail of these calculations in the cited references. However, some steps in the process have been included in the Appendix as exercises.

Entropy of an isolated chain

a. Gaussian model

In order to facilitate the calculations, in a first approximation, the chain molecule is replaced by a series of links which can rotate freely between themselves, in this way some restrictions imposed by the stiffness of a molecular chain are eliminated.

It is assumed that the chain is formed of n links each of length l_0. One end of the chain is located at the origin of coordinates, O, (as shown in figure 2.12). It is assumed that, as a result of thermal agitation, the molecular chain changes its configuration randomly with time. In order to obtain the value of any parameter of the chain (its length, for example) we must obtain a time average value. This is done using the function $p(\mathbf{r})$, which provides information about the probability of the various configurations of the chain over a sufficiently long period.

The Gaussian model states that the probability $p(\mathbf{r})\,dV$ of finding an end Q in a volume element, $dV = dx\,dy\,dz$, about point $\boldsymbol{r}$ is given by (see, for example, Flory 1989 and the posed exercise):

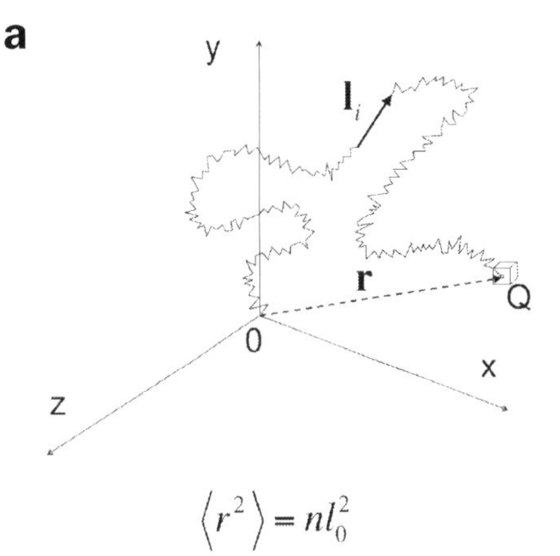

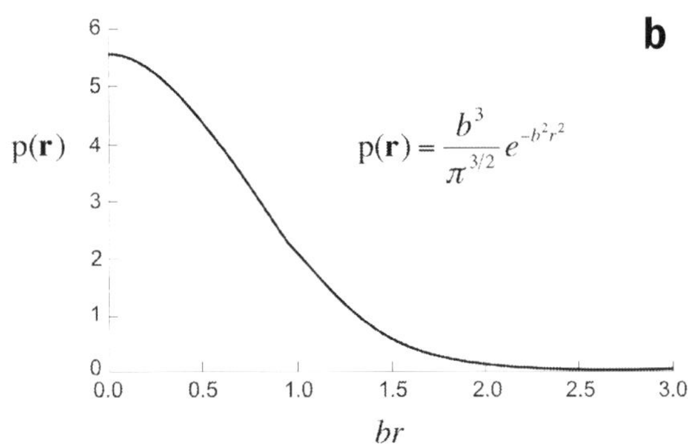

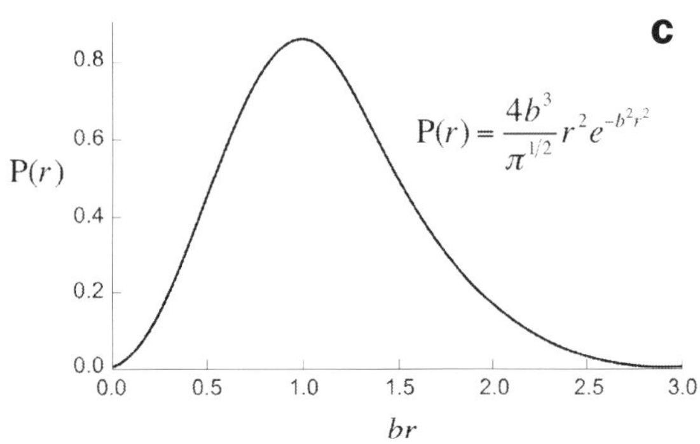

Fig. 2.12. *a. Idealised configuration of a polymer (Gaussian chain), together with functions $p(\mathbf{r})$ and $P(r)$.*
b. $p(\mathbf{r})$ – Probability of finding end Q inside dV, located at $\boldsymbol{r}$.
c. $P(r)$ – Probability of finding end Q inside a volume located between r and $r + dr$.

$$p(r)dV = \frac{b^3}{\pi^{3/2}} \exp\left(-b^2 r^2\right) dV \tag{26a}$$

where $b^2 = 3/(2nl_0^2)$. Function $p(\mathbf{r})$ takes the form of a Gaussian probabilty distribution[9], symmetric about the origin, $\mathbf{r} = \mathbf{0}$, at which the function is a maximum (see figure 2.12). Function $p(\mathbf{r})$ is normalised (its integral across space is 1) and it is only zero at infinity, which is contrary to the fact that the maximum chain length is nl_0, and the probability of exceeding that value must be zero. From a practical point of view, since the function $p(\mathbf{r})$ is practically zero for sufficiently large values of the exponent, it is sufficient to restrict its application to situations when $r << nl_0$.

In order for $p(\mathbf{r})$ to be symmetric about the origin, the most likely value –and also the average of the radius vector $\mathbf{r}$– is $\mathbf{0}$, and the average position of Q is point O. However, the mean value of the distance to the origin, r, is not zero. The probability $P(r)dr$ that Q falls within a volume located within a spherical shell between r and $r + dr$, independent of the direction of r, is the product of $p(\mathbf{r})$ and the volumen $4\pi r^2 dr$. That is, the probability is given by:

$$P(r)dr = 4\pi\, r^2\, dr\; p\left(\mathbf{r}\right) = \frac{4b^3}{\pi^{1/2}}\, r^2 \exp\left(-b^2\, r^2\right) dr \tag{27}$$

Function P(r) is shown in figure 12 and it can be seen that it has a maximum when $r = b^{-1}$, a value that depends upon the number of links n and the length of each one of them, l_0. The mean distance between the ends of the chain $< r >$, is obtained (see the exercise) as

$$< r > = \int_0^\infty r\, P(r)dr = \left(2/\sqrt{\pi}\right)b^{-1} = \sqrt{\frac{8}{3\pi}}\, n^{\frac{1}{2}}\, l_0 \simeq 1.13\, b^{-1} \tag{28}$$

and the mean of the distance squared between the ends of the chain, $< r^2 >$, (which will be used in later calculations) is (see the exercise)

$$< r^2 > = \int_0^\infty r^2\, P(r)dr = 3/(2\, b^2) = n{l_0}^2 \tag{29}$$

Note that $< r^2 > \neq < r >^2$.

The entropy S of the chain in a particular configuration can be calculated using Boltzmann[10], statistics, which yields a value that is proportional to the logarithm of the number of possible configurations (microstates), Z, compatible with that macroscopic configuration (macrostate). Its value is given by:

$$S = k \ln Z \tag{30}$$

where k is the Boltzmann constant, equal to $1.3806488(13)\times 10^{-23}$ Joules/K. Then the number of possible configurations compatible with a value $\overline{OQ} = \mathbf{r}$ of the chain must be proportional to the probability density $p(\mathbf{r})$ of that macroscopic configuration, given by equation (26), and the entropy of the chain becomes

$$S = c - k\, b^2\, r^2 \tag{31a}$$

where c is a constant.

b. FJC model

The freely articulated chain model (FJC, Freely Jointed Chain) avoids the limitation of the Gaussian model which considered an infinite chain length (and led to the restriction that it could only be applied to cases where $r << nl_0$).

In this model, the probability density function $p(\mathbf{r})$ is the result of a random path of n steps each of length l_0 Function $p(\mathbf{r})$ turns out to be (Ward, Sweeney, 2004)

$$p(\mathbf{r}) = C \exp\left[-n\left\{\frac{r\beta}{nl_0} + \ln\left(\frac{\beta}{sinh(\beta)}\right)\right\}\right] \tag{26b}$$

where C is a constant and parameter β is given by the Langevin[11] '$\mathcal{L}(\beta)$

$$\mathcal{L}(\beta) = \coth(\beta) - \frac{1}{\beta} = \frac{r}{nl_0}$$

in which $\beta = \mathcal{L}^{-1}\left(r/(nl_0)\right)$

The development of equation (26b) for large values of n yields the Gaussian chain model (26) (see exercise 12 and figure 2.13).

Applying equation (30), the entropy of the chain becomes

$$S = c - kn\left[\frac{r\beta}{nl_0} + \ln\left(\frac{\beta}{sinh(\beta)}\right)\right] \tag{31b}$$

where c is a constant.

9 Johann Carl Friedrich Gauss (1777-1855) was a German mathematician and physicist, who used Equation (26a) widely in the analysis of astronomical data, as did Abraham de Moivre, the French mathematician, who first derived it in 1733.

10 Ludwig Edward Boltzmann (1844-1906) was an Austrian physicist and pioneer of statistics, who proposed Equation (30) for the calculation of entropy. That equation can be found engraved on his tomb in the central cemetery in Vienna.

11 Paul Langevin (1872 -1946) was a French physicist who derived the function $\mathcal{L}(\beta)$ during a study of the magnetisation of paramagnetic materials.

c. WLC model

In order to obtain a better fit to the experimental data, more complex models have been developed.

One of these is the worm chain model (WLC, Worm-Like Chain model), which provides a good representation of chains that have some degree of bending stiffness. The model assumes that the average angle formed between two links of the chains is not zero –as in the FJC model– but its cosine decreases as $\exp(-s/l_p)$, where s is the distance between two links and l_p a distance particular to the model, known as the persistence length. The response of the model when the length of the chain nl_0 is much greater than l_p is

$$F = \frac{kT}{l_p}\left[\frac{1}{4(1-r/nl_0)^2} - \frac{1}{4} + \frac{r}{nl_0}\right]$$

which, as can be seen in figure 1 (Bustamante *et al.* 2000), is in good agreement with experimental results.

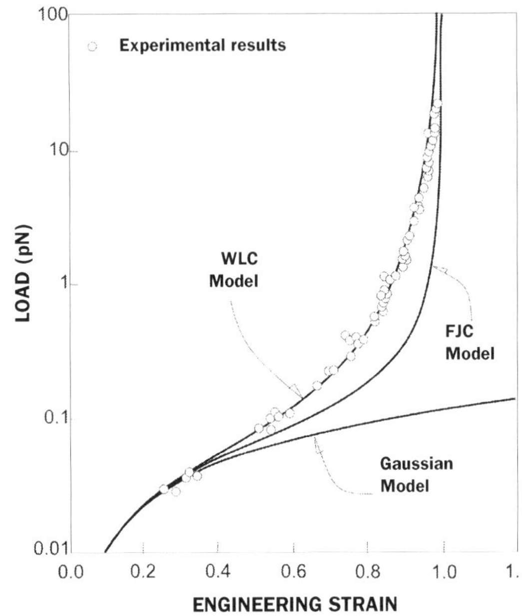

Fig. 2.13. *Load versus engineering strain curves for dsDNA (48,502 bp) of phage λ obtained by traction with magnetic microspheres. Comparison with the worm chain model WLC, which was obtained with an lp f 53 nm. The FJC model assumes a chain length l_0 of 106 nm. (Bustamante et al. 2000).*

Commentary on experiments with DNA molecules: When tensile tests are performed on DNA molecules using optical tweezers[12], or AFM[13], for example, the data shown in figure 2.13 are obtained. In this figure it is apparent that the Gaussian model (Hooke' law) only fits for small deformations, while the FJC model is good at the two asymptotic extremes: for small deformations and at very large ones. This suggests that the molcular chains (of DNA, in this case) do not behave as rigid links which rotate freely

[12] Optical tweezers use the confinement effect produced near the focal point of a spatially inhomogeneous beam of light on very small objects (microns). If the object is just outside the focus, the resulting refractive and dispersive forces of the light rays (of the order of pN) create a resultant that attracts the object to the focal point.

[13] Atomic Force Microscopy. An atomic force microscope (AFM) allows a very sharp pointer (~nm) to closely track a surface with very high precision, while measuring the force (~nN) between the surface and the tip of the pointer. It is also possible to attach the end of a molecule to the tip of the AFM and to pull it with great precision, performing a micro-tensile test.

EXERCISE 12

Using equations (25) and (26b), calculate the force on a chain as a function of its length r. Determine, also, the limits for large and small displacements.

Solution:

The force comes from equation (25) and the entropy from equation (31b). Therefore:

$$F = -T\left(\frac{\partial S}{\partial r}\right)_T = kTn\left[\frac{\beta}{nl_0} + \left(\frac{r}{nl_0} + \frac{1}{\beta} - \coth(\beta)\right)\frac{d\beta}{dr}\right]$$

and, bearing in mind that

$$\mathcal{L}(\beta) = \coth(\beta) - \frac{1}{\beta} = \frac{r}{nl_0}$$

we obtain

$$F = kTn\left[\frac{\beta}{nl_0}\right] = \frac{kT\beta}{l_0}$$

or

$$\mathcal{L}\left(\frac{Fl_0}{kT}\right) = \frac{r}{nl_0}$$

For small values of the argument (F and β, smalls), $\mathcal{L}(x) \approx x/3$, recovering the result for a Gaussian chain, which is no more than the linear relationship between force and displacement of Hooke's law

$$F = \frac{3kT}{nl_0^2}\,r$$

When the chain is highly stretched (F and β, large), F increases rapidly. For large values of the argument $\mathcal{L}(x) \approx 1 - 1/x$, giving

$$F = \frac{kT/l_0}{1 - r/nl_0}$$

In Figure 2.13, both results agree with experimental results.

Entropy of a molecular network

In this section an approximate form of the entropy of a network of polymer chains connected by cross-links will be calculted. To facilitate the calculations, the following assumptions are made.

Each chain between two anchor points behaves as a *Gaussian chain* with n links each of length l_0.

The chains are randomly distributed and in their initial state (at rest and unloaded) their mean values can be approximated as a free chain.

When the material deforms, the network of anchor points undergoes an *affine deformation*[14], as shown in figure 2.14. This type of deformation is usually considered and occurs whenever the element remains homogeneous (same composition at all points) and isotropic (with the same properties in all directions) during the deformation.

Consider, in the first instance, the entropy of one of the chains in the initial state, without load, which is, according to equation (31a),

$$S_{b0} = c - kb^2\left(x_0^2 + y_0^2 + z_0^2\right) \tag{32}$$

where it has been assumed, for simplicity, that one end of the chain is at the origin of coordinates and the other is at point $r_0\ (x_0, y_0, z_0)$.

After deformation of the material, the new contribution of this chain will be

$$S_b = c - kb^2\left(\lambda_1^2 x_0^2 + \lambda_2^2 y_0^2 + \lambda_3^2 z_0^2\right) \tag{33}$$

in which, again, we have eliminated the position of the origin of the chain–which has no relevance– in order to write the coordinates of the deformed end $(\lambda_1 x, \lambda_2 y, \lambda_3 z)$, see figure 2.14. The coefficients $(\lambda_1, \lambda_2, \lambda_3)$ are the elongations (ratios of the final and initial lengths) which produce the affine deformation in each of the three coordinate directions (x, y, z).

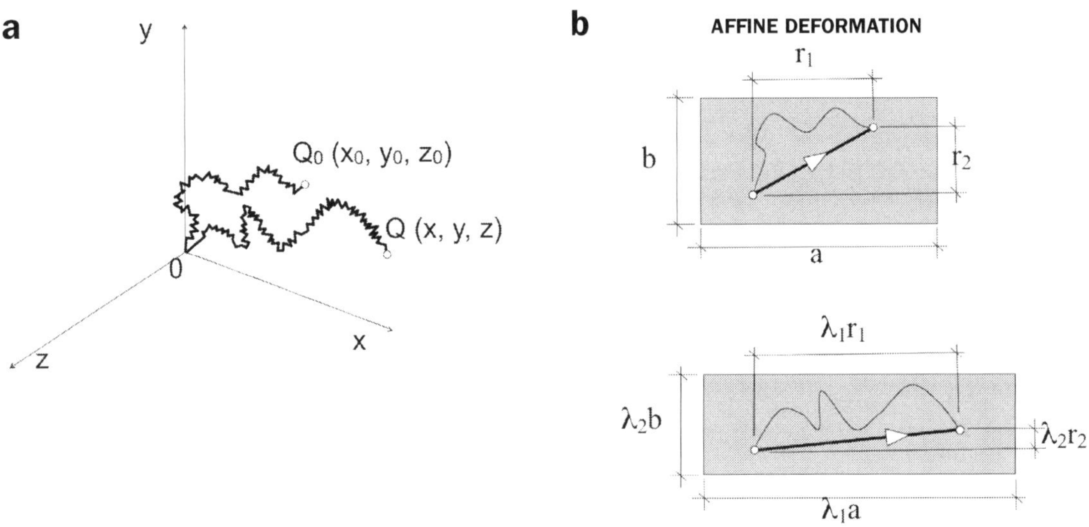

Fig. 2.14. *a. Idealised configuration of a polymer chain before and after deformation.*
b. The concept of an affine deformation. *The components in direction (1) change in the ratio λ_1 while those in direction (2) with λ_2.*

[14] In an affine deformation, the components of a vector that join two points in a material change in the same proportion as the corresponding components of the macroscopic strains. It is a linear transformation in which points initially situated on a straight line, or in a plane, remain in a straight line or forming a plane after deformation, although the lines and planes are usually different. Similarly, the transformation preserves parallelism between the lines, although not, generally, the angles or distances.

If N_b is the number of chains, the entropy due to all of them will be

$$S = \sum_{N_b} S_b$$

In more detail

$$S = cte - kb^2 \sum_{N_b} \left(\lambda_1^2 x_0^2 + \lambda_2^2 y_0^2 + \lambda_3^2 z_0^2 \right) =$$

$$= cte - kb^2 \left(\lambda_1^2 \sum_{N_b} x_0^2 + \lambda_2^2 \sum_{N_b} y_0^2 + \lambda_3^2 \sum_{N_b} x_0^2 \right) =$$

$$= cte - kN_b b^2 \left(\lambda_1^2 \overline{x_0^2} + \lambda_2^2 \overline{y_0^2} + \lambda_3^2 \overline{z_0^2} \right) \quad (34)$$

where use has been made of the fact that the deformation is the same in all of the chains and the mean values of x_0^2, y_0^2 and z_0^2 have been introduced. In the calculation it is assumed that the change in entropy due to the interaction between chains is negligible so that its only source is that due to the form of the chains (configurational entropy).

If it is assumed that the spatial average of the rest state coincides with that over the space of the free chains and that this, in turn, coincides with the time average[15], it is possible to substitute the mean values over all the chains at a given time with that over one of them, represented by the symbol < >, calculated in equation (29).

Since the chains are distributed randomly, the mean values must be equal in all directions:

$$\sum_{N_b} r_0^2 = N_b \left\langle r_0^2 \right\rangle = \sum_{N_b} \left(x_0^2 + y_0^2 + z_0^2 \right) = 3 \sum_{N_b} x^2 = 3N_b \left\langle x_0^2 \right\rangle$$

and

$$\left\langle x_0^2 \right\rangle = \left\langle y_0^2 \right\rangle = \left\langle z_0^2 \right\rangle = \frac{\left\langle r_0^2 \right\rangle}{3} = \frac{1}{2b^2} \quad (35)$$

Therefore

$$S = cte - \frac{kN_b}{2} \left(\lambda_1^2 + \lambda_2^2 + \lambda_3^2 \right) \quad (36)$$

This equation indicates that, although S_{tb} is proportional to N_b (as would be expected) it does not depend explicity on b and, therefore, the change in entropy per unit volume s, is given by

$$s = cte - \frac{kN}{2} \left(\lambda_1^2 + \lambda_2^2 + \lambda_3^2 \right) \quad (37)$$

where N is now the total number of chains per unit volume of the network.

The *constitutive equation*, that is the relationship between stress and strain, can be deduced from equation (25), since the total entropy of the network is obtained by adding a constant, independent of the strain, to equation (37).

For a tensile test in direction x, where $\lambda_1 = \lambda$, and $\lambda_2 = \lambda_3 = \lambda^{-1/2}$ (assuming the material is incompressible $\lambda_1 \lambda_2 \lambda_3 = 1$), the following result is obtained for the applied force F (see the exercise)

$$F = NkT\,(\lambda - \lambda^{-2}) \cdot A_0 \quad (38)$$

where A_0 is the initial cross-section of the specimen. Recalling that the true stress is given by $\sigma = F/A$, equation (3), and $A = A_0/\lambda$, due to concentration of volume, then:

$$\sigma = NkT\,(\lambda^2 - \lambda^{-1}) \quad (39a)$$

or, since $\sigma = s\,\lambda$

$$s = NkT\,(\lambda - \lambda^{-2}) \quad (39b)$$

This result provides reasonable values for moderate and medium strains (upto about 200%, corresponding to λ=3) as can be seen in the example in figure 2.15, despite the simplifications that have been made.

Perhaps, the most satisfactory result is that equation (39) gives a physical meaning to the parameters of the equation, something that does not occur with empirical constitutive equations. In this case it can be seen that the initial elastic modulus is directly proportional to the absolute temperature T and, also, the number of cross-links, characterised by N. (For an ideal network, N is equal to double the density of cross-links).

[15] Systems in which the temporal average of certain quantities can be obtained as averages over the state space are termed ergodic and are of great importance in statistical physics.

A similar development of the entropy in equation (31b) for the FJC model for a spatial network of three families of chains leads to a value of the true stress in a tensile test equal to

$$\sigma = \frac{NkT}{3}\,\lambda\sqrt{n}\left[\mathcal{L}^{-1}\left(\frac{\lambda}{\sqrt{n}}\right) - \lambda^{-\frac{3}{3}}\mathcal{L}^{-1}\left(\frac{1}{\sqrt{\lambda}\sqrt{n}}\right)\right] \quad (40)$$

which is shown in figure 2.15, where the engineering stress $s = \sigma/\lambda$ is depicted. The agreement is excellent even for large strains. In this case, the stress depends explicitly on the number of cross links in the chains, n. It can be readily shown (see the exercise) that for small elongations λ equation (40) converges –as would be expected– to equation (39) and the dependence upon n disappears.

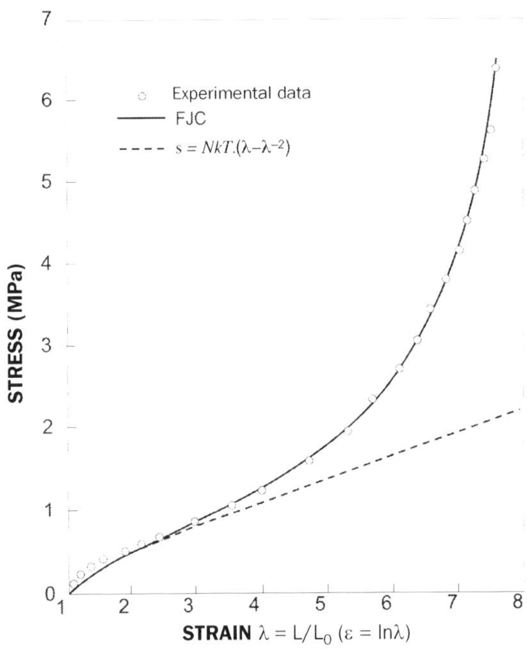

Fig. 2.15. *Engineering stress-strain curve for vulcanised rubber.*
Symbols - Experimental data (Treloar 1944).
Solid curve -FJC model (James and Guth 1943).
Dashed curve - Gaussian chain model, from equation (38).

2.4.4. Final remarks

Biological materials are, in general, quite complex as has been mentioned repeatedly. The arguments above –the thermodynamic approach and the molecular chain model– are only a first approximation in order to understand them when they exhibit elastomeric behaviour. It is, therefore, not surprising that the results obtained with this hypothesis have a limited value, which is usually confined to a regime of small deformations.

These concepts were developed in the middle of the last century and the interested reader will find a good summary in Treloar's book (Treloar, 1975). Since then there have been many attempts to refine this model: attempts to generalise the concept of cross-links, removing the restriction that they should be fixed and with some able to slide or even disappear (if they are caused by the entanglement of chains). Gaussian statistics has been generalised by taking into account the structure and mobility of real chains. And the contibution of the internal energy has also been incorporated.

In particular, the influence of different types of cross links and the modelling of more realistic molecular chains have given rise to more sophisticated constitutive equations, whose results can be found in Ball *et al.* 1981) and Edwards and Vilgis (1986), and which better reflect the stress-strain curves for large deformations. Generalisations of the non-Gaussian models have been produced for molecular chains perpendicular to each other (models with three, four, eight or more) (Arruda *et al.* 1993 and Wu and Giessen 1993) which can also reproduce the behaviour of the material for large deformations.

The objective of this section was to obtain a constitutive equation for fibres which exhibit elastomeric behaviour, that is to say, an expression which relates the force to the length of the fibre and the temperature, $F = f(l, T)$. The effects of humidity and other external factors have not been considered. The weak influence of the variation of internal energy has been justified, and it has been concluded that the constitutive equation can be obtained from the variation of entropy of the fibre with length, that is:

$$F = -T\left(\frac{\partial S}{\partial l}\right)_T \quad (25)$$

The problem has been reduced to calculation of the entropy for various models of the fibre stucture.

EXERCISE 13

Show that a material subjected to tension and whose constitutive equation is given by equation (39) does not neck before fracture.

Solution:

In a tensile test, the true strain ε, is related to the elongation λ through equation (4)

$$\varepsilon = \ln\frac{L}{L_0} = \ln\lambda$$

which can be inverted to give

$$\lambda = e^{\varepsilon}$$

Equation (8) gives the necking condition:

$$\sigma = \frac{d\sigma}{d\varepsilon}$$

introducing the constitutive equation of the material equation (39)

$$\frac{d\sigma}{d\varepsilon} = \frac{d\sigma}{d\lambda}\frac{d\lambda}{d\varepsilon} = NkT\left(2\lambda + \frac{1}{\lambda^2}\right)e^{\varepsilon} = NkT\left(2\lambda + \frac{1}{\lambda^2}\right)\lambda = \sigma = NkT\left(\lambda^2 - \frac{1}{\lambda}\right)$$

which leads to

$$\lambda^3 + 2 = 0$$

this has a solution $\lambda = -1.26$, which has no physical meaning.

The absence of necking can be confirmed graphically with Considère's construction:

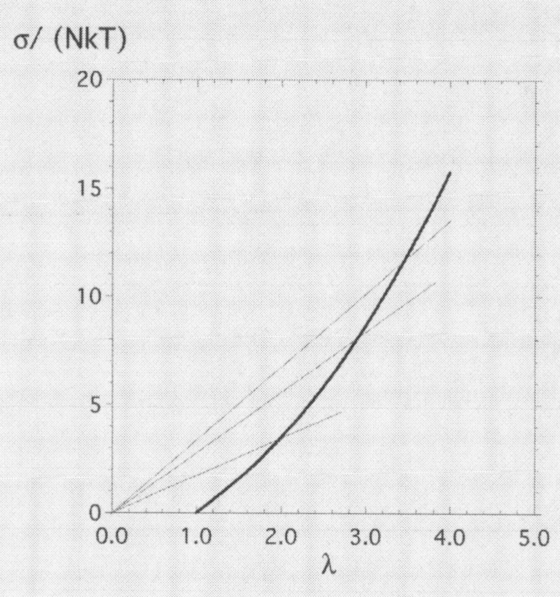

EXERCISE 14
Show that throughout the deformation of an incompressible material $\lambda_1\lambda_2\lambda_3 = 1$ and deduce equations (38) and (39).

Solution:
Consider an infinitesimal element $dV_0 = dx\, dy\, dz$ before deformation, and whose edges are orientated along directions (1), (2) y (3). After deformation, its dimensions become $dV = \lambda_1 dx\, \lambda_2 dy\, \lambda_3 dz = \lambda_1\lambda_2\lambda_3\, dV_0$

From a simple inspection of the above equation, we see the incompressibility condition (= constant volume, $dV = dV_0$) requires $\lambda_1\lambda_2\lambda_3 = 1$.

Suppose that the element is incompressible, homogeneous and isotropic, and subjected to tensile loading, with initial lengths and cross-sectional area L_0 and A_0, respectively. If the material is stretched to length L it will suffer an elongation

$$\lambda_1 = \lambda = \frac{L}{L_0}$$

The elongations in the other two orthogonal directions (those of the cross-section A_0) confirm that $\lambda_1\lambda_2\lambda_3 = 1 = \lambda\,\lambda_2\,\lambda_2$ then the material is incompressible, and $\lambda_2 = \lambda_3$ by symmetry. From this last equation we see that $\lambda_2 = \lambda_3 = \lambda^{-1/2}$

The value of the initial cross-sectional area A_0 changes under deformation depending on the values λ_2 y λ_3, giving

$$A = \lambda_2\lambda_3\, A_0 = \lambda^{-1}\, A_0$$

So that the true stress will be equal to (see equations (3) and (7)) $\sigma = \dfrac{F}{A} = \lambda \dfrac{F}{A_0}$

where F is the applied force. In order to calculate its value, we use equation (25):

$F = -T\left(\dfrac{\partial S}{\partial L}\right)_T$ with $S = s\, A_0 L_0$

noting that s is given by equation (37) and $V_0 = A_0\, L_0$ the volume of the specimen. Derivation yields

$$F = -TA_0L_0\left(\frac{\partial s}{\partial L}\right)_T = -TA_0L_0\sum_{i=1}^{3}\left(\frac{\partial s}{\partial \lambda_i}\right)_T \frac{\partial \lambda_i}{\partial L}$$

The derivatives of the elongations are

$$\frac{\partial \lambda_1}{\partial L} = \frac{\partial \lambda}{\partial L} = \frac{\partial}{\partial L}\left(\frac{L}{L_0}\right) = \frac{1}{L_0} \quad \text{and} \quad \frac{\partial \lambda_2}{\partial L} = \frac{\partial \lambda_3}{\partial L} = \frac{\partial \lambda^{-1/2}}{\partial L} = -\frac{1}{2}\lambda^{-3/2}\frac{\partial \lambda}{\partial L} = -\frac{1}{2}\lambda^{-3/2}\frac{1}{L_0}$$

resulting in $F = -TA_0L_0\left[Nk\lambda_1\dfrac{1}{L_0} + Nk\lambda_2\left(-\dfrac{1}{2}\lambda^{-3/2}\dfrac{1}{L_0}\right) + Nk\lambda_3\left(-\dfrac{1}{2}\lambda^{-3/2}\dfrac{1}{L_0}\right)\right] = NkTA_0\left[\lambda - \lambda^{-2}\right]$

And the true stress is

$$\sigma = \lambda\frac{F}{A_0} = NkT\left[\lambda^2 - \frac{1}{\lambda}\right]$$

Which are equations (38) and (39).

The engineering stress s (not to be confused with the entropy per unit volume!) is simply

$$s = \frac{F}{A_0} = NkT\left[\lambda - \lambda^{-2}\right]$$

and which is represented in figure 2.15.

2.5. VISCOELASTIC BEHAVIOUR

2.5.1. Introduction

Figure 2.16 shows the results from three tensile tests on a natural rubber fibre, vulcanised with 8% sulphur (Treloar, 1944).

For small strains, up to ε of about 0.40 (λ = 1.5), **Hookean elastic** behaviour can be assumed (Figure 2.16a).

Even as the maximum strain approaches ε = 1.7 (λ = 5.5) the response is still reversible –the loading line coincides, more or less, with that of unloading. The behaviour is nonlinear elastic and the material considered as **elastomeric** (Figure 2.16b).

For larger deformations the unloading curve no longer coincides with that on loading, as is apparent in figure 2.16c. This effect, known as *hysteresis*, complicates the analysis of the tensile test since, for example, for a particular value of strain there are two values of stress. This result means that a new variable, time, must be introduced in order to characterise this behaviour which, in what follows, will be called **viscoelastic;** *visco*, to emphasise the influence of time and *elastic* because it will be assumed that the deformation will be recovered when the specimen is unloaded.

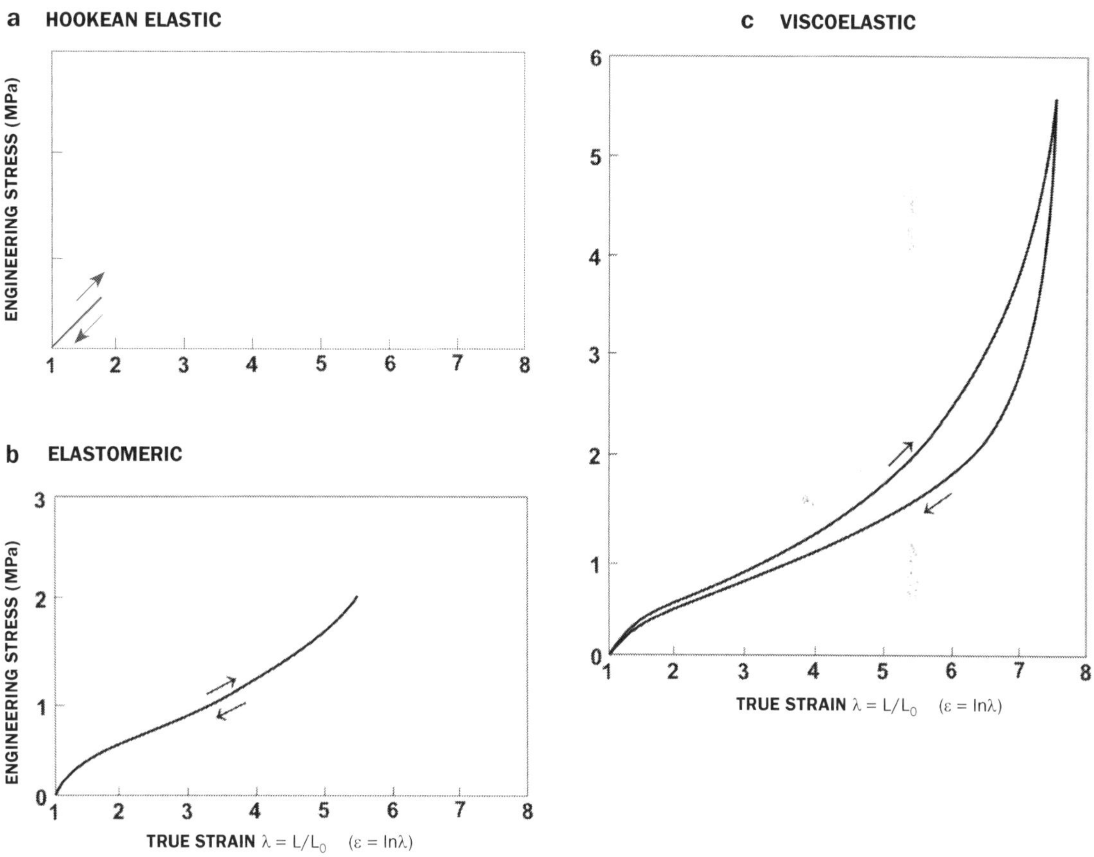

Fig. 2.16. *Engineering stress-elongation curves for a vulcanised rubber fibre (Treloar 1944).*
a. Hookean elastic up to λ = 1.5
b. Elastomeric behaviour up to λ = 5.5.
c. Viscoelastic behaviour when λ_{max} = 7.5.

2.5.2. Mechanical tests

Various forms of tensile test allow characterisation of the viscoelastic behaviour of a material. The simplest are the *creep* test, that of *relaxation*, and *cyclic* tests.

The **creep** test consists of subjecting the specimen to a constant stress and recording its deformation as a function of time.

Creep results for the mesoglea jellyfish *Cyanea capillata* (Alexander, 1964) are shown in figure 2.17a. The stress has been kept constant at 125 Pa, and the deformation increases with time (note the difference in the two curves when the same phenomenon is represented using a natural scale and a logarithmic one).

When the specimen deforms a lot, such as happens with the example cited, the cross-section may change significantly, compared to the initial value, so it is necessary to change the applied load in order to achieve a constant (true) stress experiment. When there is no information available about the variation of cross-section, the specimen constant volume assumption is usually made for the purpose of calculations. This assumption is not always accurate for biological materials and it must be confirmed on a case by case basis.

In a creep test the *flexibility* of the specimen, J, is proportional to the strain ε because the stress is kept constant; $J(t) = \varepsilon(t)/\sigma_0$.

The variation of the flexibility that occurs in creep tests has the characteristics shown in figure 2.17b, in a very general schematic form. The viscoelastic response is only shown for the duration of the experiment. For very short times, or after a long time, the material does not flow but has glassy and elastomeric behaviours, respectively. The creep test helps characterise a section of this curve.

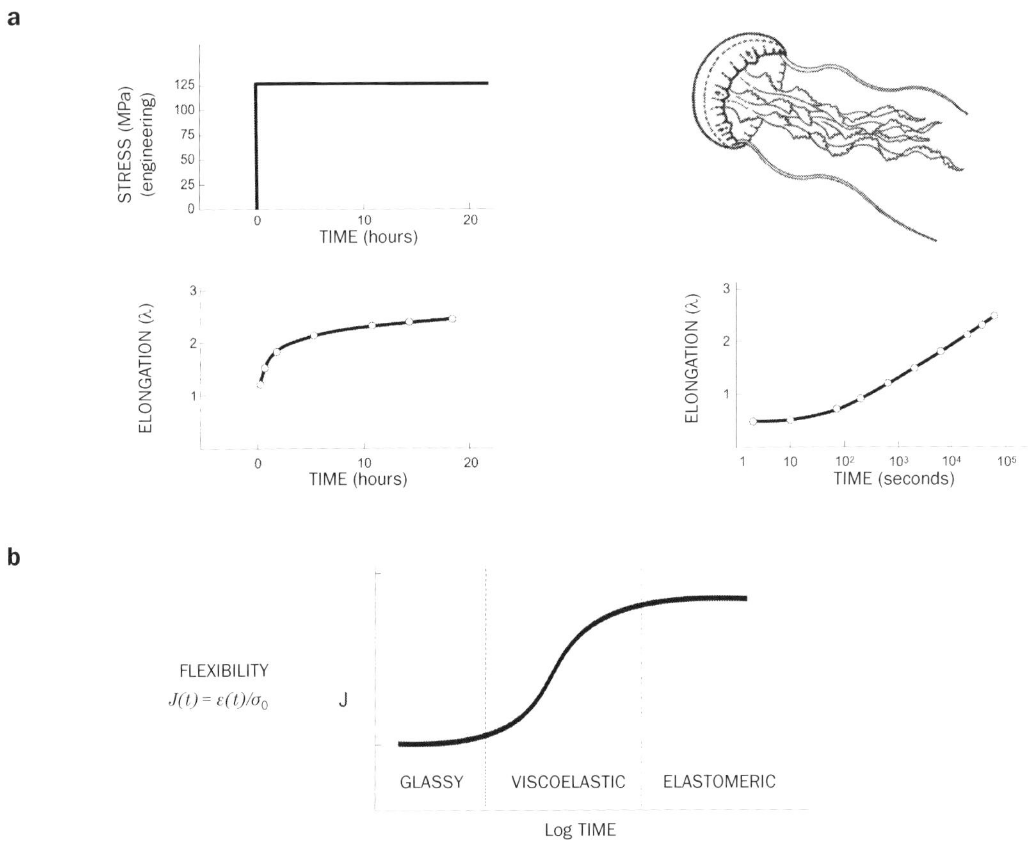

Fig. 2.17. *a. Creep test on a fibre of the mesoglea jellyfish* Cyanea capillata *(Alexander, 1964). The creep curves are quite different when drawn with natural and logarithmic scales.*
b. Variation of flexibility in a creep test as a function of time. In the figure all possible states have been represented; the experiment takes place over a particular time interval.

The **relaxation** test consists of subjecting the specimen to a constant deformation and recording the stress as a function of time.

Results for elastin are shown in figure 2.18a. The specimen is a ring taken from the porcine aortic wall from which everything except the elastin has been removed (Dorrington 1975). The initial strain –about 30%– was kept constant throughout the duration of the experiment and the load relaxation monitored (due to difficulties in estimating the cross-section at any time, only the reduction in force is represented).

In a relaxation test the *rigidity* of the specimen, G, is proportional to the stress, σ, since the strain is kept constant; $G(t) = \sigma(t)/\varepsilon_0$.

The evolution of the rigidity in relaxation experiments, figure 2.18b, exhibits similar stages to those reported above for creep. Viscoelastic relaxation is shown only for the duration of the test. For very short times the material does not relax and behaves like a glassy solid. For very long times, the behaviour is elastomeric.

a

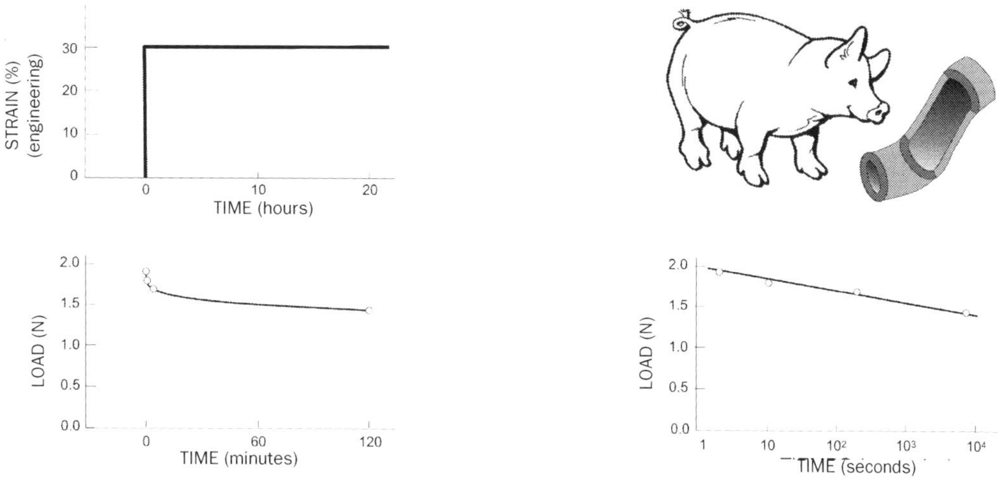

b

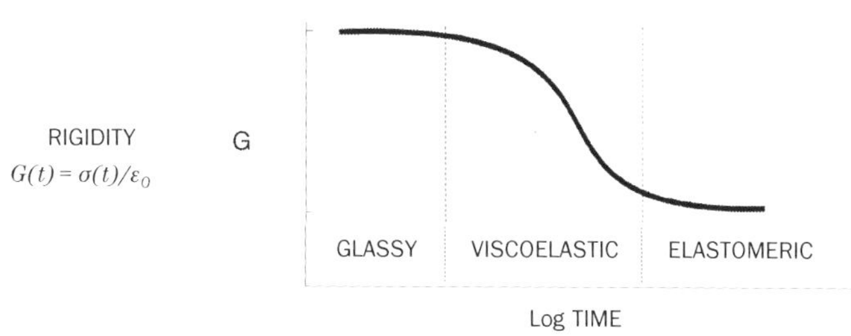

Fig. 2.18. *a. Relaxation test on a ring of aorta (Dorrington, 1975). The relaxation curves are shown with natural and logarithmic time scales.*
b. Variation of the rigidity in a relaxation test as a function of time. In the figure all possible stages have been represented, the experiment only runs for a fixed duration.

The **cyclic strain** test (also called dynamic, or forced vibration) consists of loading the specimen to an initial strain, $\varepsilon_{inicial}$, and then deforming it cyclically, with an amplitude ε_0, which is generally small, so:

$$\varepsilon = \varepsilon_{initial} + \varepsilon_0 \sin \omega t$$

where ω is the frequency of oscillation, and t time.

Only when the material is *linear viscoelastic* (a concept introduced later) the stress response –after a transitory time of material accommodation– is of the same form although out of phase, as shown in figure 2.19, that is

$$\sigma = \sigma_{initial} + \sigma_0 \sin(\omega t + \delta)$$

where δ is the phase angle, and $\sigma_{initial}$ and σ_0 are constants.

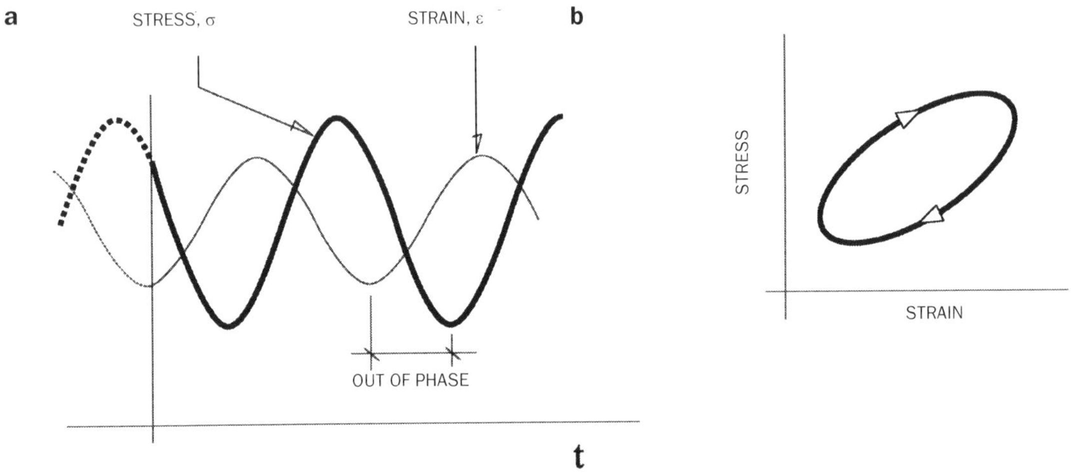

Fig. 2.19.
a. Cyclic test for a fibre with a linear viscoelastic behaviour.
b. Hysteresis loop.

BIOLOGICAL MATERIALS AND VISCOELASTIC MODELS

Time is an essential factor in understanding the mechanical behaviour of biological materials and must be taken into account when formulating models of these materials. When subjected to a constant load, the materials deform with the passage of time –a phenomenon known as *creep*– if, on the other hand, they are subjected to a constant strain, the stresses relax with time –a phenomenon known as *relaxation*. In a process of loading and unloading the paths, in a force-displacement diagram, are not coincident and form a loop, known as *hysteresis*. Time also influences the restructuring of biological materials, because they age, but they can be *rejuvenated* and reorganise themselves with time, under the appropriate conditions. The tool used for studying these phenomena –at the macroscopic level– is viscoelasticity, an essential chapter in the study of the mechanics of continuous media

In animal tissue, for example, knowing how the response to mechanical loading can play a major role in understanding their functioning –under both normal and pathological conditions– and the influence of time must be considered in interpreting its behaviour: Synchronised contractions of the cardiovascular system allow the circulation of blood. Contractions of the diaphragm and the intercostal muscles facilitate breathing. Peristalsis of the intestines helps the passage of food through the digestive tract. Muscles and tendons contract and stretch during physiological activities. Examples which, while only from animal tissue, would alone justify these studies. Other reasons for outlining the role played by time in the relationship between structure and biological properties are the advantages that this knowledge can bring to the design of new intelligent materials.

The mathematical formulation of viscoelastic models (or better, perhaps, *visco-elasto-plastic,* because plastic phenomena also occur) is more complex than that of purely elastic or elastomeric materials and, in general, viscoelastic models require much more experimental data in order to make reliable predictions. However, in certain circumstances and for small deformations, it is possible to use simplified models –based on linear viscoelasticity– which are useful, as a first step, on the way to the study of more complex situations. In this book only linear models are considered and the interested reader must consult specialist books, some of which are mentioned in this text.

In order to characterise the viscoelastic behaviour of biological materials several techniques are used: The simplest are based upon *tensile and torsion tests.* (For this reason, the section on viscoelasticity has been placed in this volume in which these tests are discussed in some detail. Another reason is that the results of these experiments are sufficient to create a *one-dimensional formulation* of linear viscoelasticity, which is the commonly used approach). The above procedures allow the performance of *creep, relaxation* and *fatigue* experiments, with the ability to vary, in all of these, the rate of loading, the temperature and other ambient parameters.

Other more specialist techniques, not discussed here, are based upon rheometers for studying how time intervenes in the relationships between forces and shear deformations. Measurements based on indentation techniques (see Section 3) provide information on the microscopic scale. For example, for animal tissue, and at the nanoscale, they provide information on the viscoelastic behaviour between cells and the extracellular matrix. More recent techniques, based on magnetic resonance (MRE) (Magnetic Resonance Elastography) or on ultrasonics (SDUV) (Shearwave Dispersion Ultrasound Vibrometry), have the advantage of not being invasive but the disadvantage, currently, of a low signal-to-noise ratio.

2.5.3. Linear viscoelasticity

1. Assumptions about linear viscoelasticity

The simplest description of viscoelastic behaviour is that of linear viscoelasticity, based two hypotheses (Boltzmann 1876). In order to define the concepts reference is made to a *relaxation* experiment but the same steps could be taken for a *creep* test.

Linearity hypothesis

Figure 2.20 illustrates the concept. Two identical specimens are used in two **relaxation** experiments; one with a strain value of ε_A (figure 2.20a) and the other with a strain ε_B (figure 2.20b). If the corresponding stresses are measured at the same time t_1 in both tests, σ_A and σ_B, the quotient of the stresses and strains must be the same (figure 2.20c), that is, $\sigma_A / \varepsilon_A = \sigma_B / \varepsilon_B = G(t_1)$. As a result, in a general form, for a relaxation test with strain ε_0:

$$\sigma(t) = \varepsilon_0 G(t) \tag{40}$$

where $G(t)$, is called the **modulus of relaxation.** Evidently, $G(t)$ is a function of time, depending only upon the material, but neither the test nor the specimen is used. This function should not be confused with the Hookean elastic shear modulus, defined in linear elasticity.

Superposition hypothesis

It is now supposed that equation (40) is not only applicable the total values of stress and strain, but also to their increments, that is:

$$\Delta\sigma(t) = G(t\text{-}t')\ \Delta\varepsilon(t') \tag{41}$$

which gives the stress increment in time, t, due to a strain increment suffered in t'.

Figure 2.21 illustrates the hypothesis. In this figure, a relaxation experiment in which the strain is applied in the form of a series of steps is shown. The deformation is decomposed into steps $\Delta\varepsilon_i$ which give rise, according to equation (40), to stress increments $\Delta\sigma_i$, as shown in figure 2.21b. The stress response of the material at some time t is given by:

$$\sigma(t) = \sum G(t - t_i)\ \Delta\epsilon_i \tag{42}$$

The result can be generalised for each applied strain history, $\varepsilon(t)$, as far as the limit of infinitesimal strain increments, as illustrated in figure 2.21c. The stress can then be evaluated from the following integral expression:

$$\sigma(t) = \int_0^{\varepsilon} G(t-\tau)d\varepsilon = \int_0^{t} G(t-\tau)\frac{d\varepsilon(\tau)}{d\tau}d\tau \tag{43a}$$

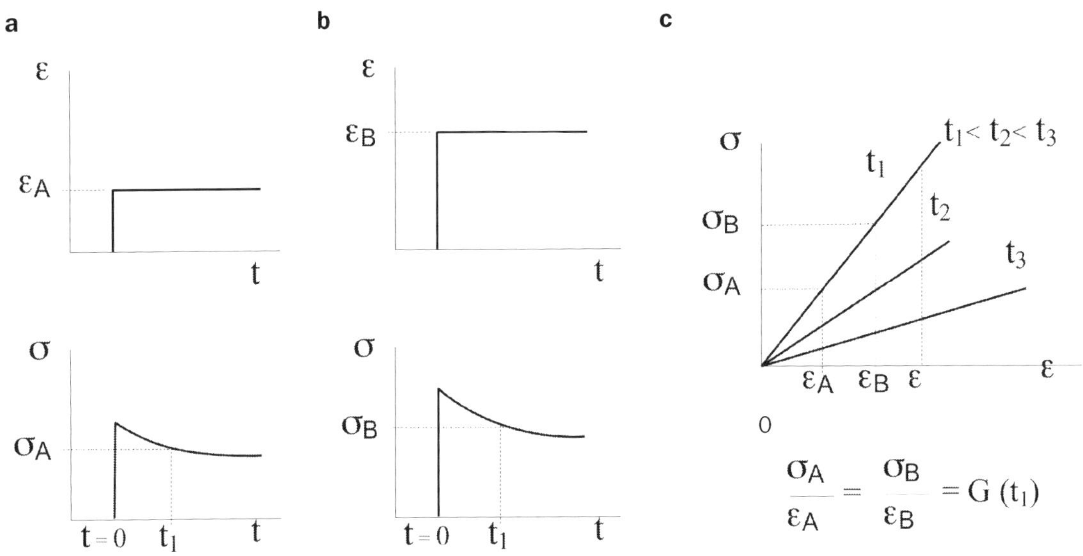

Fig. 2.20. *Linearity hypothesis.*
a. Relaxation with ε_A. Variation of strains and stresses with time
b. Relaxation with ε_B. Variation of strains and stresses with time
c. Isochronic stress-strain curves. For a particular time, the relaxation stress-strain is always linear and independent of the test

Therefore, the stress response to general strain loading $\varepsilon(t)$, can be determined if relaxation modulus function of the material, $G(t)$, is known.

In the case that the initial strain, ε_0, is not zero it is necessary to take into account the jump in strain that is produced at $t=0$. Equation (43a) then becomes (Wineman and Rajagopal, 2000):

$$\sigma(t)=\varepsilon_0 G(t)+\int_{0^+}^{t} G(t-\tau)\frac{d\varepsilon(\tau)}{d\tau}d\tau \qquad (43b)$$

in which the integral is evaluated from the positive side of 0 (that is, 0^+) without including the jump ε_0.

Creep tests can be handled in an analogous manner, producing the following result: the strain response to an applied stress, given by $\sigma(t)$, can be determined, if the material **creep modulus** $J(t)$ is known, from the expression

$$\varepsilon(t)=\int_0^{\varepsilon} J(t-\tau)d\sigma=\int_0^{t} J(t-\tau)\frac{d\sigma(\tau)}{d\tau}d\tau \qquad (44)$$

In the same way, as with equation (43), when the initial stress, σ_0, is not zero the jump produced at $t=0$ must be considered. Equation (44) then becomes (Wineman and Rajagopal, 2000):

$$\varepsilon(t)=\sigma_0 J(t)+\int_{0^+}^{t} J(t-\tau)\frac{d\sigma(\tau)}{d\tau}d\tau \qquad (44b)$$

in which the integral is evaluated from 0^+, without including the jump σ_0.

Note that both $G(t)$ and $J(t)$ on their own completely characterise the linear viscoelastic response of the material and are not, therefore, independent, so that each one can be deduced from the other. The equations relating them can be found in Wineman and Rajagopal (2000).

The restrictions imposed by linear viscoelasticity are so severe that practically no material, biological or otherwise, complies with them. It is sufficient to calculate the ratio $\sigma(t)/\varepsilon_0=G(t)$ in a relaxation test and to check that it is not independent of the imposed strain level, as would be expected. The same happens if $J(t)=\varepsilon(t)/\sigma_0$ is calculated in a creep test, which turns out to be dependent upon σ_0. However, the linear viscoelastic model is frequently used with all types of material, restricting it to a very limited range of stresses and strains so that its predictions are not too inaccurate.

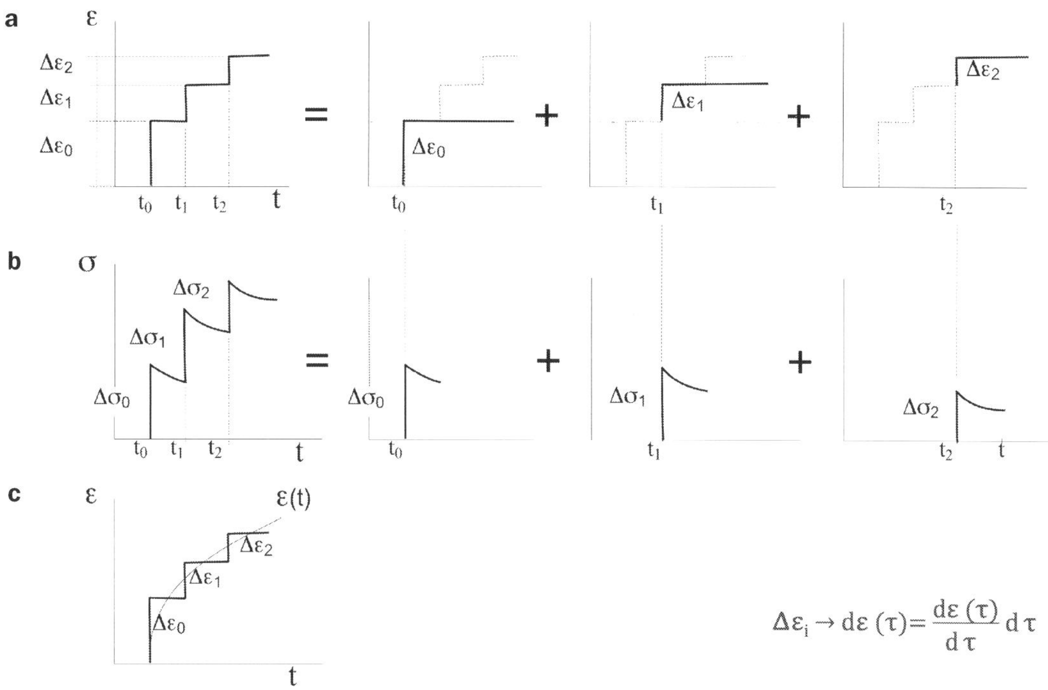

Fig. 2.21. *Superposition hypothesis.*
a. Stepped relaxation test and decomposition of the strain into steps
b. Obtaining stresses as a sum of the responses to each strain step
c. Decomposition of general loading $\varepsilon(t)$ into strain steps.

LINEAR VISCOELASTICITY

The experiments performed by Smart and Williams (1972) with polypropylene fibres serve to illustrate the concepts above, although not strictly based on natural fibres.

In figure **a** isochronic curves corresponding to *relaxation* tests are presented for several values of strain (curves analogous to those shown in figure 2.20c)

For small strains (less than 0.5%) it can be seen that a set of straight lines are obtained. This indicates that the **linear** hypothesis is valid in this range of strains. In an experiment of loading and unloading –within the cited interval– a straight line is obtained, as shown in figure **a**.

For large strains, the isochronic curves are no longer straight lines and the *linearity* assumption is not valid. If a loading and unloading test w.re performed the phenomenon of hysteresis would occur, as shown in figure **c**, for a strain of 4%.

The properties of *linearity* and *superposition* are independent. It is possible that the experimental results satisfy the conditions of linearity but not those of superposition, and the contrary. Only, when a viscoelastic material obeys both hypotheses, for different loadings, can the material be considered as *linear viscoelastic.*

In general, viscoelastic biological materials only show *linear viscoelastic* behaviour for small strains (less than 1%) and in certain temperature ranges.

When characterising the large strain behaviour of a viscoelastic material behaviour, the experimentation is more complicated and the mathematical treatment more arduous. See, for example, Lakes (2009) or Wineman and Rajagopal (2000).

a

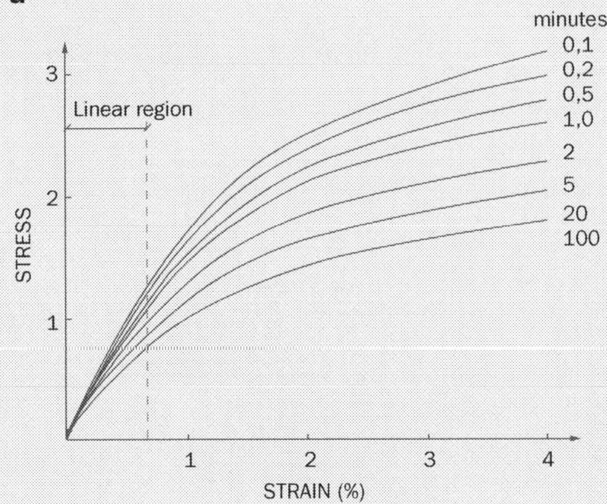

Isochronous relaxation curves

b

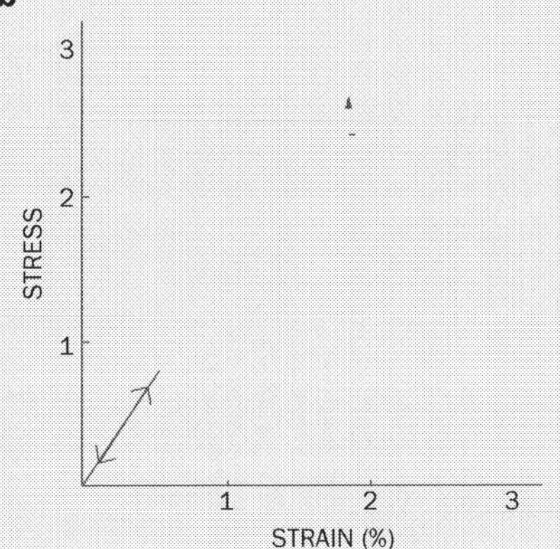

Loading/unloading curve until 0,7% strain

c

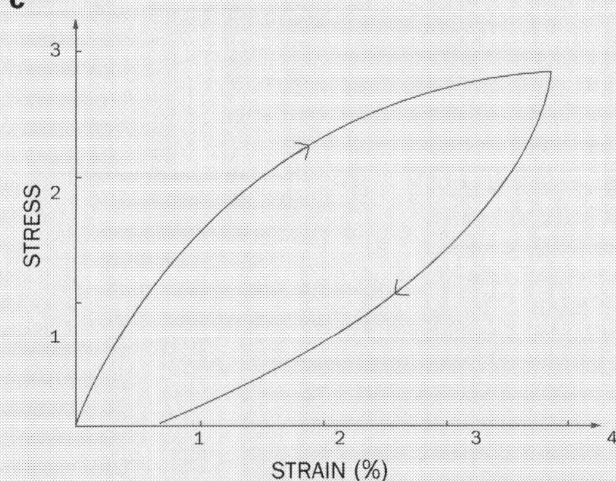

Loading/unloading curve until 4% strain

2. Mechanical models

The search for constitutive equations within the framework of linear viscoelastity is reduced, as mentioned above, to the determination of expressions for the relaxation or creep modulus. Ideally these expressions must come from a knowledge of the microstructure of the material and its response to mechanical stimuli, a complex task because biological materials are, in general, composite materials and very hierarchical, and it is difficult to unravel the role that each component plays in this process.

An attempt to simplify viscoelastic behaviour can be made by means of phenomenological models based on ideal springs and dampers (dashpots). These elemental components are represented in figures 2.22a and 2.22b.

The simplest model for a solid, capable of reproducing creep and relaxation phenomena, is called the *standard linear solid* (Zener, 1948), shown in figure 2.22c and is composed of a Hookean spring (of modulus E_H) in parallel with a Maxwell fluid model (figure 2.22d), which is an element formed of a Hookean spring, E_V, in series with Newtonian damper of viscosity η.

The corresponding constitutive equation for this model is obtained by noting that the stress is equal in the damper and spring collocated in series, and the total stress σ is the sum of the stresses in the isolated spring of the assembled damper+spring element, with both devices having the same strain ε since they are in parallel. The resulting equation is:

$$\sigma + \tau_R \frac{d\sigma}{dt} = E_H \epsilon + (E_H + E_V) \tau_R \frac{d\epsilon}{dt} \qquad (45)$$

where the time $\tau_R = \eta / E_V$ is usually called the *relaxation time*.

The responses of the model to different basic loadings: relaxation, creep, or cyclic are presented below, leaving their deduction to the reader (see, for example, Wineman and Rajagopal 2000).

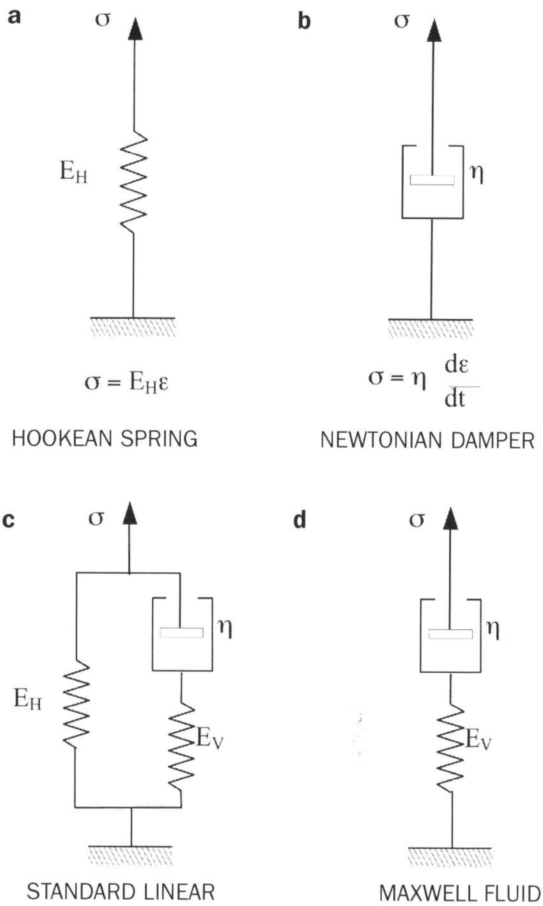

Fig. 2.22.
a. *Hookean spring; characterised by elastic modulus* E_H.
b. *Newtonian damper; characterised by viscosity* η.
c. *Standard linear model; characterised by elastic moduli* E_H, E_V *and viscosity* η.
d. *Maxwell fluid; characterised by elastic modulus* E_V *and viscosity* η.

EXERCISE 15
Derive equation (45).

Solution:
Consider the solid in figure 2.22c, subjected to a stress σ *and strain* ε. Let the strain in the Hookean spring be $\varepsilon_H=\varepsilon$, that in the spring of the Maxwell element ε_V and that in the damper ε_η. Compatibility of strains leads to $\varepsilon=\varepsilon_V+\varepsilon_\eta$

The stress in spring E_H is σ_H. In the Maxwell element the stress in the damper is σ_η and in the spring, σ_V.

Equilibrium of forces means that: $\sigma=\sigma_H+\sigma_V$ y $\sigma_\eta=\sigma_V$, since the damper and spring of the Maxwell element act in series.

The constitutive equations of the elements requires that

$$\sigma_H=\varepsilon_H E_H=\varepsilon E_H; \quad \sigma_V=\varepsilon_V E_V=\sigma_\eta=\dot{\varepsilon}_\eta\eta$$

In order to obtain the constitutive equation of the solid we combine the above equations from the equilibrium of forces. Differentiating with respect to time gives:

$$\dot{\sigma}=\dot{\sigma}_H+\dot{\sigma}_V=E_H\dot{\varepsilon}+E_V\dot{\varepsilon}_V=E_H\dot{\varepsilon}+E_V(\dot{\varepsilon}-\dot{\varepsilon}_\eta)=(E_H+E_V)\dot{\varepsilon}-E_V\frac{\sigma_\eta}{\eta}$$

since $\varepsilon_V=\varepsilon-\varepsilon_\eta$. Substituting $\sigma_\eta=\sigma_V=\sigma-\sigma_H$

$$\dot{\sigma}=(E_H+E_V)\dot{\varepsilon}-\frac{E_V}{\eta}(\sigma-\sigma_H)=(E_H+E_V)\dot{\varepsilon}-\frac{E_V}{\eta}(\sigma-\varepsilon E_H)$$

and rearranging the terms

$$\dot{\sigma}+\frac{E_V}{\eta}\sigma=(E_H+E_V)\dot{\varepsilon}+\varepsilon\frac{E_V E_H}{\eta}$$

which is equation (45).

Use of mechanical analogue models is relatively widespread not only because their equations are easy to handle, but also because the results are particularly helpful in idealising and better understanding linear viscoelastic behaviour and its peculiarities. Thus, in a relaxation test it is apparent that the initial response is only due to the springs E_H and E_V, because, with the infinite loading rate (the initial step from 0 to ε_0) the viscoelastic damper works like a rigid solid (just as if a load is suddenly applied to a real damper). Therefore, the initial stiffness must be $E_H + E_V$. Similarly, for an infinite time the strain rate in the damper is extremely slow –theoretically zero– it can be seen that the damper provides no resistance to motion (as in applying a very slow load to a real damper) and the stress is borne only by the spring E_H, whose stiffness determines the behaviour.

On the other hand, it is often useful to associate the different strain micromechanisms of a material to the mechanical analogue models characterised by their relaxation time. For example Johnson *et al.* (2010) modelled the viscoelastic behaviour of cortical bone using spring-damper elements with a relaxation times of 30ms and 10µs, corresponding to two very different mechanisms: the sliding between lamellae and tissue fluid flow in the haversian canals, respectively. In this way, mechanical analogue models help to isolate the different resistance mechanisms of each material and allow separate consideration of their effects.

Relaxation test *(standard model)*
The loading and the response are shown schematically in figure 2.23a. The values of strain and stress are given by (see the exercise):

$$\epsilon\,(t) = \epsilon_0$$
$$\sigma\,(t) = \epsilon_0\left(E_H + E_V\,e^{-t/\tau_R}\right) = \epsilon_0\,G(t) \qquad (46)$$

where τ_R, the characteristic relaxation time, is given by

$$\tau_R = \eta/E_V \qquad (47)$$

From this result which determines $G(t)$, it is possible to calculate the variation of the stress $\sigma(t)$ as a function of any strain history $\varepsilon\,(t)$, using equation (43b).

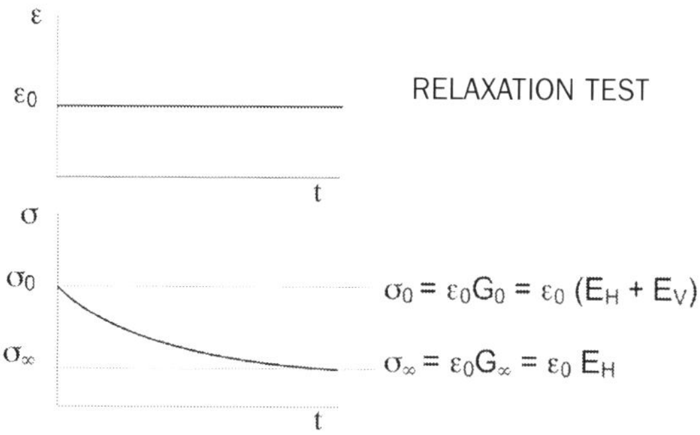

Fig. 2.23a. *Standard linear solid. Relaxation test.*

EXERCISE 16
Derive equation (46).

Solution:
Since $\varepsilon = \varepsilon_0$, a constant, then $d\varepsilon/dt = 0$ at all times, which when substituted into equation (45) gives

$$\dot{\sigma} + \frac{E_V}{\eta}\sigma = \varepsilon_0\frac{E_V E_H}{\eta}$$

This a differential equation which can be integrated with the initial condition $\sigma(0) = (E_H + E_V)\,\varepsilon_0$, since right at the beginning, and for an infinitely rapid loading, the damper behaves like a perfectly rigid material, and the solid behaves like two springs in parallel, with stiffnesses E_H and E_V.

The homogenous equation is

$$\frac{d\sigma}{dt} = -\frac{E_V}{\eta}\sigma$$

which has the solution

$$\sigma = Ce^{-\frac{E_V}{\eta}t}$$

where C is a constant. The particular integral is $\sigma = E_H\,\varepsilon_0$, which when combined with the solution to homogeneous equation gives the general solution

$$\sigma = Ce^{-\frac{E_V}{\eta}t} + E_H\varepsilon_0$$

Imposing the boundary condition $\sigma(0) = (E_H + E_V)\,\varepsilon_0$ yields

$$C = E_V\varepsilon_0$$

and

$$\sigma = E_v\varepsilon_0 e^{-\frac{E_V}{\eta}t} + E_H\varepsilon_0 = \varepsilon_0\left(E_H + E_v e^{-\frac{E_V}{\eta}t}\right)$$

Creep test *(standard model)*

The loading and response are shown in figure 2.23b, in a simplified form. The values of stress and strain are given by:

$$\sigma(t) = \sigma_0 \tag{48}$$

$$\epsilon(t) = \sigma_0 \left[\frac{1}{E_H} + \left(\frac{1}{E_H + E_V} - \frac{1}{E_H}\right) e^{-t/\tau_C}\right] = \sigma_0 J(t)$$

where τ_C, the creep time characteristic, is defined by

$$\tau_C = \frac{E_H + E_V}{E_H E_V}\eta \tag{49}$$

With this result, knowing $J(t)$ from equation (48) it is possible to calculate the evolution of strain $\varepsilon(t)$ as a function of any stress history $\sigma(t)$ using equation (44b)

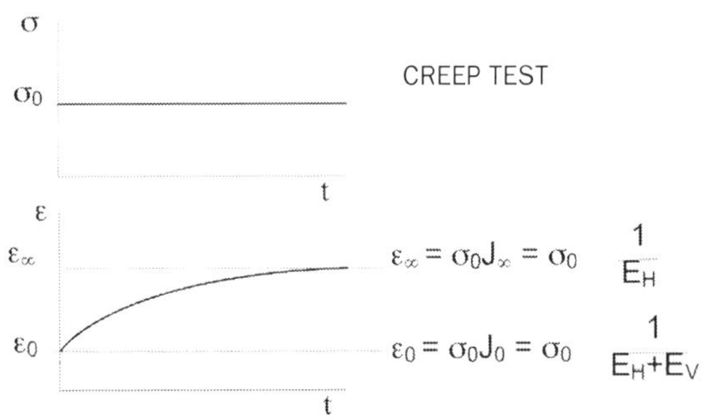

Fig. 2.23b. *Standard linear solid. Creep test.*

EXERCISE 17

Derive equation (48). Justify the values of the flexibility obtained at times t=0 and t=∞ in a creep test for a standard linear solid.

Solution:

In the case we have $\sigma = \sigma_0$, a constant, and $d\sigma/dt = 0$ at all times. These conditions, substituted in equation (45) yield

$$(E_H + E_V)\dot{\varepsilon} + \varepsilon\frac{E_V E_H}{\eta} = \frac{E_V}{\eta}\sigma_0$$

which is again a first-order differential equation that can be integrated with the initial condition $\varepsilon(0) = \sigma_0/(E_H + E_V)$, since, initially, and for an infinitely rapid loading, the damper behaves like a perfectly rigid body, and the standard solid behaves like two springs in parallel, with stiffnesses E_H and E_V.

The homogeneous equation is

$$\frac{d\varepsilon}{dt} = -\frac{E_H E_V}{\eta(E_H + E_V)}\varepsilon$$

whose solution is

$$\varepsilon = Ce^{-\frac{E_H E_V}{\eta(E_H + E_V)}t}$$

where C *is a constant. The particular integral of the complete equation is* $\varepsilon = \sigma_0/E_H$, which when combined with the solution to the homogeneous equation gives the general solution

$$\varepsilon = Ce^{-\frac{E_H E_V}{\eta(E_H + E_V)}t} + \frac{\sigma_0}{E_H}$$

Imposing the initial condition $\varepsilon(0) = \sigma_0/(E_H + E_V)$ gives C

$$C = \frac{-\sigma_0 E_V}{(E_H + E_V)E_H}$$

and

$$\varepsilon = \frac{\sigma_0}{E_H}\left(1 - \frac{E_V}{E_H + E_V}e^{-\frac{E_H E_V}{\eta(E_H + E_V)}t}\right)$$

Cyclic tests
Using the notation defined above and assuming a sinusoidal loading (see figure 2.19) with a small amplitude ε_0, the strain is: $\epsilon(t) = \epsilon_i + \epsilon_0 \sin \omega t$ the response, after reaching **stable test,** will be (see the exercise):

$$\sigma(t) = \sigma_i + \sigma_0 \sin(\omega t + \delta) =$$

$$= \epsilon_i G_\infty + \epsilon_0 G_1 \sin \omega t + \epsilon_0 G_2 \cos \omega t \qquad (51)$$

where G_1 and G_2 are known as *storage* and *loss* moduli respectively, and G_∞ is the value of $G(t)$ at $t = \infty$.

From the two previous results it can also be deduced that

$$\tan \delta = G_2 / G_1 \qquad (52)$$

If the energy dissipated in, one cycle is calculated, then

$$\Delta E = \int_{t=T+0}^{t=T+2\pi/\omega} \sigma(t)\, d\epsilon(t) = \pi \sigma_0 \epsilon_0 \sin \delta =$$

$$= \pi G_2 \epsilon_0^2 \qquad (53)$$

which is the area under the curve $\sigma - \varepsilon$ and which accounts for the name of loss modulus for G_2.

For the standard *linear solid*, G_1 and G_2 are given by

$$G_1 = \frac{G_\infty + G_0 (\omega \tau_R)^2}{1 + (\omega \tau_R)^2} \qquad (54)$$

$$\text{and } G_2 = \frac{\omega \tau_R (G_0 - G_\infty)}{1 + (\omega \tau_R)^2} \qquad (55)$$

where the values of G_0 and G_∞ are obtained from equation (46) for $t = 0$ and $t = \infty$, respectively. Both limiting values are shown in figure 2.24.

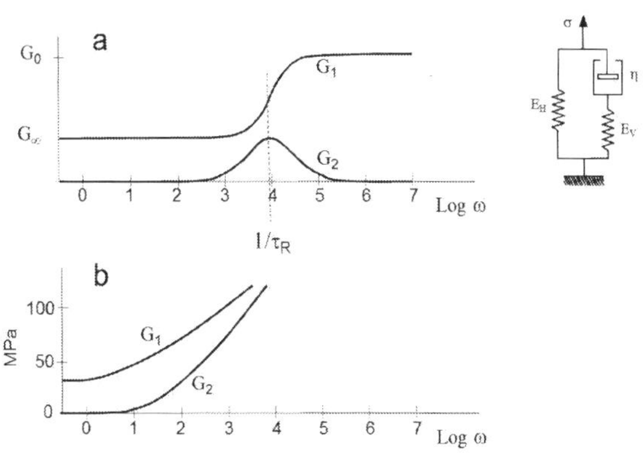

Fig. 2.24. *Standard linear solid.*
a. Response to forced vibrations of frequency ω; moduli G_1 and G_2.
b. Experimental results for elastin fibres (after Gosline and French 1979).

Equations (54), (55), (46) and (48) provide, for the special case of the standard linear solid, a set of relationships which yield the moduli G_1 and G_2 from $G(t)$ or $J(t)$. This is always possible, no matter what linear viscoelastic model is used, since the values of $G(t)$, $J(t)$ and the pair $G_1(\omega)$ and $G_2(\omega)$ are uniquely related (see Wineman and Rajagopal, 2000).

Figure 2.24a contains a schematic representation of the values of moduli G_1 and G_2 as a function of the frequency, ω. For very high frequencies, the material behaves as a purely elastic solid since the damper acts like a rigid solid, and there is hardly any dissipation of energy (G_2 is practically zero). For very low frequencies (very long deformation times), the damper offers no resistance and the material also responds elastically, exhibiting hardly any hysteresis and the energy dissipation is practically zero. The material only exhibits a viscoelastic behaviour in an intermediate range of frequencies in which G_2 reached a maximum.

Experimental results for cyclic tests on elastin are shown in figure 2.24b (Gosline and French 1979). In that figure it can be seen that G_1 and G_2 follow the expected trends in the achievable range of frequencies.

In general, a model as simple as the standard linear solid is not capable of reproducing the behaviour of complex biological materials (see, for example, figure 2.25), but it does serve as a starting point other more satisfactory models.

When the loss modulus G_2 is measured a bell-shaped curve centred around the relaxation time τ_R is not always obtained. On the contrary, the measurements often produce curves with a rather flat maximum, or *plateau*, in the central region, as shown in figure 2.25b, corresponding to the test on the vascular wall of sheep. That figure also contains data for the differing viscoelastic responses of veins and arteries.

In order to try to replicate this relatively frequency-insensitive behaviour the mechanical model is enhanced by inserting in parallel various Maxwell modules and a Hookean spring (generalised Maxwell model). Each Maxwell module gives rise to its own relaxation time τ_i with the objective of summing the bell-shaped curves resulting in a curve with a plateau. The relaxation function $G(t)$ is given, in this case, by an expression of the type

$$G(t) = E_0 + \sum E_i \, e^{-t/\tau_i} \qquad (56)$$

which is known as a Prony series. E_i are the discrete values of the relaxation time spectrum. These ideas can be generalised by introducing a continuous distribution $E(t)$ of relaxation times (or creep) giving rise to

$$G(t) = E_0 + \int_0^\infty E(\tau) e^{-t/\tau}\, d\tau \qquad (57)$$

EXERCISE 18
Derive equation (51).

Solution:
Let $\varepsilon = \varepsilon_{initial} + \varepsilon_0 \sin(\omega t)$ be strain applied to a material whose relaxation is described by $G(t) = G(\infty) - \Delta G(t)$. Note that the function $\Delta G(t)$ *decreases monotonically* and that for sufficiently long times $\Delta G(t) \approx 0$.

Since that loading occurs rapidly at t=0, we calculate $\sigma(t)$ using equation (43b)

$$\sigma(t) = \varepsilon_{initial} G(t) + \int_{0^+}^{t} G(t-\tau)\frac{d\varepsilon(\tau)}{d\tau}d\tau = \varepsilon_{initial} G(t) + \omega\varepsilon_0 \int_{0^+}^{t} G(t-\tau)\cos(\omega\tau)d\tau$$

Making the substitution $u = t-\tau$, the equation becomes

$$\sigma(t) = \varepsilon_{initial} G(t) - \omega\varepsilon_0 \int_0^{t'} G(u)\cos[\omega(t-u)]du$$

and applying $G(t)$

$$\sigma(t) = \varepsilon_{initial} G(t) - \omega\varepsilon_0 G(\infty)\int_0^{t'} \cos[\omega(t-u)]du + \omega\varepsilon_0 \int_0^{t'} \Delta G(u)\cos[\omega(t-u)]du =$$
$$= \varepsilon_{initial} G(t) + \varepsilon_0 G(\infty)\sin\omega t + \omega\varepsilon_0 \int_0^{t'} \Delta G(u)\cos[\omega(t-u)]du$$

since

$$\cos[\omega(t-u)] = \cos\omega t \cos u + \sin\omega t \sin u$$

$$\sigma(t) = \varepsilon_{initial} G(t) + \varepsilon_0 G(\infty)\sin\omega t + \omega\varepsilon_0 \cos\omega t \int_0^{t'} \Delta G(u)\cos(\omega u)du + \omega\varepsilon_0 \sin\omega t \int_0^{t'} \Delta G(u)\sin(\omega u)du$$

which can be rearranged as

$$\sigma(t) = \varepsilon_{initial} G(t) + \varepsilon_0\left[G(\infty) + \omega\int_0^{t'} \Delta G(u)\sin(\omega u)du\right]\sin\omega t + \varepsilon_0\left[\omega\int_0^{t'} \Delta G(u)\cos(\omega u)du\right]\cos\omega t$$

If the integral of ΔG is bounded, that is to say

$$\int_0^{\infty} \Delta G(u)du \leq M$$

since both sine and cosine functions are less than or equal to one, the integrals in the expression for $\sigma(t)$, for times tending to infinity, approach certain limiting values which allow definition of the storage modulus (G_1), and loss modulus (G_2).

$$G_1 = G(\infty) + \omega\int_0^{\infty} \Delta G(u)\sin(\omega u)du$$
$$G_2 = \omega\int_0^{t'} \Delta G(u)\cos(\omega u)du$$

With these values we have

$$\sigma(t \to \infty) = \varepsilon_{initial} G(\infty) + \varepsilon_0 G_1 \sin\omega t + \varepsilon_0 G_2 \cos\omega t$$

which corresponds to the steady-state solution, equation (51). From a practical point of view, convergence is always rapid, achieving a stable behaviour after a few cycles, generally less than 10.

In these cases the relaxation functions, or creep, can no longer be derived from mechanical models with a finite number of modules. Expressions (56) and (57) and their analogues for creep are a particularisation of the integrands which appear in equations (42) and (43).

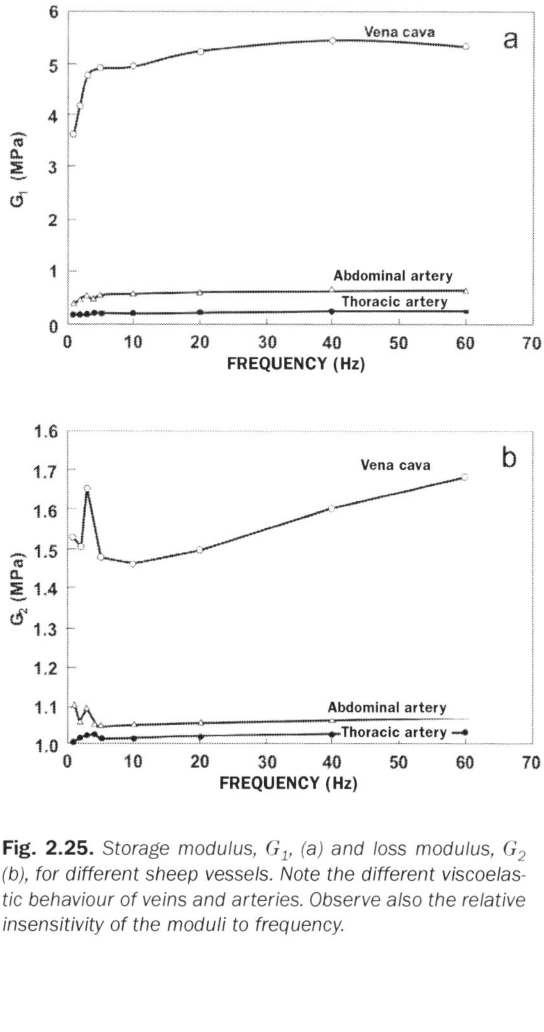

Fig. 2.25. *Storage modulus, G_1, (a) and loss modulus, G_2 (b), for different sheep vessels. Note the different viscoelastic behaviour of veins and arteries. Observe also the relative insensitivity of the moduli to frequency.*

2.5.4. Final remarks

The *linear viscoelastic* model is well founded and has an elegant well-developed mathematical foundation. For example, starting from tensile loading –load history– it is possible to calculate the strain response –the corresponding strain history– if the creep modulus is known. Likewise, knowing the relaxation modulus it is possible to calculate the tensile stress response to an applied strain. The model allows calculation of one modulus from a knowledge of the other, providing predictions of the energy dissipated in cyclic processes and many other things.

However, the linear viscoelastic model has its limitations when attempts are made to apply it to biological materials which are capable of withstanding large deformations, because the assumptions on which it is based are, in general, only good for small deformations.

Experimental measurement of the moduli (creep, relaxation, loss, *etc.*) is not always easy and often requires several complementary processes. In figure 2.26 the time scales of different experimental techniques are shown (note that the time scale is logarithmic).

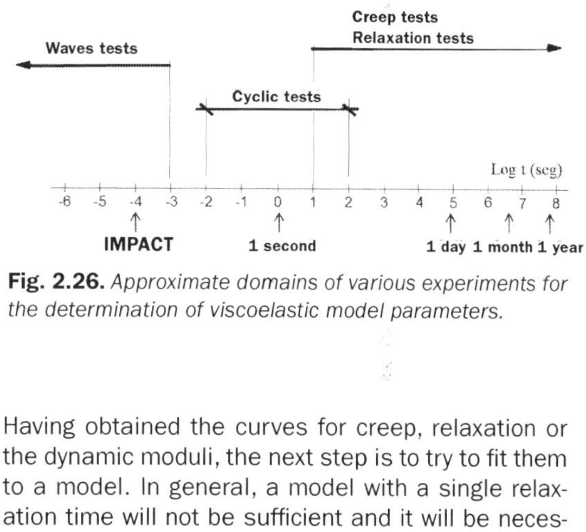

Fig. 2.26. *Approximate domains of various experiments for the determination of viscoelastic model parameters.*

Having obtained the curves for creep, relaxation or the dynamic moduli, the next step is to try to fit them to a model. In general, a model with a single relaxation time will not be sufficient and it will be necessary to use models with a spectrum of relaxation (or creep) times reflecting the specific deformation micromechanisms of each material. If successful, the material tested can be described by the model and predictions made about the mechanical behaviour. Furthermore, as already indicated, viscoelastic models assume that the deformation is recoverable (can be cancelled) when the applied load is removed, a situation that does not occur in real materials, which can be classified more precisely as *visco-elasto-plastic*.

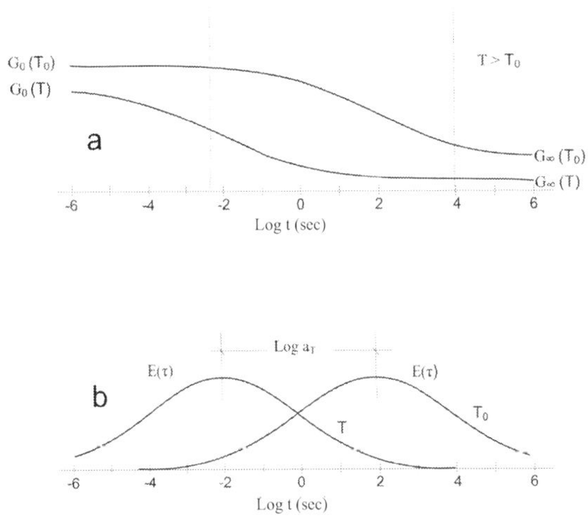

Fig. 2.27. *Effect of temperature on*
a. relaxation modulus,
b. distribution function of relaxation times.

Temperature is another variable which influences the viscoelastic behaviour of biological materials. For example, if the temperature is increased in a relaxation test, the relaxation will accelerate and the modulus of relaxation curve will move to the left, when the horizontal scale represents time (figure 2.27a). If the test temperature is decreased, the opposite occurs.

In many cases this displacement of the curves is associated with a shift in relaxation time function $E(\tau)$ (figure 2.27b). The temperature also influences the extreme values of the moduli (G_0 and G_∞, in the current example), as shown in figure 2.27a. It is usually assumed that the temperature has an equal effect upon the two moduli, that is to say, for two temperatures T and T_0:

$$\frac{G_0\ (T)}{G_0\ (T_0)} = \frac{G_\infty\ (T)}{G_\infty\ (T_0)} \qquad (58)$$

In these circumstances it is possible to make the curves, obtained for temperatures T and T_0, coincide and represent them on a logarithmic scale, by means of a horizontal translation, $log\ a_T$, and a vertical displacement $log\ b_T$, that is to say, $G(t,\ T) \cdot b_T = G(t \cdot a_T, T_0)$. In general, $log\ a_T >> log\ b_T$ and the vertical displacement is usually ignored. The same considerations can be made for the creep modulus and the dynamic moduli G_1 and G_2 (51), and $tan\ \delta$ (equation 52).

In 1955, Williams, Landel and Ferry proposed an empirical expression for the calculation of a_T, applicable to many amorphous polymers, and known as the WLF equation. This relation is:

$$\log a_T = - \frac{C_1\left(T - T_g\right)}{C_2 + \left(T - T_g\right)}$$

where C_1 and C_2 are positive constants and T_g is the glass transition temperature. The displacement obtained with this expression is relative to the curve corresponding to the glass transition temperature. This empirical recipe must be applied with care.

The existence of a relationship such as $G(t,T) \cdot b_T = G(t \cdot a_T,\ T_0)$ allows an easy expansion of the measurement interval in the time scale –always difficult experimentally– without performing experiments at different temperatures.

The complexity of the mechanical behaviour of biological materials does not end with the inclusion of time and temperature in their constitutive equations, other variables also have influence; the water content of the material or the relative humidity during the test are parameters that must be taken into account.

For example, the glass transition temperature of dry elastin is about 200 °C. If 10% by weight of water is added, that temperature reduces to 70°C and if the amount of water is increased to 38% the glass transition temperature reaches 0°C (Kakivaya and Hoeve 1975).

The shapes of the moduli curves as a function of time for different water contents are similar to those at different temperatures and attempts have also been made to correlate them using appropriate translations (shifts).

The effect of humidity, temperature and time on mechanical tests are factors to take into account when interpreting experimental results. It has already been noted that, according to the values of these parameters, it is possible to obtain a behaviour which is elastic, viscoelastic or elastomeric. What is more, with a certain combination of these values it is possible to return the specimen to its original state and to erase the memory of its mechanical history. So, for example, with wool fibres, if strained less than 30% and tested for less than 1 hour, it is possible to recover the initial state if they are submerged in water, at 52 °C, for 1 hour (Feughleman 1997).

2.6. FRACTURE BEHAVIOUR

2.6.1. INTRODUCTION

When a fibre is loaded in tension beyond a certain limit, its microstructure and internal organisation change irreversibly, giving rise to permanent deformation, as has already been noted. If the loading continues, the material separates and breaks, first on a microscopic scale, and ultimately fracturing and breaking the fibre. The process of deformation and degradation prior to rupture is complex and depends as much on the type of material and its microstructure as the geometry and applied load. Therefore, the measured failure loads and strains often vary widely, much more than other mechanical properties (elastic modulus, for example)

A detailed analysis of the principal fracture mechanisms and their mechanical models, which are practically unaffected by microstructure, is presented in Section 3. For the moment it is enough to differentiate between the two main types of fracture behaviour: *brittle* fracture and *ductile* fracture, and identify the implications for testing and characterisation of biological fibres.

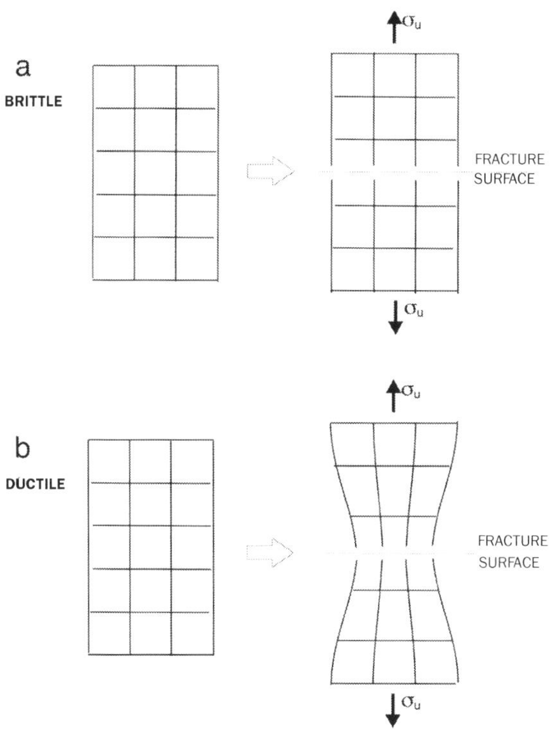

Fig. 2.28. *Fracture behaviour*
a. Brittle
b. Ductile

Brittle fracture is characterised by the absence of permanent deformation prior to rupture. That is, the material behaves elastically up to failure and it suddenly fractures. The material suffers large, very localised strains which have hardly any effect upon the microstructure away from the fracture surface. These deformations happen so rapidly relative to the time scale of the test that they can be considered instantaneous (Figure 2.28a).

Ductile fracture, on the other hand, occurs when the material suffers large plastic deformations during the fracture process (Figure 2.28b). The appearance of irreversible deformations usually induces significant changes in the microstructure beyond the fracture zone and even the geometry of the specimen. These changes alter the stress distribution and, if not taken into account, may give rise to serious errors in estimating the values of the load and/or failure strain.

The boundary between brittle and ductile fracture is not well defined and depends not only upon the type of material but also the geometry of the specimen, its size and the manner in which the load is applied and the surrounding environment. The phenomenon is complex and the interested reader can consult specialist texts on the fracture mechanisms of fibres (for example Elices and Llorca 2002). Simplistically, as a general rule, it is possible to indicate that–for the same material– the larger the size of the specimen relative to its microstructure, the more likely the behaviour will be brittle.

As can been seen, the type of fracture–brittle or ductile– is linked to the (microscopic) deformation which takes place in the fracture zone (Figure 2.28) and may bear little relation to the elongation (macroscopic deformation) of the specimen. It is wrong to establish –as is sometimes done– that a fracture is ductile when there is a large elongation before rupture. An obvious example is the rupture of an elastomer, whose elastic elongation can be more than 100% before brittle fracture. By the same token, the definition of tenacity as the area under the macroscopic stress-strain curve up to the point of rupture can give an erroneous estimate of the ductility of the material.

In the particular case of a test on fibres, ductile behaviour brings with it the appearance of a localised deformation, a process known as *necking*, which was described in Section 2.2.1. It has to be borne in mind that once necking has occurred, the values of fibre stress and strain are not the same throughout the specimen, so that it becomes difficult to define a value of rupture stress or strain.

When the behaviour of the fibre can be treated as brittle, or while ductile the plastic deformation is limited and failure is rapid, it is possible to approximate fracture behaviour by means of a probabilistic model based on the Weibull theory of the weakest link. This model, which will be explained later, is useful in handling the variability of the experimental values of the rupture stress or strain and to predict the behaviour of specimens of different sizes.

2.6.2. Weibull model

Weibull theory[16] assumes that the failure of a structure happens as a sudden random event with the appearance of the first macroscopic crack and whose probability increases with the volume of the structure and the level of the applied stress. It is also based on the supposition that the material is brittle, so that there are no changes in geometry, and therefore the stress distribution, before failure. The model does not consider time-dependent effects.

Figure 2.29a shows a fibre of volume V_T subjected to a stress σ. Its probability of fracture $P_f(\sigma, V_T)$ will depend as much on the level of the applied stress as the volume and is, in principle, an unknown fuction that needs to be determined.

If the volume of the fibre is notionally divided into two parts V and $V_T - V$ respectively (Figure 2.29b), the probability of survival of the fibre is $1-P_f(\sigma, V_T)$ which can be expressed as the product of the probabilities of survival of the two parts into which the specimen was divided, given that the fibre fails when one of its parts fail and that these are statistically independent events, then

$$1-P_f(\sigma,V_T)=[1-P_f(\sigma,V)][1-P_f(\sigma,V_T-V)] \quad (59)$$

in which the same probability function has been used for each part of the fibre because it is the same material.

Equation (59) must be true for any chosen division, that is, any V *between* 0 and V_T, which yields an equation which is a differentiable function of V, leading to:

$$\frac{P_f'(\sigma,V)}{1-P_f(\sigma,V)}=\frac{P_f'(\sigma,V_T-V)}{1-P_f(\sigma,V_T-V)} \quad (60)$$

In which (′) indicates the partial derivative with respect to V. Since V and $V_T - V$ can take on arbitrary values between 0 *and* V_T, the above fractions must be independent of the volume, so that

$$\frac{P_f'(\sigma,V)}{1-P_f(\sigma,V)}=c(\sigma) \quad (61)$$

where $c(\sigma)$ is a function only of the stress, called the *defect concentration function*. Equation (61) can be integrated to obtain:

$$P_f(\sigma,V)=1-e^{-c(\sigma)V} \quad (62)$$

which is the Weibull probability of failure (1939, 1951). $c(\sigma)$ usually interpreted as the density of microdefects present whose failure stress is less than or equal to σ.

The function $P_f(\sigma,V)$ provides that probability that a fibre of volume V loaded to stress σ will suffer brittle fracture or, in other words, the probability that the rupture stress of the fibre is less than or equal to σ. Therefore, $P_f(\sigma,V)$ is the probability distribution function of the random variable "rupture stress" of the fibre.

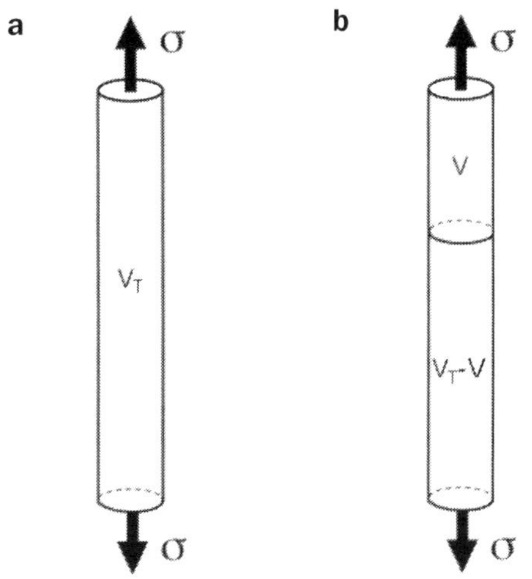

Fig. 2.29. *Tensile test on a fibre*
a. Original fibre volume V_T
b. Notional division of the fibre into two parts $V_T - V$ *and* V

[16] Ernst Hjalmar Waloddi Weibull (1887-1979) was a Swedish engineer interested in the fatigue and fracture of materials and who in 1951 published the statistical distribution that bears his name. The concept was suggested earlier by Fréchet (1927) and by Rosin and Rammler (1933).

The *probability density* function $\phi(\sigma,V)$ is obtained by differentiating equation (62) with respect to σ:

$$\phi(\sigma,V) = V\frac{dc}{d\sigma}e^{-c(\sigma)V} \tag{63}$$

The quantity $\phi(\sigma,V)d\sigma$ represents the probability of failure of a fibre of volume V subjected to a stress between σ and $\sigma+d\sigma$ or, in other words, the probability that the value of the rupture stress of the fibre will be between σ and $\sigma+d\sigma$. Weibull proved that a good fit with experimental data is obtained by assuming a possible defect concentration function of the form.

$$c(\sigma) = \frac{1}{V_0}\left\langle\frac{\sigma-\sigma_{th}}{\sigma_0}\right\rangle^m \tag{64}$$

where V_0 and σ_0 are two constants denoted reference volume and stress, σ_{th} is the rupture threshold stresses and m is a nondimensional constant known as the Weibull modulus. The brackets $\langle\rangle$ indicate that only positive values of the argument are taken, and otherwise zero. If the applied stress is less than the threshold stress σ_{th} the probability of failure is zero.

With the concentration function in equation (64), the probability of failure and the probability density become:

$$P_f(\sigma,V) = 1 - e^{-\frac{V}{V_0}\left\langle\frac{\sigma-\sigma_{th}}{\sigma_0}\right\rangle^m} \tag{65}$$

$$\phi(\sigma,V) = \frac{Vm}{V_0\sigma_0}\left\langle\frac{\sigma-\sigma_{th}}{\sigma_0}\right\rangle^{m-1} e^{-\frac{V}{V_0}\left\langle\frac{\sigma-\sigma_{th}}{\sigma_0}\right\rangle^m} \tag{66}$$

which are represented in figures 2.30 and 2.31 for various values of the parameters in them. In these figures it is clear that the Weibull modulus, m, controls the distribution of the failure stresses, with them being more defined the greater the modulus. The Weibull modulus for several fibres are presented in Table 2.1.

TABLE 2.1. WEIBULL MODULUS FOR SEVERAL FIBRES

FIBRE	WEIBULL MODULUS	REFERENCE
Spider silk	3 - 4	Pérez-Rigueiro *et al.* (2001)
Silkworm silk	5.8	Pérez-Rigueiro *et al.* (1998)
Semolina (dry)	10-15	Guinea *et al.* (2004)
Glass*	10-30	Le Bourhis (2008)
Jute	2.7	Fidelis *et al.* (2013)
Pita	3.7	Fidelis *et al.* (2013)
Flax	3-5	Andersons *et al.* (2005)

* For specimens which are not of fibre form the Weibull modulus lies between 3 and 7

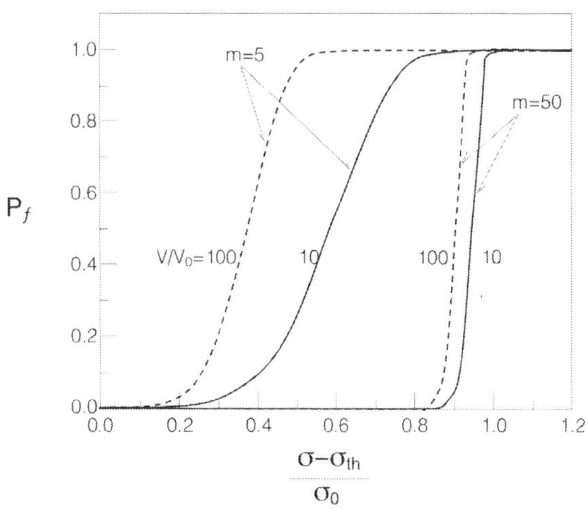

Fig. 2.30. *Weibull model for the probability of failure*

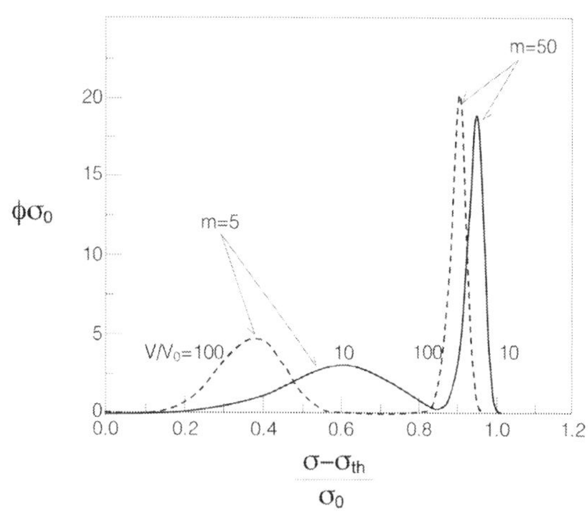

Fig. 2.31. *Weibull model for the probability density of failure.*

It is usual to employ the Weibull modulus without the value of the threshold stress σ_{th} setting it to zero. This parameter normally has very little influence on the results, and the assumption simplifies the calculation of the statistical parameters from equation (66). The mean failure stress σ_u then becomes

$$\sigma_u = \int_0^\infty \sigma\ \phi(\sigma)d\sigma = \Gamma(1+1/m)\ \sigma_0\left(\frac{V_0}{V}\right)^{1/m} \quad (67)$$

and the coefficient of variation of the failure stress, ω, is given by

$$\omega^2 = \frac{\int_{\sigma_{th}}^\infty (\sigma-\sigma_u)^2\ \phi(\sigma)d\sigma}{\sigma_u^2} = \frac{\Gamma(1+2/m)}{\Gamma^2(1+1/m)} - 1 \quad (68)$$

where $\Gamma(z)$ is Euler's Gamma function which is defined as

$$\Gamma(z) = \int_0^\infty t^{z-1}e^{-t}\,dt$$

which, for values $5<m<50$, can be approximated to $\Gamma(1+1/m) \approx 0.6366^{1/m}$ with which

$$\sigma_u = \sigma_0\left(0.6366\frac{V_0}{V}\right)^{\frac{1}{m}}$$

with an error of less than 0.5%. Equally

$$\omega = \frac{1}{0.462+0.783m}$$

With an error of less than 0.25% for the same range of values of m (Bazant and Planas 1997).

Equations (67) and (68) provide a simple means of calculating the Weibull parameters from statistical values of rupture and making estimates of the mean stress for fibres of different lengths.

EXERCISE 19

Integrate equation (61) and obtain equation (62).

Recall the model hypothesis, for V tending to 0, P_f must also tend to zero.

Solution:

If we write equation (61) in differential form and separate the variables

$$\frac{dP_f(\sigma,V)}{1-P_f(\sigma,V)} = c(\sigma)dV$$

the equation can be easily integrated

$$-\ln\left(1-P_f(\sigma,V)\right) = c(\sigma)V + C(\sigma)$$

where C is a constant of integration. The condition $P_f(\sigma, V=0)=0$ yields the value $C=0$, so that

$$1-P_f(\sigma,V) = e^{-c(\sigma)V}$$

which is equation (62).

EXERCISE 20

Pérez-Rigueiro *et al.* (1998) recorded the values for the rupture of silk fibres presented in the Table A. Calculate the Weibull modulus of silk thread assuming that there is no stress threshold: a) from the probability of failure, using equation (65) and the given data; b) from equations (67) and (68).

TABLE (A) : FAILURE STRESS (MPa)

203	280	280	319	327	336	346	366	375	375	392	401
415	419	426	433	436	454	455	457	464	471	471	523

Solution:

a) Equation (65) with σ_{th} =0 gives: $P_f = 1 - e^{-\frac{V}{V_0}\left(\frac{\sigma}{\sigma_0}\right)^m}$ or $\ln\left(\ln\left(\frac{1}{1-P_f}\right)\right) = m\ln(\sigma) + \ln\left(\frac{V}{V_0 {\sigma_0}^m}\right)$

Plotting ln(ln(1/(1- P_f))) against a ln(σ) gives a straight line of slope m.

Estimation of the probability exceeding a determined value from a series of experimental data is a subject of great engineering interest, and there are many expression for this, as well as the two referenced in the text. An extensive discussion of the best method can be found in Barnett 1975.

In order to assign probabilities $P_f(\sigma)$ to the measured values of failure stress Hazen's equation (1930) is usually employed.

$$P_f(\sigma) = \frac{i(\sigma) - 0.5}{N}$$

where i is the ordinal of each value of the failure stress, once ordered from smallest to largest and N is the total number of measurements. In Table (B) the values obtained are presented, together with value of the probability calculated with the alternative expression (Weibull 1939)

$$P^*_f(\sigma) = \frac{i(\sigma)}{N+1}$$

which is also often used. The previous expressions differ above all in the assignment of probabilities at the extremes (large and small values) and are more precise approximations as the size of the specimen increases, converging for $N \to \infty$.

TABLE B. STRESS σ (MPa)	Order (i)	P_f	$P_f{}^*$
203	1	0.020833	0.04
280	2	0.062500	0.08
280	3	0.104170	0.12
319	4	0.145830	0.16
327	5	0.187500	0.20
336	6	0.229170	0.24
346	7	0.270830	0.28
366	8	0.312500	0.32
375	9	0.354170	0.36
375	10	0.395830	0.40
392	11	0.437500	0.44
401	12	0.479170	0.48
415	13	0.520830	0.52
419	14	0.562500	0.56
426	15	0.604170	0.60
433	16	0.645830	0.64
436	17	0.687500	0.68
454	18	0.729170	0.72
455	19	0.770830	0.76
457	20	0.812500	0.80
464	21	0.854170	0.84
471	22	0.895830	0.88
471	23	0.937500	0.92
523	24	0.979170	0.96

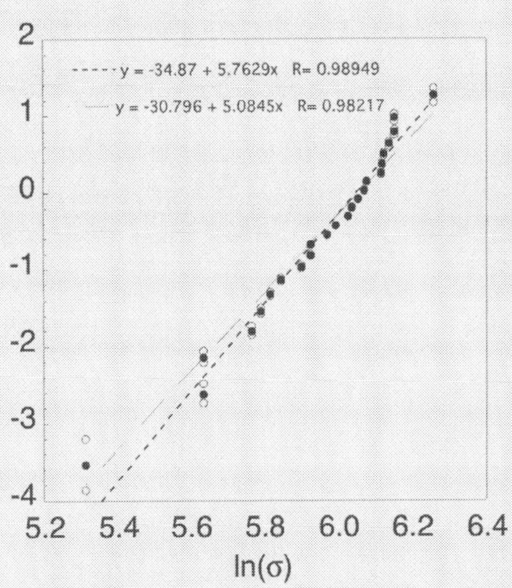

The graph presents the data with two estimated probabilities P_f *and* $P_f{}^*$ and their corresponding straight-line approximations. The slope of the lines (Weibull modulus) lies between **5.76** for P_f and **5.08** for $P_f{}^*$.

b) With the data in Table (A) we can calculate the mean value (= 392.67 MPa) and standard deviation (= 75.133 MPa) of the failure stress.

The coefficient of variation of the failure stress is

ω = standard deviation / mean value =
= 75.133 / 392.67 = 0.1913

and since $\omega \approx (0.462+0.783m)^{-1}$ we see that m= **6.09**.

EXERCISE 21

Tensile tests on a batch of fibres have provided a value for the mean failure stress of 203MPa, with a standard deviation of 17.3 MPa. Calculate the value of the Weibull modulus of the material. Neglect the threshold stress.

Solution

The ratio of the standard deviation and the mean value,ω gives ω = 17.3/203=0.0852. Using the equation of Bazant & Planas $\omega \approx (0.462+0.783m)^{-1}$ we see that m= 14.4.

EXERCISE 22

If the fibres in the previous exercise are divided in half before testing, what will the mean stress to failure be according to the Weibull model?

Solution

Equation (67) provides the average breaking stress $\sigma_u = \Gamma(1+1/m)\, \sigma_0 (V_0/V)^{1/m}$

Then, for a length L and diameter $\emptyset$ the volume is $V=\pi L\emptyset^2/4$ and the mean stress to failure will be

$$\sigma_u = \Gamma(1+1/m)\, \sigma_0 (4V_0/(\pi L \emptyset^2))^{1/m}$$

For a length L' of the same diameter $\emptyset$ the volume will be $V'=\pi L'\emptyset^2/4$ and

$$\sigma_u' = \Gamma(1+1/m)\, \sigma_0 (4V_0/(\pi L' \emptyset^2))^{1/m}$$

dividing both expressions

$$\sigma_u' / \sigma_u = (L/L')^{1/m} = (2)^{1/m} = 1.05$$

If σ_u=203 MPa then σ'_u=213 MPa

2.6.3. Final remarks

The rupture of a material is a complex phenomenon in which the deformation processes in the fracture zone and its surroundings play an important role. The models used to describe failure depend upon the type of deformation produced (reversible, irreversible, time-dependent etc.) and the extent of the affected zone.

Brittle fracture, in which ideally it is assumed that the material does not suffer any plastic (irreversible) deformation, is perhaps the simplest case to analyse. Even so, the appearance of discontinuities inside the material enormously complicates the mechanical analysis and requires the use of empirical information (location and geometry of defects, for example) which are rarely easy to obtain.

Probabilistic failure models in, for example, that of Weibull simplify the situation assuming that the defects are randomly distributed in the material, and when one of these reaches a critical stress the entire structure collapses. However, this is not normally the case for most structures (which can usually sustain a certain level of defect, and a resulting redistribution of stresses and strains). The assumption is more realistic in the case of brittle fibres, and moderate success is obtained in these models, including for biological materials. A study which used the Weibull model to characterise the properties of serrano ham (*jamón serrano*) has been recently published (Romero de Ávila *et al.* 2014).

When a deeper analysis is required, and when an adequate prediction of the rupture load and stress are needed, it is essential to delve further into the material, its deformation and failure mechanisms. Section 3 of this book deals with this.

3 CHAPTER Examples of Fibres

3.1. COTTON FIBRES

3.1.1. Introduction and fibre structure

Cotton fibres are polysaccharides. They come from the hairs that cover the seeds of plants of the genus *Gossypium*. The structure of the fibres is shown in Figure 3.1. The fibres have a diameter of about 25 µm and lengths which vary between 1 and 5 cm. In their natural state the cross-section of the fibre is circular, exhibiting daily growth rings (Figure 3.1a). Once picked, they lose water and collapse, adopting the twisted form with the kidney-shaped cross section shown in Figures 3.1b and 3.1c.

From plant to fabric, cotton fibres pass through the following stages: *harvest* (*picking*), in which the fibres are extracted from the seed capsule, manually or by machine; *ginning*, the process by which the fibres are separated from the seeds; *yarn*, manipulation when the fibres are stretched, twisted, folded and pressed against each other in order to create threads with distinct characteristics; and, finally, the *fabric* made from the fibres. In some cases the fibres are subjected to a preparation process for later manipulation, called *mercerisation* (due to John Mercer, 1844), which consists of dousing them in an 18% solution of NaOH which causes a swelling of the fibres. By the end of the treatment the fibres are thicker, less twisted, stronger, more lustrous and absorb colour better.

The use of cotton goes back to prehistoric times. There is evidence that it was cultivated in the Indus valley, in India, around 3000 BC. The ancient Egyptians and Chinese spun and wove cotton. In the Middle Ages, the Arabs introduced cotton to Europe, among other routes through Spain. They called it *qutn*, from which it became *algodón* in Spanish, *cotó* in Catalan, or *cotton* in English. The invention of the mechanical cotton gin in 1793, by the American Eli Whitney, revolutionised its production;

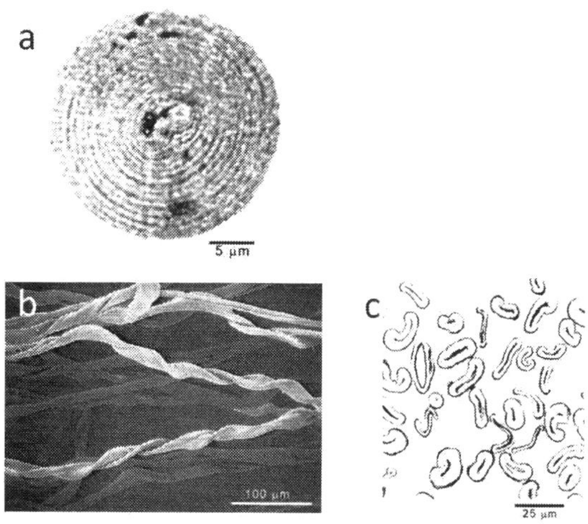

Fig. 3.1. *Form of cotton fibres.*
a. Cross-section of a fibre with water, in which the growth rings are apparent (Kerr 1937).
Dry fibres; longitudinal view, and transverse view.

it took production from 3,000 bales per year to 73,000, in 1800. According to some historians, cotton production played a role in triggering the American Civil War, because of the need for slave labour. Currently more than 26 million tons of cotton are produced annually, a clear indication of the importance of this fibre.

Cotton fibres have a hierarchical structure– as do all natural fibres– represented, in a simplified form, in figure 3.2. In a mature cotton fibre six levels can be identified. The first is the *cuticle*, a fine shell –of several molecules– which contains waxes, pectin and proteins. The second is the *primary wall*, with a thickness of about 0.1 μm, formed of a network of cellulose fibrils arranged helically with an angle between 20 and 30° to the fibre axis. The thickness of this wall depends upon the maturity of the fibre; the more mature, the thicker, and, in general, the more mature, the stronger the fibre. The third level is known as the *layer S1*, it is another, but simpler, network of fibrils aligned to the axis of the fibre forming angles between 40 and 70°. The thickness is about 0.1 μm. The fourth level, called *layer* S2, is the *secondary wall* which has a greater thickness, of about 4 μm, and is, to a large extent, responsible for the stiffness of the fibre. It is also composed of cellulose fibrils which lie, approximately, at an angle of 45° to the fibre axis. The fifth level is the *lumen wall*, or *layer* S3. And the sixth level is the *lumen* itself, a channel which runs down the centre of the fibre and which, in a mature fibre when the calyx of the plant is opened, the interior dries, the lumen collapses, and the cross-section takes on the kidney-shaped form shown in figure 3.2.

The purest form of cellulose that nature produces is in cotton fibres, with a content of about 90%. The remaining components (waxes, proteins, *etc.*) are located in the external shell or inside the lumen.

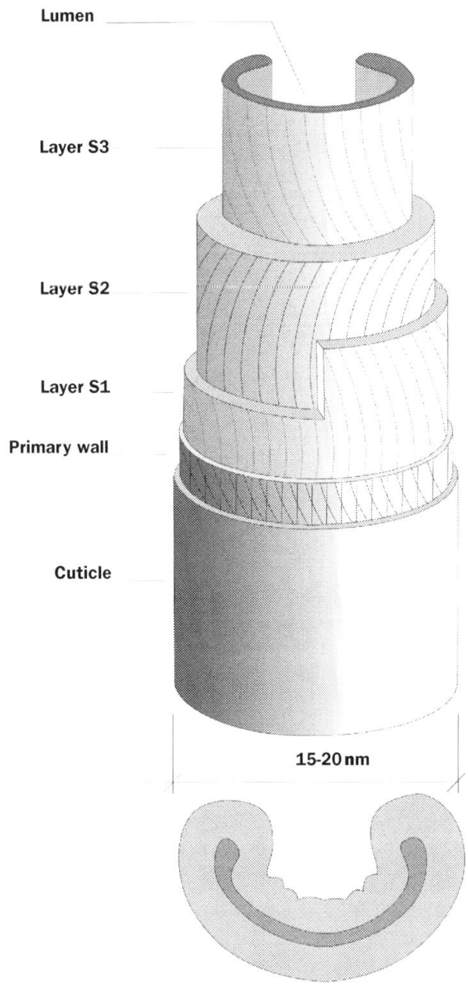

Fig. 3.2. *Hierarchical structure of cotton fibres (Jefferies et al. 1969).*

3.1.2. Mechanical properties

The mechanical properties of cotton fibres are derived from tensile tests, tests which are not without their difficulties because of the small dimensions of the fibres; the diameter in particular. The characteristics of these tests and their peculiarities are not the purpose of this chapter, but the interested reader can consult the articles of Sasser *et al.* (1991), Hebert *et al.* (1995), or Thibodeaux *et al.* (1998).

When testing natural, rather than artificial, fibres a significant spread in the results is obtained, a factor that must be taken into account when modelling their behaviour. Figure 3.3 is a typical example of the distribution of the results when the strength of fibres and threads (a bundle of fibres) are measured.

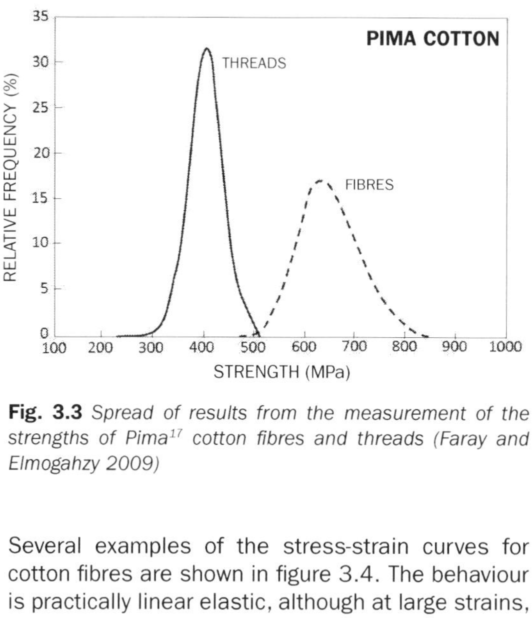

Fig. 3.3 *Spread of results from the measurement of the strengths of Pima[17] cotton fibres and threads (Faray and Elmogahzy 2009)*

Several examples of the stress-strain curves for cotton fibres are shown in figure 3.4. The behaviour is practically linear elastic, although at large strains, on unloading, the deformation is not completely recovered, as will be discussed later.

Strength values can range from 200 to 800 MPa. The strength of these fibres is attributed to the cellulose chain stiffness, the fibrillar structure and the multiple hydrogen bonds; as much intermolecular as intramolecular. However, there is no reliable model for the prediction of the failure load, owing to the hierarchical structure of the fibre, shown in figure 3.1.

Strains in tensile tests may reach as much as 6 to 14%. These values are small in absolute terms, allowing the use of small deformation mechanical models. Strain measurement is tricky since the fibre is initially wavy, as can be seen in figure 3.1, and during straightening, in the first phase of loading, an erroneous elongation is recorded as a result of this process. For this reason it is common to "comb" the fibres a little to try to unravel them prior to testing. Fibres with larger elongations before rupture produce more flexible threads and, also, larger failure strains. Both characteristics are desirable for fabric production because the fibres are subjected to a conditioning process before weaving in order to increase their stiffness.

The elastic modulus of cotton fibres typically lies between 5 and 12 GPa. Given the shape of the stress-strain curves, which are essential straight lines, the initial modulus is valid for the whole curve. Measurement of the initial modulus presents the same difficulties mentioned above with measurement of the initial strains, because the fibres are wavy.

As has already been mentioned, the mechanical response of the fibres is not completely elastic, beyond a certain strain the fibres do not recover their original length on unloading. This characteristic is very important because the fibre suffer repeated loading and unloading cycles during handling, weaving and

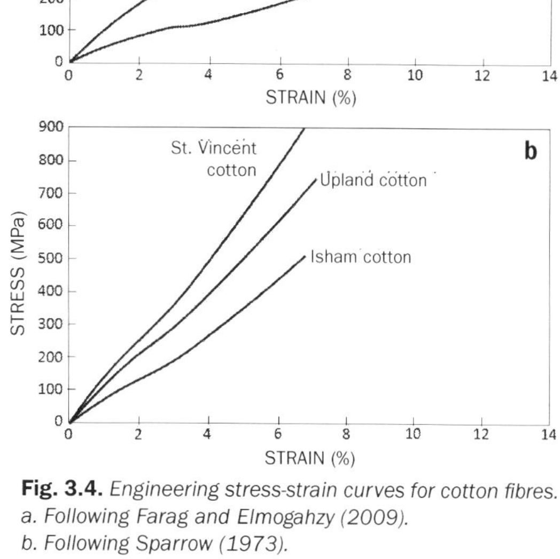

Fig. 3.4. *Engineering stress-strain curves for cotton fibres.*
a. Following Farag and Elmogahzy (2009).
b. Following Sparrow (1973).

[17] Pima cotton is type of cotton with a longer fibre than conventional cotton. Pima cotton fibres measure between 35 to 50 mm in length, while conventional cotton fibers can be up to 28 mm long. Pima cotton fabric comes out softer and more durable than if it were made from a shorter staple cotton.

later use. Linen fibres are more elastic and, for large deformations, recover up to 80%. Cotton fibres, under the same circumstances, only recover 50% (elastic recovery is measured as the quotient of the difference between the maximum and final strains after one cycle of loading and unloading, and the maximum strain to which the fibre was subjected). An approximate picture of the elastic recovery of cotton fibres can be obtained from the work of Meredith (1945); if the strain is less than 2% the recovery will be typically about 80%. For strains of about 5% recovery is about 50% and for large strains, of the order of 10%, recovery does not exceed 30%.

Another factor which influences the mechanical properties of the fibres is humidity. It affects the dimensions, strength and flexibility. The effects of humidity on the stress-strain curves are shown in figure 3.5. Increasing the humidity increases the strength, strain to failure and the flexibility.

The tensile test on cotton fibres, as well as providing data on their quality, can give an indication of the strength of threads manufactured from the fibres. The contribution of the fibres to the strength of threads is due to their own strength, and the friction between them. A detailed analysis of the strengths of threads can be found in Farag (2009) and in Schwartz (2008). The conclusion is that the strength of threads is less than that of the fibres, as indicated in figure 3.3, although the dispersion in the data is usually less than that of fibres (as is also apparent in the figure).

One conclusion of this study is that the value of the elastic modulus of a thread E_h is a function of the modulus of the fibre E_f and the angle α which the fibre forms with the axis of the thread. The result is given by

$$E_h = E_f\,[\cos^2\alpha\,(1 - k\,\mathrm{cosec}\,\alpha)]$$

where k is a constant which depends upon the geometry of the fibre and the coefficient of friction

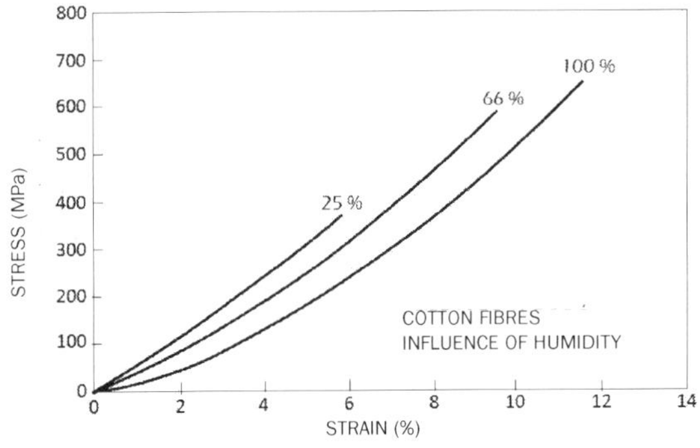

Fig. 3.5. *Engineering stress-strain curves of cotton fibres. Influence of humidity (R. Meredith, in Fibre Science, J.M. Preston (Ed.), The Textile Institute 1953, p. 246).*

between the fibres (see Exercise 24).

If all of the fibres have the same length as the thread, only the first factor appears $-\cos^2\alpha-$, which indicates that the stiffness can be reduced by increasing the angle. The second factor $-(1 - k\,\mathrm{cosec}\,\alpha)-$ takes into account the dimensions of the fibres and the friction between them through the parameter k, which is given by

$$k = A\,(a/\mu)^{1/2}\,L^{-1}$$

where A is a constant, a and L the diameter and length of the fibres, respectively, and μ is the coefficient of friction between them.

3.1.3. Concluding remarks

One of the reasons for the continued use of cotton is its large number of applications. From a 230 kg bale of cotton it is possible to make about 700 shirts, 1,200 T-shirts, 6,000 pairs of knickers, 22,000 handkerchiefs, 4,000 socks, 200 pairs of jeans, 240 sheets, 600 towels, 1,200 pillowcases or 3,000 nappies.

The vast range of cotton products is due, in great part, to its physicochemical properties and the organoleptics of this fibre. The smooth texture of cotton makes it very suitable for underwear, because it feels like skin. It is absorbent, durable against abrasion, and resistant to many chemical agents.

Nonwoven cotton products also have a future –for example, wipes, cloths, and cleaning gauzes– because of their great powers of absorption and wet strengths, and they have many application in daily use, in hospitals and in cosmetics. Other articles fabricated by punching, such as serviette, tablecloths or sheets, are economical and can be used as disposable products, although some can be washed as many as six times. The challenge is to make them economically competitive by means of the appropriate selection of fibres and manufacturing processes.

Biotechnology has also played a part in the production of cotton fibres and may have an even greater role in the future. Currently, varieties of cotton have been produced that are more resistant to certain herbicides and insects, and attempts will be made to produce longer, finer and stronger fibres and, if possible, of different colours.

Another challenge for the cotton industry is to optimise production and to be able to recycle the waste products. Transformation of cotton products into cellulose, by dissolving or chemical processing, is another challenging technology.

EXERCISE 23

The curve in the figure show the degree of **elastic recovery** (in Isham cotton fibre) measured by Meredith (1945), as a function of the load, and strain reached in each cycle of loading and unloading.

Derive the engineering stress-strain curves for the fibres and the value of the elastic modulus of the unloading-reloading branches as a function of the maximum strain reached in the cycles. Assume that the behaviour is linear during unloading and reloading.

Data for Isham cotton fibres: mass = 2.4 denier; density of cotton =1.54g/cm³

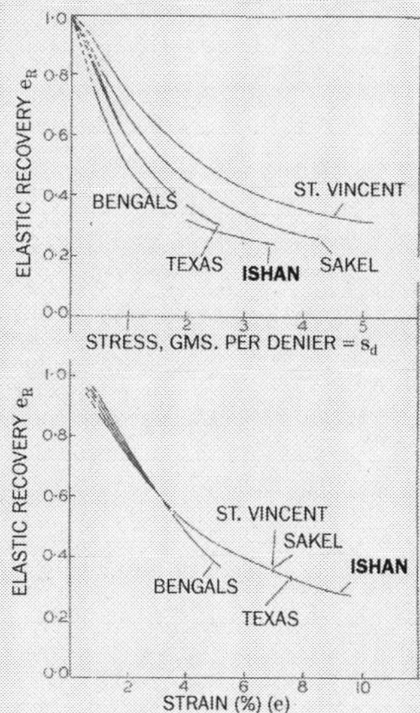

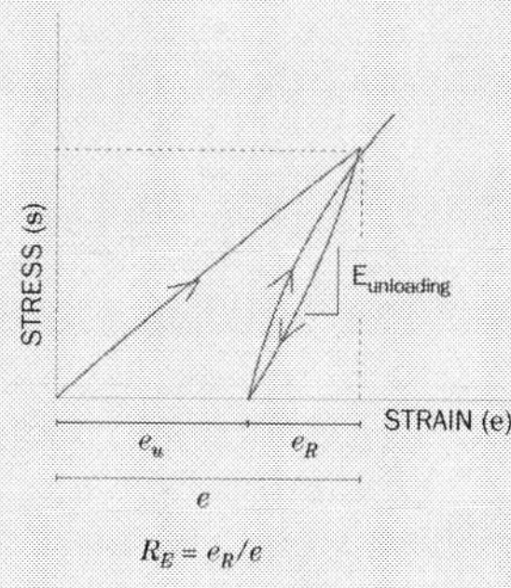

Solution:

The engineering stress-strain curve is obtained by plotting pairs of stress-strain points corresponding to the same value of the elastic recovery. This will be the new curve of the fibre.

The resulting curve is shown in **figure a** with the stress expressed in MPa. In order to obtain it, the definition of *denier* given in Section 2.2.1 has been used. If *A* is the cross-sectional area of the fibre and ρ its density, the linear density will be $\mu = \rho A$ so the engineering stress s is then

$$s = \frac{F}{A} = \frac{\mu F}{A \mu} = \rho \frac{F}{\mu} = \rho s_d$$

stress expressed in units of force per linear density. In Meredith's chart we see that s_d is given in grammes-force/denier and, for that, we will write

$$s = 1540 \frac{kg}{m^3} s_d \left(\frac{g_f}{denier} \right) = 1540\, s_d \frac{kg}{m^3} \frac{10^{-3} kg_f}{(2.4 g/9km)} \frac{9.81N}{kg_f} = 56.65\, s_d MPa$$

In order to obtain the elastic modulus of the unloading branch E_u, we simply divide the stress s at each point by the value of the elastic strain $e - e_u = e.e_R$:

$$E_u = \frac{s}{e.e_R}$$

Then, the strain recovered on unloading is the product of the elastic recovery $e_R = (e - e_u)/e$ and the strain e reached in each cycle. The results are shown in **figure b** below, in which the secant elastic modulus is also shown, defined as the ratio of the stress and the maximum strain reached in each cycle s / e.

a

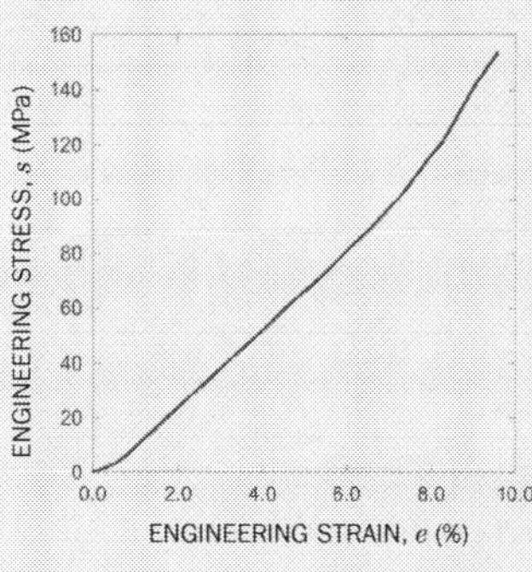

b

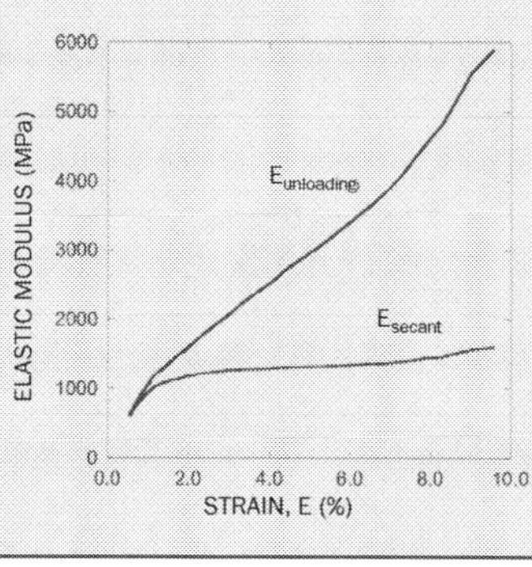

As can be seen in **figure b** (left), when the fibre undergoes larger strains, the response hardens –as much in new loading as in the unloading and reloading branches– probably as a result of the forced alignment of its microfibrils.

If we compare the curve in figure a with that of figure 3.4b, obtained by Sparrow (1973) we see significant differences, probably attributable to the different threading and moisture content during the experiments.

EXERCISE 24
In the case of fibres of equal length, determine the value of the elastic modulus of a thread E_h from the elastic modulus of a fibre E_f

Solution:
Consider a thread, of length L, consisting of a bundle of N fibres which form a cylindrical helix of angle α with the axis of the thread, as shown in the figure. A straight line element of a single fibre is shown in the figure. We shall assume that all of the fibres have the same length L_f and cross-sectional area A_f. We shall also assume that the strain in the thread is small enough that any change in the angle α can be ignored, and that there is no friction between the fibres.

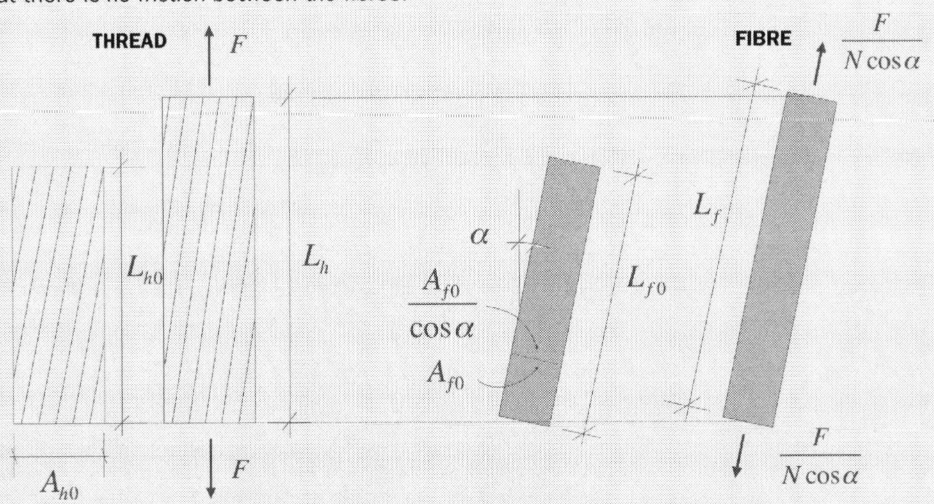

Under these conditions, if the thread is subjected to a tensile force F, the engineering stress in the thread will be:

$$s_h = \frac{F}{A_{h0}} = \frac{F\cos\alpha}{NA_{f0}}$$

where A_{h0} and A_{f0} are the initial cross-sectional areas of the thread and fibre, respectively. In order to obtain A_{h0} we note that the cross-sectional area of the thread will be the sum of the oblique cross-sections of the fibres $A_{f0}/cos\alpha$.

If the initial length of the thread is L_{h0}, the engineering strain will be $e_h = \frac{L_h}{L_{h0}}$

In the fibres, the stress state will also be tensile and the force carried by each of them will be $F/(N\ cos\alpha)$ since there is no friction. Note that the vector sum of the forces on all of the fibres must be equal to the external force and that the components perpendicular to the axis of the thread must cancel out.

The engineering stress in a fibre will be $s_f = \frac{F}{NA_{f0}\cos\alpha}$ and its engineering strain $e_f = \frac{L_f}{L_{f0}} = \frac{L_h}{L_{h0}} = e_h$

since $cos\alpha = L_h/L_f = L_{h0}/L_{f0}$, and the ratio of the height of a helix and its length is the cosine of the angle which the helix forms with its axis.

From these results, the elastic modulus of the thread and fibre must be related by

$$E_h = \frac{s_h}{e_h} = \frac{F\cos\alpha}{NA_{f0}}\frac{1}{e_h} = \frac{F}{NA_{f0}\cos\alpha}\frac{1}{e_h}\cos^2\alpha = \frac{s_f}{e_f}\cos^2\alpha = E_f\cos^2\alpha$$

which is the required relationship.

3.2. WOOL FIBRES

3.2.1. Introduction and fibre structure

Wool fibres are proteins; amino acid polymers. They come from the hair of domesticated sheep of the genus *Ovis aries*. The name has been generalised and is also used for the fine hair of some goats (*cashmere*, from Himalayan goats, and *angora* or *mohair*, from goats of Ankara), from camels, vicunas, alpacas and yaks, and also Angora rabbits.

From wool to fabric, the fibres pass through the following stages: *cleaning*, since they contain impurities (grease, earth, excrement and vegetable material); *mechanical* treatment consisting of disentangling the fibres by carding, combing and drawing; *chemical* treatments, such as carbonising (this process destroys the vegetable residues) bleaching (the process of removing the natural colouring of the fibres) and, sometimes, dyeing. The final product, made from the fibres, may be subjected to moth and stabilisation treatments. For more detailed information see, for example, Simpson and Crawshaw (2002).

Wool was, probably, the first fibre used for clothing by mankind, initially in the form of felt, from cut or torn hair. Spinning of fibres and then fabric came later. It is difficult to track down the origins of the use of woollens because they do not survive well, but it is known that in ancient Babylon (which according to some sources means "land of wool") there was an active market in this product. One of the oldest woollen fabric remains, dating from around 1800 BC, comes from a tomb in Denmark, which has been preserved in a reasonable state thanks to an acid soil (Barber, 1992).

The Phoenicians introduced sheep from Asia Minor as far as North Africa. They entered Spain around the twelfth century by the hand of the Beni-Merines tribe and, quite possibly, from this comes the name of sheep (merino) which produce very fine wool of excellent quality. Spain had a monopoly of wool between the twelfth and sixteenth centuries, and the export of merino sheep was severely punished. The Napoleonic wars almost ended the Spanish merino breed. At the beginning of the nineteenth century, this breed began to proliferate in Germany, the United States, Australia and South Africa.

An ideal wool fibre has a circular cross-section, whose diameter ranges from 20 to 40 μm, a length which varies from 5 to 50 cm, and is curly.

The actual structure of wool fibres is very complicated, as shown in figure 3.6a. The components of the fibre are: the cortical cells, the extracellular complex (*cell membrane complex* CMC) and the cuticle that surrounds the fibre. The three components are shown schematically in figure 3.6b.

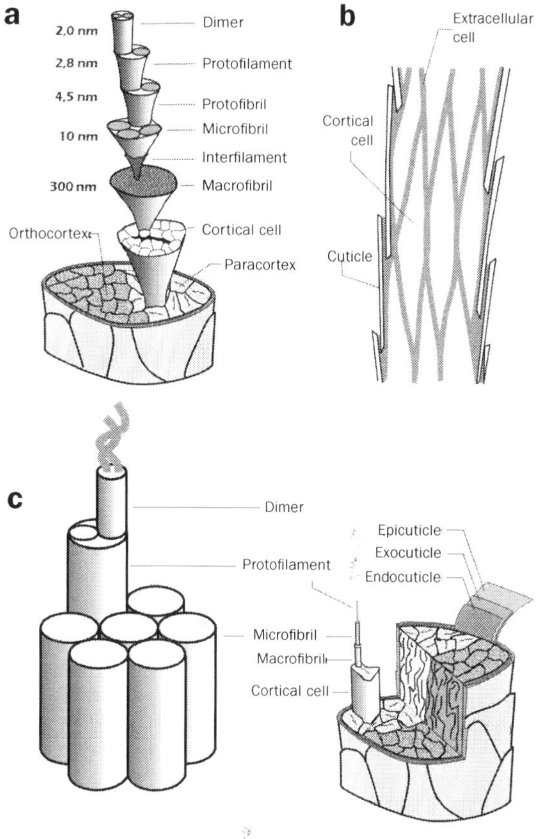

Fig. 3.6. *Hierarchical structure of wool fibre (after Eichner et al. 1985).*
b. Longitudinal section of a fibre with its principal components: cuticle, cortical cells and extracellular matrix.
c. Schematic of the hierarchical structure.

The cuticle is formed of flattened cells, interleaved like (roof) tiles, whose tips point towards the fibre end. The approximate dimension are 20x30x0.8 μm. In each cell four layers can be distinguished: the surface, or *epicuticle*, which is a thin membrane (about 5 nm) strong and hydrophobic, the α *layer*, the *exocuticle* and the *endocuticle* (figure 3.6c). In figure 3.6c the α *layer* is not shown.

The extracellular complex covers all of the cortical cells and ensures its stability by means of intercellular links via proteins known as *desmosomes* (binding bodies or *maculae adherentes*).

Examining the cross-section of a cortical cell it is possible to identify *macrofibrils* which are aligned in the direction of the fibre axis (figure 3.6c) and embedded in an intermacrofibrilar matrix. The macrofibrils, in turn, are formed of hundreds of *microfibrils*, a sort of protein which is present in the cell cytoskeleton and which is usually called *intermediate keratin filaments* (IKF), embedded in a *keratin associated protein* (KAP) matrix. It is possible to identify two classes of cortical cells by their different appearances after staining with silver nitrate: the *orthocortical* cells, which appear clearer compared to the darker stained *paracorticals*.

The cortical cells are spindle-shaped (figure 3.6b) with lengths between 45 and 95 μm and a width of between 2 and 6 μm. The combined cortical cells occupy 90% of the fibre, the *orth-* type being more abundant than the *para-*.

In each cortical cell between 5 and 20 macrofibrils can be observed. Each microfibril has a diameter of between 0.1 and 0.3 μm and is composed of a stack of 500 to 800 microfibrils (figures 3.1a and 3.1c).

Microfibrils are formed of 7 or 8 protofilaments and each protofilament is composed of four chains of IKFs grouped in two helically wound pairs (Simpson and Crawshaw 2002) (figure 3.6c).

When observed by x-ray diffraction microfibrils show peaks which indicate a certain order. If the fibres are stretched to about 50% and immersed in a steam bath, the x-ray diffraction pattern changes and other peaks appear. For this reason the undeformed fibres are called *α-keratins* and the deformed ones, *β-keratins*.

In α-keratin the peptide chain forms a right-handed helix whose stability is ensured by intramolecular hydrogen bonds (Pauling *et al.* 1951). Two helices of this type are wound and form a left-handed superhelix with hydrophobic amino acid residues facing outwards (Letai and Fuchs, 1995). These dimers are the smallest units of the microfibril, as shown in figure 3.6a.

In β-keratin, the peptide chain is more stretched and forms a different structure called β-leaf in which the chains are joined transversely by means of hydrogen bonds (Pauling, 1953).

3.2.2. Mechanical properties

The density of wool is about 1.31 g cm^{-3}, at 25 °C and relative humidity of 65%.

Moisture absorption

The protein in wool contains hydrophilic groups which attract water molecules, so that water absorption depends upon the ambient relative humidity, the temperature and, also, the history of the wool. The process of water absorption is exothermic and the water absorption curves as a function of relative humidity are almost reversible, that is there is very little hysteresis, barely 2%.

Although wool is hygroscopic, the surface of the fibre is hydrophobic, as has already been observed, which makes the penetration of water in the liquid phase difficult. The shrinkage caused by water absorption is not the same in the radial direction as the longitudinal; with a relative humidity of 100%, the fibre may extend by only 2% but increase in radius by 16% (Onions, 1962).

Tensile test

The mechanical properties of wool fibre depend upon the ambient relative humidity (as would be expected), the temperature and prior history. For example, a dry wool fibre, at room temperature and which has not previously been deformed might have a breaking load of between130 and 210 MPa, but, if tested wet, these values may decrease by 20%. The extension to failure has the opposite tendency; when dry this can vary between 28 and 48%, but if wet the value increases by between 40 to 60%.

Figure 3.7 shows two stress-strain curves for a wool fibre tested at 20 °C, with a relative humidity of 65%, and another tested in water at the same temperature. The general shape of the curves is typical of tensile response of these fibres, in which distinct regions are apparent.

In the first phase (OA), where the stress increases rapidly with strain, up to values of 1 to 2%, the initial elastic modulus decreases with humidity. This first tranche is usually called "Hookean" but this is not strictly correct, since, as will be seen later, although the behaviour is linear, it is viscoelastic.

The second phase (AB) (note that point B is not on the curve, rather it is a reference point) is known as the flow region; the load increases very slowly with strain, reaching strains of the order of 25 to 30%.

In the third phase (BC) the stress increases more quickly with strain, until rupture at point C. In general the curves are quite reproducible up to point B. Tranche BC is more variable because it is affected by imperfections which may precipitate breakage. A good fibre can reach strains as great as 60%.

X-ray diffraction spectra indicate that after point A, that is the initiation of flow, the lines corresponding to the α-helices weaken and those of the β-leaves develop (Bendit, 1960). This transition is, to some extent, reversible, as noted by Feughelman (1997) showing that if the fibre is deformed up to point B, at room temperature, it can recover its mechanical properties completely if the duration of straining is less than an hour. If the fibre is unloaded and immersed in water at room temperature for 10 hours (or one hour at 50 °C), not only is the initial length recovered, the same stress-strain response is obtained on reloading.

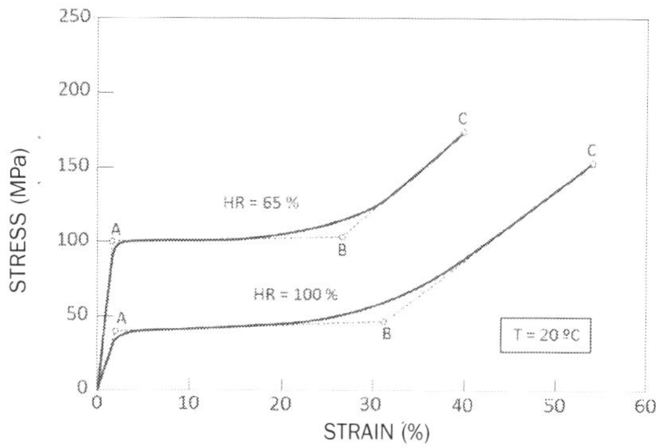

Fig. 3.7. *Engineering stress-strain curves for wool fibre (Zahn et al. 1996).*

EXERCISE 25

The figure shows a typical load (in grammes)-engineering strain curve for wool fibre in water at 20°C subjected to tensile loading in a test lasting 25 days (Feughelman 1997). The strain rate was 0.375%/hour from the beginning to the end of the Hookean region/initiation of the flow plateau, and 1.875%/day from that point on.

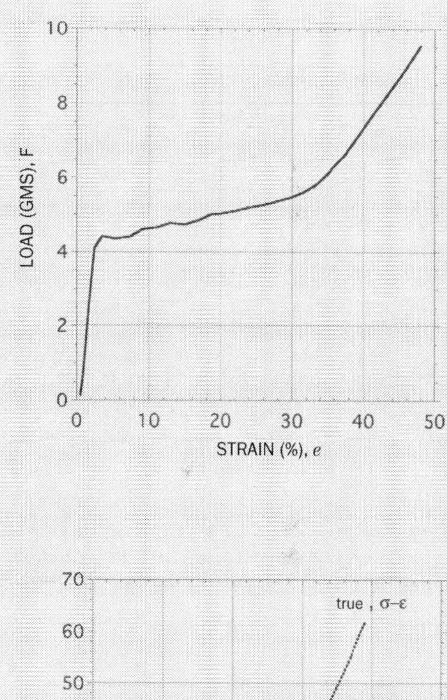

If the fibre has an average diameter of 30 μm derive the engineering and true stress-strain curves and determine salient values.

Solution:

The engineering stress is calculated using equation (1) of Section 2.2:

$$s = \frac{F}{A_0} = \frac{4F}{\pi D_0^2}$$

where F is the force on the specimen and A_0 is its cross sectional area, which we calculate from its diameter D_0. The resulting s-e curve is sketched in the figure below. From equations (6) and (7) we obtain the true strains and stresses $\varepsilon = \ln(1+e)$ and $\sigma = s(1+e)$ which are also drawn in the figure.

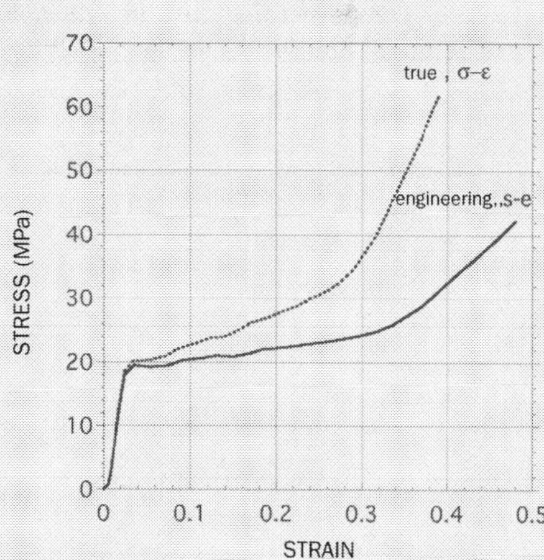

The salient values of these curves are:

	Engineering	True
Initial modulus E(GPa)	970	1013
Flow stress, s_y or σ_y (MPa)	19.4	20.2
Modulus in flow region E_2(GPa)	20	52
Modulus post-flow E_3(GPa)	120	300

Note that when values for the parameters of the stress-strain curves are given it is important to specify from which curves they have been obtained, whether from true or engineering, the data will be different, especially for strains above 5%.

Influence of humidity, temperature and ambient conditions

Figure 3.8 shows the effect of **humidity** on the stress-strain curves, according to data from Speakman (1927). The drier the fibre, the greater the elastic limit, the slope of the flow region, the initial modulus and the breaking stress, but the smaller the strain to failure.

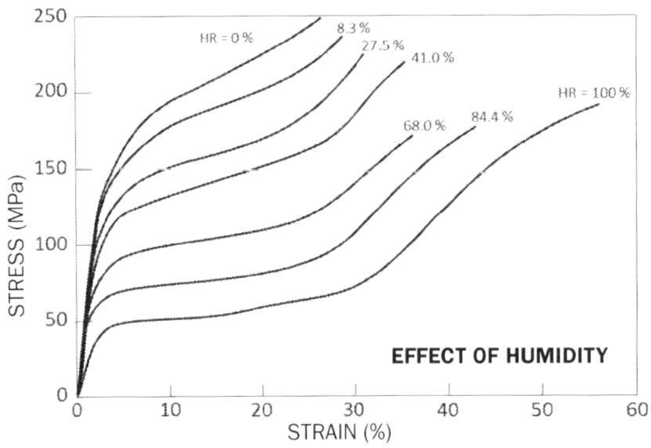

Fig. 3.8. *Engineering stress-strain curves for wool fibres at 24.4 °C as a function of the relative humidity (after Speakman 1927).*

Figure 3.9 shows the effect of **temperature** on the stress-strain curves, again using data from Speakman (1927). When the test temperature increases, the elastic limit decreases, as do the slope of the flow region and the breaking stress. Note that these changes are more pronounced between 40 and 80 °C.

Figures 3.8 and 3.9 justify the use of water and temperature to facilitate combing, although this assistance is temporary since the hairs recover their initial properties when the humidity and/or temperature are reestablished. Thanks to the presence of the amino acids with sulphur (cistein), it is possible to form permanent bonds between the keratin fibres through disulphur bridges (–S–S–), which allow *permanent* changes to the form and properties of the hairs. This is precisely the name that hairdressers have given to the treatment, in which a reducing agent first (generally tyoglicolic acid) and then another oxidant (oxygenated water) break the existing disulphur bridges and then reform them in hair with the desired shape.

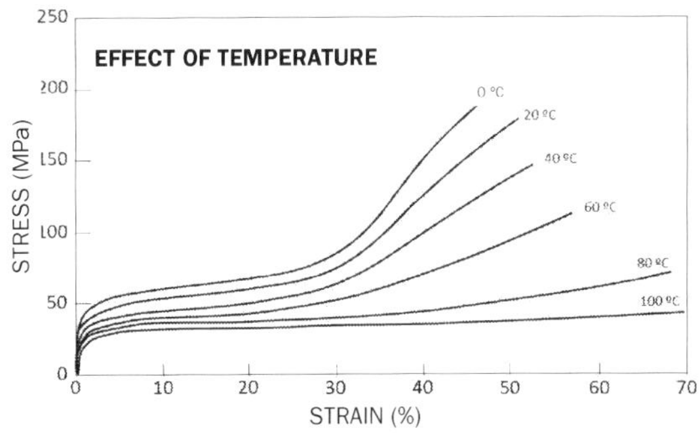

Fig. 3.9. *Engineering stress-strain curves for wool fibres tested in water as a function of temperature (after Speakman 1927).*

Figure 3.10 shows the effect of various ambient media **(acids and alcohols)** on the stress-strain curves, using data from Peters and Woods (1956). Figure 3.10a illustrates the effect of alcohols. Since the alcohol molecules are larger than that of water they do not cause the reduction in the stiffness evident in Figure 3.8. However, the larger the alcohol molecule the smaller the plastifying effect, and the mechanical properties improve. The effect of alcohols is similar to that of humidity; the larger molecules behave as if the environment were dryer. Figure 3.10b shows the effect of acids. Everything indicates that if the molecules are small they have less difficulty in penetrating the fibre, and therefore, weaken the material.

a

b

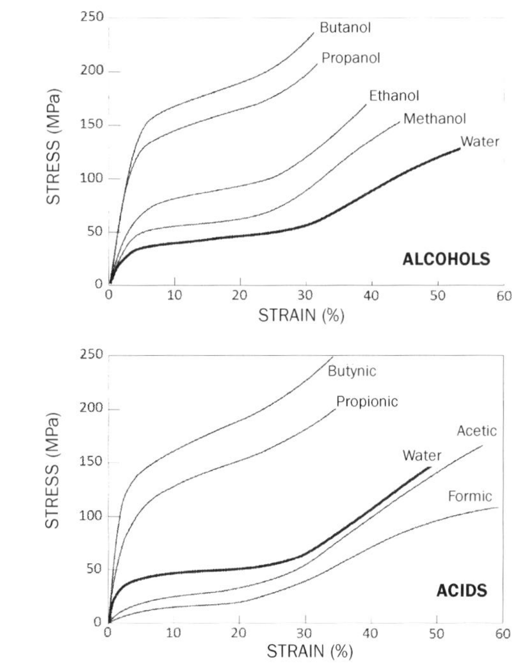

Fig. 3.10. *Engineering stress-strain curves for wool fibres immersed in several media: a. alcohols, b. acids (after Peters and Woods 1956).*

In 1894 the Spaniard A. Borrás produced *"El fraile del tiempo"* a weather forecasting device which is still on sale. This is based on the sketch of a monk on cardboard (see the figure) who has a wand and mobile hood. These items rise and fall according to whether or not it is going to rain.

Actually it is a hair hygrometer, which measures the ambient humidity thanks to the extension induced in a fibre –usually human hair– by the atmospheric moisture. The wand and hood are usually rigidly connected and rotate together about a point. In order to keep the fibre in a state of tension at all times, a counterweight is usually included.

Weather forecasting is based on the correlation between humidity and atmospheric pressure, since the latter is sensitive to the presence of anticyclones (high-pressure phenomena) which are normally accompanied by dry stable weather, and low pressure systems (troughs) which usually bring changeable weather and squalls.

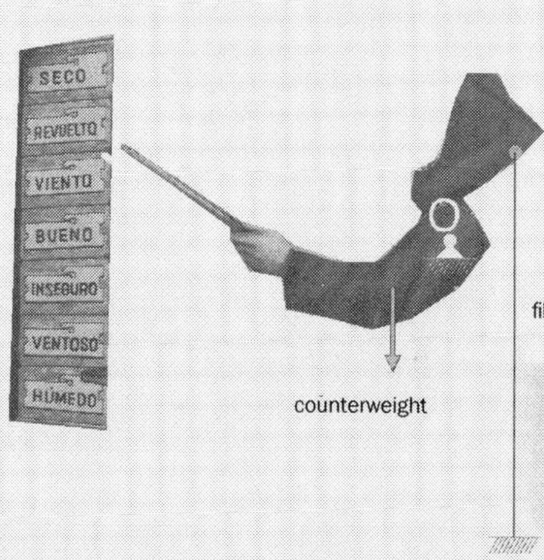

The influence of humidity on human hair has been known since ancient times (people with long hair notice that their hair curls and expands in bulk when the humidity is high) but it was not until 1783 when the Swiss physicist and geologist Horace Bénédict de Saussure built the first hygrometer from human hair very similar to the one described above.

If we take as reference values of figure 3.8 (which are for wool fibres) we see that the initial modulus changes from ~2 to ~6GPa when the relative humidity changes from 100 to 0%. This implies that elongation of the fibres (under the same load) will be (6/2 = 3) three times bigger when saturated compared to that in the absence of moisture. The hair hygrometer takes advantage of this change to precisely measure humidity on a graduated scale.

Influence of time

The stress-strain curves of wool fibres are like those of many other polymers, but only during the loading process. On unloading, if the elastic limit has been exceeded, a permanent deformation occurs in many polymers, but not in the case of wool fibres; if tested wet and before the strain has surpassed 30%, as illustrated in figure 3.11, due to Morton and Hearle (1993), the strain is completely recovered. In this example, the unloading curve does not follow that of loading, and only rejoins it for values of about one third of the elastic limit.

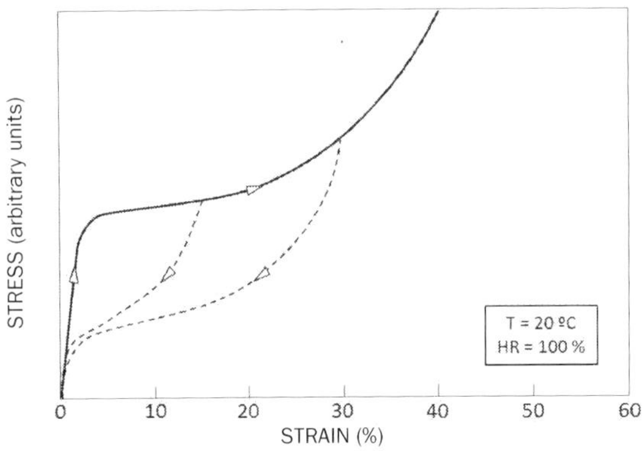

Fig. 3.11. *Loading and unloading curves for a wet wool fibre (after Morton and Hearle 1993. The stress is in arbitrary units).*

Other experiments have demonstrated this recovery capacity, which decreases with the degree of relative humidity (Meredith, 1945; Beste and Hoffman, 1950). All this indicates that **time** is a factor to take into account in the deformation of these fibres.

Figure 3.12 shows the results of several **relaxation** experiments. This phenomenon occurs even in the "elastic" region. For this reason it was noted above that it is not correct to call this first part of the curve the "Hookean region", even though it is straight. Figure 3.12a indicates how the stress relaxes from an initial strain of 0.8% (which in this fibre is in the initial straight section). The data are from Feughelman and Robinson (1971), and the stresses are normalised with the respect to the value limited by the fibre humidity. The effect of humidity on the relaxation is also shown in the figure.

Figure 3.12b shows the relaxation in stress for different values of strain beyond the elastic zone, in experiments on human hair (which behaves similar to wool), wet and at a temperature of 35 °C. The shape of the curves seem to indicate that there are several different relaxation mechanisms; initially there is a rapid reduction in the stress (in less than 1 second), then there is a slow reduction lasting about 100 seconds, after which there is another rapid decline, between 100 and 10,000 seconds, and, finally, it seems to approach an asymptote (Peters and Wood, 1956).

Figure 3.12c shows how the relaxation, in wet human hair, varies with temperature for a deformation of 40%, according to Peters and Wood (1956).

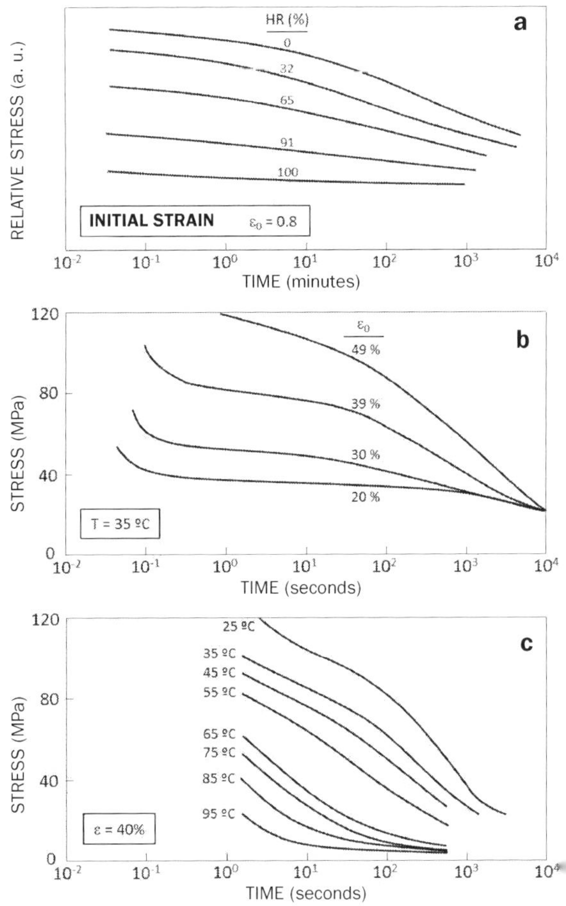

Fig. 3.12. *Relaxation curves of engineering stresses*

a. *Influence of humidity in wool fibres at 20°C (initial strain in the linear zone; 0.8%) (after Feughelman and Robinson 1971).*

b. *Influence of the initial strain. Wet human hair at 35 °C (after Peters and Woods 1956).*

c. *Influence of temperature. Wet human hair with an initial strain of 40% (after Peters and Woods 1956).*

Creep test results are presented in figures 3.13a and 3.13b for short duration tests (2.5 minutes) and in figure 3.13c for tests which lasted a week. The effect of the initial stress is shown in figure 3.13a, for wet wool at 18.5°C. The influence of temperature, for the same wet wool tested subjected to an initial stress of 38.5MPa is illustrated in figure 3.13b (Feughelman, 1954). These results –obtained with a low initial stress, of the order of the elastic limit– seem to suggest that, after a short while, an asymptote is reached and the strain stabilises. This does not happen when the experiments are performed with higher initial stresses –about 70MPa instead of 40MPa– and longer duration, as is apparent in figure 3.13c (taken from Peters and Woods, 1956), in which there seems to be a new creep mechanism coming into play after roughly 100 minutes.

Viscoelastic behaviour of wool fibres is also observed in the **dynamic** behaviour, even in the "Hookean" region. Figure 3.14 presents dynamic data for wool fibre tested under low amplitude vibrations, with a frequency of 116Hz, at 25°C and different levels of relative humidity (from Danilatos and Feughelman, 1979). Figure 3.14a shows the variation in the storage modulus, G_1; note that the maximum occurs in the Hookean region, then it decreases in the flow region in order to increase again in the hardening zone. Figure 3.14b represents the loss angle δ and is a mirror image of the previous one; minimum values in the Hookean and hardening regions, and maximum in the flow region.

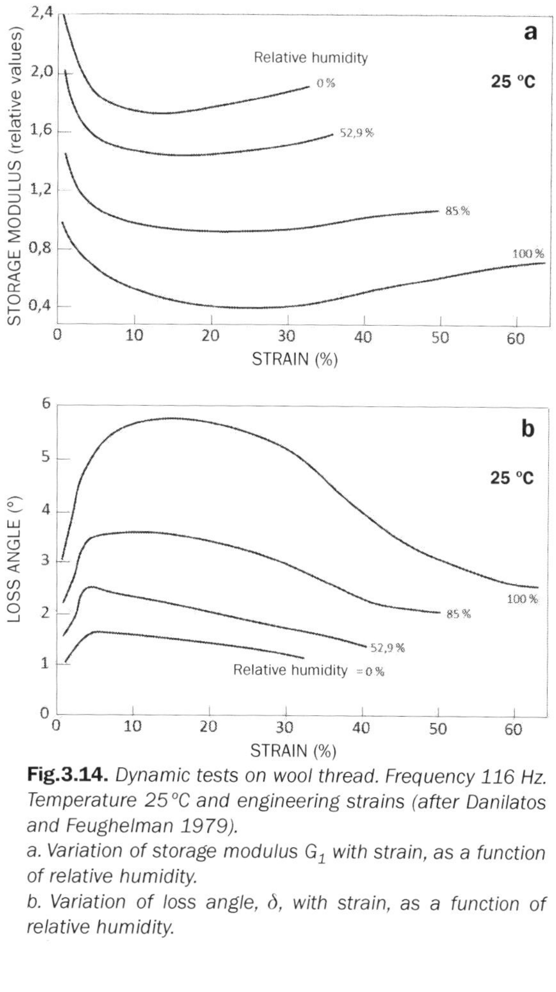

Fig.3.14. *Dynamic tests on wool thread. Frequency 116 Hz. Temperature 25°C and engineering strains (after Danilatos and Feughelman 1979).*
a. Variation of storage modulus G_1 with strain, as a function of relative humidity.
b. Variation of loss angle, δ, with strain, as a function of relative humidity.

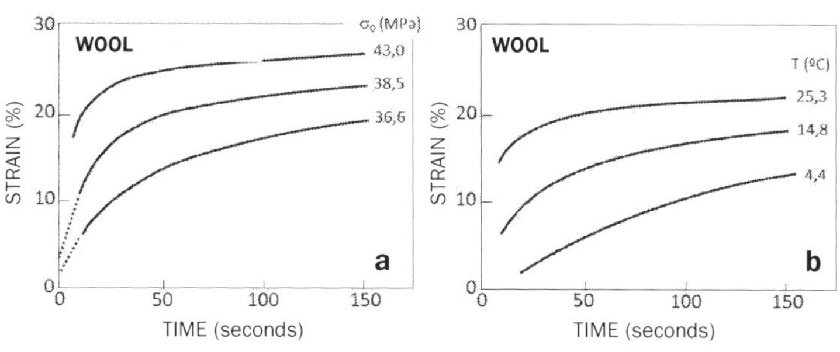

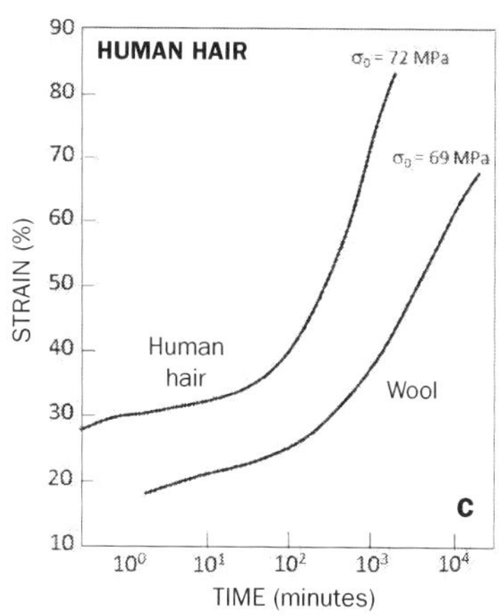

Fig. 3.13. *Creep curves (Engineering stresses and strains)*
a. Effect of initial stress on wet wool fibres at 18.5°C (after Feughelman 1954).
b. Effect of temperature on wet wool fibres with an initial stress of 38.5 MPa (after Feughelman 1954).
c. Effect of test duration. (Note that in figures a and b the test time is short compared to c). Tests in water (after Peters and Woods 1956).

EXERCISE 26
Assess the validity of the linear viscoelastic model for the study of wool fibres in a short duration test from the curves shown in figure 3.13a. Investigate the influence of temperature using the data in figure 3.13b.

Solution:
The linear viscoelastic hypothesis proposes that in a creep test under a load σ_0 the creep function is $J(t) = \varepsilon(t)/\sigma_0$, a function which is a material property and independent of the test conditions. Therefore, if we divide the data in the curves of figure 3.13a by the values of the applied stress in each case all the curves must collapse onto a single curve which gives the creep function $J(t)$.

As we see in figure (a), this is not the case and the curves $\varepsilon(t)/\sigma_0$ remain quite separate. This is common in the majority of viscoelastic materials, particularly biological materials. We shall revisit this in more detail in the next section, when we examine the viscoelastic behaviour of collagen.

In order to study the effect of temperature, we draw the curves in figure 3.13b in figure (b) with a logarithmic time scale and strain values divided by the creep stress (38.5MPa) with the aim of obtaining the creep modulus $J(t) = \varepsilon(t)/\sigma_0$. We see that it is possible to identify a time shift which superimposes the curves. Taking the curve for T=14.8°C as a reference we see that $J(t;14.8°) = J(6.4t;25.5°) = J(0.22t;4.4°)$. The dashed line corresponds to the displaced curves. It can be seen that the fit between the two higher temperatures is better but, even so, it seems that for T=14.8° we can deduce the creep curves for times ranging from somewhat less than 10s to more than 15 minutes (900s).

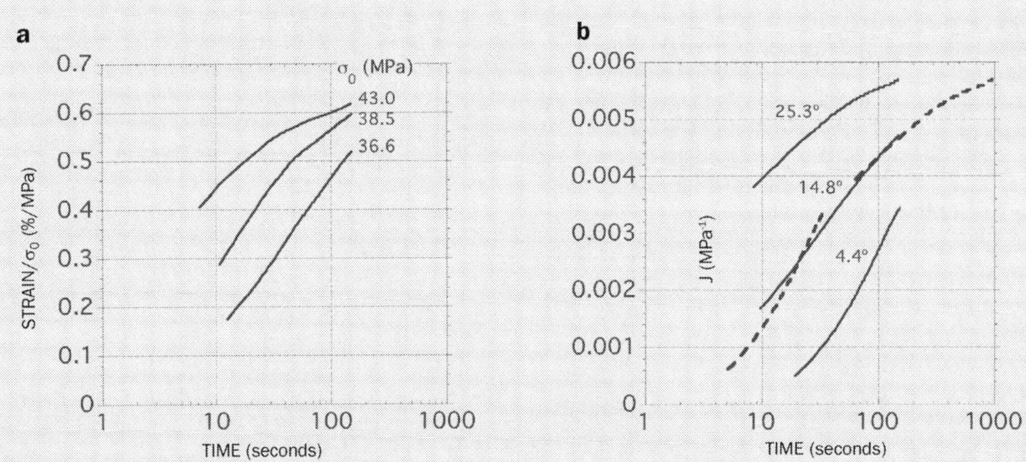

Although attempts have been made to introduce models that provide a better representation of experimental behaviour (such as Fung's linear quasi-viscoelastic model (Fung, 1972)) the linear viscoelastic model and parameters are still widely used for the characterisation of biological materials.

Currently there is extensive information on the chemical composition and mechanical behaviour of wool fibres as a function of humidity, temperature, time and environment, but much remains to be done to achieve a model that provides a detailed description of their behaviour. This is not surprising given the complex hierarchical structure shown in figure 3.6. Nonetheless, approximations have been made in this direction and one of the simplest is a two-phase model, due to Feughelman, which explains some of the experimental features discussed above. The interested reader will find details in Feughelman (1997). A summary of some ideas extracted from this book follow.

Feughelman's model, described schematically in figure 3.15, provides reasonably satisfactory results in the small deformation regime; up to values of 1% strain. The model of a keratin-water system assumes that the fibre is composed of three interpenetrating polymer networks: a) a microfibril network (IKF) formed of α-keratin fibres orientated parallel to the fibre axis, b) an amorphous globular protein (KAP) matrix with a high sulphur, glycine and tyrosine content, c) water, in the form of a network of hydrogen bonds, which interact with the residual surface hydrofils of the proteins of a and b. Increasing the water content (humidity) produces a constriction in the matrix and the model accounts for the larger increase in radial strain compared to the axial strain.

For small deformations the strength is due to the strain energy in the hydrogen bonds with the α-helices more than the strength provided by the matrix. However, the shear strength, or value of the shear modulus, is principally due to the properties of the matrix.

For large deformations the model is more complicated because there are three phases: the α-helices destabilise and are transformed into β-leaves, the *gel* type matrix structure collapses into a state more like a *sol*, but when the fibre is unloaded, the structures recover again to form a polymer with the three networks. It is not the aim of this chapter to analyse the model in detail. In the reference cited above (Feughelman 1997) calculations based on this model, which coincide with the published experimental data, can be found.

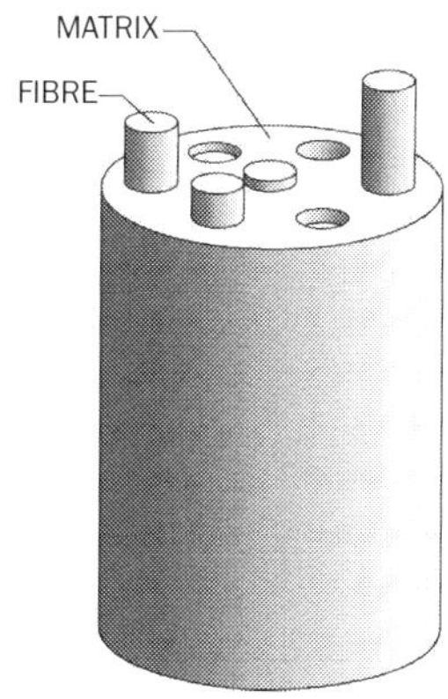

Fig. 3.15. *Feughelman's model (1997).*

3.2.3. Concluding remarks

The worldwide production of clean wool in 1992 was 1.7 million tonnes (Mt), in 1995 1.5 Mt, in 2000 1.4 Mt and in 2017 it did not reach 1.2 Mt. The trend is decreasing although it seems that in the last few years it has tended to stabilise. The production of wool is modest in comparison with cotton and synthetic fibres, the percentage of wool fibres is about 3% of all fibres. However, the current trend is towards quality rather than quantity.

In the last few years, with the introduction of synthetic fibres, there has been increased emphasis on producing smoother and lighter wool fibres, and in a simpler manner, by reducing the fibre diameter.

Wool fibres are classified by their diameter; *fine* fibres have a diameter between 17 and 23μm, *medium* fibres between 23 and 33μm, and *thick* between 33 and 42μm. The preference for fine fibres has been notable in the last few years; in Australia, for example, in 1994 only 8.5% of fibres had a diameter of less than 19μm, 12 years later the percentage was 31%. The finest fibres have strength problems; if a fibre has an intrinsic strength of 200MPa, for example, and a dimeter of 18μm it will only support 50mN. In order to improve the strength of fine fibres several techniques have been used, principally employing genetic engineering (Rogers, 2000) but so far progress has been slow.

3.3. SILK FIBRES

3.3.1. Introduction and fibre structure

Silk fibres –like wool fibres– are proteins; amino acid polymers. Silk is usually associated with the fibre produced by the larva of the silkworm *Bombyx mori* but it is also produced by many arthropods, in particular some insects and spiders. There are more than 30,000 species of spiders and 100,000 species of insects (of the order *Lepidoptera)* which produce silk. In general, any proteic fibre produced by extrusion of a viscous liquid through a fine orifice can be considered as silk.

Silk fibres can be obtaincd directly from spider webs, from silkworm cocoons or the yarn forced through the glands of the creatures. From a cocoon it is possible to obtain about 1,000 metres of silk, while that from a spider's forced yarn about 100 m – in several extractions– and only 10 m from the larger cobwebs.

Silks have multiple uses: to produce cocoons which protect and thermally insulate the larvae, to fabricate lairs, or to construct webs to catch prey. The most studied silks are those that come from domesticated silkworms *Bombyx mori* and the silk from some spider webs, particularly those from the genera *Nephila*, *Argiope* and *Araneus*

Silkworm silk

About 5,000 years ago, according to Chinese legend, Princess Xi Ling Shi was walking in her garden with a cup of tea in her hand. As luck would have it a silkworm fell into her cup. While trying to remove it, she saw that she could get a delicate shiny and iridescent thread from the soft mass formed by the heat of the infusion. This accident was the origin of the silk art and industry.

It is not surprising that this interesting polymer attracted the interest of peoples throughout history. For centuries the Chinese jealously guarded the secret of sericulture, an imperial law decreed death to those who revealed it. Silk was unknown in Europe 2,000 years ago, and until the sixth century that the techniques of sericulture reached the West, thanks to industrial espionage by one of the monks encouraged by the emperor Justinian (Feltwell, 1990). Currently, the textile industry consumes, annually, more than 400,000 tonnes of silk cocoons.

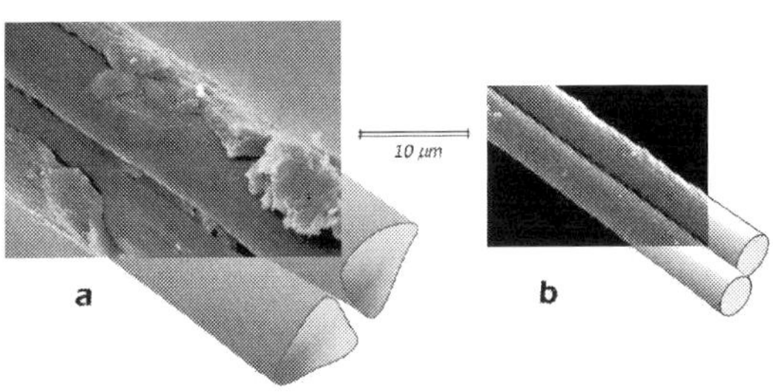

Fig. 3.16. *a. Silkworm silk; fibres of fibroin coated sericin. b. Spider silk; fibres of spidroin.*

Most of the silkworms belong to the domesticated species *Bombyx mori* and produce silk during one stage of its life cycle. Domestication of the *Bombyx mori* has reached the point that the insect is totally dependent on humans for food and protection, perhaps the only insect that has reached this extreme.

The thread that silkworms spin –moving their head in a figure of eight– is formed of two filaments of fibroin coated with a sericin adhesive (see figure 3.16a). The thickness of the fibres is about 10 μm. Silk fibres are extracted from the cocoons which have been previously boiled in soapy water to remove the sericin coating.

From each cocoon, 300 to 1200 metres of fibre are obtained, with the Japanese holding the record which they achieved at the start of the twentieth century from a variety of silkworm whose cocoon produced two kilometres of fibre. In order to make a tie about 100 cocoons are needed, 650 for a shirt and more than 3,000 for a high quality kimono; for the last the silkworms must consume about 60 kilos of mulberry leaves.

The microstructure of silk thread is not known in great detail; the threads are formed by two fibres of triangular cross-section, covered with a layer of sericin (figure 3.16a). Little is known about the hierarchical structure inside the fibres.

The protein of the silkworm –also called *fibroin*– contains two proteic chains (a heavy one, H, and the another lighter one, L) joined by sulphur crosslinks, together with a glycoprotein called P25. The molecular weights of these chains are, approximately, 350 kDa for H, 26 kDa for L, and 30 kDa for P25. These chains are connected to form a single large molecule, with respective proportions 6:6:1. The percentages of amino acid which comprise the fibroin are presented in Table 3.

The primary structure of H chain is known (Zhou *et al.* 2000) and the GlyAlaGlyAlaGlySer motif is repeated many times. It is believed that this amino acid chain is folded on itself in a structure known as β folding and which ends with the formation of nanocrystals.

Silk fibres can be modelled as a semi-crystalline material in which the macromolecules are, more

TABLE 3.1. PERCENTAGE OF AMINO ACID IN SILK FROM B. MORI, N. CLAVIPES AND A. TRIFASCIATA

AMINO ACID	B. MORI[a]	N. CLAVIPES[b]	A. TRIFASCIATA[c]
Ala	30.0	22.71	22.8
Gly	42.9	49.96	33.7
Tyr	4.8	2.99	3.4
Ser	12.2	2.24	8.7
Asp	1.9	1.06	0.5
Arg	0.5	1.76	1.4
His	0.2	0.21	0.1
Glu	1.4	11.02	1.0
Lys	0.4	0.10	0.5
Val	2.5	0.89	1.9
Leu	0.6	4.26	3.1
Ile	0.6	0.07	1.4
Phe	0.7	0.26	0.9
Pro	0.5	2.04	5.7
Thr	0.9	0.34	1.4
Met	0.1	0.04	0.5
Cys	0.0	0.06	0.1

a. Y. Nakazawa (2002), b. R.W. Work (1987), c. Average values of MaSp1 and MaSp2 for Gatesy *et al* (2001) and Motriuk *et al.* (2005).

or less, orientated in the direction of the fibre axis. In this model the crystalline regions play a dominant role in the stiffness and strength, while the amorphous regions contribute to the elasticity and deformation of the fibre. The crystallinity of *B. mori* is estimated to be about 50% (Drummy *et al.* 2007).

The textile industry is a unique consumer of silk fibres, but currently it is thought that the use of silk is expanding –dissolved and respun silk– for multiple applications, in particular in the field of medicine (Murphy and Kaplan, 2009).

Spider silk

Spider silk is, possibly, the natural fibre with the greatest mechanical performance. However, so far, it has not been possible to use it in the textile industry because spiders are, by their nature, solitary predators, and it is not known how to domesticate them, as the Chinese did with the silkworm.

Until the middle of the twentieth century spider silk fibres –because of their fineness– were only used to make reticules in optical instruments. The natives of New Guinea, New Hebrides and the Solomon Islands used plaited strings with silk threads from spiders of the species *Nephila* to make nets and fishing implements. In 1709, Bon de Saint-Hilaire made gloves and stockings from silk taken from the ovigenous sacks, but the French Academy thought that silk based on siders would never be viable. Three hundred years later, because of the extraordinary properties of this material, industrial production of silk proteins from genetically modified organisms is being considered.

The production of spider silk threads is quite different from that of silkworms, since, as mentioned earlier it has not been possible to domesticate spiders; the threads obtained directly from webs or extracting the rows situated in the abdomen of spiders by forced silking (Guinea *et al.* 2005), which is how the spider makes it using its legs. For this reason spider silk thread has not been exploited industrially and is used only for research with the aim of determining the causes of their excellent mechanical properties, and to try to make it artificially.

Spiders can make, at least, seven types of silk threads (Vollrath and Porter, 2006): that from the web spokes (the strongest), the circumferential (elastic and sticky), that which envelops the prey, that from the cocoon, *etc*. One of the most studied is that which is made for the web spokes, which is also used for the mooring network and as a safety line for moving. Most ampullate is secreted from the gland, conveniently designated *MAS* (major ampullate silk). In the following discussion attention is centred around silks of the genera *Nephila* and *Argiope*, since they are the best documented.

The microstructure of these threads is also not known in much detail; the MAS threads are formed of two fibres of almost circular cross-section with a diameter of about 3.5 μm, without the coating, as shown in figure 3.16b. Some authors believe that they have detected a hierarchical structure in the fibres consisting of a multilayer coating (Sponner *et al.* 2007).

MAS fibres contain at least two types of protein which are usually called *spidroins*; *MaSp1* (for major ampullate spidroin) and *MaSp2*. The molecular weight of these spidroins is several hundred kDa, of the same order as the heavy H chain of the fibroin of the *B. mori* (Hakimi *et al.* 2007). The motifs which appear in these spidroins are typically grouped in four modules: $(Gly\ Ala)_n(Ala)_m$, ... GlyGlyX, ... GlyProGlyXX, and spacers. The two motifs of the first module are believed to be composed of β-nanocrystals, just like the fibroin motif mentioned above. The second module probably adopts a helicoidal configuration of type 3_1. The third is believed to involve more flexible helicoidal structures than the previous ones, and which are responsible for the elastic behaviour and large deformations in these fibres. The percentage of amino acid spidroins of *Nephila clavipes* and *Argiope trifasciata* are given in Table 3.1.

Like silkworm silk, spider silk fibres can be modelled as a semi-crystalline material in which the macromolecules are, more or less, aligned with the fibre axis. The estimated crystallinity of the MAS threads of *Nephila* and *Argiope* varies between 11 and 15% (Hakimi *et al.* 2007), rising to more than 20% (Plaza, G. R. *et al.* 2012).

3.3.2. Mechanical properties

The tenacity of silk fibres (measured approximately as the area under the stress-strain curve) is far higher than that of the best synthetic fibres, as shown in figure 3.17. This property is a consequence of a good combination of strength and deformation; the strength is due, in part, to the β-nanocrystals and the deformability of the amorphous regions between these crystals. The tenacity depends not only upon the chemical composition of the fibres but also the processing: that is, how they are generated, since the orientation of the macromolecules during spinning has a great influence.

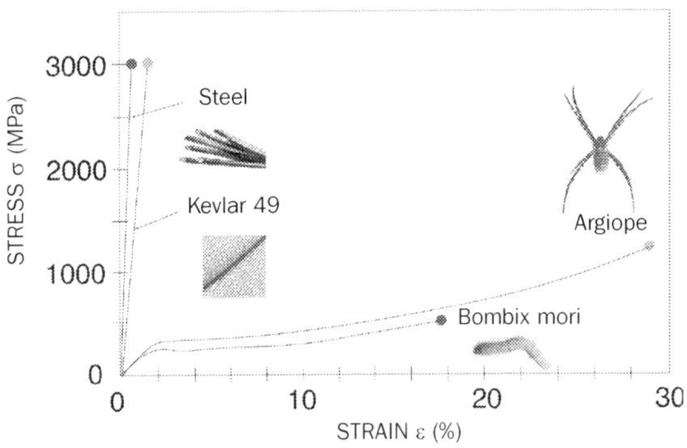

Fig. 3.17. *True stress-strain curves for different types of fibre.*

Silkworm silk

Silk fibre can be obtained from the cocoons of the silkworm or by extracting it directly from the worm when it begins to spin. Both techniques are described in detail in Pérez-Rigueiro *et al.* (2002), in which the usual procedure for tensile testing these fibres is also presented

A typical stress-strain curve of silk thread is represented in figure 3.17. Such curves depend upon the relative humidity, temperature and strain rate; the data shown are for a test at room temperature, 60% relative humidity and a strain rate of 0.0002 s^{-1}.

The means of extraction has relatively little effect upon the behaviour of the fibres, as indicated in figure 3.18, which depicts force displacement curves for fibres obtained by forced silking and those produced directly by the worm.

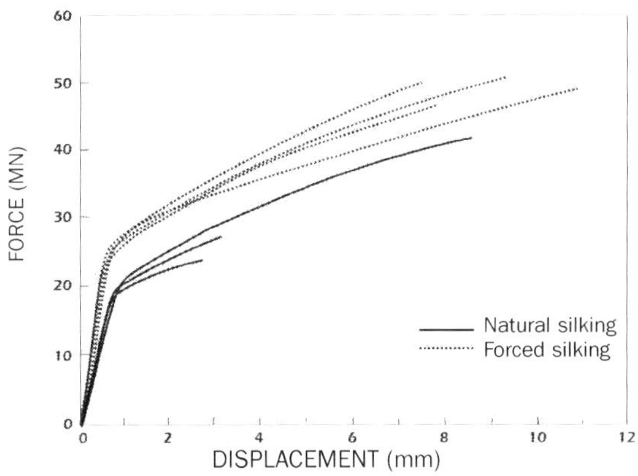

Fig. 3.18. *Force-displacement curves for silk fibres of B. mori obtained naturally by the worm and by forced silking (both from the same worm).*

Fibres obtained by degumming –boiling the cocoons in a detergent solution– have a similar behaviour, although the initial modulus is less, as are the elastic limit and the strain to failure (Pérez-Rigueiro *et al.* 2002). The degumming process reduces the size of the protein chains and worsens the properties of the fibre. Nonetheless, its use is essential for the elimination of sericin which binds the fibres and can also produce an allergy in contact the skin.

Martel *et al.* (2007) details the effect of temperature on tensile tests. It seems that the initial modulus increases gradually up to 227 °C, the temperature at which the tiny crystals degrade. On the other hand, the failure load and strain to failure remain constant up to 127 °C, the temperature above which both parameters begin to decrease.

EXERCISE 27

Sketch the stress-strain curves corresponding to the data in figure 3.18, and comment on the results. All of the fibres have a diameter of 10 µm and the initial tensile test specimen length is 30mm.

Solution:

Taking the two curves with the greatest stress for each fibre, we calculate the true stresses and strains using equations (4) and (7), based on the usual constant volume assumption. In this way

$$\varepsilon = \ln\frac{L}{L_0} = \ln\frac{L_0+\Delta L}{L_0} \quad \text{and} \quad \sigma = \frac{F}{\pi\,{}_0^2/4}\,\frac{L_0+\Delta L}{L_0}$$

with $L_0 = 30$mm and $D_0 = 10\mu$m, where F and ΔL are the force and displacement, respectively, presented in figure 3.18. Hence we obtain the curves in figure a.

If that has happened the curve is no longer representative of a material property beyond that point since the geometry of the fibre is not uniform. Since it is difficult to determine the condition $\sigma = d\sigma/d\varepsilon$ at every point, because numerical differentiation would be needed, we will use Considère's construction (see the exercise in Section 2.2).

The procedure is based on writing equation (8) in terms of the true stress σ and the elongation $\lambda = L/L_0$. Since $d\varepsilon = d\lambda/\lambda$ equation (8) become $\sigma = (d\sigma/d\lambda)\lambda$ which can be written as

$$\frac{\sigma}{\lambda} = \frac{d\sigma}{d\lambda}$$

an equation which indicates when the slope of the (σ,λ) curve coincides, at the point of instability, with the tangent to the curve drawn from the origin. The construction is shown in figure b

As we see from the curves instability is never achieved, so that the specimens break while maintaining a uniform geometry.

a **b**

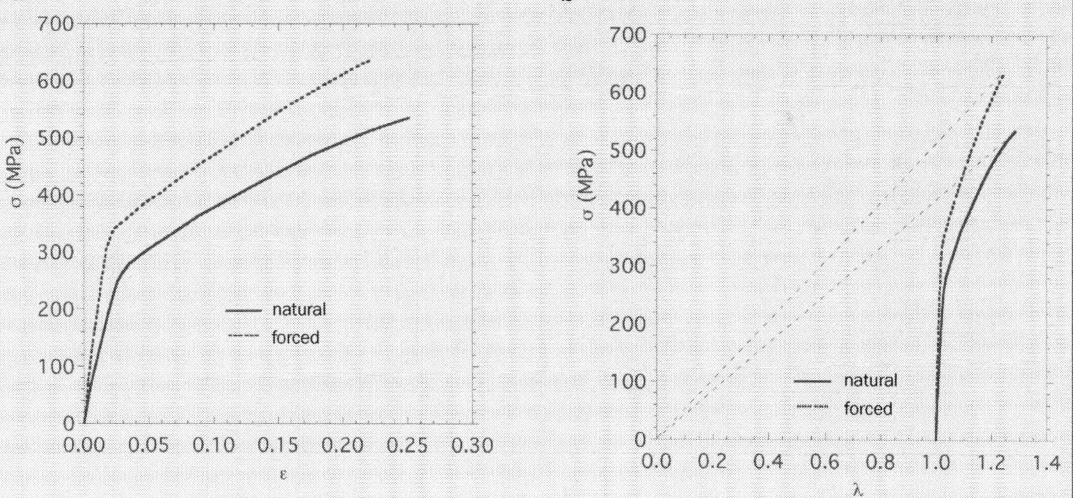

Experimental data on the influence of strain rate on the stress-strain curves are not available.

The influence of various aqueous media on tensile tests is shown in figure 3.19 (Pérez-Rigueiro *et al.* 2000). Testing in water reduces the stiffness of the fibre and increases the deformability. Ethanol, on the other hand, increases the rigidity and reduces the strain to failure. These results seem to indicate that water breaks the hydrogen bonds present in the amorphous phase, while other solvents (ethanol, acetone, isopropanol) remove water and promote the formation of new hydrogen bonds in the amorphous phase.

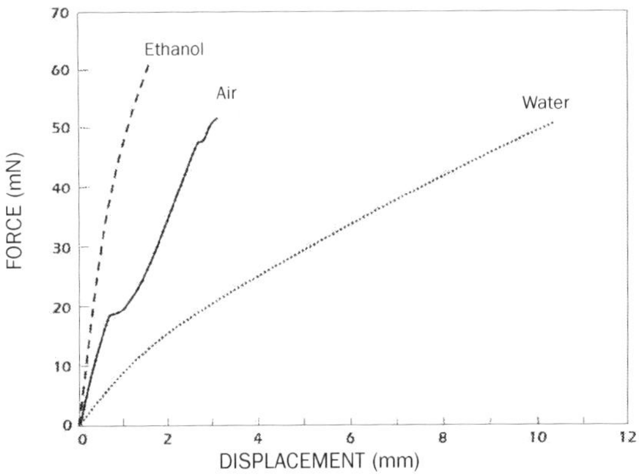

Fig. 3.19. *Force-displacement curves for silk fibres of B. mori in different media: ethanol, air and water.*

Spider silk

Silk fibres can be obtained directly from the web or also by forced silking. This technique is described in detail by Pérez-Rigueiro *et al.* (2005). The tensile test method is the same as that used for silkworm silk. For calculation of the true stress it is assumed that the volume is constant during the test (Guinea *et al.* 2006).

The mechanical behaviour of the fibres obtained from the two processes is rather different, as shown in figure 3.20 (Guinea *et al.* 2005a). Fibres obtained by forced silking are stiffer and have a lower strain to failure. It is thought that these differences are due to the orientation of the polymer chains during processing; this will be greater in forced silking than in the natural fibre. Fibres which the spider spins have a larger dispersion between stress-strain curves since the spider exerts active control during spinning according to the end use of the fibre; the fibre for weaving a web is not the same as that for when the spider needs to jump off quickly to flee from a threat (Garrido *et al.* 2002).

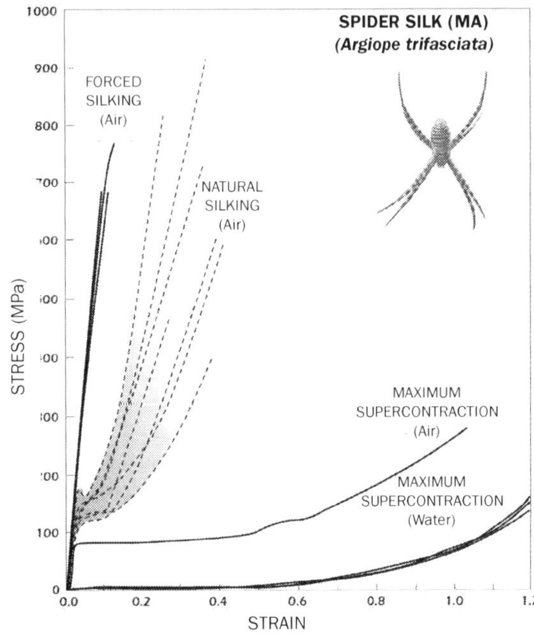

Fig. 3.20. *Engineering stress-strain curves of MAS silk fibres of A. trifasciata obtained under various conditions (see the text).*

The test conditions influence to stress-strain curves. In figure 3.21 (Plaza *et al.* 2006) the effects of **temperature** on tests performed with a relative humidity of 50% are shown. It is apparent that increasing the temperature softens the fibres and reduces the rupture stress.

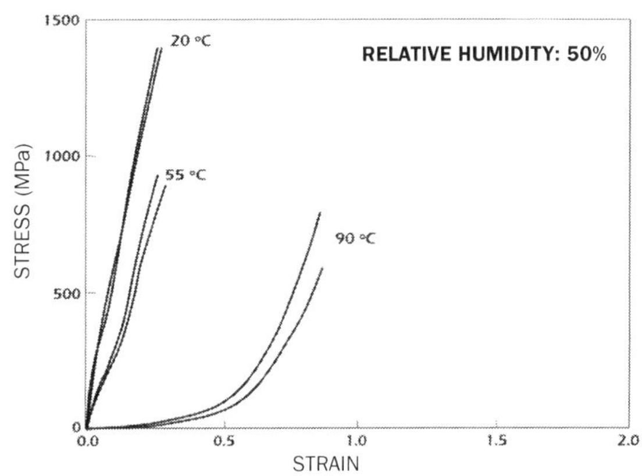

Fig. 3.21. *Engineering stress-strain curves for silk fibres (MA of* A. trifasciata*) for different temperatures (at a constant 50% relative humidity).*

In figure 3.22 (Plaza *et al.* 2006) the effect of **humidity** on tests carried out at 55 °C are shown. Here, the effect of an increase in the humidity on fibre softening and rupture stress are more marked than in the case of temperature.

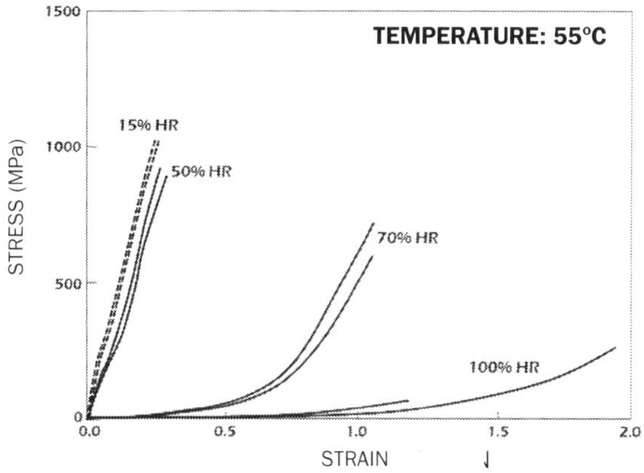

Fig. 3.22. *Engineering stress-strain curves for MA silk from* A. trifasciata *for different values of relative humidity HR (at a constant temperature, 55 °C).*

The effect of the **loading rate** is illustrated in figure 3.23. For moderate velocities –between 10^{-5} and 10^{-1} s^{-1}– a small increase in all of the characteristic parameters of the stress-strain curve is produced, as may be observed in figure 3.23a (Elices *et al.* 2005). For greater values, between 10^{-3} and 10^{3} s^{-1}, the increase is more marked and there is a change in the shape of the curve (a plateau appears in the last section) as shown in figure 3.23b (Hudspeth *et al.* 2012).

One of the most striking characteristics of the spider MAS thread is the huge contraction which it experiences when submerged in water, a phenomenon known as **super-contraction** and whose microscopic mechanisms are not fully understood (Elices *et al.* 2011). This phenomenon can be quantified by means of the super-contraction index SC, defined as $(L_0 - L_C)/L_0$, where L_0 is the initial length of the fibre and L_C is the length of the fully contracted unstressed fibre. Table 3.2 contains values of the maximum observed contraction (usually called the super-contraction) for the MAS fibres of *Argiope* and *Nephila*.

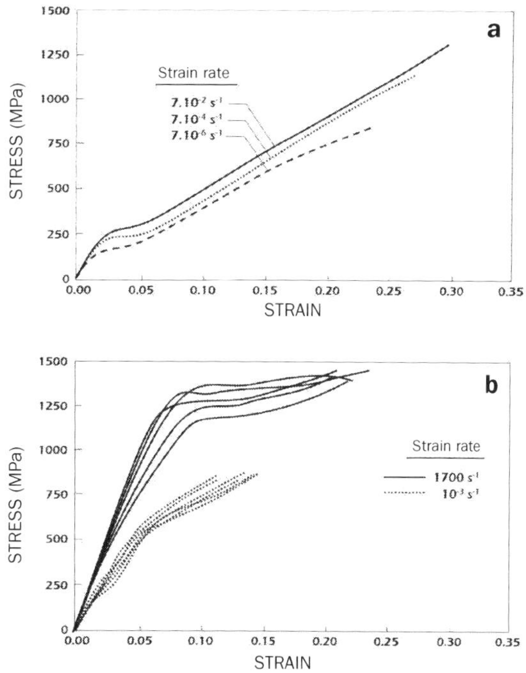

Fig. 3.23. *Engineering stress-strain curves of MA silk fibres for different values of strain rate.*
a. Moderate speeds (A. trifasciata).
b. High speed (N. clavipes)

TABLE 3.2 SUPER-CONTRACTION INDICES FOR SOME SPIDER MAS FIBRES

TYPE OF SPIDER	SC INDEX = $(L_0-L_C)/L_0$	REFERENCE
Argiope trifasciata	0.31	Pérez-Rigueiro *et al.* 2003
Natural thread	0.52	
Forced spun		Elices *et al.* 2009
Nephila sp	0.20	Work 1985
Natural thread	0.40	Elices *et al.* 2009
Forced spun	0.44	Liu *et al.* 2005
		Jelinski *et al.* 1999

The tensile behaviour, in **water**, of the super-contracted fibres is typical of an elastomer (see figure 3.20), characterised by small values of the initial elastic modulus and a marked increase in the stiffness for large deformations. When the super-contracted fibre is dry the behaviour is totally different, as illustrated in figure 3.20 (fibre Maximum Super-contraction tested in **air**).

Super-contraction modifies the tensile behaviour of the fibre and provides ways of understanding the mechanical properties (Guinea *et al.* 2005a, b, and Elices *et al.* 2011). The stress-strain curves of a dry super-contracted fibre (figure 3.20, curve MS) are independent of the previous loading history; that is, the curve is the same after the super-contraction, independent of whether the fibre has been loaded or unloaded several times, or if the fibre was obtained from the web by forced silking. This property allows the use of the MS fibre curves as **reference curves** since, from these, it is possible to obtain all of the others (that which the spider spun and those obtained by forced silking) by means of a process of controlled loading in water (Guinea *et al.* 2005a) characterised by an alignment parameter α; in the initial state, unstretched, $\alpha = 0$ and its value increases with the deformation of the fibre during treatment. It has been observed that for the fibres that spiders weave, α ranges from 0.45 to 0.9, approximately. The largest values correspond to fibres obtained by forced silking. These results are presented in figure 3.24 (Elices *et al.* 2011), for MAS fibres of the *Argiope trifasciata*.

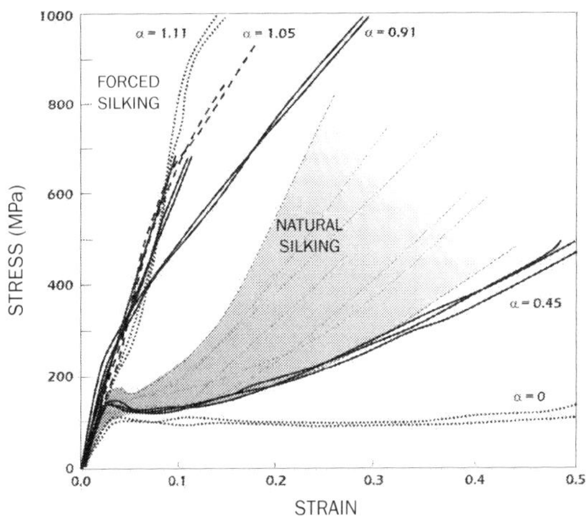

Fig. 3.24. *Engineering stress-strain curves of MA silk fibres of* A. trifasciata *obtained from MS fibres (see text) by straining under humidity (different values of* α*).*

The mechanical behaviour of silks can be modelled by a semi-crystalline composite material, as has already been mentioned.

At the molecular level, Termonia's model (Termonia, 2000) gives results that provide reasonable agreement with experiment; for small strains (less than 0.2%) it predicts a linear response, with a modulus of 10 GPa, agreeing with experimental observation. Beyond this strain, the hydrogen bonds between the chains in the amorphous regions begin to break and a yielding phenomenon appears accompanied by a plateau. After a strain of approximately 0.1%, the stress begins to increase with strain until it reaches about 0.3%, at which it breaks. The model is also capable of reproducing the effect of water by eliminating the hydrogen bonds between the amorphous phase chains.

In a continuum mechanics-based treatment, the fibre can be modelled as a composite of microfibrils dispersed in a deformable matrix (Planas *et al.* 2007). The microfibrils are modelled as nonlinear springs in parallel with a system formed of a slider and a linear spring; the nonlinear spring represents the behaviour of the molecular chains in the elastomeric state (with the network of hydrogen bonds deactivated), while the slider-spring system attempts to reproduce rupture of the hydrogen bonds and sliding of the chains when the fibre is above the glass transition temperature. This model is not only able to reproduce tensile tests it can also simulate tests on super-contracted fibres and stressed with different values of α.

EXERCISE 28

A simple way of understanding the behaviour of spider thread is to consider the protein chains which form it as tangled creating an elastic (nonlinear) network which is stable thanks to the weak bonds between the chains. These bonds break easily in the presence of humidity. Therefore, when the fibre is submerged in water the network unlocks and tends to shrink to its maximum value. If the network is stretched in **water** up to a certain value and the fibre dried in this new position, the bonds recombine to provide stiffness, so that we obtain a longer fibre than the initial one but with a certain internal "pre-stress", which can be relieved if it is wetted again.

With these ideas, explain figures 3.20 and 3.24, and the effect of the alignment parameter α, defined as $\alpha = L_0 / L_C - 1$ (Guinea *et al.*, 2005a) where L_0 is the length of the relaxed dry fibre L_C is relaxed dry length when it is totally constrained (state of maximum contraction, MS).

Posing the problem:

As explained, the stress-strain curve in water, shown in figure 3.20 **(maximum super-contraction-water)** represents the behaviour of a molecular network, free from the effects of the bonds between the chains, and therefore a material characteristic, same in all of the fibres. We shall call this curve $s_{net}(e)$ (Figure a).

If we let the fibre contract **freely** in water, and then test it dry (in air) we will obtain the **maximum super-contraction-air** (curve MS air) which will be the combined effect of the elastic molecular network and the network of weak bond. The subtraction of the two maximum super-contraction curves in Figure 3.20 shows the behaviour of the network of weak bonds. The result is shown in figure a. This engineering stress-strain curve will be called $s_{bonds}(e)$.

We now consider a test in which we **stretch** a fibre of initial length and area L_C and A_C, in **water,** until the strain $e_S = 0.6$ the corresponding stress is ($s_S \approx 13.5$MPa) and at this point allow it to dry.

What will the shape of the stress-strain curve be, for a new tensile test performed on this fibre?

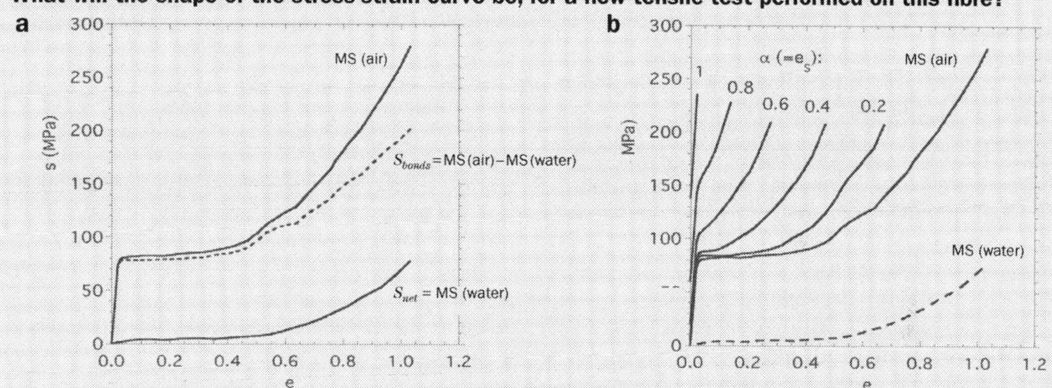

Solution:

To answer this question, suppose that by drying the fibre the residual strain is e_S, without producing any contraction during drying, which is not strictly true, but which we can take as a first approximation.

The stretched dry fibre will now have a length $L_S = (1+e_S) L_C = 1.6\ L_C$ and a cross-sectional area $A_S = A_C/(1+e_S)$, which we obtained from conservation of volume. Note that L_S and A_S become the initial reference length and area for the new tensile test.

To stretch the fibre we must apply a force greater than the internal residual forces created when drying the fibre F_S. Its value can be calculated from the stress at the point of drying $F_S = s_{net}(e_S)\ A_C$. Note that this value of the force now gives, in the new test, an engineering stress $s_{net}(e_S)\ A_C / A_S = s_{net}(e_S)\ (1+e_S)$, having to refer the force to a smaller area A_S.

The force exerted in the test must overcome the behaviour of the polymer network, represented by the load curve in water and also must overcome the forces in the network of weak links, whose stress-strain curve we assume to be approximated by the calculation shown in figure a. In both cases, the forces must be referred to the initial cross-sectional area in the new test, that is A_S and not A_C.

Bringing this together, the force due to the molecular chain **network** will be

$$F_{net}=A_C s_{net}\left(\frac{L}{L_C}-1\right)=A_C s_{net}\left[\frac{L}{L_S}\frac{L_S}{L_C}-1\right]=A_C s_{net}\left[(1+e)(1+e_S)-1\right]$$

So now $e=(L-L_S)/L_S$, and $L_S/L_C=1+e_S$.

On the other hand, the force due to the **weak links** will be $F_{bonds}=A_S s_{bonds}\left(\frac{L-L_S}{L_S}\right)=A_S s_{bonds}(e)$

in this case we consider that the links are completely restored when the water is eliminated, with no memory of its previous state.

Therefore, the engineering stress becomes

$$s=\frac{F_{net}+F_{bonds}}{A_S}=\frac{A_C}{A_S}s_{net}\left[(1+e)(1+e_S)-1\right]+s_{bonds}(e)=(1+e_S)s_{net}\left[(1+e)(1+e_S)-1\right]+s_{bonds}(e)$$

The result for different values of $\alpha=e_S$ is represented in figure b. Note that, according to the definition given for the alignment parameter, this result will be equal to e_S.

3.3.3. Concluding remarks

Silkworm fibres have been principally destined for the textile industry although recent interest has developed in biomedical applications, particularly in regenerated fibroin threads.

Silkworm fibres are covered with a layer of sericin which provokes an inflammatory response *in vivo*. For this reason when used for stitches, they must be de-gummed, eliminating this coating. In these circumstances it seems that the inflammatory response is less than with traditional biomaterials, such as collagen or polylactic acid.

Another form of use of the fibre is regenerating; the de-gummed fibres dissolve at high temperature (between 60 and 80 °C) in various watery solutions, dialysed in water and the fibroin solution is processed in the form of fibres, films or sponge structures which can be used as scaffolds in engineering of textiles. The potential of these scaffolds to function chemically makes them attractive for controlling the interaction between fibroin and living systems. The surface chemistry of the fibroin scaffolds can greatly influence the cellular response in many ways; from changes in adhesion to modifying the cellular proliferation processes, differentiation or survival (Murphy and Kaplan, 2009).

Spider silk threads have many attractive features, among them its huge tenacity (energy-absorbing capacity when strained, represented by the area under the stress-strain curve) as can be inferred from figure 3.17. The tenacity of a MAS thread from *Argiope* is about 130 kJ/kg compared to 30 kJ/kg of Kevlar® and 4 kJ/kg of steel. However, they have the disadvantage that it is not possible to obtain large quantities for the reasons noted above, a motive for the growing interest in obtaining these fibres through genetic engineering.

The route forward has two stages: in the first, attempts are made to identify the spider genes responsible for the production of silk proteins, synthesise them and express them in other organisms. The second stage consists of producing a protein solution, spinning it and fabricating the fibre.

With this strategy, several related amino acid sequences with the composition of silk threads have been identified, and artificial genes which mimic the most characteristic aspects of the natural fibre have been prepared. These genes have been expressed in bacteria, in several microorganisms, in plants such as sunflowers and in the mammary cells of genetically modified goats from which milk rich in spider protein has been produced. After spinning the protein solutions, the mechanical properties of these artificial fibres are quite different from those of the natural fibres, possibly because the spider protein has not been completely reproduced, and because the spinning process has not recreated the natural spinning strategy.

The spinning difficulties could be avoided if it were possible to insert in the genome of the worm the genes responsible for the enhanced properties of spider silk. This process has already begun in several research laboratories but, as of now, the results obtained are inferior to those of natural fibres.

Despite not achieving the excellent mechanical properties of natural silk, the silk thread obtained from these processes may have important applications in the medical field (Kluge *et al.* 2008, Elices *et al.* 2011), in particular in the textile technology and in pharmacology (in the form of microcapsules for the administration of medication).

3.4. COLLAGEN FIBRES

3.4.1. Introduction and fibre structure

Collagen is the most abundant fibrous protein in vertebrates and is present throughout the animal kingdom, except for the simplest eukaryotes (protists). About 25% of all mammal proteins are some sort of collagen molecule (Smith 1991). It is the protein that provides strength and resistance to biological tissues and appears in practically all organs from bones to skin, tendons, cornea, blood vessels, uterus or epithelia.

The name collagen comes from the Greek κόλλα (cola, glue) and literally means "generator of adhesion". Since ancient times, glue has been made by cooking skin, cartilage, tendons and ligaments of horses and other animals, and also the skin and bones of fish. Collagen glue was used to give strength and impermeability to everyday utensils, and also in ancient mortuary practices. Up to now, the oldest preserved remains, more than 8,000 years old, come from the Nahal Hemar cave near the Dead Sea (Nissenbaum 1997). Nowadays musical instrument makers still use animal glue in more elaborate string instruments in order to be able to reopen them without damaging the wood when needed, since it is a thermoplastic adhesive which softens with heat.

Currently 29 types of collagen are known and are codified by 44 different genes (Soderhall 2007). All of them are trimers (generally called tropocollagen) composed of three levorotatory chains. In each chain, the most common amino acid repeating unit is Gly-X-Y, in which the proline amino acids or hydroxyproline and, to a lesser extent, lysine or hydroxylysine occupy positions X and Y (Ramshaw 1998). Each type possesses a characteristic structure in which the chains that form the triple helix can be the same or different. Table 3.3 contains the composition of the most common types of collagen, their preferred location, the cells that secrete them and their principal function.

Type I collagen is by far the most important in vertebrates, by constituting about 90% of the total and being the major part of organs such as the skin, tendons, ligaments, bones, blood vessels and the cornea. Its microstructure is shown in figure 3.25. Each molecular chain consists of about 1,000 amino acids which take on the form of a levorotary helix of 3.6 amino acids per turn. Two $\alpha1[I]$[18] chains and one $\alpha2[I]$ chain link together to form a triple helix with a right-hand rotation. The molecular weight of the $\alpha1[I]$ chain is about 139 kDa and that of the $\alpha2[I]$ chain 129kDa, yielding a value approaching 400 kDa for the trimer (Cowan 1955).

TABLE 3.3 PRINCIPAL TYPES OF COLLAGEN IN VERTEBRATES (SOURCE: BRINCKMANN 2005)

TYPE	COMPOSITION*	STRUCTURE	TISSUES	PRODUCTION	FUNCTION
I	$\alpha1[I]_2\ \alpha2[I]$	Fibrillar	Dermis, tendons, ligaments, bones, blood vessels, cornea, scar tissue	Fibroblasts, chondroblasts and osteoblasts	Thick structural fibres: tension
II	$\alpha1[II]_3$	Fibrillar	Cartilage, vitreous humour	Chondroblasts	Very fine structural fibres: tension and compression
III	$\alpha1[III]_3$	Fibrillar	Accompanies Type I, in arterial veins, skin, intestine and uterus	Smooth muscle cells, fibroblasts	Spongey mass support and support of other cells: expandible bodies
IV	$\alpha1[IV]_2\ \alpha2[IV]$ $\alpha3[IV]\ \alpha4[IV]$ $\alpha5[IV]$ $5[IV]_2\ \alpha6[IV]$	Reticulated	Basal lamina under the epithelia	Epithelial cells and endothelia	Support and filtration

*) Type of chain–same or different– which forms the tropocollagen molecule

[18] Although the symbol α is used for helices which form collagen molecules, it must be remembered that its structure is different from the alpha helix which forms the secondary structure of proteins, and is dextrorotatory.

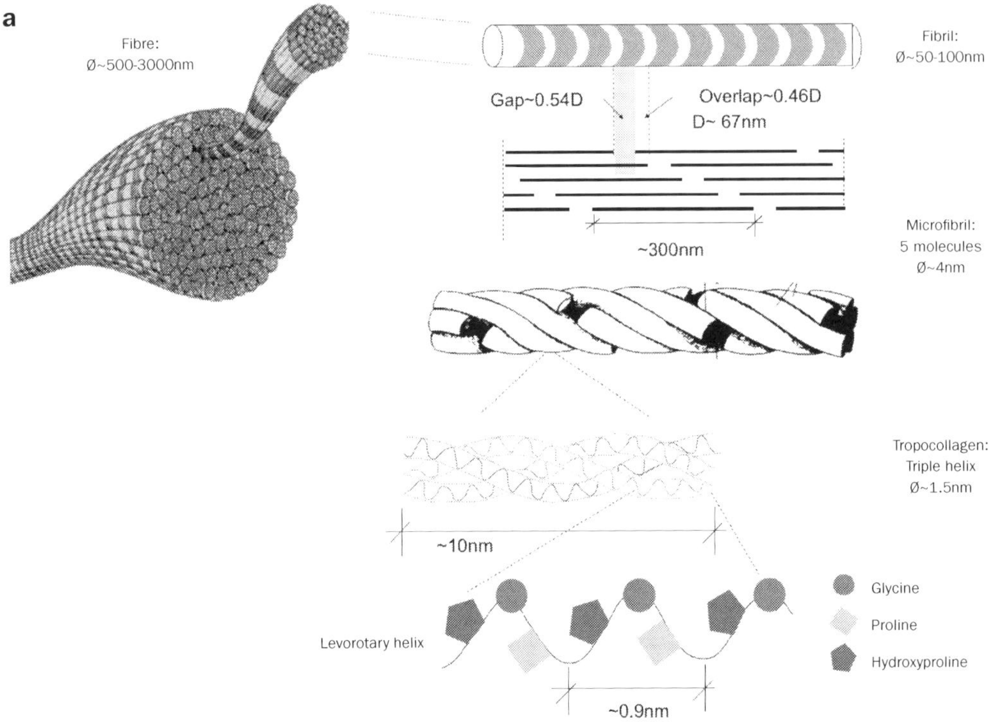

Fig. 3.25. *a) Typical microstructure of Type I collagen fibres. b) Electron microscopic image which shows the typical striated fibres.*

A single α chain is not stable and must combine with two similar ones to form the tropocollagen triple helix. Nor is Type I tropocollagen stable at body temperature (Leikina 2002), so five molecules combine (figure 3.25a) to form a stable structure (Orgel 2006). The packing is such that adjacent molecules are shifted by approximately one-quarter of their length (67nm), and this arrangement produces the striated effect which can be observed under an electron microscope (figure 3.25b).

The stability of tropocollagen molecules is achieved by means of hydrogen bonds, principally between the proline residues and hydroxyproline. The lack of this last amino acid destabilises the chains and causes weakening, with grave consequences for the tissues (malformations, weakness and bleeding). Human beings, like the rest of primates, are incapable of producing their own vitamin C–essential for the synthesis of hydroxyproline– and a diet deficient in this vitamin triggers the serious disease of scurvy. It is calculated that about 2 million sailors died between 1600 and 1800 due to a lack of vitamin C, resulting from the difficulty of transporting fresh fruit and vegetables in ocean crossings (Drymon 2008).

Some genetic illnesses directly affect the production and assembly of collagen causing very severe syndromes which are difficult to treat. For example, Ehlers-Danlos syndrome originates in an extreme loosening of the skin and tendons (the so-called rubber men) and promotes the formation of aortic aneurisms. The Osteogenesis Imperfecta syndrome impedes the correct assembly of the triple helix and causes bone brittleness (the so-called glass children). Nor is the overproduction of collagen synonymous with good health, since fibrosis (for example in the liver and lungs) produced as a response of the organ to an attack on its tissues can inhibit the function of the affected organ.

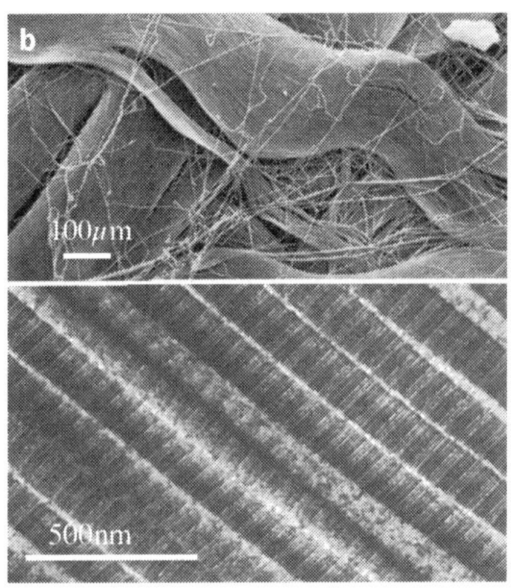

3.4.2. Mechanical properties

Collagen represents a paradigm example of the hierarchical structure of biological materials, in which –as mentioned above– it is difficult, and always a matter of judgement, to distinguish between the structure and the material. Figure 3.26 shows the structural organisation of tendon, in which continuity is not observed from the atomic and molecular up to the macroscopic level of the insertion into the bone or muscle. Therefore, to talk in general about the mechanical properties of collagen is very difficult and it must always be done with care to specify clearly the structure studied. In this section attention is focussed on the two extremes of the hierarchical organisation: fibril and microfibrils, almost molecular collagen, and tendons and ligaments.

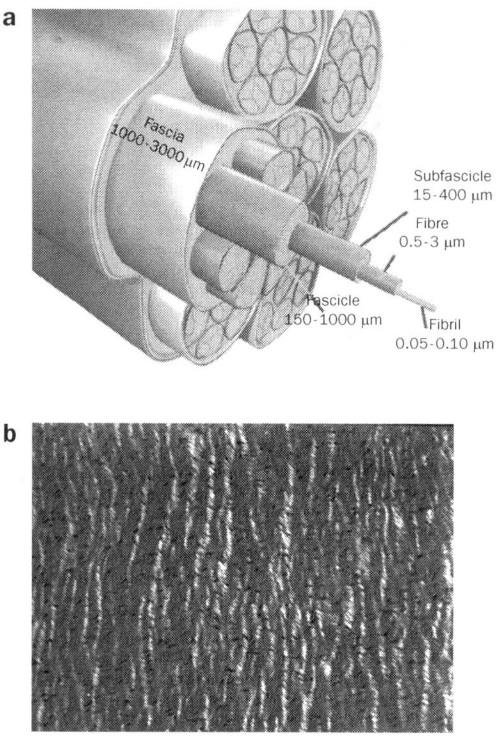

Fig. 3.26. *Typical structure of a tendon.*
a. Hierarchical organisation.
b. Wave pattern on collagen fibres (Weiss 2001).

Mechanical behaviour of collagen fibrils

Collagen fibrils are formed by the addition of tropocollagen microfibrils (figure 3.25a) and their diameter varies depending upon the tissue and the production. In figure 3.27 the stress-strain curves of *the skin of the holothuria* are shown. The fibril diameter is about 300 nm and the length 10μm. The tests were performed using a micro-electro-mechanical system (MEMS) *(*Shen 2008).

In general it can be said that the behaviour of the fibrils is linear elastic until the yield stress, probably due to sliding between the microfibrils. It can also be seen that unloading of the fibre is linear elastic, and that there is a small recovery of the plastic strain if the fibre rests in a humid environment (*Shen 2008)*. Figure 3.27c shows that the value of the yield stress depends upon the volume of the fibril, decreasing with size.

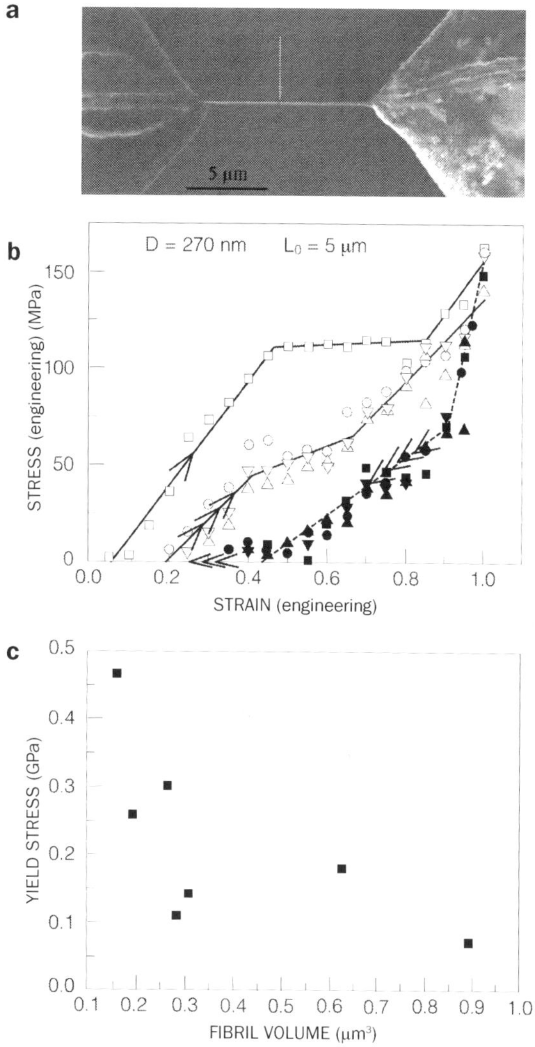

Fig. 3.27. *Tensile tests on collagen fibrils (Shen 2008).*
a. Photograph of the test.
b. Cyclic stress-strain curves loading (open symbols), unloading (filled symbols). Each symbol represents one test.
c. Yield stress.

TABLE 3.4. ELASTIC MODULUS OF COLLAGEN FIBRILS

TYPE OF FIBRIL (REF.)	TECHNIQUES USED	STATE	E(GPa)
Rat tail tendon *(Wenger 2007)*	Nanoindentation	Dehydrated	3.75–11.5
Rat tail tendon *(Harley 1977)*	Brillouin spectroscopy	0.15 M NaCl 30% humidity 0% humidity	9.0 14.7 21.5
Bovine Achilles tendon *(Sasaki 1996)*	X-ray diffraction	0.15 M NaCl	2.9
Bovine Achilles tendon *(van der Rijt 2006)*	Brillouin spectroscopy	0% humidity PBS	2-7 0.2–0.5
Holothuria *(Eppell 2006)*	Tensile test (MEMS)	water	12
Holothuria *(Heim 2006)*	Nanoindentation	45% humidity	1-2
Holothuria *(Shen 2008)*	Tensile test (MEMS)	30-60% humidity	0.86

Table 3.4 contains the values of the elastic moduli of collagen fibrils measured by different researchers. The values are quite dispersed and go from about 200 MPa to somewhat more than 21 GPa. The elastic modulus is, in general, high indicating that the fibrils are quite stiff, especially in the absence of moisture. Experimental data show a wide degree of variability owing to the different conditions in which the fibrils were tested, the difficulty in preparing the test specimens and the test methods themselves, which include micromechanical tests, optical techniques and x-rays.

In the tests mentioned, as much as because of the techniques used –some nondestructive– as the handling difficulty and size of the fibrils, the strength has not been measured. The determination of this parameter is much more difficult than the elastic modulus but the reference data cited and numerical simulations performed with molecular models show that the strength of collagen fibrils is more than 0.6GPa and probably greater than 1 GPa. On the other hand, the estimated failure stress of the tropocollagen molecule is more than 10GPa *(Buehler 2006, 2008)*.

Mechanical behaviour of tendon

Tendon is a very uniform fibrous connective tissue which joins the bones and muscles in order to transmit the mechanical forces that permit movement. Its organisation is complex (figure 3.26a) being composed of collagen fibres arranged in fascicles. The composition of tendon is by mass 90% Type I collagen, although this varies in the vicinity of its insertion into the bone –osteotendinous junction– or to the muscle –miotendinous junction (Amiel 1984).

Collagen fibres in the interior of tendons adopt a position parallel to the axis a typical undulation of wavelengths between 100 and 300μm (figure 3.26b) which has clear repercussions on the mechanical behaviour, as will be seen below.

Tests on tendon are usually done immersed in serum and at 37°, in order to reproduce, as closely as possible, the physiological condition. Figure 3.28 shows the stress-strain curve of human tendon of the extensor digitorum longus muscle (Schechtman 1997). In that figure four distinct zones are apparent (labelled from I to IV in the figure) corresponding to different fibre strain states. In the initial zone (I) the curve takes on a concave shape and the stresses increase slowly with the strain. In this zone, whose limits lie between 1 and 4% strain, the collagen fibres gradually lose their initial rippled texture, and they regroup and re-orientate themselves in the loading direction. From this point in which all the fibres are stretched the behaviour of the tendon is practically linear elastic, as shown in zone (II) of figure 3.28, up to strains of around 10%. The first fibre failures signal the start of zone (III), in which there is a gradual loss of stiffness by the weakening of the tendon up to its rupture in zone (IV).

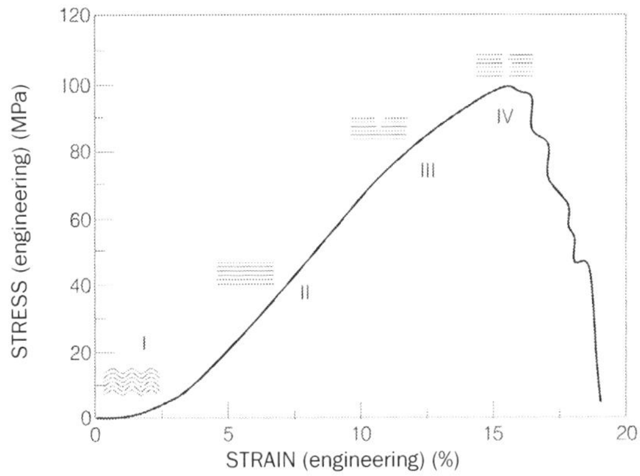

Fig. 3.28. *Tensile test on the tendon of the extensor digitorum longus muscle (Schechtman 1997).*

The average values of the elastic modulus in the linear zone are quite variable, possibly due to the different experimental techniques used and the difficulty in estimating the cross sectional area of the specimens, given that tendons are soft tissues which deform under pressure. In general the elastic modulus ranges from 0.3 to 1GPa (Cowin 2007) and is sensitive– like the rest of the loading curve– to the strain rate.

Tests with strain rates from 0.05 to 150 %/s have shown that the initial zone (I) hardly changes while the slope of the linear zone (II) and also the rupture stress increase with the loading rate. The strain to failure, however, remains essentially constant (Abrahams 1967, Ng *et al.* 2004). Figure 3.29 shows the experiments of Abrahams (1967) for the human Achilles tendon. The values obtained by Cowin (2007) can be taken as a reference for the rupture stress (70-100MPa) and the strain to failure (~11%). The values of the elastic modulus of tendon are generally an order of magnitude less than those of collagen fibrils, and a similar ratio is maintained in the values of the rupture stress (figure 3.29a).

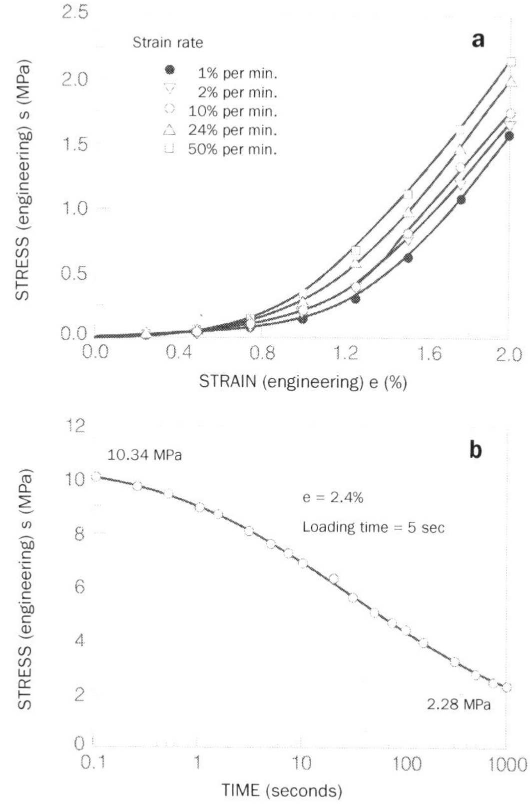

Fig. 3.29. *Experiments on the human Achilles tendon (Abrahams 1967).*
a) Effect of strain rate. The experiments were stopped at a strain of 2%
b) Relaxation test. The tendon was loaded to e=2.4% in 5 seconds.

EXERCISE 29
Estimate the elastic modulus of collagen fibrils from holothurian skin using the curves of figure 3.27b.

Solution:
Using figure 3.27b we can identify several sections in which we can easily calculate the slope, as shown in the figure below. The values are:

section	E(MPa)
1	275
2	204
3	150
4	82
5	212
6	933
7	275

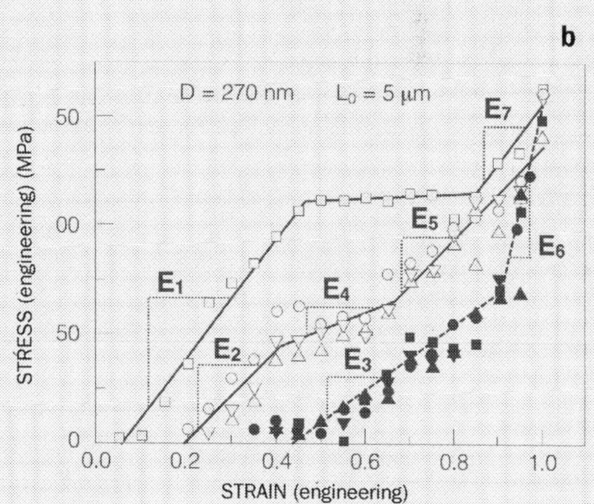

Given the difficulty in the testing, and the dispersion in the values, care must be taken when considering the values above since, for example, section 4 could be influenced by spurious effects associated with the experimental technique such as microslip in the glue regions of the grips. However, the values obtained (with the exception of section 4) lie reasonably within the range observed in other experiments.

Tendon exhibits a viscous behaviour, illustrated in figure 3.29, which is for a *relaxation test.* Unfortunately, and like many other soft tissues, tendon is not linear viscoelastic. In figure 3.30 experimental results are presented for relaxation tests performed by Hingorani (2004) on the lateral median ligament of rabbit at different strain levels Although not for tendon these results are perfectly extrapolatable since ligaments and tendons share the same structure of aligned collagen fibres.

It can be observed, in the figure, that the relaxation is not proportional to the level of the applied strain since the relaxation modulus $G(t)$ (obtained by dividing the stress by the initial strain, $G(t) = \sigma(t)/\varepsilon_0$) does not remain constant in all of the tests (figure 3.30b).

The tendon strain is not elastic much beyond zone (I) (Abrahams 1967). Even so, and neglecting the different unloading behaviour and time dependency, hyper-elastic models have been proposed to represent the nonlinear behaviour of tendon. Details can be found in Cowin (2007). In general, three-dimensional transversely isotropic models are used, in which the contribution of the collagen fibres is separated from that of the matrix which is assumed isotropic. The study of these models lies outside the scope of this chapter and the interested reader is left to consult the cited reference.

When it is required to model time-dependent behaviour, given the *nonlinearity* of the viscoelastic response and its dependency of the load level, it is common to adopt an *ad-hoc* linear viscoelastic model for the range of desired loadings. Obviously the validity of the approximation and its accuracy are compromised by the underlying hypothesis. In this case the modelling is exclusively one-dimensional.

Another possibility is the use of more elaborate models developed to account for the dependency of the relaxation (or creep) on the loading state, such as the quasi-linear viscoelastic model of Fung (1972). The interested reader can find its application in Cowin (2007).

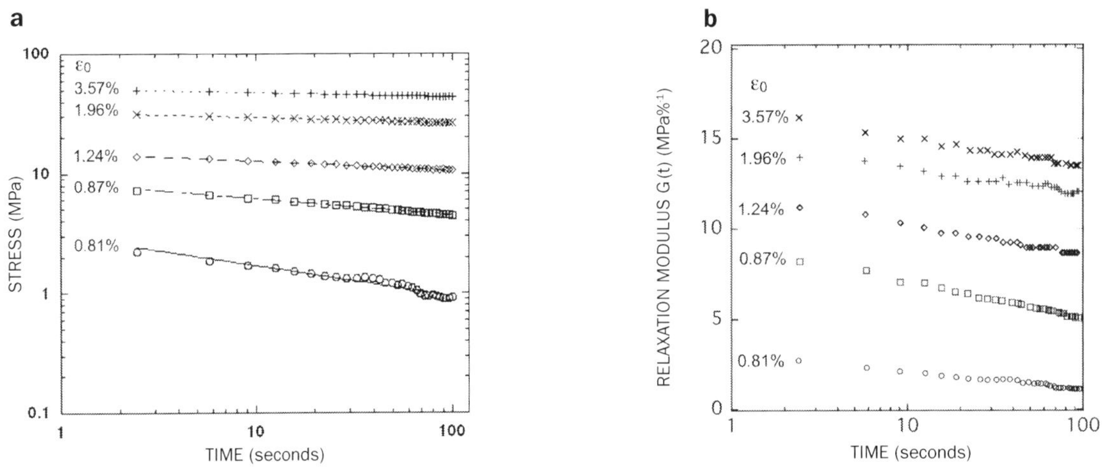

Fig. 3.30. *Relaxation tests on the lateral median ligament of rabbit and on the human Achilles tendon for different levels of strain* ε_0 (Hingorani *et al.* 2004)
a) Stress as a function of time $\sigma(t)$
b) Relaxation modulus $G(t) = \sigma(t)/\varepsilon_0$.

3.4.3. Final remarks

Collagen is a biological material of huge mechanical capacity and with numerous applications in medicine, bioengineering and in the cosmetics and food industries.

Using processes of alkaline, acid or enzyme hydrolysis collagen is obtained industrially from tendons and skin of bovines, equines and pigs and it is commercialised in liquid solution or –more commonly– in powder. Its molecular weight varies from 1kDa to 10kDa and the molecules have a globular disordered structure.

Being a component of the extracellular matrix, collagen possesses haemostatic wound-healing properties and promotes cell growth. Therefore it is used in the treatment of severe burns and is a frequent ingredient of scaffolds used in tissue engineering and as an implant coating (Lee 2001). It can be used on its own or in combination with silicones, glycosaminoglycans, fibroblasts, growth agents and other substances (Parenteau-Bareil 2010).

Hydrolysed collagen is also commercialised for use in cosmetics as a skin moisturiser and conditioner, and as a dietetic supplement to promote bone growth and maintain articular joints (Oesser 2008).

If reconstituted with water, hydrolysed collagen forms gelatine, a viscoelastic birefringent gel below its melting point whose uses are many and varied; from the preparation of desserts to pharmaceutical capsules, supplements for low calorie products, adhesives and photographic emulsions (Ward 1977).

When it is required to take advantage of the mechanical potential of collagen fibres it is necessary to use animal tissues directly, changing their elaborate hierarchical structure as little as possible. In these cases it is possible to opt to use the entire organ, as has been done for years with simple elements like porcine blood valves, and there are more advanced studies with such sophisticated organs as the kidneys or the heart (Maher 2013). The strategy used is simple: it aims to eliminate completely the cellular component of the donor preserving the collagen matrix, so that that can be colonised by the cells of the patient. For the first time, a human has been given a transplant of a pig´s heart that had been genetically modified to boost the chances of acceptance in a human body. (New Scientist, 2022).

Otherwise tissues rich in collagen are used, generally membranes such as the bovine pericardium or dura, to repair veins or hernias, or as prosthetic elements like heart valve vanes. In these cases, the collagen tissues are usually treated with glutaraldehyde to crosslink and delay its biodegradation (Nimni 1987).

Collagen is one of the most versatile and useful natural fibres, with applications in very diverse sectors. Moreover, their excellent mechanical and biological properties mark them out as one of the most important biomaterials of the present and with great future development potential.

SECTION 2

BIOLOGICAL MEMBRANES

4 CHAPTER Mechanical Behaviour of Membranes. General Concepts.

4.1. INTRODUCTION

The objective of this section is to outline the particular properties of biological materials which have two preponderant dimensions –those which characterise the surface rather than the thickness. In the previous section, dedicated to fibres, there was only one relevant dimension –the length of the fibre– which greatly simplified the mathematical treatment in order to understand mechanical tests and to model their behaviour. Bidimensional materials can be grouped into two families: *membranes* and *laminas*. Laminas, due to their small thickness, have a certain flexural rigidity, while in membranes this can be considered negligible. This is the case with soft biological materials, which are always very flexible.

The analysis of bidimensional materials is more complicated than that of fibres because, with the presence of the additional dimension, there is often *anisotropy*; the material behaves differently in different directions in the surface.

The structure of these materials is essentially that of a material composed of fibres embedded in an organic matrix. P*olysaccharide* fibres form the leaves of plants (if the fibres are cellulose) or the cuticles of insects (when chitin fibres are used). The epithelia are constructed from *polypeptide* fibres as are blood vessels (with elastin and collagen fibres). Therefore, **Chapter 4** focuses on two important biological materials: the skin of mammals and blood vessels. A detour into the membranes of marines creatures is made, because of their growing importance in the field of biomimetics.

An additional difficulty when materials are very flexible is the mathematical treatment of the tensile state, because the Hookean approximation is no longer valid and it is necessary to define stress and strain tensors within the framework of the theory of finite deformations. In order to familiarise the reader with these new concepts, **Chapter 5** outlines the most important aspects of the kinematics and stress state which are used to characterise the mechanical behaviour of biological membranes. These concepts are necessary for the understanding of the most common mechanical tests used to characterise bidimensional biological materials, which will be described below. Some tests are special cases of more general tests developed for three-dimensional materials (such as fracture or indentation tests), and for this reason they are only mentioned in passing, since they will be covered in more detail in a later section.

In **Chapter 6** two important bidimensional biological materials in biomedical engineering are discussed in detail: the pericardium and blood vessels. Also, elasmoid scales are mentioned as a source of inspiration for the design of armours.

4.2. SKIN OF MAMMALS

Skin is the coating which totally covers the body of animals. It is a tissue with multiple functions; it is the first defensive barrier, has sensors and excretory organs, and plays a critical role in regulating body temperature and in containing fluids. Furthermore, it is the source of several symptoms, some corresponding to specific skin diseases and others to internal maladies which appear on the skin.

Skin is the defensive element that protects traumas, from ultraviolet radiation, toxic agents and microorganisms. Skin plays a part in regulating body temperature by increasing or decreasing the flow of blood reaching it. Sweat can also be involved the thermoregulation, since when it evaporates, it takes away heat from the organism. Some nerve ends reach the skin, terminating in tactile corpuscles which capture heat or cold, or pressure or pain, information and transmit it through the nervous system to the brain. The production of keratin (keratinisation) hardens the outer layers of the skin and contributes to replacement of its cells.

According to Lyman (2019),"The skin –despite being our largest and most visible organ, despite us seeing and touching it, indeed living in it, every moment in our lives– is the organ most overlooked by the medical profession. Weighing nine kilograms and covering two square meters, skin wasn't even recognized as an organ until the eighteenth century."

4.2.1 Structure

In the skin of mammals it is possible to distinguish three superimposed layers which, from outside to inside, are: the *epidermis*, the *dermis* and the *hypodermis* (figure 4.1).

The **epidermis** is the outermost part. It is a thin epithelium (in humans of a similar thickness to a sheet of paper) composed of cells, keratinocytes, which are in a state of continuous renewal. Dispersed between these are the melanocytes, cells which produce melanin and the Langerhans cells which have a defensive function. The epidermis has neither blood vessels nor lymphatics, but it does have nerve endings. In the epidermis it is possible to observe four strata (see the detail in figure 4.1): corneum, granulosum, spinosum and basale or germinative.

The **dermis** is rich in blood vessels and lymphatics, and it has nerve endings which collect tactile sensitivities (Meissner corpuscles), the sense of cold (Kraus corpuscles) or pressure (Pacini corpuscles). Sweat glands are situated in this layer, the sebaceous glands and hair roots. The dermis also contains collagen and elastin fibres which confer its strength and elasticity, and defence cells.

The **hypodermis** is the deepest part of the skin. It is a zone rich in fat tissue and acts like a cushion to protect the organs beneath it.

The skin which covers the natural orifices of the organism (mouth, nasal passages, *etc.*) changes its structure a little when it passes into the interior and is called the *mucosa*. The most notable difference between the structure of ordinary skin and the mucosa is that these, under normal conditions, lack keratinised layers, that is to say, its epidermis lacks the corneum stratum and the granulosum.

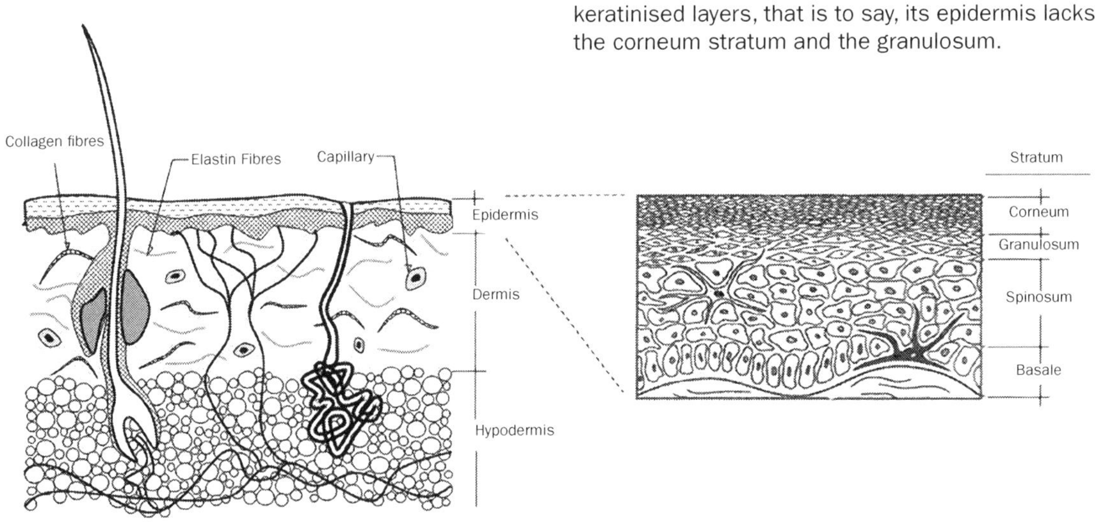

Fig. 4.1 *Structure of the skin of mammals.*

4.2.2 Mechanical properties

The study of the mechanical properties of skin has centred around the first two layers: the epidermis and the dermis, which, in a simplified form, can be modelled as follows: In the epidermis, the superficial layer –the corneum stratum– is a keratin membrane and the rest, the raw material of leather, an amorphous material. The dermis is, roughly, a mat composed of collagen and elastin fibres in a protein and polysaccharide matrix (principally hyaluronic acid).

Later, some notable aspects of the mechanical behaviour of human skin (epidermis and dermis) will be summarised in relation to age; the variation of thickness and elasticity (Escoffier *et al.* 1989). Both properties are illustrated in figure 4.2. The experiments were performed *in vivo,* on the ventral forearms. The thickness is measured with ultrasonic techniques and the elasticity determined from a torsion test, described in the cited reference, which evaluates the recovery of skin once the applied stress is removed. The elasticity was estimated by means of the quotient of the instantaneous strain after unloading, ε_R, and the initial instantaneous strain, ε_o, as indicated schematically in figure 4.2b. The results show that skin maintains its thickness up in people to their 70s, while the elasticity decreases progressively with age.

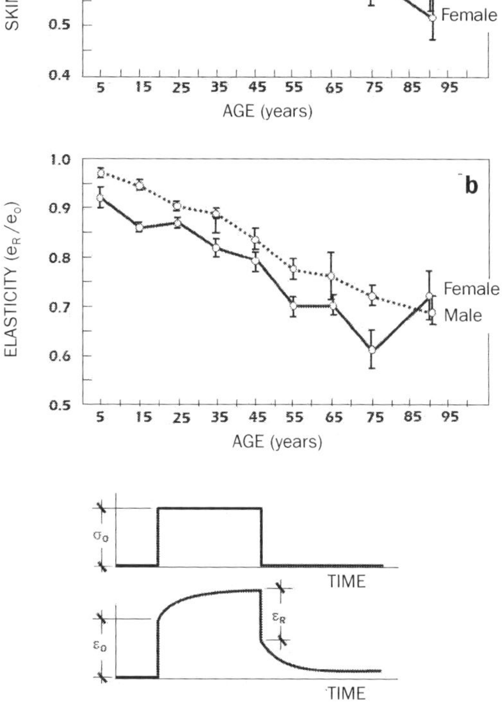

Fig. 4.2. *Variation of thickness and elasticity of human skin with age (Escoffier et al. 1989).*

A study of the mechanical behaviour of the human epidermis is interesting not only for a knowledge of its resistance to abrasion and penetration by foreign objects, but also due to the increasing use of patches and adhesives for the administration of transdermal medication. For these reasons it is important to know mechanical properties, in particular the effect of temperature and humidity on the adhesion strength of the exterior layer of the epidermis; the *stratum corneum*. The appropriate tests for this are delamination tests (which are described in detail below). Figure 4.3 shows the delamination energies and the stress necessary to separate the skin (from *in vitro* tests performed by Wu *et al.* in 2006). The adhesion strength (or delamination) (measured as the work required to separate an area of 1 m^2) decreases with temperature from 8 to 1 J/m^2 in totally hydrated specimens, while in those with a relative humidity of 45% it is practically constant at about 3 J/m^2 (figure 4.3a). The adhesion stress falls with temperature and hydration (figure 4.3b). The cited article also contains a study of the influence of lipids on the mechanical properties, finding that specimens in which the lipids have been partially removed are more resistant.

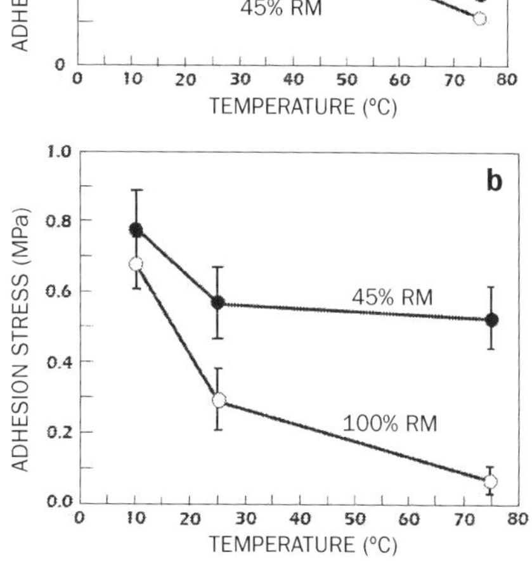

Fig. 4.3. *Variation of the energy and adhesion stress with temperature and humidity in the human epidermis (Wu et al. 2006).*

The effect of loading rate on the mechanical response of skin is another issue that deserves attention. H.G. Vogel studied the influence of the rate at which load was applied to the skin of rats; the speeds went from very slow movement ($5 \cdot 10^{-3}$ cm/minute) to the those which corresponded to free fall from about 1 metre high (~$3 \cdot 10^{4}$ cm/minute) (Vogel 1972). The results of this study showed that the failure load and the elastic modulus increased almost linearly with the log of the strain rate, while the strain to failure was practically constant. Vogel also investigated the effect of age and corticosteroids, which were known to be two parameters that influenced the mechanical properties.

The location of the skin, in this case the dorsal or ventral parts, also influences the mechanical properties. In particular, many large herbivores have very thick areas of skin for defensive purposes. R. Shadwick performed a detailed study of the mechanical properties of rhinoceros skin, one of the mammals with the thickest skin (Shadwick *et al.* 1992). Skin from the back and sides of the white rhinoceros has a very high rupture stress, 30 MPa, and a small failure strain of about 0.24. As a result, the stress-strain curve is quite steep, unlike curves for the epithelia discussed above, and its average elastic modulus can reach 240 MPa. The fracture toughness is of the order of 70 kJ/m^2. On the other hand, the ventral skin, on the stomach, is weaker, less stiff and more deformable; the average rupture stress is about 14 MPa and the corresponding strain is about 0.33. The elastic modulus lies at around 100 MPa.

Although rhinoceros skin is stronger and stiffer than that of other mammals, it should be noted that the tear resistance of the skin is small, or at least, it is not in keeping with the tensile or compressive strengths. This observation can be related to an adaptive advantage: In fights between rhinoceroses, the horns can puncture or tear (scratch) the skin, while the most frequent wounds produced are scratches, which are less dangerous because punctures could cause damage to internal organs. The design of the skin of these animals improves survival by favouring tear breaks and leads to the end of the fight before more serious harm is caused.

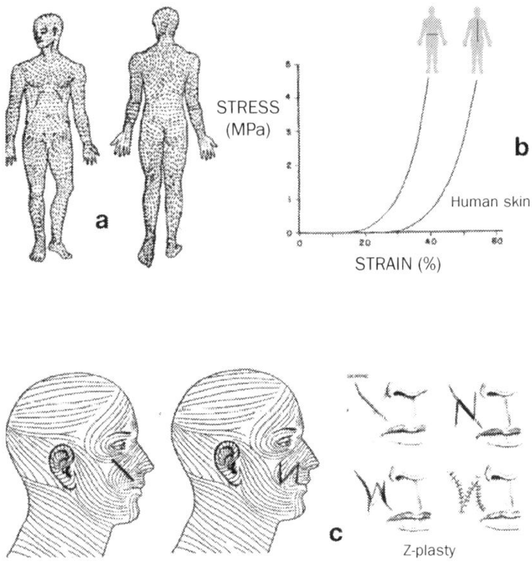

Fig. 4.4. *a) Langer lines.*
b. Traction curves. Greater stiffnesses are observed in the Langer directions.
c. Z-plasty, a surgical technique to conceal scars.

4.2.3 Concluding remarks

In a very simplified form, and in order to understand the mechanical behaviour, it is possible to assume that the principal components of the *dermis* of mammals are collagen fibres, which are embedded into a soft tissue matrix. Collagen fibres are naturally arranged in preferential directions in the skin, which in the case of human beings are known as Langer lines (figure 4.4a). The greater density of collagen fibres and their alignment make the skin more rigid when subjected to tension in the direction of the Langer lines, generating a pronounced anisotropy (figure 4.4b). Scar formation induces an accumulation of collagen fibre points and produces, as experience confirms, a zone of much more rigid skin. Some surgical techniques take advantage of the Langer lines in order to conceal, as far as possible, the scars produced by surgery, orienting the position of the suture (that is, the scar) along the length of one of these lines. In this way, the increase in the stiffness due to the presence of the scar is masked by the grater overall stiffness of the Langer lines. When it is not possible, and the line of the suture crosses the Langer lines, a technique known as Z-plasty is used (figure 4.4c) in which a Z-shaped scar is created, which is accommodated in a possible Langer network.

When embedded in a highly deformable matrix, the position and orientation of collagen fibres change with the deformation of the tissue. The gradual take-up of load in these fibres is responsible for the J-shaped stress-strain curves. As a result, the mechanical properties of the skin depend essentially on the structural arrangement –on the architecture– of these fibres. In figure 4.5 the various stress-strain curves show distinct behaviour depending upon the orientation of the collagen fibres in different tissues: Tendons and ligaments are tissues with a high collagen content and fibres that are already highly aligned before load introduction. Therefore, these structures are already very stiff from the beginning and, although very strong, the strain to failure is very low. The dermis of the rhinoceros is probably the mammal epithelium with collagen fibres having the greatest degree of alignment. This does not demonstrate such a marked J-curve as that of the epithelium of a cat, in which the collagen fibres are initially quite disordered, as shown schematically in the figure.

Disturbance of the network of collagen fibres and elastin can have serious consequences; some diseases are due to breaking the cross-links between them. One of these cases is lathyrism, which can be caused by ingestion of grass pea seeds (*Lathyrus sativus*), or its principal agent β-aminopropionitrile, which reduces the formation of cross-links in the collagen by inhibiting the enzyme lysyl oxidase. Another serious illness is scurvy, caused by vitamin C deficiency. Vitamin C is essential for the synthesis of the hydroxyproline and hydroxylysine amino acids, which form the collagen chain. Human beings and some higher primates lack the capacity to synthesise vitamin C, which they must obtain from their diet. This was a very common sickness among sailors in previous centuries. Their diets often excluded perishable products, rich in vitamin C, such as fruits and vegetables. Defective fabrication of collagen weakens the structures of all organisms, producing haemorrhages and, in extreme cases, leading to death.

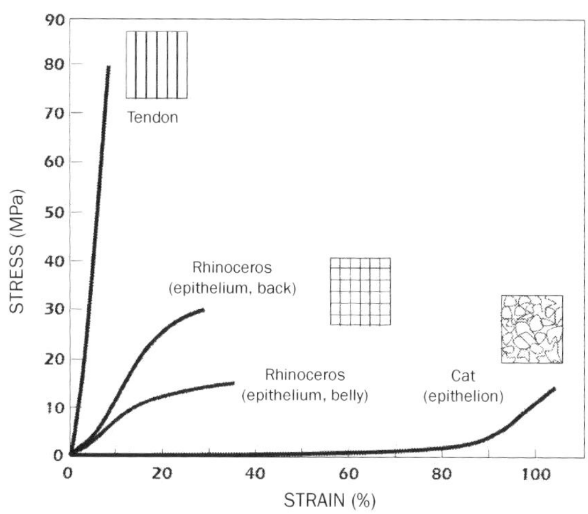

Fig. 4.5. *Schematic of stress-strain curves as a function of the orientation of collagen fibres.*

In concluding this introduction to the epithelium, it is worth elucidating some aspects of the structure of the *stratum corneum* because, although it is an excellent barrier for avoiding water loss, it is also a superb biosensor capable of initiating a series of biological processes to repair wounds and restore a homeostatic barrier.

The structure of the stratum corneum can be likened, in a simplified form, to a brick wall, in which the bricks are keratinocytes (cells which produce keratin) and the mortar a matrix of lipids. The keratinocytes play the role of protector against abrasion from external agents, and the lipids are essential barrier elements which avoid water loss. The understanding of the biochemistry of the stratum corneum was facilitated by the diagnosis and treatment of various skin diseases which range from cosmetic problems –dryness, peeling– to more serious conditions, such as psoriasis or atopic dermatitis (Harding 2004).

Tissue engineering has achieved major success in the production of epithelia and has commercialised several products. One of the first, in 1997, was *TransCyte*, based on a degradable polymer scaffold on which skin keratinocytes grow. This product was a temporary protection until the skin regenerates. Nowadays, the fabrication of skin from a few donor cells is a reality. In the Unidad de Biomedicina Epitelial Ciemat Madrid, a team led by Dr. Jorcano produces skin for severe burns by cultivating epithelial cells on a fibrin matrix. This technique has been perfected in recent years, and the ultimate objective is to try to regenerate skin; which is functional and with all of its components, including glands and capillary follicles. The new technique of 3D bioimpression allows the design of structural scaffolds which, as well as being biocompatible and dissolvable, can be functionalised so that the cells that are seeded on them grow and differentiate properly (Cubo *et al.* 2016).

4.3. BLOOD VESSELS

This section is focussed on the human cardiovascular system, although reference will also be made to the blood vessels of other mammals.

The cardiovascular system is the means by which blood runs throughout an organism. The heart is the engine; it acts like a pump that pushes the blood through the arterial system and sucks through the venous system.

The arterial system is composed of a network of blood vessels called arteries which start in the heart and extend throughout the organism carrying oxygen and essential nutrients for the functioning of the cell, by means of the arterial blood. Moving further from the heart, the calibre of the vessels become smaller; initially called arteries, then arterioles and, finally, arterial capillaries.

The venous system consists of a network of vessels which, following a path parallel to the arterial network but in the opposite direction, carries deoxygenated blood loaded with waste products –venous blood– back to the right cavities of the heart from which it will pass to the lungs to be oxygenated and converted to arterial blood. The veins gradually run together, progressively forming vessels of larger calibre, called capillary veins, venules and, finally, veins.

4.3.1 Structure

The structures of arteries and veins are shown, schematically, in figure 4.6. The arteries must support greater pressures and pressure fluctuations than veins due to the way in which the cardiovascular system works, and this is reflected in the their structure. Veins have less elastic and muscular walls because the blood circulates along their length due to the suction effect of the thorax and the pumping muscles in which they are embedded. Inside the veins there are valves which restrict the return of blood. In figure 4.6 it is shown that the structure of arteries and veins is formed of three layers or tunica: the *intima*, the *media* and the *adventitia* or exterior.

The *intima* layer is principally formed by endothelial cells, the *media* layer contains smooth muscular tissue cells, collagen fibrils and elastin, and the *adventitia* includes collagen fibres, fibroblasts, nerves and capillaries (except in the finest veins).

The composition and structure of blood vessels varies with size and type of vessel. The thickness of veins is, approximately, one-tenth of its diameter, while the thickness of arteries is usually double that, about one-fifth of their diameter. Thicker arteries, such as the aorta, contain elastic tissue that allows them to stretch during the systole (contraction of the heart, which launches blood into the arterial system) and contracts during the diastole (relaxation of the heart, during which it is filled with the venous blood). The arteriole walls contain less elastic tissue and deform less. The capillary walls are formed by a layer of endothelial cells, and their diameter is so small that red blood cells can only just pass through them.

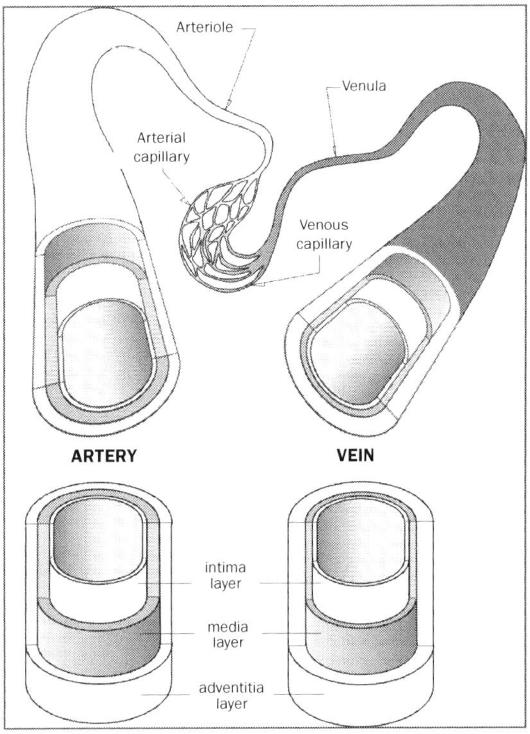

Fig. 4.6. *The structure of arteries and veins.*

The walls of arteries and veins consist of, approximately, 70% water, 15% collagen and elastin and the rest is smooth muscle fibre. In the aorta, there is more elastin than collagen (more than 50%), and less elastin than collagen in the arterioles and veins (the proportions can be 50% in the arterioles and 30% in the veins). Veins always contain less elastin than the arteries. The percentage of smooth muscle is less in the aorta and increases in the more peripheral arteries and arterioles.

4.3.2 Mechanical properties

Knowledge of the mechanical properties of blood vessels should provide an understanding of their function and help prevent any dysfunction. Techniques for measuring the mechanical properties, modelling behaviour and describing the performance of the vessels will be discussed as this section progresses. For the moment, in order to pave the way, the itinerary that will be followed is indicated in the paragraphs below.

Due to the difficulties in testing blood vessels *in vivo* almost all experiments have been performed *in vitro*, with specimens extracted during surgical interventions or from cadavers, aiming to reproduce physiological loadings, a task that is not always straightforward because the vessels are subjected simultaneously to axial tension and internal pressure (which varies as shown in figure 4.7), and, in addition, having residual stresses in their walls.

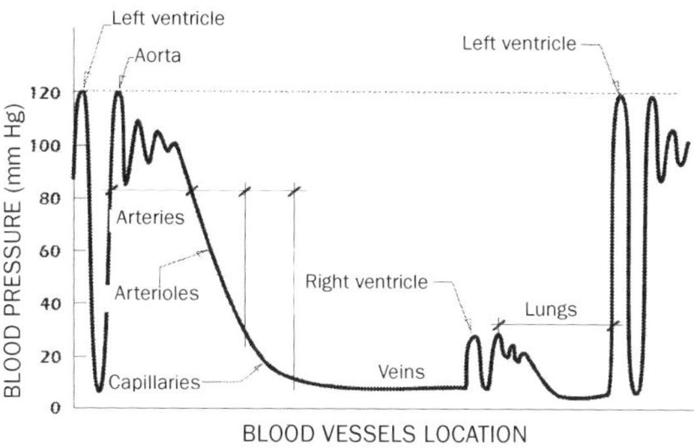

Fig. 4.7. *Pressure (mm Hg) in blood vessels as a function of their location.*

In order to understand the mechanical behaviour, to a first approximation, it can be assumed that blood vessels behave like a *membrane*; with no bending stiffness. Tensile tests, with specimens extracted in the longitudinal and circumferential directions (as indicated in figure 4.8a, are the simplest. Stress-strain curves are obtained from these tests, and from relaxation or creep tests, as a function of time. Dynamic tests have also been performed on these types of specimens. For example, a study of the mechanical behaviour of the aorta of a dog, following these procedures, can be found in the article by Tanaka and Fung (1974). This information is not completely satisfactory because, in reality, the vessel tissue is subjected to biaxial loading, due to the blood pressure and the longitudinal tension, so it would be worth performing more representative tests. To this end, various devices have been used to apply traction to the vessels and, at the same time, introduce internal pressure (see, for example, Froneck *et al.* 1976, or Guinea *et al.* 2005). Other devices, as well as loading the tube in tension and subjecting it to pressure, can simultaneously twist it (Fung 1993, page 337). Biaxial tests performed on a coronary artery subjected to traction and internal pressure are shown in figure 4.8b in which the development of the diameter is shown as a function of the pressure.

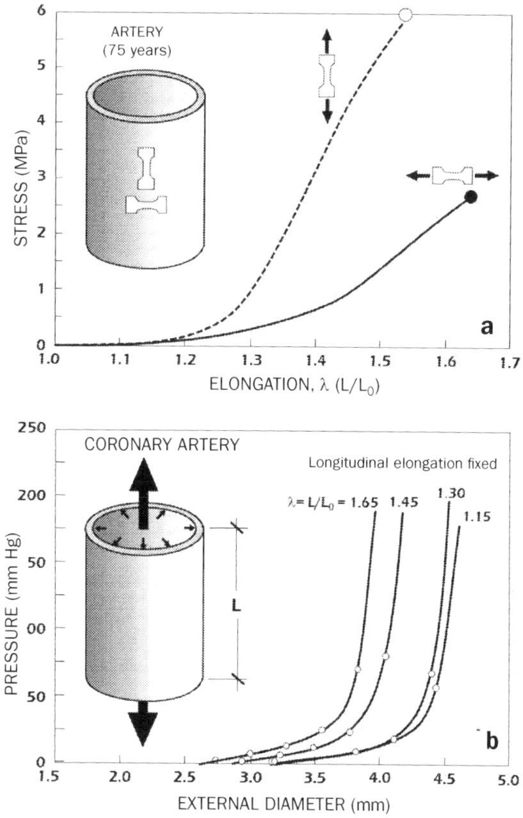

Fig. 4.8. *a. Engineering stress-elongation curves for an arterial blood vessel, in the radial and longitudinal directions. b. Internal pressure-external diameter curves for a coronary artery pulled to different values of λ, where $\lambda = L/L_0$, and L_0 is the initial length of the vessel.*
Note the dog-bone specimen geometry (Claes, 2010).

More refined studies can be performed if the thickness and flexibility of the vessels need to be taken into account, that is to say they can be modelled as *laminas* subjected to traction, internal pressure and torsion. For these cases, experiments have been proposed that simulate a beam (the lamina) on two supports or bending tests on a cantilevered beam (lamina) with an end load (Fung 1993, page 352, 355).

4.3.3 Final comments

All of this information is very useful for the diagnosis and treatment of diseases of the cardiovascular system. It is also helpful for modelling the mechanical behaviour of blood vessels and formulating the constitutive equations that allow reliable predictions of their functioning.

There are two types of accidents that occur, unfortunately frequently, in human blood vessels. These will be reviewed and discussed in more detail below: *Ruptures* by longitudinal tearing and *aneurysms*, a biological degeneration which leads to a mechanical instability. The information derived from experiments indicates that the vessels possess a structure and are made of materials which are especially adapted to withstand these accidents thanks to the nonlinear nature of the stress-strain curve, and the presence of residual stresses.

If blood vessels had a mechanical response in which the stiffness (slope of the curve), within the physiological range of pressures, decreases with increasing diameter, any local changes in the diameter of the vessel would provoke an instability, which gives rise to an increase in the diameter and the appearance of a widening zone known as *aneurysm*.

Indeed, if the tube is not perfectly uniform, in a region where the diameter was a little larger than those nearby, a swelling will be formed –an aneurysm– because it is easier to dilate locally than become less stiff, leaving the adjacent regions (of smaller diameter, stiffer) unaltered. This is the behaviour of **elastomers** such as rubber, which is shown in figure 4.9, in which the phenomenon is clearly visible, for example the swelling of a cylindrical rubber balloon. On the other hand, if an increase in diameter of the tube increases its stiffness, as happens in arteries, instability will not be produced. The design of blood vessels, unlike that of rubber, protects it from that eventuality.

The shape of the stress-strain curve (or pressure-diameter, as will be discussed in detail later) is due to the structure of the vessels, formed of three layers of different mechanical properties, and the characteristics of their constituent materials; fundamentally elastin and collagen fibres. At very low pressures, of the order of 70 mmHg, only the elastin fibres work. The collagen fibres come into play little by little and are those that carry the greater part of the load at high pressures.

Residual stresses in blood vessels make the stress in the wall more uniform and allow support of greater pressures with less risk of rupture. The presence of residual stresses becomes apparent through a lateral cut in a circular section (as shown in figure 4.10) because the section opens spontaneously to a certain angle α, known as the *angle of aperture*, being greater for greater residual stresses.

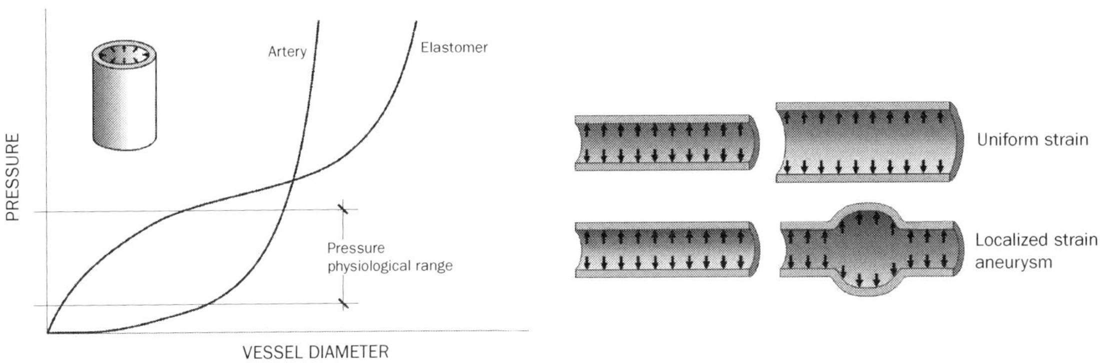

Fig. 4.9. *Schematic of pressure-diameter curves for tubes of different materials; an elastomer and an artery. Inside the pressure physiological range, stiffness* ***(slope of the curve)*** *increases with the diameter of the artery, whereas it decreases for elastomers.*

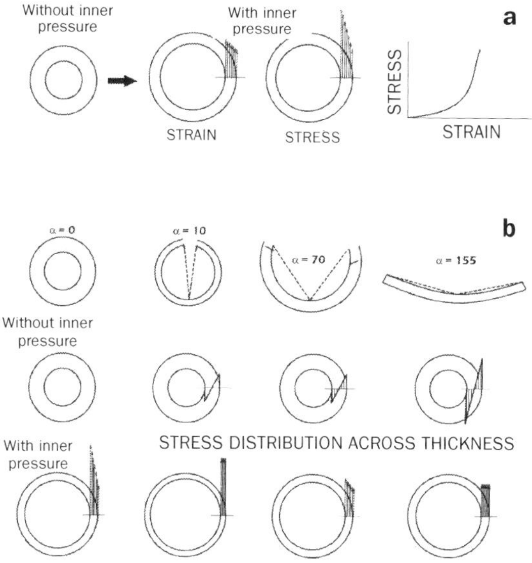

Fig. 4.10. *Distribution of through-the-thickness stresses in a blood vessel.*
a. Without residual stresses.
b. With residual stresses (which increase with the value of α).

The beneficial role of these stresses can be appreciated, using the following reasoning due to Fung (1993). Initially the occurrence of the residual stresses is due to the fact that the arteries have an appreciable thickness (the ratio of the radius to thickness is between 2 and 4) and the stress-strain relationship for the tissue walls is not linear, the stress increases much more rapidly than the strain (see the curve in figure 4.10a). When the arteries support the blood pressure (between 110 and 130 mmHg), the radius increases and the thickness decreases. In these circumstances, the strain in the internal layers of the aorta is greater than that in the external ones (see figure 4.10a) and, since the stresses grow more rapidly than the strains, this effect is more marked in the distribution of stresses through the thickness. The beneficial role of residual stresses is shown in figure 4.10b, in which it can be seen that the distribution of stresses through the thickness becomes more uniform as the angle α increases. Depending upon the presence of the residual stresses (first column, figure 4.10b), the stress distribution through the thickness is quite ununiform when there is internal pressure, reaching very high values in the internal layers. Due to the increasing residual stresses (with increasing α) the superposition of the residual stresses and those due to internal pressure renders the total distribution more uniform through the thickness and the maximum values are reduced, so that the risk of rupture also reduces.

Pressure changes in the cardiovascular system can cause changes in the geometry, structure and mechanical properties of blood vessels, a phenomenon known as continuous structural adaption or *remodelling*. The causes of this process can be many; diseases, drugs, surgery, high salt diet, lack of oxygen (hypoxia), overpressure, lack of gravity, *etc*. An example of remodelling of the pulmonary artery of a rat due to increased blood pressure is described in Fung and Lin (1991). The authors observed that, initially (up to 12 hours), the *media* grew thicker while the *adventitia* hardly changed. After 95 hours, the *adventitia* had achieved the same thickness as the *media*, its thickening continuing over the next few days (between 10 and 20). Remodelling was also observed in the walls of the arterioles. In summary, they found that remodelling of the vessels is not uniform and proceeds quite rapidly; histological changes can be detected after a few hours and most of the remodelling had taken place after a couple of days. They also observed the variation of the α angle (and as a consequence the residual wall stresses) with change of pressure.

Tissue remodelling adds new parameters to the prediction of mechanical behaviour. Everything seems to indicate that the tissue cells live under stresses which modify their metabolism, internal structure, and the production or reabsorption of extracellular structures. Under normal conditions, the organisms exist in dynamic equilibrium, called a *homeostatic state*. When there are changes in the external pressure, the homeostatic state must re-balance through remodelling, that is to say by means of changes in the mass, structure and mechanical properties of the tissue.

Surgical repair of blood vessels consists of replacing the blocked or deteriorated section with an implant, or diverting the blood flow using a bypass. The grafted tissue may be natural, such as a piece of saphenous vein or internal mammary artery, from the actual patient or a donor, or artificial, such as Dacron. In both cases, a knowledge of the mechanical properties is fundamental. Tissue engineering offers the possibility of constructing blood vessels *to-measure* for each patient, aiming to reproduce the mechanical and biological behaviour of the original material. This requires a scaffold which is biocompatible, biodegradable and functionalisable, to seed it with the appropriate cells of the patient and to allow them to grow and proliferate. Some success has already been achieved in the fabrication of blood vessels using this technique (Pashneh-Tala *et al.* 2016).

4.4. ANEMONES AND MEDUSAS

The apparent simplicity of the tissue of medusas and other marine animals has encouraged their study, with the aim of unravelling the relationships between structure and mechanical properties. The resistance of anemones and medusas to the ravages of the sea, despite their apparent fragility, has also motivated the study and the attempts to create fabrics with new features inspired by them.

4.4.1 Structure

Medusas and anemones are *cnidarians* –or flower animals– and, although they exist in different forms (figure 4.11), their organism has the same general arrangement. The body, with axial symmetry, has a wall formed of three layers: the external or ectoderm, the intermediate or *mesoglea*, and the internal or endoderm. In its interior, there is a gastric cavity with only one opening which functions as a mouth and an anus. This aperture is surrounded by a series of stinging tentacles (hence the name of the *cnidaria* phylum, from the Greek κνιδη, nettle).

Two forms can be observed inside the phylum: the *polyp*, which is adapted to the sessile, and the *medusa,* to the free pelagic life. Some aspects of the mechanical behaviour of the mesoglea of the polyp *Metridium senile* (a common anemone off the coasts of northeast Europe) and of the mesoglea *Rhopilema esculenta* (an edible medusa abundant in the Pacific Ocean) are discussed in this section.

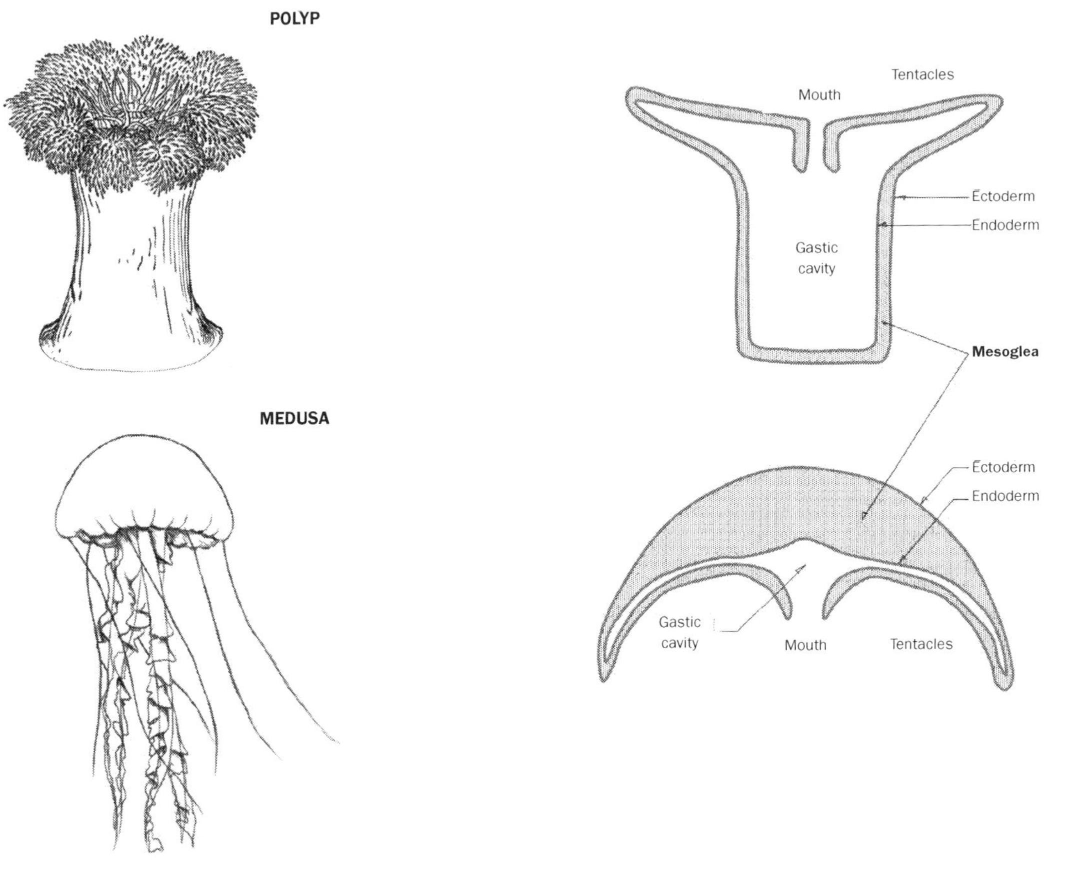

Fig. 4.11. *Transverse sections of an actinia and a jellyfish, in which the same general organisation is apparent. The mesoglea is shown in both cases.*

4.4.2 Mechanical properties

Chapman (1953) observed that the mechanical properties of the body of anemones can be attributed only to the mesoglea, barely depend upon the muscle fibres. Alexander (1962) performed creep experiments on the mesoglea of the actinia *Metridium* and observed that under small loads the material deformed up to two or three times the initial length, and that it took about 15 hours to stabilise. On unloading the test specimen, and after a similar period of time, the material returned to its original length. Gosline (1971) continued these studies seeking a relationship between the structure and composition of the mesoglea and its mechanical properties. The following commentary has been extracted from the work of Gosline.

The mesoglea of *Metridium* can be considered as a composite of collagen fibres embedded in a matrix of polysaccharides and proteins (a gel with 2.4% solid contents). The composition of the mesoglea is, approximately, 6.7% collagen, 2% matrix, 5% salt and the rest, 86%, water. In terms of structure, (see figure 4.12) it is possible to distinguish two layers: an exterior in which the fibres are intertwined and another in which the fibres are more ordered radially and circumferentially. This structure suggests that the mechanical behaviour may be anisotropic, indeed this is observed in the tests discussed below.

Figure 4.12 shows the results of relaxation tests, at 23°C, performed in different directions: longitudinal and circumferential. In both cases the stresses decrease with time reaching stability after about 15 hours. The previously mentioned anisotropy is clearly apparent in the experimental results.

Gosline has proposed a model for the mesoglea in which the intertwined (but not transversely connected) collagen fibres are embedded in an elastomeric matrix formed of a high molecular weight amorphous polymer. The matrix is responsible for the deformation and the elasticity of the mesoglea. The collagen fibres, whose role is to reinforce the matrix, are responsible for the initial stiffness, but the large deformability and the elastic recovery are due to the changes in the configurational entropy of the matrix polymer network. According to Gosline (1971), if the collagen fibres were transversely joined, the mesoglea could not deform more than 30% strain in the longitudinal and 55% in the circumferential direction. Alexander (1962) recorded strains of 200% in creep tests.

Another notable aspect of the mesoglea of the polyp *Metridium* is its scratch resistance: 1.2 kJ/m^2 (similar to that of bone, 1.7 kJ/m^2, or of the pig's aorta 1.0 kJ/m^2, but much lower than that of the epithelium of a mouse, 20 kJ/m^2). The measurements were made by Purslow (1983) by means of rupture by tear experiments which are described in detail in Sections 5.2.2 and 5.3.3. The rupture strength of the mesoglea, like other highly deformable materials, is small, due in part to the elastic energy stored by deformation –which can cause rupture.

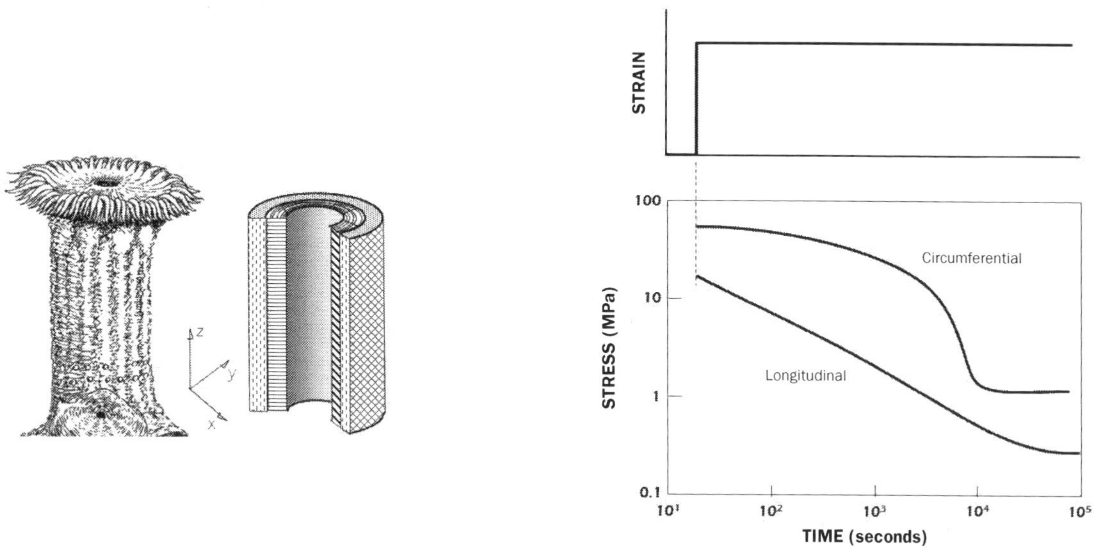

Fig. 4.12. *Relaxation experiments on the mesoglea of the polyp* Metridium s., *which illustrates the anisotropic character of the material (Gosline, 1971).*

Another biological material with these characteristics –simple soft membranes– is the mesoglea of medusa. It is curious that this material –which is 97% water– has mechanical properties that withstand the ravages of the sea without suffering harm. Once again, with this type of relatively simple gel, the challenge is to study the relationship between the structure and properties, and to design and fabricate hydrogels with better mechanical properties.

For this purpose, Zhu *et al.* (2012) investigated the mesoglea of the edible medusa *Rhopilema esculentum*, of the rhizostomatidae family. The bell –or umbrella– of the medusa is composed of three layers, as shown in figure 4.11; the exterior or ectoderm, the intermediate or *mesoglea,* and the interior or endoderm.

The structure of the mesoglea is outlined in figure 4.13. It is formed of fine laminas, stacked parallel to the surface of the bell, connected by nanofibres whose diameters range from 20 to 50 nm. In terms of chemical composition, it is 97% water and the rest is composed of (type II) collagen fibres and unsaturated fatty acids.

The mechanical properties of the mesoglea were obtained from uniaxial tensile and compression tests, whose details can be found in the cited reference. A schematic representation of the specimens used in each case is shown in figure 4.13, as are the direction of loading and the results obtained. It can be seen that the tensile rupture load is around 0.20 MPa and the corresponding strain is about 100%. In the compression test, the rupture load approaches 2.0 MPa, and the strain is around 86%. These results are better than those for most synthetic hydrogels with similar water content, and even less. Despite its apparently simple microstructure –thin sheets of interlocked fibres– there is no satisfactory model for prediction of the mechanical properties of the mesoglea. On the other hand, and bearing in mind the considerations made in previous sections, it would be very useful to have biaxial experiments for a more reliable database.

4.4.3 Final comments

Hydrogels are networks of three-dimensional hydrophilic polymers which can retain considerable amounts of water. Synthetic hydrogels have numerous applications in biomedicine and pharmacy, but their mechanical properties are poor and seldom appropriate for applications that involve support of mechanical forces. To this end, it would be desirable to develop more resistant hydrogels inspired by the structure of the mesoglea of medusas.

The locomotion of medusas is quite unique; it is based on successive contractions and expansions of the bell, produced by striated muscles similar to those of vertebrates. Even though it has not yet been possible to fabricate a synthetic mesoglea, it has been possible to make an artificial medusa capable of replicating its movements (Nawroth *et al.* 2012). For that, a sheet of cells from the heart muscle of a rat was grown –which contracts under the stimulus of an electric current– combined with a silicone polymer layer –which is elastic and allows the recovery of the initial geometry when the current stops. On applying an electric field to the *artificial medusa*, it contracts imitating the impulse of real medusas. It even creates a stream of water which could attract food near where the mouth would be.

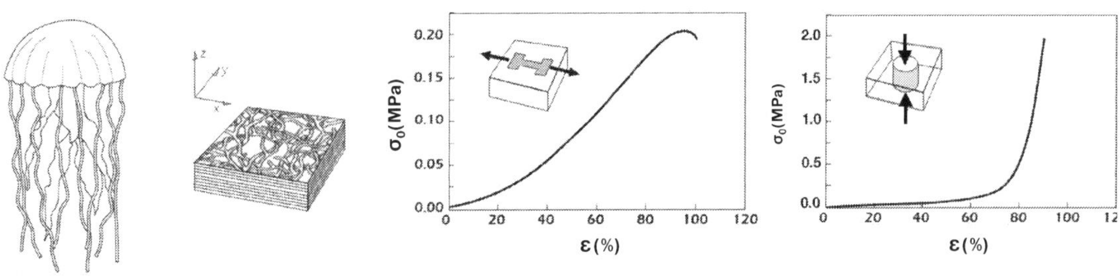

Fig. 4.13. *Stress-strain curves of the mesoglea of the medusa* Rhopilema esculentum, *in the radial and transverse directions, which illustrate their anisotropic character (Zhu et al., 2012).*

4.5. NATURAL ARMOURS

4.5.1 Evolution of armours

Natural armours are another type of biological membranes and laminas. Not as flexible as those mentioned above, natural armours have evolved over time to provide living beings with protection against external threats.

Fossils of fish covered in hard plates (*placoderms*) have been found. These creatures lived in, and dominated, the seas 420 million years ago, and became extinct towards the end of the Devonian period. These creatures had large plates which were juxtaposed, and, although they provided a protective shell, the plates impeded mobility and locomotion. Fossils of these fish of more than 8 metres long have been found.

Armour plates have evolved to improve mobility while maintaining their protective qualities. Within the gnathostomata (jawed mouth), there are two classes: the chondrichthyes (with cartilaginous endoskeleton) and osteichthyes (with skeletal bone) (figure 4.14).

The chondrichthyes (sharks and rays) appeared 400 million years ago and have bodies covered in *placoid* scales partially embedded in the skin. As well as the protective function, placoid scales reduce the drag resistance and accommodate various cutaneous and light organ sensor systems.

Osteichthye fish appeared in the late Silurian period, also around 400 million years ago. Osteichthyes are divided into two subclasses: sarcopterygians (lobe-finned) and actinopterygians (ray-finned) (figure 4.14).

Most fish are now actinopterygii and their scales differ from those mentioned above. Primitive ones, called *ganoids*, are typical of the most basal bony fish, such as sturgeons. They are composed of bone, enamel and ganoine, they look like rhomboidal shields and are arranged in the skin like pavement slabs. The other bony fish have *elasmoid* scales (due to the presence of elasmoidine) of two types: *cycloid* and *ctenoid*. The former have smooth edges and the latter jagged, which gives them a rough texture. These two types of scales are not arranged like mosaic tiles but are imbricated, like roof tiles. The free edge is directed towards the rear of the fish to reduce water resistance during swimming.

In the past, sarcopterygii occupied every aquatic ecosystem on the planet, but they now are restricted to the freshwaters of Africa, South America and Australia, as well as the marine waters of the Indian Ocean, where there are only two species. The coelacanth, the famous "living fossil", belongs to one of them. One group derived from the sarcopterygii, the tetrapods, (figure 4.14), has colonised all terrestrial environments of the planet; these are amphibians, reptiles, birds and mammals. These, also, have developed natural armours.

Turtles are reptiles; more accurately, chelonia. Mammals which appeared towards the end of the Triassic period, about 200 million years ago. The shell of turtles is formed of hard bony plates which protects them against the attacks of predators: cats, crocodiles, sharks and, even, birds which can scratch, bite or peck them. They must also withstand hard impacts, when eagles seize them and drop them on rocks to break their shells. This armour is very strong but not very flexible.

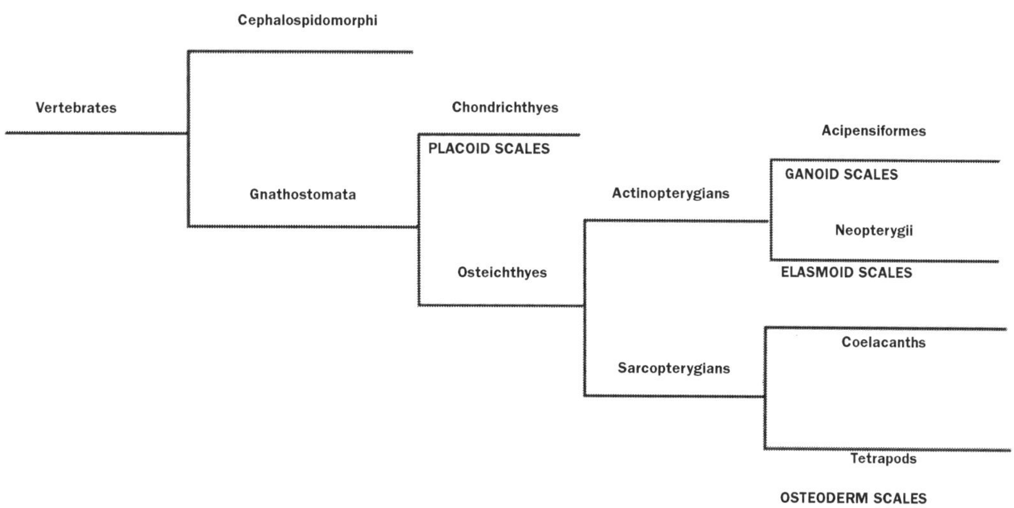

Fig. 4.14. *Types of scales and their position in evolution.*

Other reptiles which are coated in armour are crocodiles, whose ancestors appeared in the Triassic period, about 180 million years ago, and who lived alongside the dinosaurs. This long lineage has now been reduced to just three families: gavials (gharials), crocodiles and alligators. They all have armoured skin. Although they rarely encounter dangerous enemies, they need their armour to defend their territory from members of their own species because their bites can be fatal. Compared to turtles, their armour is more flexible and it gives them greater agility to capture prey. In former times, some warriors created armoured breastplates from crocodiles to protect themselves from arrows and pointed objects.

Mammals have also developed natural armour during their evolution. In particular, within the order of Edentata, a group of armadillos (whose name is Spanish, already indicates that function) and pangolins. Some edentates in the relatively recent past were much more impressive than current ones. Glyptodonts, which became extinct about 10,000 years ago, were a mammalian version of land turtles and reached weights of several tons. They were equipped with a golf ball-like shell, made of dermal bones. They achieved great success and diversity; originating in South America, they spread into southern North America.

Pangolins are one of the few remaining representatives of armoured edentates. They live in tropical and subtropical regions of Africa and Asia. Their scales are *imbricated* and cover the major part of their skin. This arrangement provides great flexibility than other reptiles whose plates are *juxtaposed* and connected by collagen fibres. When threatened, the pangolin curls up into a ball, and it sticks out its exterior scales, like a half-open pine cone, protecting it and hurting the attacker. The distribution of the scales on the pangolin has inspired the design of armour since ancient times. In 1820, the United Kingdom's King George III was presented with a coat made of pangolin scales.

Armadillos are another living group of armoured edentates. They live in South America and the lower part of North America. Armadillos are covered in bony plates, called *osteoderm*, which protect them from predators, in particular, when the animal rolls up into a ball.

4.5.2 Composition and structure

The ingredients available for living species to fabricate armours have been limited: cell secretions (which provide collagen fibres) and minerals (principally calcium, in the form of carbonates and phosphates).

To a large extent, the elements of natural shields are formed of two layers: the very hard exterior provides protection against impact, and the tougher interior provides strength. However, the strength of the plates is provided, mainly, by the hierarchical structure of the components.

The exterior layer is formed of highly mineralised tissues, which are hard, avascular and with little or no collagen. The most common tissue is *enamel,* and another is *ganoin* which covers scales of some bony fish. The enamel which covers teeth is the same as that which we share with other tetrapod animals and fish. A recent study (Qu *et al.* 2015) suggests that enamel may have its origin in primitive fish scales, because a genetic analysis identified a link between *ganoin* and *enamel* matrices.

The interior layer is formed of mineralised collagen fibres within a mineral matrix. Depending on the grade of mineralisation and the distribution of the collagen fibres, several types of structures have been identified: *dentin, elasmoidine,* and *lamellar bone*, among others (Sire *et al.* 2009).

The microstructure and composition of the plates vary according to the species considered. The three examples selected –*ganoid* scales, *elasmoid* scales, and *osteodermal* plates– give an idea of the structures which nature has developed to protect animals (Yang *et al.* 2013a).

The fish *Atractosteus spatula* provides an example of the structure of *ganoid* scales. It is one of the largest fresh water fishes in North America, reaching a length of 3 metres and weighing 100 kg when fully grown. The scales are formed of two layers (figure 4.15a). The external layer, called *ganoine*, which has a thickness of 0.5 mm in adults, is hard and composed of hydroxyapatite crystals. The internal layer is thicker, about 3.5 mm, and formed of sheets of dentine, elasmodine and bone tissue. Native Americans value these scales and use them as adornments.

The *Arapaima gigas* is another large freshwater fish; reaching lengths of 2.5 metres and weighing 150 kg, it inhabits the Amazon river and has elasmoid scales. In the adult, the scales can approach a length of 100 mm and width of 40 mm.

They are also formed of two layers (figure 4.15b). The external layer can have a thickness of 0.6 mm and the internal one 1.0 mm. The external layer is highly mineralised and rough. The internal layer contains collagen fibres which are arranged in parallel planes. The planes can be stacked in a helical form, creating a structure called *Bouligand* (Bouligand, 1972) which confers great rigidity. The scales can be considered as a multifunctional protector; the external layer provides hardness and wear resistance and the internal layer flexibility and toughness.

A mammal, the armadillo *Dasypus novemcintus* is covered with hexagonal plates called *osteoderms*, which can be translated as "bone skin". It is the same protection used by some reptiles (crocodiles and turtles). *Osteoderms* in turtles form part of their skeleton, in other animals they form part of the skin and are vascular. Once again, two layers can be identified in these plates: The exterior, or epidermis, is made of α-keratin which provides stiffness and sealing. In the interior layer three regions can be observed: the first is dense and bony, the middle is porous bone and the inner is, again, dense bone (figure 4.15c). This structure –dense bone, porous bone, dense bone– is frequently seen in animals which need strength and energy absorbing capability combined with lightness, that is to say low density.

It is interesting to compare the similarities between the structure and properties of armours of turtles and armadillos because they fulfil a common function –both provide protection against predators– despite important differences, in turtles the dorsal plates include vertebrae and ribs, and in armadillos both are inside and separate from the plate. Even so, the connections between the plate and its structure are very similar. In both creatures the plates are interconnected by collagen fibres to provide a degree of flexibility to the shell, although the arrangement of the structures in turtles is more complicated than that of armadillos. The structure is similar in both cases: dense bone, porous bone, dense bone. Since turtles and armadillos belong to different families –reptiles and mammals– it is interesting to note that their evolutions converged on similar solutions to the problem of protecting these animals from predators.

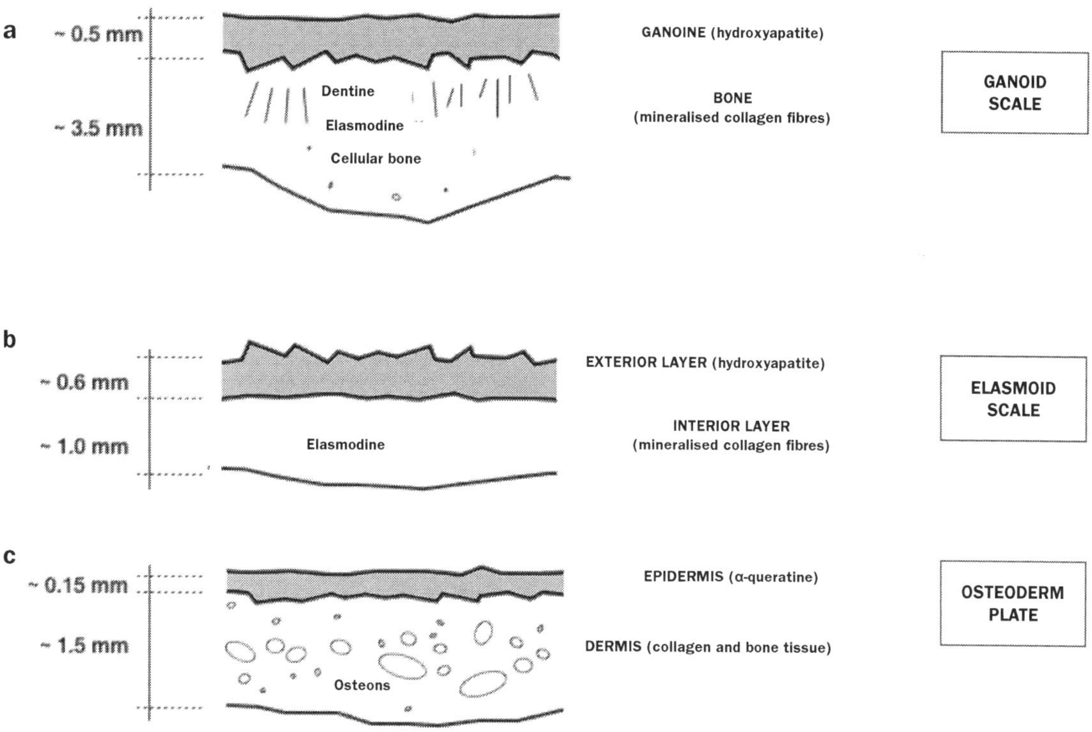

Fig. 4.15. *Greatly simplified schematic sections of two types of scales (ganoid and elasmoid) and a bony plate.*

4.5.3 Mechanical properties

Living beings have developed materials for protection against external threats over millions of years. These materials –which form plates and scales– have excellent mechanical properties, even though their components are soft and brittle (hydroxyapatite and collagen, for example).

The function of natural armours is two-fold: to provide protection and, at the same time, sufficient flexibility to allow movement. In particular, the components of the armour are materials which possess high hardness, resistance to indentation and tenacity. It is difficult to characterise these properties because the plates and scales are small, thin, and heterogeneous. All this means that the traditional techniques for measuring strength, fracture toughness or resistance to indentation provide results that are difficult to interpret.

The tensile strengths of scales or bony plates have been measured by extracting dog-bone specimens in the longitudinal (direction of movement of the fish, for example) and transverse directions, with the aim of detecting anisotropy. Humidity affects the results; tensile tests with dry specimens produce higher rupture values and lower strains. Temperature also has its influence on the test results.

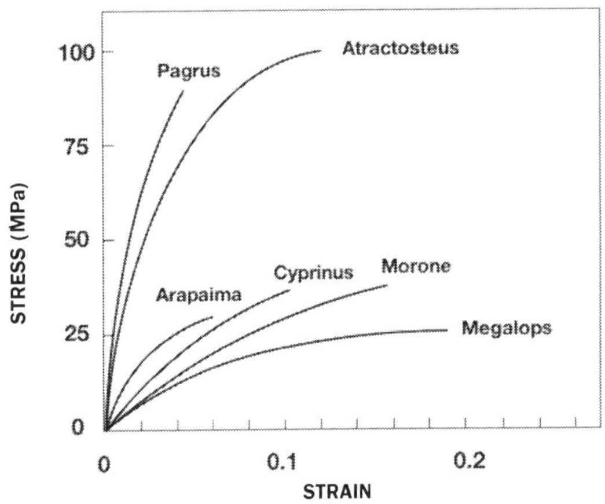

Fig. 4.16. *Tensile tests on fish scales, in a humid environment, at ambient temperature and in a direction parallel to the body. Approximate mean values are due to the large spread in results. Engineering stresses and strains: Pagrus (Ikoma, 2003). Atractosteus (Yang, 2013b). Arapaima (Yang, 2014). Cyprinus (Cugno, 2012). Morone (Zhu, 2012). Megalops (Murcia, 2018).*

Figure 4.16 contains the tensile test results for various scales, performed in a humid environment and at ambient temperature. The aim of the graph is to provide an idea of the values, but these should be treated with caution, since the tests produce results with a great deal of scatter (consult the cited references for details). Tensile tests with specimens taken from bony plate provide different values according to the degree of humidity, for example, in a humid environment 40 MPa (pangolin) or 13 MPa (turtles and armadillos), in a dry medium 70 MPa (pangolin) or 23 MPa (turtles and armadillos). Once again these are only indicative values.

The resistance to penetration and the toughness of scales has caught the attention of researchers. For example, Meyers and co-workers studied the protection capacity of the scales of *Arapaima gigas* against bites from the piranha *Sarrasalmus manueli*, because both fish live in the Amazon basin and coexist in harmony (Meyers *et al.* 2012). The aforementioned investigators performed penetration experiments with the teeth of the piranha on the scales of the arapaima. The results showed that the teeth of the piranha were not capable of penetrating the exposed part of the scales; the teeth broke and remained embedded in the scales without the power to penetrate further. The external layer of the highly mineralised and very hard scales played a fundamental role in protecting the arapaima from the bites of the piranha. In some cases, the external layer could break, owing to its hardness and the pointed form of the teeth, leaving the collagen laminae exposed. If the teeth continued to penetrate, they would separate the collagen fibres in the layer but meet the next layer with collagen fibres in another direction (the Bouligand structure, mentioned above) which slow down further progress. This is the basic mechanism which provides toughness to ctenoid scales.

Attempts have been made to estimate the toughness of some scales using standard tests with pre-cracked specimens (these tests will be discussed later in this book). As these are anisotropic, heterogeneous and hierarchical materials, it has only been possible to obtain indicative values which are difficult to interpret.

Microhardness tests suffer from the same difficulties mentioned above, but they provide comparative values for various types of scales and predator teeth, such as in the example cited for the piranha and the arapaima.

4.5.4 Artificial armours

Since ancient times, hardness of scales and their distribution over the skin –imbricated or juxtaposed, providing some degree of flexibility to movement– have inspired the fabrication of body armour for human beings.

The origins of the these artificial armours go back more than 4,000 years; remains were found in the tomb of Kenamon, who lived in Egypt during the reign of Amenotep II (1436-1411 B.C.). The Seythians had coats with iron or bronze scales stitched onto leather vests. The best documented ancient armours are those of the Roman Empire (Bishop and Coulston, 2006). Armour scales (*lorica squamata*) were formed from small metal plates sewn into a supporting fabric. The scales, fabricated from a copper alloy, had dimensions of about 15 mm long by 13 mm wide, joined together by wires and then sewn to a linen cloth using threads of the same material. While being less flexible than a chain mail, they were, however, popular throughout the Roman period. In the Far East, fish scales also inspired a means of protection; Japanese armours (*gyorin kozane* or *kozane do*) were similar to those described above although the laminas were made of leather, *kozane* means scale, and their early versions date to the Fujiwara period (eleventh century). In these cases, the scales were fabricated from materials which were more available: copper alloys, iron, leather (raw or boiled), horn, seeds or pangolin scales.

Nowadays, the strategies developed by nature for making armours continue to inspire the fabrication of artificial armours. Armours based on polymers have the advantage of flexibility and lightness but lack the required hardness for some situations (Rudykh *et al.* 2015, Funk *et al.* 2015). When hard elements are incorporated, such as in *Dragon Skin* armour with ceramic plates, mobility is reduced. Other armours can be found which are more flexible, imitating an armadillo shell (Chintapalli *et al.* 2014). In that case, hexagonal glass is used on a rubber base. The mechanical properties of the skin of fish continues being an attractive model for developing more flexible light armour, particularly when scales with greater resistance to penetration, a good linking system and tough elastic substrate are available.

5 CHAPTER Membranes. Mechanical Properties.

5.1. INTRODUCTION

In order to use a biological material for structural purposes, it is ideal to have its constitutive equation. This equation allows a knowledge of the mechanical response of the material (the strain as well as the stress state) under external loading (forces, pressure, displacements, temperature, humidity, *etc.*).

This objective is not easy due to the complexity involved, and to achieve it requires experimentation. For that it is essential to define precisely the relevant variables, design the experiment properly, and know how to interpret the results correctly.

The bi-dimensionality further complicates the study and interpretation of experiments on biological materials in the form of membranes. Familiarisation with the basic concepts outlined below requires considerable effort, but this is necessary for when the reader comes across specialist articles which take for granted, for example, a knowledge of the *Green-Lagrange* strain tensor or the *Piola-Kirchhoff* stress tensors.

To this end, the structure of this chapter is: first, the concepts are presented which are required to describe the stress-strain behaviour of the material. Then, some experiments are described, which provide data for the characterisation of the material, attempting to accompany the exposition with simple examples that can facilitate the assimilation of the basic concepts of continuum mechanics. In this book there is no room to expand upon these notions, and the interested reader may consult, for example, the books of Fung (1993), Humphrey (2002), Holzapfel or Capaldi (2012).

The exposition is completed by presenting the most common experiments used to characterise these materials: biaxial tension, rupture by tearing, and simple peeling or by pressurisation tests. The information obtained from these is essential for the formulation of the constitutive equations that define the mechanical behaviour of soft materials, and very valuable for understanding the relationships between their microstructure and mechanical properties. It is also very useful in the field of medicine; for regenerative surgery, for transplants and to feed into computer programmes in robotic surgery.

5.2. MECHANICAL BEHAVIOUR

5.2.1. Pseudo-elastic materials

The mechanical behaviour of soft biological materials is far from being elastic; the deformations depend upon the loading history, do not usually recover, and when they do they depend upon time. All of this makes it very difficult to establish a one-to-one relationship between stress and strain; when subjected to cyclic loading and unloading they exhibit **hysteresis** (the unloading curve does not coincide with that of loading, and for the same value of strain there are two possible values of stress), if the strain is held constant the stress **relaxes** with time, and if the stress is kept constant the strain increases (**flows**) with time. If, in addition to these difficulties we add, among other inconveniences, anisotropy, heterogeneity and the influence of the age of the tissue, the mathematical formulation of a model of mechanical behaviour is greatly complicated.

However, under cyclic testing a soft biological tissue exhibits a phenomenon known as *accommodation* which often allows the proposition of an approximate constitutive equation. Accommodation means that, after several loading and unloading cycles, the hysteresis loop does not change, and the stresses and strains always follow the same path. When that happens and the hysteresis cycle (for a given strain rate) is stabilised, it is said that the specimen has been *preconditioned* (figure 5.1). In these cases, the relationship between stress and strain is unique, repeatable and predictable. The relationship between the stress and strain is **unique** –both for the *rising load* branch and the *unloading branch*– if the branches are considered separately.

As the stress and strain have a one-to-one relationship in each branch of the hysteresis loop it is possible to treat the material as if they were two different *elastic* materials: one for the loading process and the other for that of unloading. This idea, due to Y. C. Fung, allows use of the tools of elasticity to model the behaviour of *preconditioned* materials. In order to avoid possible confusion with truly elastic materials, Fung has proposed calling them **pseudo-elastic.**

Bear in mind that pseudo-elasticity is not an intrinsic property of the material, it is merely a convenient description of the relationship between the stress and strain in a cyclic loading process, which is valid only up to the level of the maximum load reached in a preconditioning cycle and, most often, for a given strain rate.

The key kinematic concepts and the state of stress used for characterising the mechanical behaviour of membranes subjected to biaxial loads are summarised below. The aim of this exposition is to delve a little into the concepts and definitions that the reader will meet when they analyse experimental results or constitutive equations. A rigorous development of these concepts would take us beyond the purpose of this book. Interested readers can usefully consult the already quoted books as well as the article of Sacks (2003). The following definitions are intended only as a guide to assist the reader of works related to this topic.

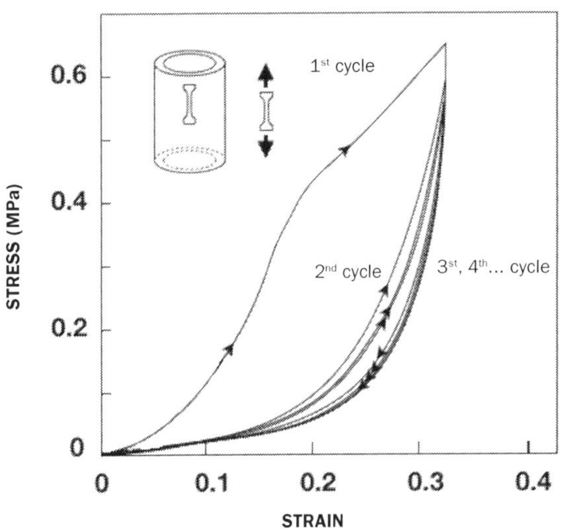

Fig. 5. 1. *Cyclic true stress-strain curves from human thoracic aorta specimens (adapted from Horry et al. 2010). The material shows a very important hysteresis loop which reduces notably after a few cycles, attaining a stable repeatable behaviour.*

5.2.2. Strain and stress

Strains and strain tensor

Consider a *two-dimensional* continuous body formed by a set of particles. The simultaneous position of all the particles is called the *configuration,* and the position $\mathbf{X}$ of all the particles which occupy a space Ω_0 at a certain time (see figure 5.2) is the *reference* configuration. In what follows, capital letters $\mathbf{X}$ are used to denote the position vectors of points in the reference configuration, which is the configuration with respect to which displacements or strains are measured.

When an external load or displacement is applied to the body, it may deform and the positions of the particles change. The *deformed* configuration, which now occupies a space Ω, is the set of the new positions $\mathbf{x}$ of the points, and lowercase letters are used for this situation (as indicated in figure 5.2).

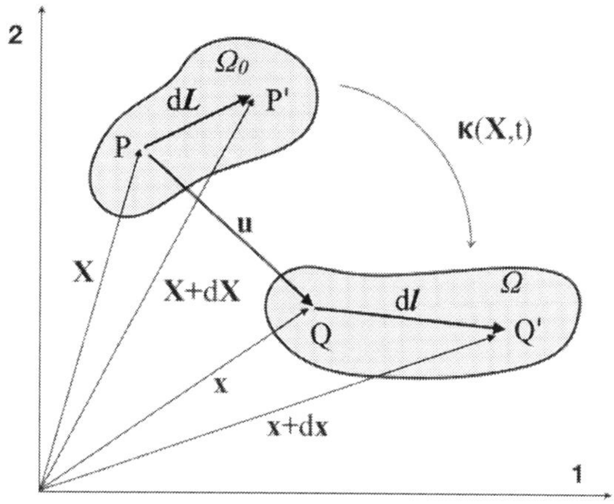

Fig. 5.2. *Notation used in the reference Ω_0, and the deformed Ω configurations.*

The relationship between the values of the points in the deformed configuration $\mathbf{x}$ and the values of the reference configuration $\mathbf{X}$ is provided through the equation of motion

$$\mathbf{x} = \kappa(\mathbf{X},t) \tag{1}$$

Each point in the reference configuration must have a corresponding point in the deformed configuration and vice versa. This implies that there must be an inverse function

$$\mathbf{X} = \kappa^{-1}(\mathbf{x},t) \tag{2}$$

which allows us to obtain the position of a point in the reference configuration when its position in the deformed configuration is known.

The *displacement,* **u**, of a material point, *P*, is the difference between the deformed position $\mathbf{x}(\mathbf{X}_p,t)$ and the reference position $\mathbf{X}_p$, that is to say:

$$u(\mathbf{X}_P,t) = \mathbf{x}(\mathbf{X}_P,t) - \mathbf{X}_P \tag{3}$$

This is the *material* (or Lagrangian) description of displacement, where the independent variables are the coordinates of the reference configuration and time.

It is also possible to write the displacement as a function of the variables in the deformed configuration and time, which is known as the *spatial* (or Euclidian) description in the form:

$$u(\mathbf{x}_P,t) = \mathbf{x}_P - \mathbf{X}(\mathbf{x}_P,t) \tag{4}$$

Figure 5.2 illustrates these concepts and allows us to advance a little further into the ideas related to the deformation of a body.

Suppose that the points $P(X_1, X_2)$ and $P'(X_1 + dX_1, X_2 + dX_2)$ (in Ω_0) are very close to each other, if the body is continuous, after deformation the corresponding points $Q(x_1, x_2)$ and $Q'(x_1 + dx_1, x_2 + dx_2)$ (in Ω) will also be close. The vector $d\mathbf{X} = (dX_1, dX_2)$ in the initial (or reference) configuration will be transformed into the vector $d\mathbf{x} = (dx_1, dx_2)$ of the deformed (or instantaneous) configuration. We can calculate the vector $d\mathbf{x}$ from $d\mathbf{X}$ by means of equation (1):

$$d\mathbf{x} = \frac{\partial \kappa(\mathbf{X},t)}{\partial \mathbf{X}} d\mathbf{X} = \mathbf{F} d\mathbf{X} \tag{5a}$$

Expression (5a) defines the *deformation gradient tensor* $\mathbf{F}$, which contains all of the information needed to determine the relative position $d\mathbf{x} = (dx_1, dx_2)$ of the material particles from their initial separation $d\mathbf{X} = (dX_1, dX_2)$:

$$\begin{bmatrix} dx_1 \\ dx_2 \end{bmatrix} = \begin{bmatrix} \frac{\partial \kappa_1}{\partial X_1} & \frac{\partial \kappa_1}{\partial X_2} \\ \frac{\partial \kappa_2}{\partial X_1} & \frac{\partial \kappa_2}{\partial X_2} \end{bmatrix} \begin{bmatrix} dX_1 \\ dX_2 \end{bmatrix} \quad \text{or} \quad dx_i = \frac{\partial \kappa_i}{\partial X_j} dX_j = \frac{\partial x_i}{\partial X_j} dX_j \tag{5b}$$

in which we have assumed a cartesian coordinate system. To simplify writing, and in what follows, we will use the Einstein summation convention in which a repeated index indicates the sum of the terms over all of its possible values.

In a similar manner, we can obtain $d\mathbf{X}$ from $d\mathbf{x}$ using equation (2)

$$d\mathbf{X} = \frac{\partial \kappa^{-1}(\mathbf{x},t)}{\partial \mathbf{x}} d\mathbf{x} = \mathbf{F}^{-1} d\mathbf{x} \quad \text{or} \quad dX_i = \frac{\partial X_i}{\partial x_j} dx_j \tag{6}$$

It is useful to know, for reasons which will become apparent later, how an infinitesimal length between two infinitely close points, such as P and P' in figure 5.2 deforms. For that we look for a relationship between the *reference* dL and the deformed dl segments, whose magnitudes are given by:

$$(dL)^2 = d\mathbf{X} \cdot d\mathbf{X} = dX_1^2 + dX_2^2 \qquad (7)$$

and

$$(dl)^2 = d\mathbf{x} \cdot d\mathbf{x} = dx_1^2 + dx_2^2 \qquad (8)$$

Taking into account equation (6) we obtain

$$(dL)^2 = \frac{\partial \boldsymbol{\kappa}^{-1}}{\partial \mathbf{x}} d\mathbf{x} \cdot \frac{\partial \boldsymbol{\kappa}^{-1}}{\partial \mathbf{x}} d\mathbf{x} = \left(\frac{\partial \boldsymbol{\kappa}^{-1}}{\partial \mathbf{x}}\right)^T \frac{\partial \boldsymbol{\kappa}^{-1}}{\partial \mathbf{x}} d\mathbf{x} \cdot d\mathbf{x} \qquad (9)$$

where the superscript (^{T}) indicates the transposed tensor. The above expression becomes, in cartesian coordinates,

$$(dL)^2 = \delta_{ij} \cdot dX_i \cdot dX_j = \delta_{ij} \frac{\partial X_i}{\partial x_l} \frac{\partial X_j}{\partial x_m} dx_l dx_m = $$

$$= \frac{\partial X_i}{\partial x_l} \frac{\partial X_i}{\partial x_m} dx_l dx_m \qquad (10)$$

where δ_{ij} equals 1 if $i=j$, and 0 if $i \neq j$. From equation (5) we also have

$$(dl)^2 = \left(\frac{\partial \boldsymbol{\kappa}}{\partial \mathbf{X}}\right)^T \frac{\partial \boldsymbol{\kappa}}{\partial \mathbf{X}} d\mathbf{X} \cdot d\mathbf{X}$$

and

$$(dl)^2 = \frac{\partial x_i}{\partial X_l} \frac{\partial x_i}{\partial X_m} dX_l dX_m \qquad (11)$$

The difference between the squares of the deformed segment dl and the reference segment dL, can be written as:

$$(dl)^2 - (dL)^2 = 2\mathbf{E} d\mathbf{X} \cdot d\mathbf{X}$$

or $(dl)^2 - (dL)^2 = 2E_{ij} dX_i \cdot dX_j$ (12)

or,

$(dl)^2 - (dL)^2 = 2\eta d\mathbf{x} \cdot d\mathbf{x}$ or $(dl)^2 - (dL)^2 = 2\eta_{ij} dx_i \cdot dx_j$ (13)

where $\mathbf{E}$ is the *Green-Lagrange strain tensor* (introduced by Green and Saint Venant) and η is the

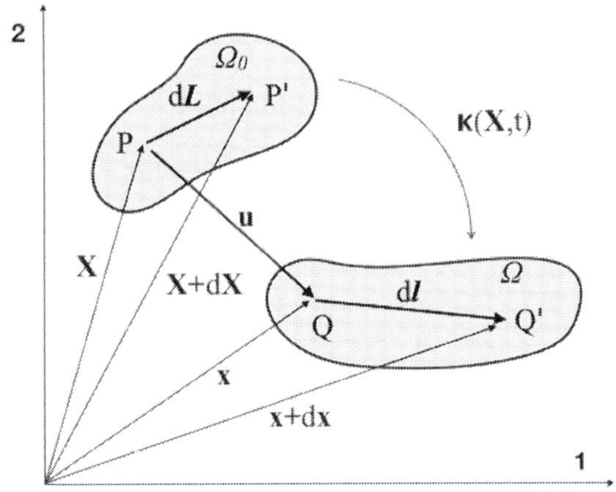

Fig. 5.2. *Notation used in the reference Ω_0, and the deformed Ω configurations.*

Almansi strain tensor (introduced by Almansi and Hamel[19]), defined as

$$\mathbf{E} = \frac{1}{2}\left[\left(\frac{\partial \boldsymbol{\kappa}}{\partial \mathbf{X}}\right)^T \frac{\partial \boldsymbol{\kappa}}{\partial \mathbf{X}} - \mathbf{1}\right] \text{ or } E_{ij} = \frac{1}{2}\left[\frac{\partial x_\alpha}{\partial X_i}\frac{\partial x_\alpha}{\partial X_j} - \delta_{ij}\right] \qquad (14)$$

and

$$\eta = \frac{1}{2}\left[\mathbf{1} - \left(\frac{\partial \boldsymbol{\kappa}^{-1}}{\partial \mathbf{x}}\right)^T \frac{\partial \boldsymbol{\kappa}^{-1}}{\partial \mathbf{x}}\right] \text{ or } \eta_{ij} = \frac{1}{2}\left[\delta_{ij} - \frac{\partial X_\alpha}{\partial x_i}\frac{\partial X_\alpha}{\partial x_j}\right] \qquad (15)$$

Expressions (14) and (15) obtained for the Green-Lagrange and Almansi tensors can be written in terms of the displacement **u** taking into account equations (3) and (4):

$$\mathbf{E} = \frac{1}{2}\left[\left(\frac{\partial \mathbf{u}}{\partial \mathbf{X}}\right)^T + \frac{\partial \mathbf{u}}{\partial \mathbf{X}} + \left(\frac{\partial \mathbf{u}}{\partial \mathbf{X}}\right)^T \frac{\partial \mathbf{u}}{\partial \mathbf{X}}\right]$$

or

$$E_{ij} = \frac{1}{2}\left[\frac{\partial u_j}{\partial X_i} + \frac{\partial u_i}{\partial X_j} + \frac{\partial u_\alpha}{\partial X_i}\frac{\partial u_\alpha}{\partial X_j}\right] \qquad (16)$$

$$\eta = \frac{1}{2}\left[\left(\frac{\partial \mathbf{u}}{\partial \mathbf{x}}\right)^T + \frac{\partial \mathbf{u}}{\partial \mathbf{x}} - \left(\frac{\partial \mathbf{u}}{\partial \mathbf{x}}\right)^T \frac{\partial \mathbf{u}}{\partial \mathbf{x}}\right]$$

or

$$\eta_{ij} = \frac{1}{2}\left[\frac{\partial u_j}{\partial x_i} + \frac{\partial u_i}{\partial x_j} - \frac{\partial u_\alpha}{\partial x_i}\frac{\partial u_\alpha}{\partial x_j}\right] \qquad (17)$$

[19] Emilio Almansi (1869-1948) was an Italian physicist and mathematician famed for his work on the theory of elasticity and the solution of nonlinear problems. Georg Karl Wilhelm Hamel (1877-1954) was a German mathematician and author of several treatises on mechanics.

The following example may help to cement the ideas and concepts mentioned above. In what follows the effect of time is ignored.

EXERCISE 30

Consider the homogeneous membrane represented in the figure, loaded uniformly by forces $\mathbf{f}_1 = f_1 \mathbf{e}_1$ and $\mathbf{f}_2 = f_2 \mathbf{e}_2$. The dimensions and the coordinate system are also shown in the figure.

It is assumed that the deformation is homogeneous, that is, the same at every point, and the thickness h changes during the deformation but the volume remains constant. Consequently the density r will be constant. It is further assumed that there are no angular deformations, and, therefore, the membrane maintains a rectangular shape during the deformation. This is the case if the material is isotropic and also if, being anisotropic, there is a material symmetry with respect to the 1 and 2 axes (since the deformation must respect that symmetry and, therefore, also the rectangular form of the plate, which is symmetric with respect to said axes).

In the *reference* configuration, the initial values are: L_1, L_2 and H.

In the *deformed* configuration, these values will be $l_1 = \lambda_1 L_1$, $l_2 = \lambda_2 L_2$ and h, where by conservation of volume, $l_1\, l_2\, h = L_1\, L_2\, H$.

Assume, for convenience, that the origin of the coordinates does not change its position with the deformation.

Under these conditions, apply expressions (1) to (15) to the deformation of the membrane.

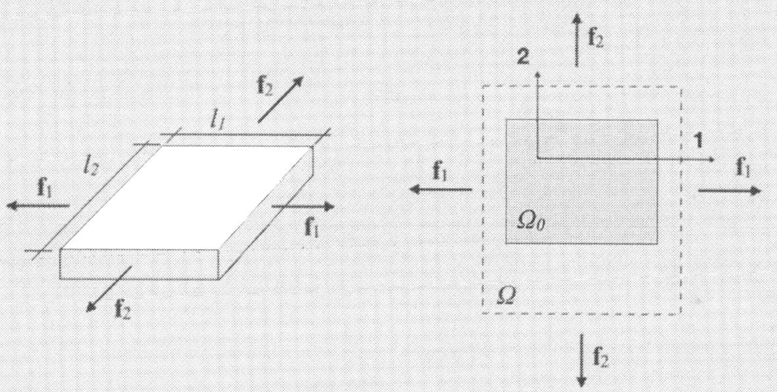

SOLUTION:

Equations (1) and (2) provide a knowledge of the *position* of a point in the membrane after the deformation. A point in the reference configuration (initial) with position (X_1, X_2) is displaced to point (x_1, x_2) in the deformed configuration according to equation (1):

$\mathbf{x} = \kappa(\mathbf{X})$ vector notation

$(x_1, x_2) = (\lambda_1 X_1,\ \lambda_2 X_2)$ coordinate notation

$$\begin{bmatrix} x_1 \\ x_2 \end{bmatrix} = \begin{bmatrix} \lambda_1 & 0 \\ 0 & \lambda_2 \end{bmatrix} \begin{bmatrix} X_1 \\ X_2 \end{bmatrix}$$ matrix notation

The position of a point in the reference configuration, when its position in the deformed configuration is known, is given by equation (2)

$\mathbf{X} = \kappa^{-1}(\mathbf{x})$ vector notation

$(X_1, X_2) = (x_1/\lambda_1, x_2/\lambda_2)$ coordinate notation

$$\begin{bmatrix} X_1 \\ X_2 \end{bmatrix} = \begin{bmatrix} 1/\lambda_1 & 0 \\ 0 & 1/\lambda_2 \end{bmatrix} \begin{bmatrix} x_1 \\ x_2 \end{bmatrix}$$ matrix notation

The value of the *displacement,* **u**, of a point P (see figure 5.18) as a function of the coordinates (X_1, X_2) of the reference configuration is given by equation (3)

$u(\mathbf{X}_P) = \mathbf{x}(\mathbf{X}_P) - \mathbf{X}_P$ vector notation

$[u_1(X_1, X_2), u_2(X_1, X_2)] = [(\lambda_1 - 1)\, X_1,\, (\lambda_2 - 1)\, X_2]$ coordinate notation

$$\begin{bmatrix} u_1 \\ u_2 \end{bmatrix} = \begin{bmatrix} \lambda_1 - 1 & 0 \\ 0 & \lambda_2 - 1 \end{bmatrix} \begin{bmatrix} X_1 \\ X_2 \end{bmatrix}$$ matrix notation

Similarly, the displacement **u** as a function of the coordinates (x_1, x_2) in the deformed configuration is given by equation (4)

$u(\mathbf{x}_P) = \mathbf{x}_P - \mathbf{X}(\mathbf{x}_P)$ vector notation

$[u_1(x_1, x_2),\, u_1(x_1, x_2)] = [(1 - 1/\lambda_1)\, x_1,\, (1 - 1/\lambda_2)\, x_2]$ coordinate notation

$$\begin{bmatrix} u_1 \\ u_2 \end{bmatrix} = \begin{bmatrix} 1 - 1/\lambda_1 & 0 \\ 0 & 1 - 1/\lambda_2 \end{bmatrix} \begin{bmatrix} x_1 \\ x_2 \end{bmatrix}$$ matrix notation

The *deformation gradient tensor,* $\mathbf{F}$, defined by equations (5a,b) becomes

$$\mathbf{F} = \frac{\partial \kappa(\mathbf{X})}{\partial \mathbf{X}} \quad \text{or} \quad F_{ij} = \frac{\partial x_i}{\partial X_j}$$

which, for this example, is $\mathbf{F} = \begin{bmatrix} \lambda_1 & 0 \\ 0 & \lambda_2 \end{bmatrix}$

for this problem $\mathbf{F}$ is symmetrical so that $\mathbf{F} = \mathbf{F}^T$. We can also obtain the inverse tensor $\mathbf{F}^{-1}$ from equation (6)

$$\mathbf{F}^{-1} = \frac{\partial \kappa^{-1}(\mathbf{x})}{\partial \mathbf{x}} \quad \text{or} \quad F_{ij}^{-1} = \frac{\partial X_i}{\partial x_j}$$

which for this example becomes,

$$\mathbf{F}^{-1} = \begin{bmatrix} 1/\lambda_1 & 0 \\ 0 & 1/\lambda_2 \end{bmatrix}$$

which is also symmetric, so that $\mathbf{F}^{-1} = \mathbf{F}^{-T}$.

If we need to know the relationship between the infinitesimal length dL between two infinitely close points (for example P and P' in figure 5.2) and its value after deformation, dl, we use expressions (9) and (11)

$$(dL)^2 = \mathbf{F}^{-T}\mathbf{F}^{-1}d\mathbf{x}\cdot d\mathbf{x} = \begin{bmatrix} 1/\lambda_1^2 & 0 \\ 0 & 1/\lambda_2^2 \end{bmatrix} d\mathbf{x} \cdot d\mathbf{x}$$

$$(dl)^2 = \mathbf{F}^T\mathbf{F}\; d\mathbf{X}\cdot d\mathbf{X} = \begin{bmatrix} \lambda_1^2 & 0 \\ 0 & \lambda_2^2 \end{bmatrix} d\mathbf{X} \cdot d\mathbf{X}$$

Thus, using equation (7)

$$(dl)^2 - (dL)^2 = \begin{bmatrix} \lambda_1^2 & 0 \\ 0 & \lambda_2^2 \end{bmatrix} d\mathbf{X} \cdot d\mathbf{X} - d\mathbf{X} \cdot d\mathbf{X} = \begin{bmatrix} \lambda_1^2 - 1 & 0 \\ 0 & \lambda_2^2 - 1 \end{bmatrix} d\mathbf{X} \cdot d\mathbf{X}$$

and also, using (8)

$$(dl)^2 - (dL)^2 = d\mathbf{x} \cdot d\mathbf{x} - \begin{bmatrix} 1/\lambda_1^2 & 0 \\ 0 & 1/\lambda_2^2 \end{bmatrix} d\mathbf{x} \cdot d\mathbf{x} = \begin{bmatrix} 1 - 1/\lambda_1^2 & 0 \\ 0 & 1 - 1/\lambda_2^2 \end{bmatrix} d\mathbf{x} \cdot d\mathbf{x}$$

which can be identified as the *Green-Lagrange strain tensor* expression (12)

$$\mathbf{E} = \frac{1}{2}\begin{bmatrix} \lambda_1^2 - 1 & 0 \\ 0 & \lambda_2^2 - 1 \end{bmatrix}$$

and the *Almansi strain tensor* expression (13) is

$$\eta = \frac{1}{2}\begin{bmatrix} 1 - 1/\lambda_1^2 & 0 \\ 0 & 1 - 1/\lambda_2^2 \end{bmatrix}$$

which can also be expressed using equations (14)

$$\mathbf{E} = \frac{1}{2}\left[\mathbf{F}^T\mathbf{F} - \mathbf{1}\right] = \frac{1}{2}\left\{\begin{bmatrix} \lambda_1^2 & 0 \\ 0 & \lambda_2^2 \end{bmatrix} - \begin{bmatrix} 1 & 0 \\ 0 & 1 \end{bmatrix}\right\}$$

and equation (15)

$$\eta = \frac{1}{2}[\mathbf{1} - \mathbf{F}^{-T}\mathbf{F}^{-1}] = \frac{1}{2}\left\{\begin{bmatrix} 1 & 0 \\ 0 & 1 \end{bmatrix} - \begin{bmatrix} 1/\lambda_1^2 & 0 \\ 0 & 1/\lambda_2^2 \end{bmatrix}\right\}$$

The diagonal components of the *Green* tensor measure the change in length of fibres orientated parallel to the reference axes, relative to the *initial* length. It can be shown that

$$E_{11} = \frac{1}{2}(\lambda_1^2 - 1) \quad \text{and} \quad E_{22} = \frac{1}{2}(\lambda_2^2 - 1) \quad (18)$$

where $\lambda_i = dl_i/dL_i = 1 + \Delta(dL_i)/dL_i$ is the unit elongation (or more simply, elongation) in the i-direction $(i=1, 2)$. λ_i measures the relationship between final length and the initial length of an element *initially* lying parallel to the i-direction. Note that $\Delta(dL_i)/dL_i$ is the nominal or engineering strain in the i-direction.

In the same manner it can be seen that the diagonal components of the *Almansi* tensor provide a measure of the relative change in length of linear elements relative to the *deformed* length:

$$\eta_{11} = \frac{1}{2}\left(1 - \frac{1}{\tilde{\lambda}_1^2}\right) \quad \text{and} \quad \eta_{22} = \frac{1}{2}\left(1 - \frac{1}{\tilde{\lambda}_2^2}\right) \quad (19)$$

now $\tilde{\lambda}_i$ is the elongation of an element which *after strain* stays parallel to the i-direction. Note that dl and dL don't have to be parallel, and in general they are not (figure 5.2).

EXERCISE 31

Derive equations (18) and (19)

SOLUTION:

If we apply equation (12) to an element dL_i *lying initially* (in the reference configuration) in the i-direction we see that:

$$(dl_i)^2 - (dL_i)^2 = (dl_i)^2 - (dX_i)^2 = 2E_{ii}(dX_i)^2$$

since $dL_i = dX_i$. The square of the elongation is given by

$$\lambda_i^2 = \frac{(dl_i)^2}{(dX_i)^2} = \frac{(dX_i)^2 + 2E_{ii}(dX_i)^2}{(dX_i)^2} = 1 + 2E_{ii}$$

from which equation (18) is immediately deduced. Note that an element initially parallel to the i-direction does not have to remain parallel to this direction after deformation.

In order to derive equation (19) we now consider an element in the deformed configuration, in the i-direction, and for that $dl_i = dx_i$. We have:

$$(dl_i)^2 - (dL_i)^2 = (dx_i)^2 - (dL_i)^2 = 2\eta_{ii}(dx_i)^2$$

the square of the elongation is

$$\tilde{\lambda}_i^2 = \frac{(dx_i)^2}{(dL_i)^2} = \frac{(dx_i)^2}{(dx_i)^2 - 2\eta_{ii}(dx_i)^2}$$

From which equation (19) is deduced. Just as above, note that although dl_i is parallel to the i-direction in the deformed configuration, dL_i does not have to have been parallel to this direction in the initial configuration.

Other relationships which are convenient to quantify in straining are:

- The value of the strain which is produced in a segment which is initially orientated in any direction
- The value of the angle created in the deformed configuration by two directions which initially form an arbitrary angle
- The change suffered by an elemental area once deformed relative to the initial area in the reference configuration
- The change in volume of the membrane once deformed relative to the initial volume in the reference configuration

All of these relationships, as will be seen, can be expressed as functions of the Green-Lagrange, or Almansi strain tensors.

If dA is the area of an enclosure in the reference configuration and da is the corresponding area in the deformed configuration, their values are related by:

$$\frac{da}{dA} = \det\left[\frac{\partial \kappa(\mathbf{X},t)}{\partial \mathbf{X}}\right] = \det \mathbf{F} = \det\left[\frac{\partial x_i}{\partial X_j}\right] \qquad (20)$$

In order to derive this expression without loss of generality let us consider the area $dA{=}d\boldsymbol{X}{\cdot}d\boldsymbol{Y}$ defined by the elemental vectors $d\boldsymbol{X}$ and $d\boldsymbol{Y}$ aligned with the axes 1 and 2 in the reference configuration. The deformed values will be $d\boldsymbol{x}$ and $d\boldsymbol{y}$, so that the deformed area $da = dx\,dy\,sin\theta_{12}$ from which we can write

$$\left(\frac{da}{dA}\right)^2 = \left(\frac{dx}{dX}\right)^2\left(\frac{dy}{dY}\right)^2(1-\cos^2\theta_{12}) =$$
$$= (2E_{11}+1)(2E_{22}+1) - 4E_{12}^2$$

in which equation (18) has been used along with the value of the angle θ_{12} obtained from exercise 33. Given that the Cauchy-Lagrange strain tensor is symmetrical and $E_{12}{=}E_{21}$, we can write

$$\left(\frac{da}{dA}\right)^2 = (2E_{11}+1)(2E_{22}+1) - 4E_{12}E_{21} =$$
$$= \det\left[2E_{ij}+\delta_{ij}\right] = \det\left[\frac{\partial x_\alpha}{\partial X_i}\frac{\partial x_\alpha}{\partial X_j}\right] = \det{}^2\left[\frac{\partial x_i}{\partial X_j}\right] \qquad (21)$$

and, making use of equation (14), we arrive at equation (20). Observe that equation (21) –like equation (20)– is independent of the choice of the coordinate system, since the determinant is a tensorial invariant. Note that in this example $da/dA = \lambda_1\,\lambda_2$

From equation (20) we can calculate the change in volume of the membrane, letting its thickness be H in the initial (reference) configuration and h in the deformed configuration. The deformed volume will be $dv{=}hda$ and the initial volume $dV{=}HdA$. Using equation (20) we see that

$$\frac{dv}{dV} = \frac{hda}{HdA} = \frac{h}{H}\det\left[\frac{\partial \kappa(\mathbf{X},t)}{\partial \mathbf{X}}\right] = \frac{h}{H}\det \mathbf{F} \qquad (22)$$

Again, for this example $dv/dV = h\lambda_1\lambda_2/H$

When the displacement and gradient of the displacements $(\partial u/\partial X$ or $\partial u/\partial x)$ are small, it can be seen that the second order terms in equations (16) and (17) can be ignored. In these circumstances the Green and Almansi tensors coincide

$$\varepsilon = \frac{1}{2}\left[\frac{\partial \mathbf{u}}{\partial \mathbf{X}} + \left(\frac{\partial \mathbf{u}}{\partial \mathbf{X}}\right)^T\right]$$

or

$$\varepsilon_{ij} = \frac{1}{2}\left[\frac{\partial u_i}{\partial X_j} + \frac{\partial u_j}{\partial X_i}\right] \qquad (23)$$

called the *infinitesimal strain tensor.*

Even though the membranes have a small thickness, the bodies are not strictly bi-dimensional, and in order to analyse their behaviour adequately it is necessary to develop a hypothesis about the deformation in the thickness direction. Since soft biological materials have a high water content, and this is nearly incompressible for most ordinary load levels, it is often assumed that the membrane itself is incompressible or, which is the same, that its density is constant.

It is also assumed that the movement of points in the plane of the membrane (axes 1 and 2) is independent of their position through the thickness (axis 3) or, in other words, the function $\mathbf{x} = \boldsymbol{\kappa}(\mathbf{X}, t)$, equation (1), does not depend upon the coordinate X_3 for the point considered $(x_1{=}\kappa_1(X_1,X_2);\ x_2{=}\kappa_2(X_1,X_2))$. In this way the problem can be treated as purely two-dimensional (with only the dimensions 1 and 2), adjusting, for each point, the value of the thickness required to conserve the volume.

EXERCISE 32

Determine as a function of the Green-Lagrange tensor the value of the elongation λ_N in an arbitrary direction $\boldsymbol{N}$ in the reference configuration. $\boldsymbol{N}$ is a unit vector.

Apply the result to the deformation of the membrane in exercise 30.

SOLUTION:

Let $d\boldsymbol{X} = dL\boldsymbol{N}$ be an elemental vector along direction $\boldsymbol{N}$ in the reference configuration. Its deformed value is $d\boldsymbol{x} = dl\,\boldsymbol{n}$ ($\boldsymbol{n}$ is a unit vector). From equation (5)

$$d\mathbf{x} = dl\,\mathbf{n} = dL\frac{\partial \kappa(\mathbf{X},t)}{\partial \mathbf{X}}\mathbf{N} \quad \text{or} \quad dl\,n_i = dL\frac{\partial x_i}{\partial X_j}N_j$$

performing the scalar multiple of this with itself, we obtain

$$(dl)^2 = (dL)^2\,\delta_{ij}\frac{\partial x_i}{\partial X_l}\frac{\partial x_j}{\partial X_m}N_lN_m$$

and, using equation (14), the elongation becomes

$$\lambda_N^2 = \frac{(dl)^2}{(dL)^2} = \delta_{ij}\frac{\partial x_i}{\partial X_l}\frac{\partial x_j}{\partial X_m}N_lN_m = (2E_{lm} + \delta_{lm})\,N_lN_m$$

or $\lambda_N^2 = (2E_{11} + 1)\cos^2\alpha + 2(E_{12} + E_{21})\cos\alpha\sin\alpha + (2E_{22} + 1)\sin^2\alpha$

where α is the angle which the unit vector N forms with the 1-axis, so $N_1 = \cos\alpha$ and $N_2 = \sin\alpha$.

When $\boldsymbol{N}$ is the unit vector in the 1-direction 1 ($\alpha = 0$) we recover the expressions in equations (18).

If we apply the above result to the membrane in exercise 30 we find that the elongation in the 1-direction is

$$\lambda_{N=\bar{1}}^2 = 2E_{11} + 1 = 2\frac{1}{2}(\lambda_1^2 - 1) + 1 = \lambda_1^2$$

since $dl_1/dL_1 = \lambda_1$, and the deformation is homogeneous (the same at every point) then $l_1/L_1 = \lambda_1$ as well.

Equally, in the 2-direction

$$\lambda_{N=\bar{2}}^2 = 2E_{22} + 1 = 2\frac{1}{2}(\lambda_2^2 - 1) + 1 = \lambda_2^2$$

and

$dl_2/dL_2 = l_2/L_2 = \lambda_2$.

EXERCISE 33

As a function of the Green-Lagrange strain tensor, determine the angle θ_{NM} formed in the deformed configuration by two arbitrary directions $\boldsymbol{N}$ and $\boldsymbol{M}$ in the reference configuration. $\boldsymbol{N}$ and $\boldsymbol{M}$ are unit vectors.

Apply the result to the deformation of the angle formed by the 1 and 2 axes in the membrane in exercise 30.

SOLUTION:

Let $d\boldsymbol{X}=dX\boldsymbol{N}$ and $d\boldsymbol{Y}=dY\boldsymbol{M}$ be elemental vectors along directions $\boldsymbol{N}$ and $\boldsymbol{M}$ in the reference configuration. Their deformed values will be $d\boldsymbol{x}=dx\boldsymbol{n}$ and $d\boldsymbol{y}=dy\,\boldsymbol{m}$ ($\boldsymbol{n}$ and $\boldsymbol{m}$ are unit vectors). In order to calculate the angle subtended by $d\boldsymbol{x}$ and $d\boldsymbol{y}$ we perform their scalar product:

$$\cos\theta_{NM}=\frac{\mathbf{dx}\cdot\mathbf{dy}}{dxdy}=\frac{\delta_{ij}\dfrac{\partial x_i}{\partial X_l}\dfrac{\partial x_j}{\partial X_m}dX\,N_l\,dY\,M_m}{dxdy}=\frac{\left(2E_{lm}+\delta_{lm}\right)N_lM_m}{\left(dx/dX\right)\left(dy/dY\right)}$$ in which we have made use of equations (5) and (14).

However, and from exercise 32, we know that $\lambda_N^2=\left(2E_{lm}+\delta_{lm}\right)N_lN_m$ and $\lambda_M^2=\left(2E_{lm}+\delta_{lm}\right)M_lM_m$

so that $$\cos\theta_{NM}=\frac{\left(2E_{lm}+\delta_{lm}\right)N_lM_m}{\sqrt{\left(2E_{ij}+\delta_{ij}\right)N_iN_j}\sqrt{\left(2E_{sr}+\delta_{sr}\right)M_sM_r}}$$

In particular, if $\boldsymbol{N}$ and $\boldsymbol{M}$ are the 1 and 2 axes, then: $$\cos\theta_{12}=\frac{2E_{12}}{\sqrt{2E_{11}+1}\sqrt{2E_{22}+1}}$$

In exercise 30, if we take directions $\boldsymbol{N}$ and $\boldsymbol{M}$ as the 1 and 2 axes, since $E_{12}=0$ we obtain the previous expressions $\cos\theta_{12}=0$ and $\theta_{12}=90°$. That is to say, the deformed angle corresponding to the right angle formed by the 1 and 2 axes remains a right angle.

EXERCISE 33bis

Determine the relative change in the area and volume of the membrane in exercise 30.

SOLUTION:

Using equation (21): $$\left(\frac{da}{dA}\right)^2=\left(2E_{11}+1\right)\left(2E_{22}+1\right)-4E_{12}E_{21}=\lambda_1^2\lambda_2^2$$

from which we obtain the relative change in area $da/dA=\lambda_1\lambda_2$, where λ_1, λ_2 are the elongations of the 1 and 2-axes, respectively.

Given the homogeneity of the deformation we can see that this result coincides with the relationship between the total areas of the membrane after and before the deformation:

$(l_1\,l_2)/(L_1\,L_2)=(l_1/L_1)\,(l_2/L_2)=\lambda_1\,\lambda_2$ since $l_1/L_1=\lambda_1$ y $l_2/L_2=\lambda_2$, as was seen in exercise 32.

On the other hand, from equation (22) we obtain the relationship between the volumes dv/dV

$$\frac{dv}{dV}=\frac{hda}{HdA}=\frac{h}{H}\lambda_1\lambda_2$$

where H and h are the thicknesses of the reference configuration and that of the deformed configuration, respectively.

Conservation of volume, established in the statement of EXERCISE 30, implies that the dimensions of the initial and deformed configurations satisfy the relationship: $l_1\,l_2\,h=L_1\,L_2\,H$ and bearing in mind that $l_1/L_1=\lambda_1$ and $l_2/L_2=\lambda_2$, we obtain $\lambda_1\,\lambda_2\,h=H$, an expression which if introduced into the equation for the relationship between the volumes dv/dV leads to

$$\frac{dv}{dV}=\frac{h}{H}\lambda_1\lambda_2=\frac{1}{\lambda_1\lambda_2}\lambda_1\lambda_2=1$$ conforming with the statement made in exercise 30.

Stresses and the stress tensor

Consider the two-dimensional body represented in figure 5.3a, subjected to forces on its boundary, and the action of gravity. Imagine a small segment of length Δl, located in its interior, together with the unit vector normal to it, $\mathbf{n}$. This vector allows us to distinguish space on either side of the segment Δl; the positive side, which is where $\mathbf{n}$ points and the negative side, which is opposite. Furthermore, we assume that the body has a thickness h, which is very small compared to the other dimensions.

Examine closely the part of the material which is on the positive side (the shaded area in figure 5.3a). This part exerts a force $\Delta \boldsymbol{F}$ on the other part. The force $\Delta \boldsymbol{F}$ will depend upon the location of the element Δl considered, and its size and orientation.

If the hypothesis is made that the quotient $\Delta \boldsymbol{F}/(h\Delta l)$, when Δl tends to zero, has a definite limit, the resulting vector

$$\boldsymbol{t} = \lim_{\Delta l \to 0} \frac{\Delta F}{h\Delta l} = \frac{dF}{hdl} \tag{24}$$

is known as the *stress vector* at that point, which acts on dl (due to the material on the positive part) defined by the normal $\mathbf{n}$. Observe that $\Delta \boldsymbol{F}$ does not have to be parallel to $\mathbf{n}$, and in general it is not.

With the aim of explaining the notation used, consider a small square of side dl orientated along the axes 1 and 2, as shown in figure 5.3b, and the stresses acting on its faces. On each one of them, the stress vector can be decomposed into two vectors: one *normal* to the face and the other *tangent* to it. In figure 5.3b this decomposition is shown on one of the two faces perpendicular to the 1 axis. The stress vector $\boldsymbol{t}$ is decomposed into the *normal stress vector* σ_{11}, and the *tangential stress vector* σ_{12}. The other faces can be treated in an analogous manner. The set of four stresses (σ_{11}, σ_{12}, σ_{21}, σ_{22}) is the *stress tensor* $\sigma_{ij}(i, j=1, 2)$, which characterises the forces which act a point in the body.

The stress tensor σ_{ij} is symmetric and $\sigma_{12}=\sigma_{21}$, so that there are only three independent components (σ_{11}, σ_{12}, σ_{22}). If this tensor is known, then the stress $\boldsymbol{t}$ which acts on some inclined segment dl^*, characterised by its normal $\boldsymbol{n}$ (figure 5.3c) is also known, through the expression:

$$\mathbf{t} = \sigma \mathbf{n} \quad \text{or} \quad t_i = \sigma_{ij} n_j \tag{25}$$

where n_j is the component of vector $\mathbf{n}$ in the j-direction.

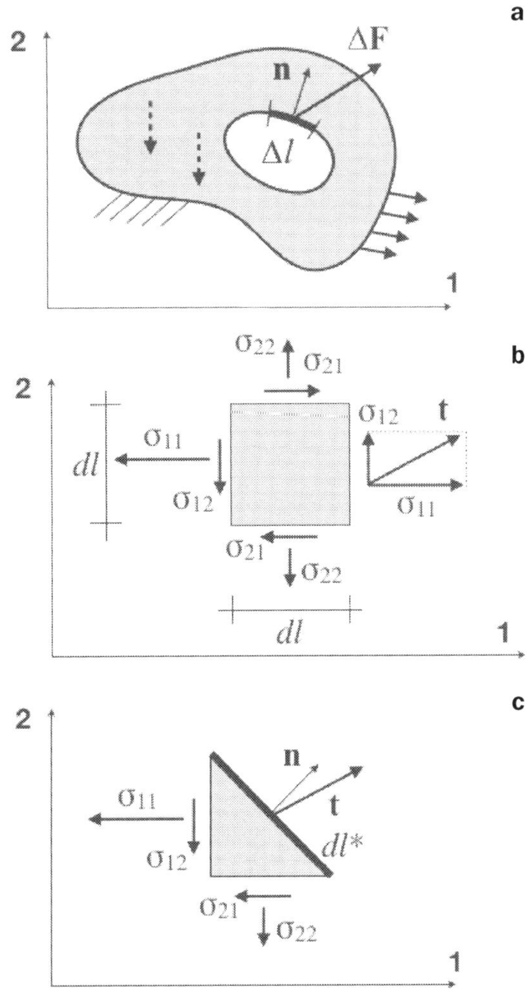

Fig. 5.3. *a) Definition of stress.*
b. Stress tensor notation.
c. Equilibrium of an infinitesimal element.

Equilibrium of the element represented in figure 5.3b requires that the following equations be satisfied,

$$\begin{aligned} \frac{\partial \sigma_{11}}{\partial x_1} + \frac{\partial \sigma_{21}}{\partial x_2} + \rho g_1 = 0 \\ \frac{\partial \sigma_{12}}{\partial x_1} + \frac{\partial \sigma_{22}}{\partial x_2} + \rho g_2 = 0 \end{aligned} \tag{26}$$

where g_1 and g_2 are the components of the gravitational acceleration along the 1 and 2 axes, and ρ is the density. The equilibrium equations (26) are none other than the statement that the sum of all of the forces on the element is zero.

The following exercises are intended to help the reader gain familiarity with the above concepts.

EXERCISE 34

Show that the stress tensor is symmetric.

SOLUTION:

If we consider equilibrium of moments about the lower right corner for the square dl x dl of figure 5.3b (and taking the thickness of the element to be h), we obtain:

$$\sigma_{12} h\,(dl)^2 - \sigma_{21} h\,(dl)^2 - \rho g_1 h (dl)^3/2 - \rho g_2 h\,(dl)^3/2 = 0$$

In the limit as $dl \to 0$ it is concluded that $\sigma_{12} = \sigma_{21}$.

EXERCISE 35

Derive equation (25).

SOLUTION:

Considering figure 5.3c, equilibrium of the elements along the 1 and 2 axes gives:

$$t_1\, h\, dl^* - \sigma_{11}\, h\, dl^*\, n_1 - \sigma_{21}\, h\, dl^*\, n_2 + \rho\, g_1\, h\,(dl^*)^2\, n_1\, n_2 / 2 = 0$$

$$t_2\, h\, dl^* - \sigma_{12}\, h\, dl^*\, n_1 - \sigma_{22}\, h\, dl^*\, n_2 + \rho\, g_2\, h\,(dl^*)^2\, n_1\, n_2 / 2 = 0$$

since (hdl^*n_1) and (hdl^*n_2) are the projections of the area (hdl^*) along the 1 and 2 axes, respectively, and $\rho h (dl^*)^2\, n_1\, n_2 / 2$ is the mass of the element. g_1 and g_2 are the components of the gravitational acceleration along the 1 and 2 axes.

In the limit as $dl^* \to 0$ for the above expressions, and exploiting the symmetry of the stress tensor, we arrive directly at

$$\begin{bmatrix} t_1 \\ t_2 \end{bmatrix} = \begin{bmatrix} \sigma_{11} n_1 + \sigma_{12} n_2 \\ \sigma_{21} n_1 + \sigma_{22} n_2 \end{bmatrix} = \begin{bmatrix} \sigma_{11} & \sigma_{12} \\ \sigma_{21} & \sigma_{22} \end{bmatrix} \begin{bmatrix} n_1 \\ n_2 \end{bmatrix}$$ matrix notation

which can be written as

$$t_i = \sigma_{ij} n_j$$ coordinate notation

or also

$$\mathbf{t} = \sigma \mathbf{n}$$ vector notation

which are the two forms of equation (25).

EXERCISE 36

Derive the equilibrium equations (26).

SOLUTION:

In order to obtain the equilibrium equation, we must consider the changes in the stresses acting on the differential element in figure 5.3b, along the 1 and 2 directions. The stresses acting on the left and lower faces will be σ_{11}, σ_{22}, σ_{12} and σ_{21}, while the stresses on the right and upper faces, which are displaced by distances $dx_1 (= dl)$ and $dx_2 (= dl)$, respectively, will be

$\sigma_{11} + (\partial\sigma_{11}/\partial x_1)\, dx_1$ and $\sigma_{12} + (\partial\sigma_{12}/\partial x_1)\, dx_1$ on the right face

$\sigma_{21} + (\partial\sigma_{21}/\partial x_2)\, dx_2$ and $\sigma_{22} + (\partial\sigma_{22}/\partial x_2)\, dx_2$ on the upper face

The corresponding equilibrium equations for the element will then be:

$$-\sigma_{11}\, hdx_2 - \sigma_{21}\, hdx_1 + (\sigma_{11} + (\partial\sigma_{11}/\partial x_1)\, dx_1)\, hdx_2 + (\sigma_{21} + (\partial\sigma_{21}/\partial x_2)\, dx_2)\, hdx_1 + \rho\, g_1 h\, dx_1\, dx_2 = 0$$

$$-\sigma_{22}\, hdx_1 - \sigma_{12}\, hdx_2 + (\sigma_{22} + (\partial\sigma_{22}/\partial x_2)\, dx_2)\, hdx_1 + (\sigma_{12} + (\partial\sigma_{12}/\partial x_1)\, dx_1)\, hdx_2 + \rho\, g_2 h\, dx_1\, dx_2 = 0$$

cancelling terms and eliminating $h\, dx_1\, dx_2$ we obtain the equations

$$\begin{bmatrix} \partial\sigma_{11}/\partial x_1 + \partial\sigma_{21}/\partial x_2 + \rho g_1 \\ \partial\sigma_{12}/\partial x_1 + \partial\sigma_{22}/\partial x_2 + \rho g_2 \end{bmatrix} = \begin{bmatrix} 0 \\ 0 \end{bmatrix}$$ matrix notation

which are equation (26), and which can also be written as

$$\frac{\partial\sigma_{ij}}{\partial x_i} + \rho g_j = 0$$ coordinate notation

or

$$div\boldsymbol{\sigma} + \rho\boldsymbol{g} = \boldsymbol{0}$$ in vector notation

Observe that the use of the updated stresses (that is, considering their variations along the 1 and 2 axes) on the right and upper faces in figure 5.3b for the calculations performed in exercise 35 does not change the result, since the terms added to the moment equilibrium equation $-((\partial\sigma_{11}/\partial x_1)dl)h(dl)^2/2$, $-((\partial\sigma_{22}/\partial x_2)dl)h(dl)^2$ and $-((\partial\sigma_{22}/\partial x_2)dl)h(dl)^2/2$ are higher order infinitesimals.

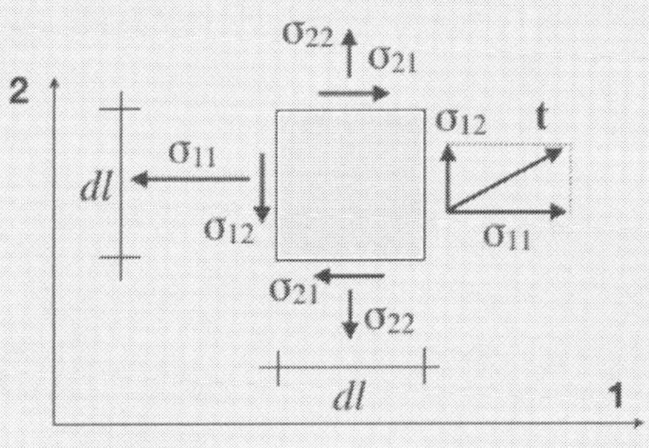

Up to now, we have only considered stresses generated by forces acting in the plane of the body. However, there can also be forces acting perpendicular to the plane, such as when the membrane of a balloon is subjected to pressure. Figure 5.4 illustrates such a situation. In this case the membrane must be curved in order to balance the pressures exerted on its faces, and an equation of equilibrium in the thickness direction must be added to the set of equations (26):

$$p = h\left(\frac{\sigma_{11}}{R_1} + \frac{\sigma_{22}}{R_2}\right) \tag{27}$$

in which R_i is the radius of curvature in the i-direction. Equation (27) is known as Laplace's law.

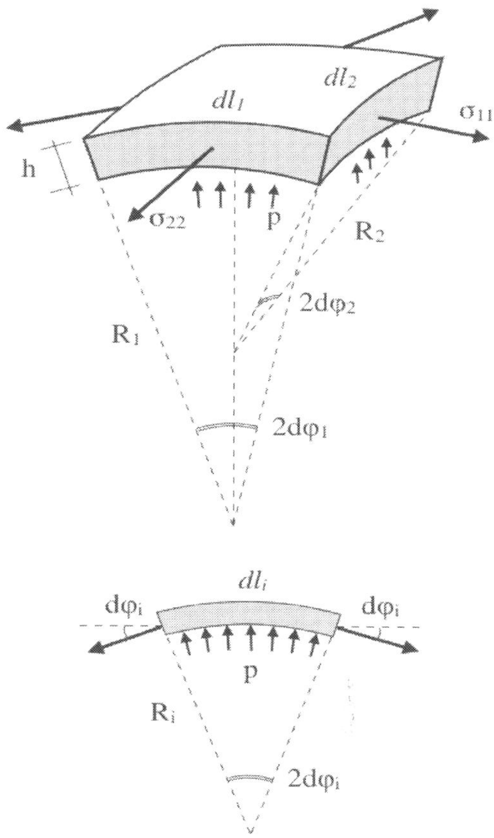

Fig. 5.4. *Equilibrium of a curved membrane element subjected to pressure.*

EXERCISE 37

Prove equation (27).

SOLUTION:

Equation (27) is obtained assuming equilibrium along the normal to the element $dl_1\, x\, dl_2$. Considering the forces shown in figure 5.4, we have

$$p\, dl_1\, dl_2 - 2\, \sigma_{22}\, h\, dl_1\, d\varphi_2 - 2\, \sigma_{11}\, h\, dl_2\, d\varphi_1 - \rho\, g_n\, h\, dl_1 dl_2 = 0$$

where g_n is the component of the gravitational acceleration along the normal to the element. In the above equation we have used the approximation $\sin(d\varphi_2) \approx d\varphi_2$. Also since $2\, d\varphi_1 = dl_1 / R_1$ and $2\, d\varphi_2 = dl_2 / R_2$ then

$$p\, dl_1\, dl_2 - \sigma_{22}\, h\, dl_1\, dl_2/R_2 - \sigma_{11}\, h\, dl_2\, dl_1/R_{12} - \rho\, g_n\, h\, dl_1 dl_2 = 0$$

which leads directly to equation (27), if the self-weight of the element is ignored. This last assumption is reasonable given that for thicknesses of less than a millimetre the term, $h\rho g_n$ is less than 10 Pa. This value is several orders of magnitude less than the usual physiological pressures such as the pressure measured in blood vessels (100 mmHg ≈ 13 kPa) or the eyeball (14 mmHg ≈ 2 kPa).

The fact that biological membranes undergo large deformations, and as a result the initial and deformed configurations may be very different, makes it necessary to clarify in which configuration (initial or deformed) the force and areas are expressed for use in equation (24) which defines the tension vector $\boldsymbol{t}$.

Depending upon the choice of one or the other, it is possible to make the following definitions:

- *Cauchy stress vector,* $\boldsymbol{t}$, or true stress vector, defined as the force acting in the deformed configuration per unit area in the deformed configuration. That is to say:

$$\mathbf{t} = \frac{d\mathbf{f}}{hdl} \qquad (28)$$

in which dl and h are the values of length and thickness in the deformed configuration.

- *First Piola-Kirchhoff stress vector,* $\boldsymbol{p}$, nominal (or engineering) stress vector, defined as the force acting in the deformed configuration per unit area in the initial configuration. That is to say:

$$\mathbf{p} = \frac{d\mathbf{f}}{HdL} \qquad (29)$$

in which now dL and H are the values of the length and thickness in the initial configuration, corresponding to dl and h. Note that in equations (28) and (29), the force vector $\mathbf{f}$ is the same.

- *Second Piola-Kirchhoff stress vector,* $\boldsymbol{s}$, also defined in terms of the unit area in the initial configuration, and using the force referred to the initial configuration

$$\mathbf{s} = \frac{d\mathbf{T}}{HdL} \qquad (30)$$

$\mathbf{T}$ *is the vector corresponding to vector* $\mathbf{f}$ (=the force acting in the deformed configuration) according to the relationship which connects the vectors $d\boldsymbol{X}$ and $d\boldsymbol{x}$, equation (6), that is to say:

$$\mathbf{T} = \frac{\partial \kappa^{-1}(\mathbf{x},t)}{\partial \mathbf{x}}\mathbf{f} = \mathbf{F}^{-1}\,\mathbf{f} \quad \text{or} \quad T_i = \frac{\partial X_i}{\partial x_j}f_j \qquad (31)$$

Vector $\boldsymbol{s}$ has no obvious physical meaning, but it is introduced, as will be explained later, because the corresponding stress tensor is symmetric, and also is the conjugate[20] of the Green-Lagrange strain tensor.

From these different stresses, and following the development in the previous paragraphs, it is possible to find the corresponding stress tensors. Therefore, we will define the *stress vectors of Cauchy* σ, and of the *first Piola-Kirchhoff,* $\boldsymbol{P}$, and the *second,* $\mathbf{S}$.

We have that:

$$\begin{aligned} \mathbf{t} &= \sigma\mathbf{n} & t_i &= \sigma_{ij}n_j \\ \mathbf{p} &= \mathbf{PN} & p_i &= P_{ij}N_j \\ \mathbf{s} &= \mathbf{SN} & s_i &= S_{ij}N_j \end{aligned} \qquad (32)$$

which correspond to equations (25), and which provide knowledge of the stress vectors in the deformed ($\mathbf{t}$) or the reference configuration ($\mathbf{p}, \mathbf{s}$) from the stress tensor and the direction normal to the segment on which the stress acts.

The *first Piola-Kirchhoff stress tensor* $\mathbf{P}$ is a function of the *Cauchy stress tensor* $\mathbf{s}$ and given by:

$$\mathbf{P} = \frac{h}{H}\det\left[\frac{\partial \kappa(\mathbf{X},t)}{\partial \mathbf{X}}\right]\sigma\left[\frac{\partial \kappa^{-1}(\mathbf{x},t)}{\partial \mathbf{x}}\right]^T =$$

$$= \frac{h}{H}\det\mathbf{F}\;\;\sigma\mathbf{F}^{-T} \qquad (33a)$$

$$P_{ij} = \frac{h}{H}\det\left[\frac{\partial x_\alpha}{\partial X_\beta}\right]\sigma_{im}\frac{\partial X_j}{\partial x_m} \qquad (33b)$$

which is not, in general, symmetric, even though σ_{ij} is.

The *second Piola-Kirchhoff stress tensor* $\mathbf{S}$ is also a function of the *Cauchy stress tensor* σ and given by:

$$\mathbf{S} = \frac{h}{H}\det\left[\frac{\partial \kappa(\mathbf{X},t)}{\partial \mathbf{X}}\right]\left[\frac{\partial \kappa^{-1}(\mathbf{x},t)}{\partial \mathbf{x}}\right]\sigma\left[\frac{\partial \kappa^{-1}(\mathbf{x},t)}{\partial \mathbf{x}}\right]^T =$$

$$= \frac{h}{H}\det\mathbf{F}\;\;\mathbf{F}^{-1}\sigma\mathbf{F}^{-T} \qquad (34a)$$

$$S_{ij} = \frac{h}{H}\det\left[\frac{\partial x_\alpha}{\partial X_\beta}\right]\frac{\partial X_i}{\partial x_l}\frac{\partial X_j}{\partial x_m}\sigma_{lm} \qquad (34b)$$

[20] In mechanics, it is said that two quantities are conjugate when their product permits calculation of the work done.

Alternatively, we can write

$$\sigma = \frac{H}{h}\frac{1}{\det \mathbf{F}}\,\mathbf{PF}^T$$

or

$$\sigma_{ij} = \frac{H}{h}\frac{1}{\det\left[\partial x_\alpha / \partial X_\beta\right]} P_{im}\frac{\partial x_j}{\partial X_m} \qquad (35)$$

$$\sigma = \frac{H}{h}\frac{1}{\det \mathbf{F}}\mathbf{FSF}^T$$

or

$$\sigma_{ij} = \frac{H}{h}\frac{1}{\det\left[\partial x_\alpha / \partial X_\beta\right]}\frac{\partial x_i}{\partial X_l}\frac{\partial x_j}{\partial X_m} S_{lm} \qquad (36)$$

which is no more than the solution for σ in equations (33) and (34). As can be seen, S_{ij} is symmetric, since σ_{ij} is, and also the product of the derivatives. The proofs of the previous equations can be found in the reference texts.

EXERCISE 38

Calculate the Cauchy and Piola-Kirchhoff stress tensors for the deformed membrane of exercise 30.

SOLUTION:

The *Cauchy stress tensor* follows from equations (28) and (32):

$$\sigma_{11} = \frac{f_1}{l_2 h} \qquad \sigma_{22} = \frac{f_2}{l_1 h} \qquad \text{and} \qquad \sigma_{12} = \sigma_{21} = 0$$

Similarly, the *first Piola-Kirchhoff tensor* follows from equations (29) and (32):

$$P_{11} = \frac{f_1}{L_2 H} \qquad P_{22} = \frac{f_2}{L_1 H} \qquad \text{and} \qquad P_{12} = P_{21} = 0$$

The same result can be achieved by applying equations (33) and bearing in mind that for this problem and that $l_1/\lambda_1 = L_1$ and $l_2/\lambda_2 = L_2$, as was seen in exercises 30 and 32.

The *second Piola-Kirchhoff* tensor follows from equations (30) and (32),

in which we have calculated the forces in the reference configuration according to equation (31):

$$S_{11} = \frac{1}{\lambda_1}\frac{f_1}{L_2 H} \qquad S_{22} = \frac{1}{\lambda_2}\frac{f_2}{L_1 H} \qquad \text{and} \qquad S_{12} = S_{21} = 0$$

The equivalent result for **S** is obtained from equations (34).

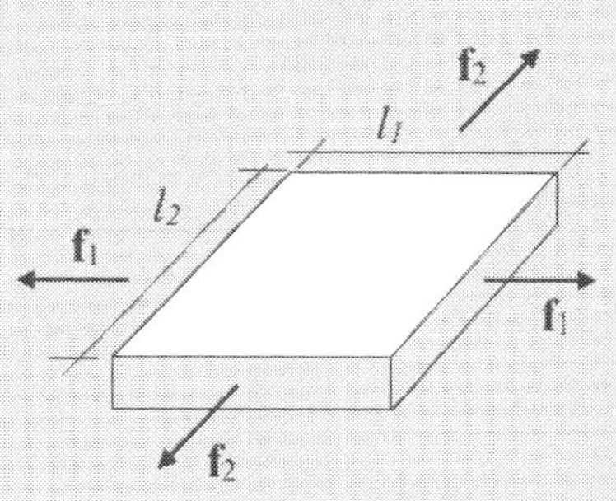

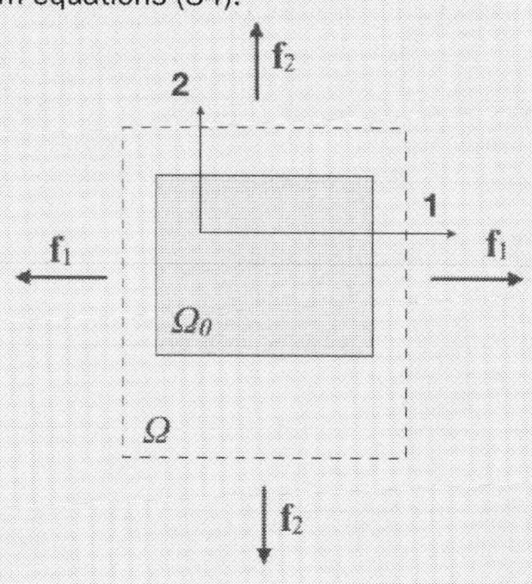

5.2.3. Example: Homogeneous biaxial test

The biaxial tensile test (figure 5.5) was illustrated in exercise 30 and analysed in the subsequent exercises. The conclusions reached are summarised in this section, grouping the results obtained, for greater clarity and use in other examples.

The experiment consists of subjecting a homogeneous isotropic membrane to a uniform load under orthogonal forces $\mathbf{f}_1$, $\mathbf{f}_2$, which produce the same strain at every point (homogeneous). Furthermore, it is assumed that during deformation the volume remains constant.

In the *reference* configuration, the initial values (length, width and thickness) are: L_1, L_2, H.

In the *deformed* configuration, the values will be $l_1 = \lambda_1 L_1$, $l_2 = \lambda_2 L_2$ and h. Satisfaction of conservation of volume requires that $l_1 l_2 h = L_1 L_2 H$.

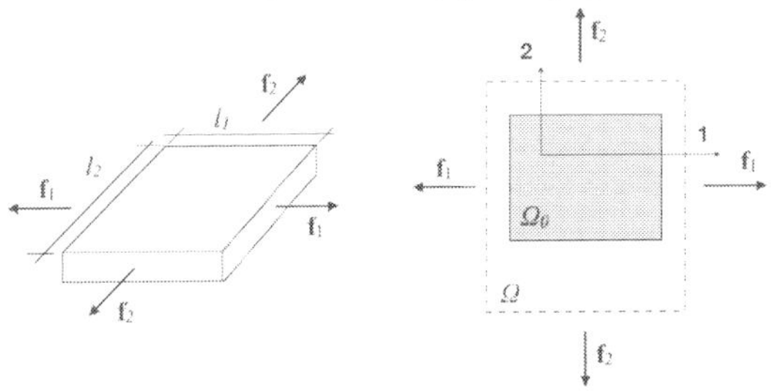

Fig. 5.5. *Homogeneous biaxial test. Geometric and loading notation.*

The displacement equations for directions 1 and 2 are, respectively,

$$x_1 = \lambda_1 X_1 \qquad x_2 = \lambda_2 X_2 \tag{37}$$

and,

$$X_1 = x_1/\lambda_1 \qquad X_2 = x_2/\lambda_2 \tag{38}$$

In which it has been assumed, for convenience, that the origin of the coordinates does not change its position with the deformation.

The *displacement* vector $\mathbf{u}_i$ is given by equation (3)

$$u_1 = (\lambda_1 - 1)X_1 \qquad u_2 = (\lambda_2 - 1)X_2 \tag{39}$$

or equation (4)

$$u_1 = (1 - \lambda_1^{-1})x_1 \qquad u_2 = (1 - \lambda_2^{-1})x_2 \tag{40}$$

The *Green-Lagrange strain tensor,* E_{ij}, the equation, in this simple example, is reduced to the diagonal (E_{11}, E_{22}) whose values become:

$$E_{11} = \frac{1}{2}\left(\lambda_1^2 - 1\right) = \frac{1}{2}\left(\frac{x_1^2}{X_1^2} - 1\right)$$

and

$$E_{22} = \frac{1}{2}\left(\lambda_2^2 - 1\right) = \frac{1}{2}\left(\frac{x_2^2}{X_2^2} - 1\right) \tag{41}$$

Similarly, the *Almansi strain tensor,* η_{ij}, the equation is also reduced to the diagonal (η_{11}, η_{22}), and given by

$$\eta_{11} = \frac{1}{2}\left(1 - \frac{1}{\lambda_1^2}\right) = \frac{1}{2}\left(1 - \frac{X_1^2}{x_1^2}\right)$$

$$\text{and } \eta_{22} = \frac{1}{2}\left(1 - \frac{1}{\lambda_2^2}\right) = \frac{1}{2}\left(1 - \frac{X_2^2}{x_2^2}\right) \tag{42}$$

In both cases, for small deformations ($\lambda \approx 1 + \varepsilon$, with ε being very small), the two tensors coincide with the tensor ε_{ij} in equation (23) which, for this example, become:

$$E_{11} = \eta_{11} = \varepsilon_{11} = \lambda_1 - 1 \qquad E_{22} = \eta_{22} = \varepsilon_{22} = \lambda_2 - 1 \tag{43}$$

The *Cauchy stress tensor* is, following equations (28) and (32),

$$\sigma_{11} = \frac{f_1}{l_2 h} \qquad \sigma_{22} = \frac{f_2}{l_1 h} \qquad \text{and} \qquad \sigma_{12} = \sigma_{21} = 0 \tag{44}$$

The *first Piola-Kirchhoff tensor* is given, from equations (29) and (32), or alternatively, equation (33), by

$$P_{11} = \frac{f_1}{L_2 H} \qquad P_{22} = \frac{f_2}{L_1 H} \qquad \text{and} \qquad P_{12} = P_{21} = 0 \tag{45}$$

The *second Piola-Kirchhoff tensor* is, according to equations (30) and (32), or alternatively equation (34)

$$S_{11} = \frac{1}{\lambda_1}\frac{f_1}{L_2 H} \qquad S_{22} = \frac{1}{\lambda_2}\frac{f_2}{L_1 H} \qquad \text{and} \qquad S_{12} = S_{21} = 0 \tag{46}$$

with the forces in the reference configuration equation (31):

$$\mathbf{T}_1 = \frac{f_1}{\lambda_1}\mathbf{e}_1 \qquad \text{and} \qquad \mathbf{T}_2 = \frac{f_2}{\lambda_1}\mathbf{e}_2 \tag{47}$$

5.2.4. Example: Extension and inflation of a thin tube

The extension and inflation of a homogeneous thin-walled tube made from an isotropic material is now considered. The process is represented in figure 5.6a.

The tube is taken to be right and infinitely long, loaded uniformly by axial forces $\mathbf{f}$ and an internal pressure p.

We assume that the tube remains right and maintains a circular cross-section, undergoing an affine deformation process. This is the case for an isotropic material and also for an anisotropic material provided it has material symmetry with respect to the 3-axis[21].

In this development two different coordinate systems will be used, the cartesian formed of the 1,2 and 3 axes (figure 5.6b) and the cylindrical (r, θ, z) (figure 5.6c), both centred on the axis of the tube. In this way, as well as familiarising the reader with the representation of both systems, we will take advantage, as the occasion arises, of the symmetry of the problem.

In the *reference* configuration we assume that the tube has a thickness H, radius R and length L. In the *deformed* configuration, the values will be h, r and l.

If $(\mathbf{e}_1, \mathbf{e}_2, \mathbf{e}_3)$ is the standard basis for the *cartesian coordinate system* (figure 5.6b), the position of a point $\mathbf{X}=X_1\mathbf{e}_1+ X_2\mathbf{e}_2+X_3\mathbf{e}_3$ in reference configuration after deformation will be:

$$\mathbf{x} = \lambda_\theta X_1\mathbf{e}_1 + \lambda_\theta X_2\mathbf{e}_2 + \lambda_3 X_3\mathbf{e}_3 \qquad (48)$$

which we can recognise as the set of equations (1). The non dimensional parameters λ_θ and λ_3 define the state of deformation of the tube at a particular instant, and reflect the fact that the tube is homogeneous. Observe also that, by the symmetry of the problem, the deformation along the axes $\mathbf{e}_1$ and $\mathbf{e}_2$ at any point must be equal. In equation (48) we call λ_θ –for reasons that will become apparent shortly– the unit elongation in both directions. The dimensions of the tube will then be $r=\lambda_\theta R$ and $l=\lambda_3 L$,

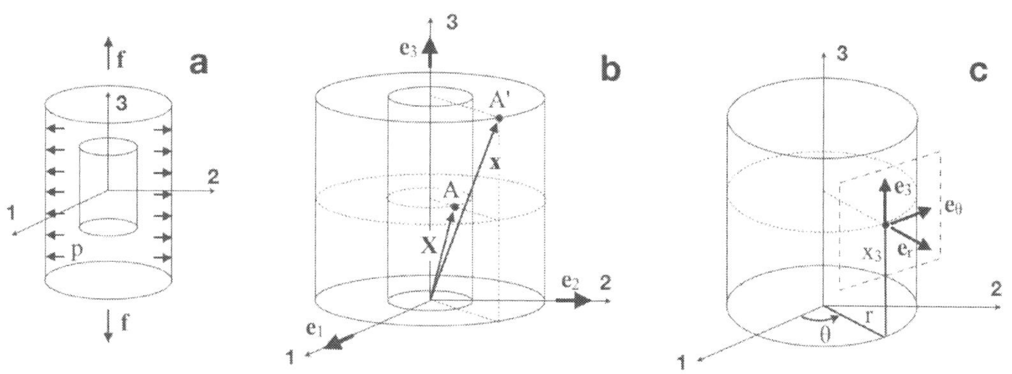

Fig. 5.6. *Extension and inflation of a thin tube.*
a. Loading.
b. Cartesian coordinates.
c. Cylindrical coordinates.

[21] This material is termed transversely isotropic, and its behaviour is the same in all directions situated in planes perpendicular to its axis of symmetry.

From equations (5), (14) and (15), it is possible to obtain the strain tensors. In our case, and given that we are using a three-dimensional coordinate system, the indices (i,j) take on the values 1, 2 and 3. A simple calculation leads to

$$\frac{\partial \kappa}{\partial \mathbf{X}} = \mathbf{F} = \begin{bmatrix} \lambda_\theta & & \\ & \lambda_\theta & \\ & & \lambda_3 \end{bmatrix} \qquad (49)$$

and Green and Almansi strain tensors

$$\mathbf{E} = \frac{1}{2}\begin{bmatrix} \lambda_\theta^2 - 1 & & \\ & \lambda_\theta^2 - 1 & \\ & & \lambda_3^2 - 1 \end{bmatrix} \quad \eta = \frac{1}{2}\begin{bmatrix} 1 - 1/\lambda_\theta^2 & & \\ & 1 - 1/\lambda_\theta^2 & \\ & & 1 - 1/\lambda_3^2 \end{bmatrix} \qquad (50)$$

The three-dimensional approach here is different from the two-dimensional situation considered previously for the membranes represented in figure 5.5, since due to the curvature of the tube no pair of coordinate axes is contained within the surface of the membrane. The strains obtained using equations (50) are referred to cartesian axes which are not parallel to the membrane, and the values will not be directly comparable to the results of other tests such as –for example– the biaxial tensile test discussed in the examples above.

Given that the tensors in equation (50) are diagonal, and also have the same values for axes 1 and 2, any other orthonormal reference system which contains the 3-axis will provide the same values, since any line perpendicular to the 3-axis is a principal axis of strain. In other words, if we rotate the 1 and 2 axes together by an arbitrary angle about the 3-axis, in order to obtain the axes 1', 2' and 3, the value of the tensors represented in the new coordinate system will be exactly the same as with the 1, 2 and 3 axes [22].

In particular, if we consider the cylindrical axes represented in figure 5.6c, the above rationale leads to the unit elongation in the circumferential (defined as $\mathbf{e}_\theta$) and the radial (defined as $\mathbf{e}_r$) directions –directions rotated from the 1 and 2 axes– being the same for the two axes, equal to λ_θ, (that is to say, $\lambda_r = \lambda_\theta$), and

$$h = \lambda_r H = \lambda_\theta H \qquad (51)$$

If, in addition, conservation of volume is considered

$$2\pi\, r\, l\, h = 2\pi\, R\, L\, H \qquad (52)$$

so

$$h = \frac{H}{\lambda_\theta \lambda_3} \qquad (53)$$

which only agrees with equation (50) in the particular case that $\lambda_3 = 1/\lambda_\theta^2$.

If we set out, the problem using the *cylindrical coordinates* shown in figure 5.6c, we define the points $\mathbf{X}$ in the reference configuration by their coordinates (R,Θ,X_3) and those of the deformed configuration $\mathbf{x}$ by (r,θ, x_3). The relationship between the coordinates of the corresponding points, according to the assumptions made, are

$$\begin{aligned} r &= \lambda_\theta R \\ \theta &= \Theta \\ x_3 &= \lambda_3 X_3 \end{aligned} \qquad (54)$$

Expressed in terms of the axes $(\mathbf{e}_r, \mathbf{e}_\theta, \mathbf{e}_3)$, the position vectors are

$$\mathbf{X}(R,\Theta,Z) = R\mathbf{e}_r + X_3\mathbf{e}_3 \qquad (55)$$

for the reference configuration, and for the deformed

$$\mathbf{x}(r,\theta,z) = r\mathbf{e}_r + x_3\mathbf{e}_3 = \lambda_\theta R\mathbf{e}_r + \lambda_3 X_3\mathbf{e}_3 \qquad (56)$$

The *deformation gradient tensor* $\mathbf{F} = \partial\kappa(\mathbf{X})/\partial\mathbf{X}$ is obtained from equation (5), expressed in cylindrical coordinates (Humphrey, 2002, appendix II), yielding:

$$\mathbf{F} = \begin{bmatrix} \dfrac{\partial r}{\partial R} & \dfrac{1}{R}\dfrac{\partial r}{\partial \Theta} & \dfrac{\partial r}{\partial X_3} \\ r\dfrac{\partial \theta}{\partial R} & \dfrac{r}{R}\dfrac{\partial \theta}{\partial \Theta} & r\dfrac{\partial \theta}{\partial X_3} \\ \dfrac{\partial x_3}{\partial R} & \dfrac{1}{R}\dfrac{\partial x_3}{\partial \Theta} & \dfrac{\partial x_3}{\partial X_3} \end{bmatrix} = \begin{bmatrix} \lambda_r = \lambda_\theta & & \\ & \lambda_\theta & \\ & & \lambda_3 \end{bmatrix} \qquad (57)$$

which, as we now know, results in the strains in equation (52).

[22] This result can be proved by applying to the tensors in equation (52) $\mathbf{Q}_\beta$ about the 3-axis by an angle β, $\mathbf{Q}_\beta = \begin{bmatrix} \cos\beta & \sin\beta & \\ -\sin\beta & \cos\beta & \\ & & 1 \end{bmatrix}$ and bearing in mind that a tensor $\mathbf{A}$ transforms under rotation as $\mathbf{A}' = \mathbf{QAQ}^T$

We see once again –this time in cylindrical coordinates– that if we consider the three-dimensional problem with the displacements in equation (54) the conservation of volume hypothesis is not satisfied. However now, given that the plane $(\mathbf{e}_\theta, \mathbf{e}_3)$ is always parallel to the surface of the tube, we can decouple the dimensions in this study and force the conservation of volume condition of equation (52). With this approach, the deformation of the tube can be analysed by decomposing the movement into that of the tangent plane $(\mathbf{e}_\theta, \mathbf{e}_3)$ (= two-dimensional displacements) together with a deformation in the thickness direction of the tube $(\mathbf{e}_r)$ which is given by the unit elongation λ_r. We can then force the condition that the unit elongation λ_r need not be equal to λ_θ as was obtained from equation (57), without requiring satisfaction of the condition

$$\frac{h}{H} = \lambda_r = \frac{1}{\lambda_\theta \lambda_3} \quad (58)$$

which is a sufficient approximation for thin tubes.

Under these conditions, the deformation gradient tensor in the tangent plane $(\mathbf{e}_\theta, \mathbf{e}_3)$ will be

$$\frac{\partial \kappa}{\partial \mathbf{X}} = \mathbf{F} = \begin{bmatrix} \frac{r}{R}\frac{\partial \theta}{\partial \Theta} & r\frac{\partial \theta}{\partial X_3} \\ \frac{1}{R}\frac{\partial x_3}{\partial \Theta} & \frac{\partial x_3}{\partial X_3} \end{bmatrix} = \begin{bmatrix} \lambda_\theta & \\ & \lambda_3 \end{bmatrix} \quad (59)$$

and the Green and Almansi tensors

$$\mathbf{E} = \frac{1}{2}\begin{bmatrix} \lambda_\theta^2 - 1 & \\ & \lambda_3^2 - 1 \end{bmatrix} \text{ and } \eta = \frac{1}{2}\begin{bmatrix} 1 - 1/\lambda_\theta^2 & \\ & 1 - 1/\lambda_3^2 \end{bmatrix} \quad (60)$$

which indicate that the circumferential $(\mathbf{e}_\theta)$ and axial $(\mathbf{e}_3)$ directions are stretched, with elongations λ_θ and λ_3 respectively. The deformed thickness h is obtained from equation (58).

The displacement vector $\mathbf{u} = \mathbf{x} - \mathbf{X}$ becomes

$$\mathbf{u} = (\lambda_\theta - 1)R\mathbf{e}_r + (\lambda_3 - 1)X_3\mathbf{e}_3 \quad (61)$$

and the small strain tensor –equation (23)– is equal to

$$\varepsilon = \begin{bmatrix} \lambda_\theta - 1 & \\ & \lambda_3 - 1 \end{bmatrix} \quad (62)$$

which coincides with the value of $\mathbf{E}$ and η for small values of λ_θ and λ_3.

The Cauchy stress in the *axial direction* σ_{33} can be calculated directly from its definition in equation (28)

$$\sigma_{33} = \frac{\mathrm{f}}{2\pi r h} = \frac{\mathrm{f}\lambda_3}{2\pi R H} \quad (63)$$

in which we have used equations (54) and (58).

The Cauchy stresses in the *circumferential direction* $\sigma_{\theta\theta}$ can be deduced from the Laplace law, equation (27) taking into account the fact that the radius of curvature in the circumferential direction is r and its value in the axial direction tends to infinity (for a right tube)

$$\sigma_{\theta\theta} = \frac{pr}{h} = p\frac{\lambda_\theta^2 \lambda_3 R}{H} \quad (64)$$

From the symmetry of the problem $\sigma_{\theta 3} = 0$, so that the Cauchy stress tensor is

$$\sigma = \begin{bmatrix} p\frac{\lambda_\theta^2 \lambda_3 R}{H} & \\ & \mathrm{f}\frac{\lambda_3}{2\pi R H} \end{bmatrix} \quad (65)$$

and the Piola-Kirchhoff tensors, deduced from equations (33) and (34) become

$$\mathbf{P} = \begin{bmatrix} p\frac{\lambda_\theta \lambda_3 R}{H} & \\ & \frac{\mathrm{f}}{2\pi R H} \end{bmatrix} \qquad \mathbf{S} = \begin{bmatrix} p\frac{\lambda_3 R}{H} & \\ & \frac{\mathrm{f}}{2\pi \lambda_3 R H} \end{bmatrix} \quad (66)$$

The stress in the *radial direction* σ_{rr} is not defined in the two-dimensional approximation that we have used. However, we can narrow down its value given that on the internal surface of the tube it must be equal to the internal applied pressure (- *p*) and zero on the exterior surface. Given the small value of the thickness of the tube, it is usually considered in the stress analysis that the radial stress does not diverge much from these two values, and frequently its value is ignored compared to the much higher values of $\sigma_{\theta\theta}$ and σ_{33}.

5.2.5. Constitutive equations

So far, we have considered the mathematical description of the deformation of a solid, the definition of the state of stress, and have derived the equilibrium equations for a differential element, but we said nothing about the nature of the material. It seems obvious that the mechanical behaviour will be very different if we examine the deformation of the mesoglea of an anemone or the dilation of an aorta, differences which come from the distinct natures of these materials.

On the other hand, the information provided by the three equilibrium equations and the conservation of mass equation are not sufficient to provide a knowledge at every point in the body of the seven variables necessary to solve the problem: the three components of the stress tensor, the two components of the displacement vector, the thickness and the density (which can depend upon position). The equations must be supplemented by other relationships which characterise the physical properties of the material in question; these equations are known as the *constitutive equations*.

In general, the theories for the development and application of constitutive equations are quite complex, and beyond the scope of this book. Sufficient to say that the constitutive equation describes a physical property of the material and must, therefore, be independent of the reference coordinate system. Consequently, it has to be a tensorial equation.

In this book we will only consider some very simple examples of constitutive equations, in order to give the reader an idea of how to proceed in other cases. Bear in mind that, in general, the deformations in soft biological materials will be large (not small, as in hard structural materials), possibly depend upon loading history, temperature, humidity and the loading rate and, in some cases, may produce rupture of the material.

Constitutive equations are significantly simplified when there is no residual deformation (when the stress is removed, the strain is also zero). Even so, its dependency on the *time* variable through the loading history (as in the viscoelastic examples discussed in our earlier study of fibres) makes them complicated because, in the stress-strain curves, the loading curve does not usually coincide with that of unloading. In these cases, the concept of *pseudo-elasticity* can be useful.

Another step towards the simplification of these equations is when they don't depend upon time, and the loading and unloading branches of the stress-strain curve coincide, that is to say, there is a one-to-one correspondence between stresses and strains. In these circumstances the material behaves *elastically,* and it can be shown that the stresses can be derived from a potential, either the internal energy U, or the free energy $F = U - TS$ per unit volume (see, for example, Capaldi, 2012).

In the particular case where the strains depend only upon mechanical causes (when there is a function for the energy which only depends upon the deformation), the constitutive equations are simplified even further. Such materials are usually called *hyperelastic*.

Until recently, the constitutive equations for soft biological materials were hardly known. This lack of information has impeded the development of this branch of Materials Science; unable to model these materials, devise experiments, or interpret their results and make predictions. Experience with the continuum mechanics of metallic materials, ceramics and elastomers has not been sufficient to address the study of tissue and blood vessels. Later we shall briefly describe some of the simpler experiments and their results, which can help model the mechanical behaviour of these complex materials.

Hyperelastic constitutive equation

If W_m is the energy, per unit mass, due to the deformation of the body and ρ_0 the density of the body in the reference configuration, $\mathcal{W}=\rho_0 W_m$ will be the energy due to the deformation per unit volume of the body in the reference configuration. This energy coincides with the internal energy per unit volume in isentropic processes and with the free energy per unit volume in isothermal processes.

If $\mathcal{W}=f(E_{ij})$ is expressed as a function of the components of the Green-Lagrange strain tensor E_{ij} equation (14) (recall that in this case the strains are taken with respect to the reference configuration), it can be shown (see, for example, Y. Fung, 1993) that:

$$\mathbf{S} = \frac{\partial \mathcal{W}}{\partial \mathbf{E}} \quad \text{or} \quad S_{ij} = \frac{\partial \mathcal{W}}{\partial E_{ij}} \tag{68}$$

which indicates that the second Piola-Kirchhoff stress tensor is the thermodynamic conjugate of the Green-Lagrange strain tensor[23]. From this result it is

[23] When this expression is used, care must be taken when writing the non diagonal terms E_{ij} and E_{ji} inside the W function, since even though $E_{ij} = E_{ji}$ in each case W must be differentiated with respect to a different argument.

possible to obtain the components of the Cauchy tensor using equation (36).

$$\sigma = \frac{H}{h}\frac{1}{detF}F\frac{\partial W}{\partial E}F^T \text{ or } \sigma_{ij} = \frac{H}{h}\frac{1}{det[\partial x_\alpha/\partial X_\beta]}\frac{\partial x_i}{\partial X_l}\frac{\partial x_j}{\partial X_m}\frac{\partial W}{\partial E_{lm}}$$

In the previous paragraphs, we have assumed that the analysis is two dimensional, and that the indices i, j, α, β, *etc...* take on the values 1 and 2, which define the axes in the plane of the membrane. The only deformation considered in the 3-direction –perpendicular to the plane of the membrane– is the change of thickness h/H. If the membrane is incompressible, the change in the thickness h/H will compensate for the change of volume produced by the deformation (= change in area) in the plane of the membrane.

It is important to note that if the strain defined by $\mathbf{F}$ does not change the area ($\det\mathbf{F}=1$, equation (20)) and the function $\mathcal{W}$ does not depend upon the area of the membrane, any hydrostatic stress $q\mathbf{1}$ defined in the plane of the membrane[24] added to the result of equation 36 will satisfy the constitutive equation, since it produces no work because the area of the membrane remains constant. In this case the correct expression for calculating the Cauchy stress tensor σ at a given point will be

$$\sigma = \frac{H}{h}F\frac{\partial W}{\partial E}F^T + q\mathbf{1} \quad \text{or}$$

$$\sigma_{ij} = \frac{H}{h}\frac{\partial x_i}{\partial X_l}\frac{\partial x_j}{\partial X_m}\frac{\partial W}{\partial E_{lm}} + q\delta_{ij} \qquad (68a)$$

where the value of the parameter q must be determined from the boundary or equilibrium conditions.

In this case, the corresponding value of the second Piola-Kirchhoff stress tensor is

$$S = \frac{\partial W}{\partial E} + qF^{-1}F^{-T} \quad \text{or} \quad S_{ij} = \frac{\partial W}{\partial E_{ij}} + q\frac{\partial X_i}{\partial x_l}\frac{\partial X_j}{\partial x_l}$$

The stresses through-the-thickness ($\sigma_{13}, \sigma_{23}, \sigma_{33}$) are not considered in the two-dimensional calculation and must be estimated separately (or neglected), as was shown in example 5.2.4.

A frequent alternative, mostly in non-planar membranes such as the walls of blood vessels, is to develop directly the three-dimensional problem in three spatial coordinates in the deformed and reference configuration $\mathbf{x}(x_1, x_2, x_3)=\kappa[\mathbf{X}(X_1, X_2, X_3)]$, and use equation (5a) to obtain the three-dimensional deformation gradient tensor $\mathbf{F}$. This process was already used in the example in Section5.2.4 to obtain equation (52).

The three-dimensional analysis leads again to equation (68), which remains valid but with the indices taking on the values 1, 2 and 3, and the stresses and strains obeying the same conventions and notation as those used in the two-dimensional case. Thus, the components of the second Piola-Kirchhoff tensor $\mathbf{S}$ will be (S_{11}, S_{12}, S_{22}) in the plane of the membrane and (S_{13}, S_{23}, S_{33}) through-the-thickness. Note that now $det\mathbf{F}$ becomes the membrane volumetric strain dv/dV, which incorporates the variation due to the thickness (included in an independent form) by means of the term h/H in equation (36). Then the Cauchy stress tensor is given by

$$\sigma = \frac{1}{detF}F\frac{\partial W}{\partial E}F^T \quad \text{or}$$

$$\sigma_{ij} = \frac{1}{det[\partial x_\alpha/\partial X_\beta]}\frac{\partial x_i}{\partial X_l}\frac{\partial x_j}{\partial X_m}\frac{\partial W}{\partial E_{lm}} \qquad (68b)$$

with components ($\sigma_{11}, \sigma_{12}, \sigma_{22}$) in the plane of the membrane and ($\sigma_{13}, \sigma_{23}, \sigma_{33}$) through the thickness.

In a parallel manner to that in the two-dimensional situation, when a deformation which conserves volume –and therefore $det\mathbf{F}=1$– and also when an energy deformation function $\mathcal{W}$ is not sensitive to the volume change of the membrane, each stress given by the expression

$$\sigma = F\frac{\partial W}{\partial E}F^T + q\mathbf{1} \quad \text{or}$$

$$\sigma_{ij} = \frac{\partial x_i}{\partial X_l}\frac{\partial x_j}{\partial X_m}\frac{\partial W}{\partial E_{lm}} + q\delta_{ij} \qquad (68c)$$

will be a solution to the constitutive equation, since an arbitrary three-dimensional hydrostatic stress[25] $q\mathbf{1}$ will produce no work on the membrane while its volume is conserved. The value of the parameter q is determined, as mentioned above, using the boundary or equilibrium conditions of the problem. In this case it is not necessary to make any estimate of, or hypotheses about, the through-the-thickness stresses, since these are all obtained directly from equation (68c).

In a similar manner, the corresponding value of the second Piola-Kirchhoff stress tensor can also be expressed as:

$$S = \frac{\partial W}{\partial E} + qF^{-1}F^{-T} \quad \text{or} \quad S_{ij} = \frac{\partial W}{\partial E_{ij}} + q\frac{\partial X_i}{\partial x_l}\frac{\partial X_j}{\partial x_l}$$

[24] $\mathbf{1}$ is the two-dimensional unity tensor in the plane of the membrane. The parameter q may be a function of position.

[25] Observe that here $\mathbf{1}$ is the three-dimensional unit tensor.

Not all elastic materials can be described by means of a strain energy function, those in which it is possible are called *hyperelastics*, as already noted above. Observe that equation (68) is biunivocal: for every state of stress, there is a corresponding strain state and vice versa.

Biological materials, and in particular tissues, are not perfectly elastic. However, in some circumstances, when it is possible to precondition them, they can be treated as *pseudo-elastic*, in the sense discussed previously. In these cases, it is possible to propose a *pseudo-strain energy* function, valid for the increasing strain branch and another for the decreasing strain of the stress-strain response. In order to illustrate and cement these ideas, we will summarise, below, the development of a *pseudo-constitutive equation* for the epithelium of a rabbit, as obtained by Tong and Fung (1976) using data from Lanir and Fung (1974).

The authors began by proposing the following empirical expression for $\mathcal{W}=\rho_0 W_m$:

$$\mathcal{W} = \frac{1}{2}\{f(\alpha,\mathbf{E}) + c\exp[F(a,\mathbf{E})]\} \tag{69}$$

where

$$f(\alpha,\mathbf{E}) = \alpha_1 E_{11}^2 + \alpha_2 E_{22}^2 + \alpha_3 E_{12}^2 + \alpha_3 E_{21}^2 + 2\alpha_4 E_{11} E_{22} \tag{70}$$

$$F(a,\mathbf{E}) = a_1 E_{11}^2 + a_2 E_{22}^2 + a_3 E_{12}^2 + a_3 E_{21}^2 + 2a_4 E_{11} E_{22} \tag{71}$$

From equations (67) and (69), the components of the second Piola-Kirchhoff tensor, S_{ij}, are given by

$$S_{11} = \alpha_1 E_{11} + \alpha_4 E_{22} + cA_1 B$$

$$S_{22} = \alpha_4 E_{11} + \alpha_2 E_{22} + cA_2 B$$

$$S_{12} = \alpha_3 E_{12} + ca_3 E_{12} B \tag{72}$$

where

$$A_1 = a_1 E_{11} + a_4 E_{22} \quad A_2 = a_4 E_{11} + a_2 E_{22}$$

and

$$B = \exp[F(a,\mathbf{E})] \tag{73}$$

Calculation of the constants α_1, α_2, α_3, α_4, a_1, a_2, a_3, a_4 and c is achieved from experimental results in the biaxial tests discussed above. In this case $S_{12}=0$, since $E_{12}=0$.

Two experimental curves (pre-conditioned, as discussed previously) are shown in figure 5.7a; one in which λ_2 is held constant ($\lambda_2 = 1$) and λ_1 varies, and the other in which λ_2 varies while keeping λ_1 constant ($\lambda_1 = 1$).

In order to calculate the constants, two points have been selected in each test: one where the force varies rapidly with the elongation (points A and C), and the other where the force varies slowly with the elongation (points B and D). At each point the values of the coordinates and slope were measured and equated to the values of S_{11} and S_{22} and their corresponding derivatives with respect to E_{11} and E_{22}. (See Tong and Fung, 1976 for details. Here, the aim is only to outline the way forward).

With the values obtained, the function $\mathcal{W}(\mathbf{E};\alpha_i,a_i,c)$ for the rising part of the force-elongation curve is now known, and from it good agreement with the two experiments is achieved, as shown in figure 5.7b. Authors Tong and Fung noted that it is not surprising that the agreement was good, since four points on the curve were used, but it is remarkable that the *whole* curve was reproduced quite well given the nonlinear behaviour of the epithelium.

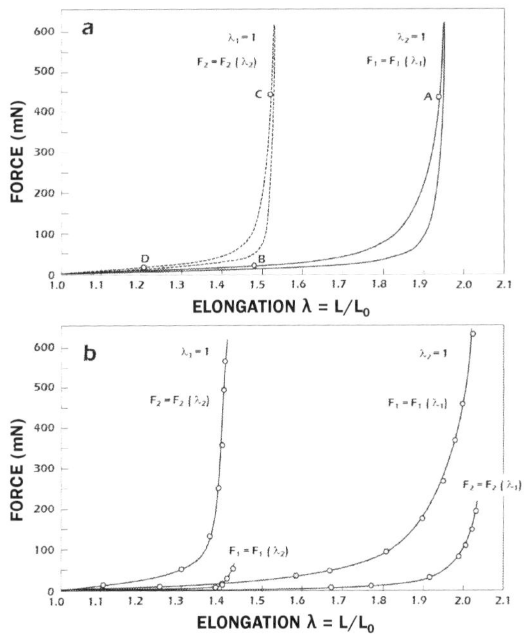

Fig. 5.7. *a. Force-elongation curves, for rabbit epithelium, corresponding to two pre-conditioned biaxial tests (Lanir and Fung, 1974).*
b. Numerical results (o) obtained from the data in figure 5.7a (see the explanation in the text) (Tong and Fung, 1976).

Pseudo-constitutive equations, similar to but even simpler than equation (69), have been developed for other tissues, by ignoring the first term, that is:

$$\mathcal{W} = c\exp\left[F(a,\mathbf{E})\right] \tag{74}$$

For artery walls and for pulmonary tissue, Fung and co-workers have developed equations like (74) and which work reasonably well (Fung 1975, Fung *et al.* 1978, Vawter *et al.* 1979).

An ambitious objective, when constitutive equations need to be formulated, is to relate the structure of the material to its properties. In general, the aim is to go from bottom to top; starting with the properties of the molecules and microscopic structure to deduce the macroscopic behaviour. Or, starting with the macroscopic behaviour, to try to understand the role played by the microscopic constituents of the material. In either case, the intention is to enhance knowledge of the relationship between the structure and mechanical behaviour.

To cite some examples along these lines, Viidik (1968, 1990) developed a tissue model based on collagen fibres which are progressively loaded by increasing the deformation. Lanir (1979) improved Viidik's first model by assuming that tissue is a material composed of elastin and collagen fibres, which come under load at different stages in the deformation process. Shoemaker and Fung (1986) have developed a similar model for human skin and the pericardium of a dog. Despite obtaining good results, the authors recognised that there was a lack of information on the microstructure of these materials and their fibres.

Following another approach, not based on microstructural data, Kwan and Woo (1989) developed models which reproduce the nonlinear relationships between stresses and strains for tendons and ligaments. With an undulating form of collagen fibres, they proposed a model in which each fibre is formed of two components: a very flexible one, which reproduces the initial part of the stress-strain curve, and the other, more rigid, which captures the final, stiffer part of the response. The model consists of a judiciously chosen bundle of fibres which is capable of reproducing the experimental results reasonably well.

EXERCISE 39

For pulmonary tissue, Fung proposed the following equation for the internal energy, based on previously expressed ideas (Fung 1975).

$$U = c\ \exp(a_1E_{11}^2 + a_2E_{22}^2 + 2a_4E_{11}E_{22})$$

in which it was assumed that the through-the-thickness strain could be ignored, compared to E_{11} and E_{22} and behaviour practically bidimensional.

The strain energy, as already mentioned, coincides with the internal energy U in an isentropic process, and with the free energy in an isothermal process.

Under these conditions, calculate the corresponding stress tensor as a function of the material constants (a_1, a_2, a_4 and c).

SOLUTION:

From equation (66) we obtain the stresses in the second Piola-Kirchhoff tensor

$$S_{11} = 2c(a_1E_{11} + a_4E_{22})\exp(a_1E_{11}^2 + a_2E_{22}^2 + 2a_4E_{11}E_{22})$$

$$S_{12} = 0$$

$$S_{22} = 2c(a_2E_{22} + a_4E_{11})\exp(a_1E_{11}^2 + a_2E_{22}^2 + 2a_4E_{11}E_{22})$$

In order to obtain the Cauchy stresses from equation (36), we need to know the deformation gradient tensor, $\mathbf{F} = \partial\kappa/\partial\mathbf{X}$, which depends upon the type of deformation $\mathbf{x} = \kappa(\mathbf{X},t)$ to which the solid is subjected.

EXERCISE 40

Calculate directly the value of the strain energy in a tensile test on an elastic material as a function of the engineering and true stresses and strains in the test. Compare its value with that obtained from equation (68). Hint: use the results of the example in Section 5.2.3.

SOLUTION:

The example analyses the case of a state of biaxial tension. For a uniaxial tensile test $f_1 = F$ and $f_2 = 0$.

If the material is elastic and there are no dissipative processes, the work Fdl_1 done by the tensile force F is stored entirely in the form of elastic strain energy. The value of the energy per unit volume in the reference configuration (= initial, undeformed) $\mathcal{W}$ will be the integral

$$\mathcal{W} = \frac{1}{L_1L_2H}\int_{L_1}^{l_1} Fdl_1 = \int_{L_1}^{l_1}\frac{F}{L_2H}d\left(\frac{l_1}{L_1}\right) = \int_1^{\lambda} sd\lambda$$

in which $L_1\ L_1\ H$ is the initial section, $s = F/(L_2H)$ the engineering stress and $\lambda = l_1/L_1$ the elongation of the specimen. Since, $\lambda = 1 + e$, and $e = (l_1 - L_1)/L_1$ the engineering strain, we have $d\lambda = de$ so that the resulting value is $\mathcal{W} = \int_1^{\lambda} sd\lambda = \int_0^{e} sde$ which is the area under the engineering stress-strain curve (s, e).

If we refer to the elastic strain energy in the instantaneous configuration, the result is

$$\omega = \frac{1}{l_1l_2h}\int_{L_1}^{l_1} Fdl_1 = \int_{L_1}^{l_1}\frac{F}{l_2h}\frac{dl_1}{l_1} = \int_0^{\varepsilon}\sigma d\varepsilon$$

where $\sigma = F/(l_2h)$ is the true stress and $d\varepsilon = dl_1/l_1$ the differential of the true strain (see the section dedicated to the tensile test in the biological fibres section). Now we recognise the value of ω, the elastic strain energy per unit volume in the instantaneous configuration, as the area under the true stress-strain curve (σ, ε).

In order to apply equation (68) $S_{ij} = \partial\mathcal{W}/\partial E_{ij}$ which relates the energy per unit volume in the reference configuration and the second Piola-Kirchhoff stresses, we will use equations (46) which lead to

$$S_{11} = \frac{1}{\lambda_1}\frac{F}{L_2H} = \frac{s}{\lambda} \qquad S_{22} = S_{12} = S_{21} = 0$$

From equation (68) we obtain

$$d\mathcal{W} = \frac{\partial\mathcal{W}}{\partial E_{ij}}dE_{ij} = S_{ij}dE_{ij} = S_{11}dE_{11} = \frac{s}{\lambda}dE_{11}$$

since S_{11} is the only nonzero term. The Green-Lagrange strain is obtained from the value just calculated

$$E_{11} = \frac{1}{2}\left(\lambda_1^2 - 1\right)$$

and differentiating it we see $dE_{11} = \lambda_1 d\lambda_1 = \lambda d\lambda$, which leads to $\mathcal{W} = \int d\mathcal{W} = \int_1^{\lambda} sd\lambda$ coincident with the result obtained directly.

If $J = l_1l_1h/(L_1L_2H)$ is the ratio of the deformed and undeformed volumes, the strain energy per unit volume in the deformed configuration will be simply
$\omega = \mathcal{W}/J$, given by

$$\omega = \frac{\mathcal{W}}{J} = \frac{1}{J}\int_1^{\lambda} sd\lambda = \int_1^{\lambda}\frac{L_1L_2H}{l_1l_2h}\frac{F}{L_2H}d\lambda = \int_1^{\lambda}\frac{\sigma}{\lambda}d\lambda = \int_{L_1}^{l_1}\sigma\frac{dl_1}{l_1} = \int_0^{\varepsilon}\sigma d\varepsilon$$

Note that when $J = 1$, both specific energies coincide.

5.3. EXPERIMENTAL AND CHARACTERISATION TECHNIQUES

The objective of this section is the summary description of the most common two-dimensional mechanical tests used to characterise biological membranes.

The essential aspects upon which the experiments are based are emphasised, without going into details of the experimental equipment, which can be found in the cited references. Besides, there is not much point in delving into these details because data acquisition and reduction techniques evolve rapidly and soon become obsolete.

The experiments discussed below are specific to bidimensional materials. This is not meant to exclude other types of tests, such as tensile tests (already mentioned in the first section) or, tests developed for three-dimensional materials, such as indentation or fracture (which will be presented in the following section).

To begin with, tensile tests in two perpendicular directions are described. Experiments on tubes are also discussed, when these are subjected to combined traction and internal pressure, and even torsion.

Several fracture tests are also outlined: The *tear* resistance is an interesting property of tissue, in particular of the epithelium. Another rupture test – the *peel* test– provides knowledge of the adhesion energy between the membrane and the substrate. To this end, two experiments are discussed: simple peeling and the tympanum test.

5.3.1. Biaxial tensile tests

Biaxial tensile tests on soft biological material tissue are difficult to perform and analyse. To begin with, gripping the specimens is complicated if damage to the material is to be avoided and the application of forces to the edges is not easy. Measurement of the deformations must be made using optical processes (contact extensometers could disturb the measurement due to the flexibility of the specimen and the reaction of the sample, if it is alive). The anisotropy, which is almost always present in these tissues, also makes the interpretation of data a complex task.

For these experiments, thin specimens are used –less than 3 mm thick and often less than 1 mm– loaded with forces acting in the plane of the specimen. The general arrangement of the biaxial test on a membrane is shown schematically in figure 5.8. In practice the sample is almost always immersed in a saline solution and tested at a controlled temperature.

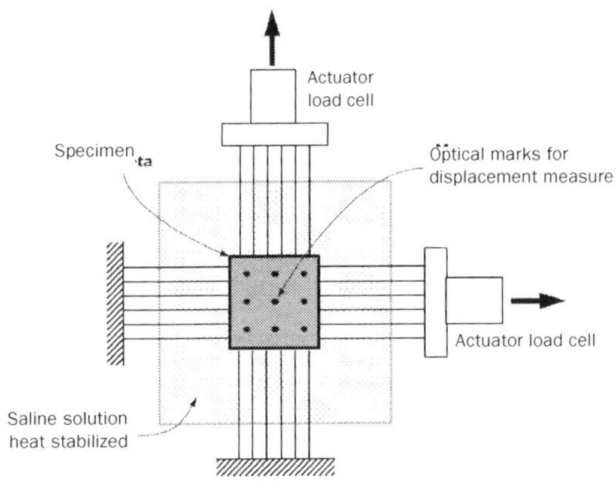

Fig. 5.8. *Schematic arrangement of a biaxial tension test.*

The loads are applied in the directions indicated in the figure and displacements measured optically using marks drawn in the centre of the membrane, in a region where it is assumed that the loading constraints have no influence, and in which it is assumed that the strains and stresses are uniform.

With this arrangement it is possible to perform a range of tests: uniaxial tension, biaxial, creep or relaxation, and dynamic. For more details, see for example Sacks (1999, 2003).

Several curves obtained from uniaxial tests on cat epithelium (Veronda and Westmann 1970), donkey and rhinoceros (Shadwick *et al.* 1992) are shown in figure 5.9. Epithelia can be modelled as if they were materials composed of elastin and collagen fibres embedded in a protein and polysaccharide matrix. In these tissues, forces are resisted primarily by the elastin and collagen fibres. The elastin fibres are much more flexible than those of collagen; they have an initial elastic modulus of about 0.6 MPa and can be strained up to 200%. The collagen fibres form a plushy mass which gradually aligns with the loading direction as the tissue deforms. Consequently, at the start of the deformation the elastin fibres are the ones that support the load, but as the deformation increases it is the stiffer collagen fibres that come under load. This behaviour is responsible for the "J-form" of the curves obtained from tensile tests.

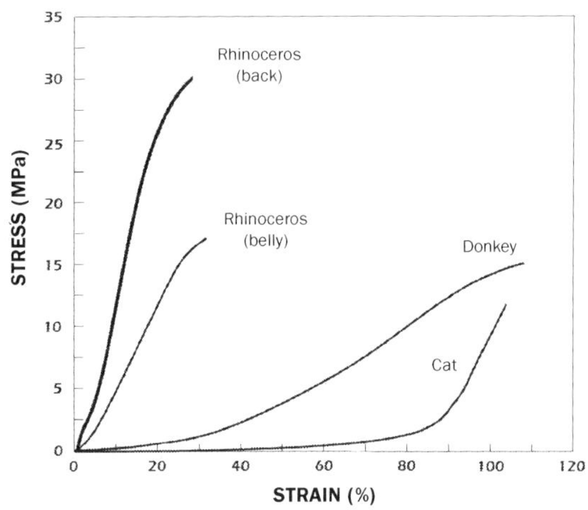

Fig. 5.9. *Typical stress-strain curves for various epithelia.*

EXAMPLE: RABBIT EPITHELIUM

An understanding of the mechanical behaviour of epithelia is of great interest to plastic surgeons, and for interpreting many pathological situations. As will be seen below, traditional tensile tests do not provide the required information, so it is necessary to resort to more complex tests, such as the biaxial ones mentioned above.

To a first approximation, skin can be considered as an incompressible material. This allows the construction of three-dimensional models –closer to reality– based on data from two-dimensional tests, in which variations in the third dimension can be deduced from the changes in the other two dimensions.

In this example, some results of Lanir and Fung's (1974), now classic, two-dimensional experiments on rabbit epithelia are summarised. Details of the equipment and specimen preparation can be found in the cited reference.

Skin specimens, measuring 35 mm by 35 mm, cut such that two edges were parallel to the axis of the body and the other two, perpendicular to it, were tested. Tensile and relaxation tests were performed, uniaxial as well as biaxial, and the influence of the rate of loading and temperature on the data were recorded.

Figure 5.10 contains the data from several tensile tests: uniaxial tests in the direction of the x-axis (Scheme A), biaxial tests with the applied load increasing in the x-direction and while keeping the displacement along the y-axis constant (Scheme B) and similar experiments with the traction in the y-direction (Schemes C and D). The direction of the x-axis is parallel to axis of the body of the rabbit.

The experiments clearly show that the epithelium is *anisotropic* since its behaviour varies with the loading direction (compare the figures corresponding to the A and C loadings, for example). The mechanical response is *nonlinear* and shows a degree of *hysteresis*. Another relevant result is that the response of the tissue depends upon the mode of loading; that under uniaxial tension (test A, for example) is not the same as in biaxial tension (test B). These results provide a warning against using uniaxial tensile tests to predict the behaviour of the epithelium, since in practice this tissue is subjected to biaxial loading.

The referenced article also includes biaxial relaxation tests, as shown schematically in figure 5.11. In this figure the relaxation force in the x-direction is reported, after the tissue was stretched in the x-direction (by a quantity denoted by $\lambda = l/l_0$) and keeping the strain in the perpendicular y-direction constant. Here, again, it was found that the stress relaxation is different when the loading is in the y-direction (these values are not shown in the figure); it is not the same if the sides are constrained or free, in which a contraction is produced in the test. All this indicates that uniaxial relaxation tests are not truly representative because in reality the sides are always restrained, and the tissue adapts itself to the loading, reorienting the collagen fibres in order to reduce both the stress in the loading direction and the strain in the transverse direction.

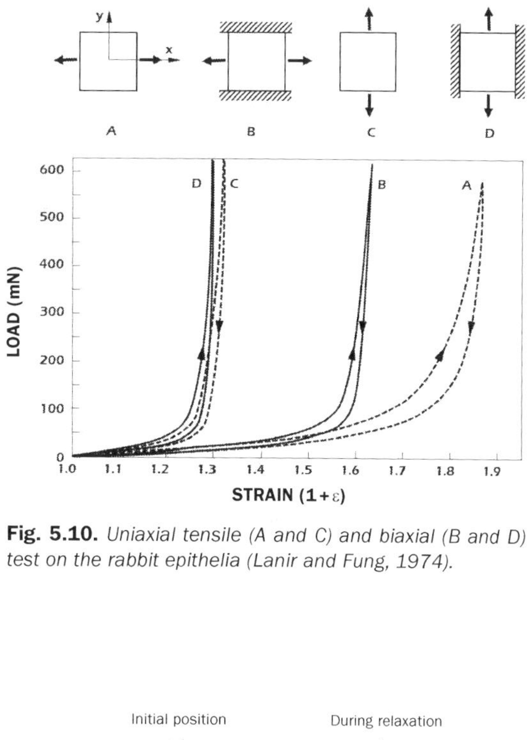

Fig. 5.10. *Uniaxial tensile (A and C) and biaxial (B and D) test on the rabbit epithelia (Lanir and Fung, 1974).*

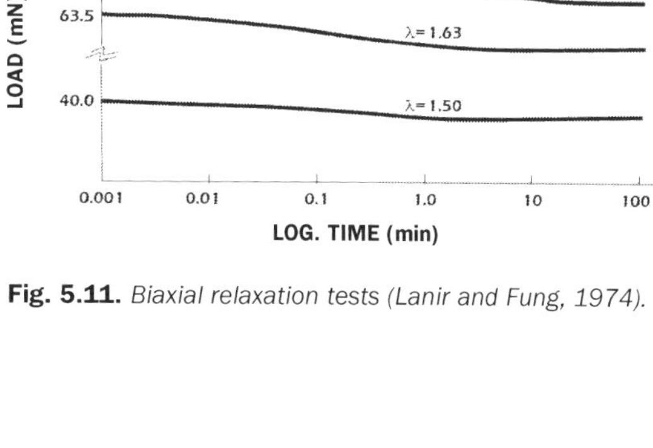

Fig. 5.11. *Biaxial relaxation tests (Lanir and Fung, 1974).*

5.3.2. Pressure test

Bulge test

An alternative means of performing biaxial tests on biological membranes consists of clamping a circular piece around its edge, subjecting it to a pressure, and measuring the displacements produced on the surface (bulge test), as shown in figure 5.12a. This is a very simple experimental technique which is, however, difficult to interpret if the membrane is not isotropic (which is almost always the case), since the anisotropy of the materials requires the application of numerical modelling (Tonge *et al.* 2013a,b). Therefore, this method is almost exclusively used to compare biomaterials and perform fatigue strength studies.

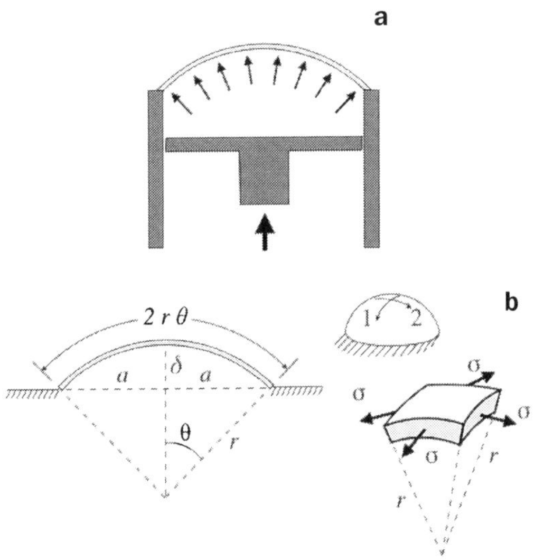

Fig. 5.12. *a. Membrane pressure test (bulge test). The membrane is subjected to a state of biaxial tension induced by the applied pressure in the interior of the apparatus. b. Simplified analysis assuming spherical deformation.*

A simplified analysis of this experiment, valid for small displacements of the membrane, which is assumed to be thin homogeneous and isotropic, can be performed by assuming that the deforming membrane takes on a spherical shape, as indicated in figure 5.12b. With the notation of the figure, we see that $sin(\theta)=a/(r-\delta)\approx a/r$ and $r cos(\theta)+\delta=r$, from which we can obtain the radius of curvature as a function of the displacement of the apex δ and the radius a of the membrane in its initial state

$$r=\frac{a^2+\delta^2}{2\delta} \tag{75}$$

With this value, we can estimate the elongations of the deformed membrane at its apex that, because of the assumed spherical deformation, will be the same in any direction. It is sufficient to refer to the initial membrane diameter *(2a)* which, in the deformed configuration, extends to an arc length equal to $2r\theta$ (figure 5.12b). Therefore

$$\lambda=\frac{2r\theta}{2a}=\frac{a^2+\delta^2}{2a\delta}arcsin\left(\frac{2a\delta}{a^2+\delta^2}\right)\simeq 1+\frac{2}{3}\frac{\delta^2}{a^2} \tag{76}$$

Taking into account equations (18) and (19) and that all of the directions on the surface of the membrane are equivalent, the strain tensor at the apex of the deformed membrane will be

$$\mathbf{E}=\eta\simeq\frac{2}{3}\frac{\delta^2}{a^2}\begin{bmatrix}1 & \\ & 1\end{bmatrix} \tag{77}$$

in which we have ignored terms of higher order than δ^2.

If we assume, as is common in biological materials, that the membrane is incompressible and volume is conserved during the deformation, we can calculate the deformed thickness h near the central apex from the initial thickness H using equation (22). If we equate this expression to 1 and solve for h taking into account that $da/dA=\lambda^2$

$$h=\frac{H}{\lambda^2}\simeq H\left(1-\frac{4}{3}\frac{\delta^2}{a^2}\right) \tag{78}$$

The Cauchy stresses σ can be obtained from the applied pressure p using Laplace's law (equation 27), since the radius of curvature is known and equal to r in every direction. The resulting value is

$$\sigma=\frac{pr}{2h}=\frac{p(a^2+\delta^2)}{4h\delta}\simeq\frac{p}{H}\left(\frac{a^2}{4\delta}+\frac{7\delta}{12}\right) \tag{79}$$

Equations (75)-(79) provide the stress-strain state of the apex of the membrane as a function of its dimensions a and H, and the displacement δ produced under the pressure p.

Pressure-diameter test

Pressure-diameter tests on blood vessels and other biological ducts are also biaxial experiments which aim to reproduce the workings of these tissues inside the body. In essence, during the test the change in diameter of the duct is measured as a function of the internal pressure. The test can be complemented by subjecting the duct to traction and, in some experiments, simultaneously to traction and torsion.

The equipment necessary for traction and internal pressure applied to blood vessels is shown, in schematic form, in figure 5.13. For more details, see Guinea *et al.* (2005). The blood vessels are tested immersed in a physiological serum at a controlled temperature. The vessel is placed in a transparent cell which allows visualisation and measurement of the diameter by optical methods. The ends of the vessel are held in the grips of a mechanical testing machine and the internal pressure is applied to the vessel by means of an auxiliary physiological serum circuit at the temperature of the experiment, driven by a compressed air pump. Pressure regulators control and stabilise the air pressure.

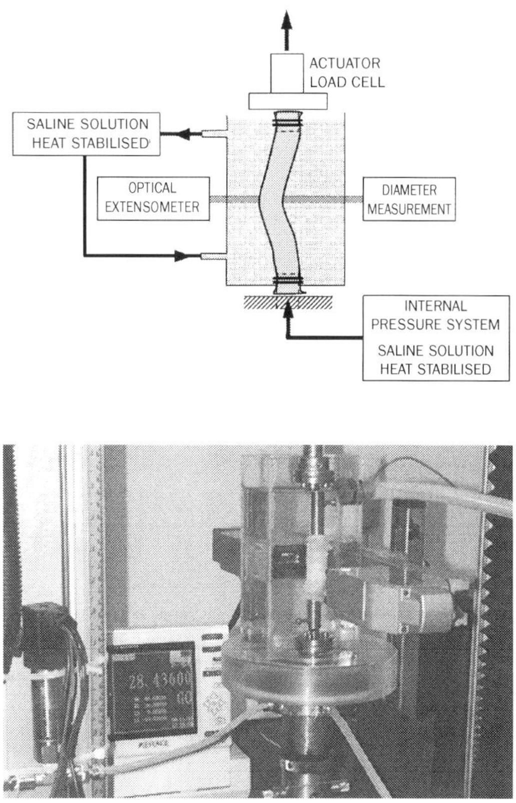

Fig. 5.13. *a) Schematic representation of an experiment for tubes subjected to traction and internal pressure. b. Experiment on an arterial aorta.*

In Fung's (1993) book, and the references cited in it, similar equipment is described for recording the variation in diameter as a function of the internal pressure, traction and torsion to which the vessel is subjected.

The process of stretching and inflation of a thin tube –which was analysed in a previous section (5.2.4)– can be applied to these experiments. If the duct wall is thick, or when residual stresses have to be considered, a complete three-dimensional analysis must be performed. This can be found in Humphrey (2002).

Two challenges in testing pressurised ducts are due to instabilities such as *buckling* or the *formation of protuberances* (known as *aneurisms* in the case of blood vessels) which change the geometry and invalidate the test. Because of their practical interest, both cases are discussed briefly below.

Buckling or the collapse of the wall of a tube generally appears at the beginning of the test, at small pressures and under axial traction. This situation is shown in figure 5.14a in a test on a carotid artery. Although buckling of the tube can only be produced for negative pressures (suction), inhomogeneities in the wall and the residual stresses may provoke its occurence with positive pressures. In general, biological ducts have only reasonably circular cross-sections above a certain minimum pressure, of around kPa (~10 mmHg) in the case of blood vessels. Figure 5.14b contains pressure-diameter curves corresponding to different values of applied axial elongation. We see that axial traction produces a stabilising effect which reduces the buckling propensity.

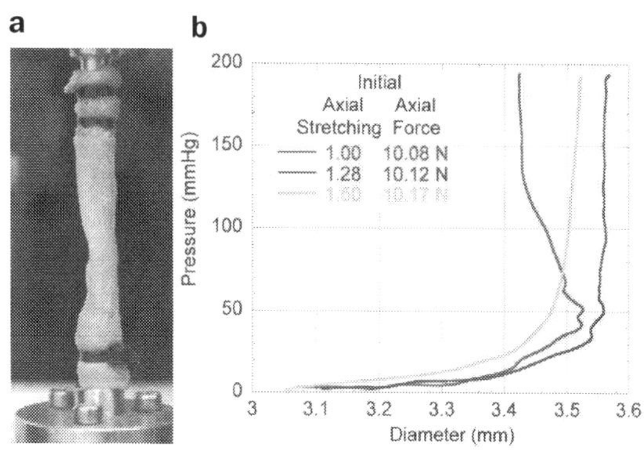

Fig. 5.14. *a. Buckling of an artery during a pressure-diameter test.*
b. Effect of the axial load on the response of the vessel.

The formation of **protuberances** in the wall is a phenomenon which is the inverse of that of buckling, because a section of the duct is dilated abnormally relative to the rest, figure 5.15a. In arterial vessels, this dilatation is known as an aneurism and is a very serious complication which can lead to incapacitation or even death. Mechanically, the formation of protuberances usually occurs because of a local weakening of the wall, which dilates due to the pressure. This is the situation of keratoconus, figure 5.15b, in which a protrusion of the cornea causes a severe loss of vision, and many aneurisms. However, it can also produce an abnormal dilatation without the presence of inhomogeneities, exclusively due to the mechanical behaviour of the material of the duct as will be seen below.

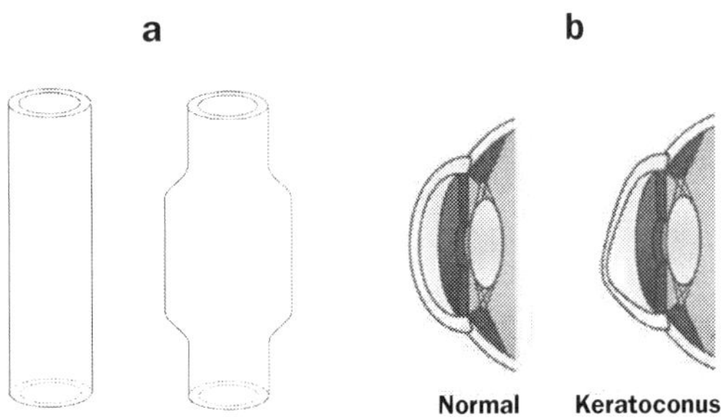

Fig. 5.15. *a. Inhomogeneous dilatation of a duct.*
b. Local dilatation of the cornea, known as keratoconus.

Figure 5.16 shows the case of a duct whose pressure-diameter curve has a local minimum. When the tube is at rest its diameter is d_O and its geometry uniform. If we apply pressure in the range $(0,\ p_B)$ the tube diameter increases uniformly d_B, which is the diameter corresponding to a unique mechanical equilibrium state. The situation changes at the instant p_B, is reached, since there are now two solutions to the mechanical problem, each one with a different value of the diameter: d_B and d_C. The points between B and C on the pressure-diameter curve are unreachable since each section of the duct has two possible diameters for a given value of pressure, and the situation is unstable.

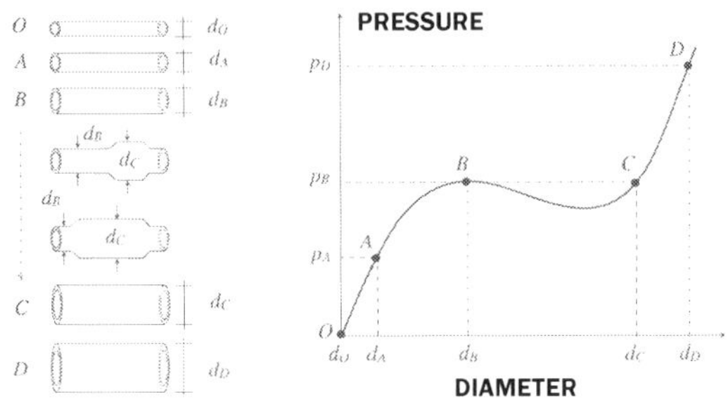

Fig. 5.16. *a) Deformation process and the appearance of protuberances in a duct.*
b. Pressure-diameter curve.

From the instant at which p_B is reached, two zones of different diameter are generated in the duct, each one corresponding to points B and C. If we try to increase the pressure in the tube, it will not exceed the value of p_B until all of the duct has transformed to the diameter d_C, corresponding to point C, after which the deformation process returns to being uniform.

We see, then, that the cause of the instability lies in the appearance of a local minimum in the pressure-diameter curve, and which, if it were to increase monotonically, the phenomenon would not occur. Certain constitutive equations, such as those of some elastomers, can give rise to this behaviour, and it is instructive to determine the pressure-diameter curve from a stretching and inflation test on these materials.

Yeoh's hyperelastic model[26] (Yeoh ,1993) is often used to model incompressible, isotropic elastomers. The strain energy per unit volume function $\mathcal{W}$ in three dimensions is

$$\mathcal{W} = \alpha tr\mathbf{E} + \beta(tr\mathbf{E})^2 + \gamma(tr\mathbf{E})^3 = \qquad (80)$$

$$= \alpha(E_{11}+E_{22}+E_{33}) + \beta(E_{11}+E_{22}+E_{33})^2 + \gamma(E_{11}+E_{22}+E_{33})^3$$

where tr is the trace operator and α, β, γ model constants.

[26] Recall that for a hyperelastic material there is a one-to-one relationship between stress and strain, which depends on neither time nor temperature.

In this example a three-dimensional analysis of the problem was performed, using expression (66) in order to obtain the second Piola-Kirchhoff stress tensor $\mathbf{S}$, since it was a hyperelastic material:

$$\mathbf{S}=\left(\alpha+2\beta tr\mathbf{E}+3\gamma\left(tr\mathbf{E}\right)^2\right)\begin{bmatrix}1&&\\&1&\\&&1\end{bmatrix}=$$

$$=\left(\alpha+2\beta\left(E_{11}+E_{22}+E_{33}\right)+3\gamma\left(E_{11}+E_{22}+E_{33}\right)^2\right)\begin{bmatrix}1&&\\&1&\\&&1\end{bmatrix} \quad (81)$$

and from equation (68c)

$$\sigma=\mathbf{FSF}^T+q\mathbf{1} \quad \text{or} \quad \sigma_{ij}=\frac{\partial x_i}{\partial X_l}\frac{\partial x_j}{\partial X_m}S_{lm}+q\delta_{ij}$$

for calculation of the Cauchy stresses σ at any point since, as discussed in section 5.2.5, the material is incompressible (= conservation of volume) and any arbitrary hydrostatic stress tension, $q\mathbf{1}$, satisfies the constitutive equation.

Applying the above results to the tube inflation and stretching experiment in section 5.2.4, we will arrive at the Cauchy stresses deduced by means of equations (81) and (68c)

$$\sigma=\left(\alpha+2\beta tr\mathbf{E}+3\gamma\left(tr\mathbf{E}\right)^2\right)\begin{bmatrix}\lambda_r^2&&\\&\lambda_\theta^2&\\&&\lambda_3^2\end{bmatrix}+q\mathbf{1} \quad (82)$$

in which we have taken into account the fact that the value of the three-dimensional deformation gradient tensor of the problem is,

$$\mathbf{F}=\begin{bmatrix}\lambda_r&&\\&\lambda_\theta&\\&&\lambda_3\end{bmatrix} \quad (82\text{bis})$$

obtained from the corresponding tensor in the two-dimensional problem –equation (59)– by adding the elongation in the radial direction λ_r.

The above stresses must coincide with those calculated from the equilibrium equation (65) in the circumferential and axial directions, which leads to:

$$p\frac{\lambda_\theta^2\lambda_3 R}{H}=\left(\alpha+2\beta tr\mathbf{E}+3\gamma\left(tr\mathbf{E}\right)^2\right)\lambda_\theta^2+q$$
$$\mathrm{f}\frac{\lambda_3}{2\pi RH}=\left(\alpha+2\beta tr\mathbf{E}+3\gamma\left(tr\mathbf{E}\right)^2\right)\lambda_3^2+q \quad (83)$$

Assuming in the radial direction $\sigma_{rr}=0$, as discussed in section 5.2.4, we will have the third equation, which allows us to obtain q

$$0=\left(\alpha+2\beta tr\mathbf{E}+3\gamma\left(tr\mathbf{E}\right)^2\right)\lambda_r^2+q \quad (84)$$

Finally, substituting into Equation (83) the values of $\mathbf{E}$ obtained from equation (82b) using equation (14), and the value $\lambda_r=1/(\lambda_\theta\lambda_3)$ given by the incompressibility condition (58) we arrive at

$$p=\frac{H}{R}\left(\frac{1}{\lambda_3}-\frac{1}{\lambda_\theta^4\lambda_3^3}\right)\left(\alpha+\beta\left(\lambda_\theta^2+\lambda_3^2+\frac{1}{\lambda_\theta^2\lambda_3^2}-3\right)+\right.$$
$$\left.+\frac{3}{4}\gamma\left(\lambda_\theta^2+\lambda_3^2+\frac{1}{\lambda_\theta^2\lambda_3^2}-3\right)^2\right) \quad (85)$$

This equation is represented in figure 5.17 which was developed for a tube of diameter D=3 mm and thickness H=0.5 mm without loading. These dimensions correspond, approximately, to those of the carotid artery. The ordinate in figure 5.17 is pressure in mmHg (1 mmHg=133 Pa, the usual unit of measurement in cardiovascular mechanics), and the abscissa has been drawn in terms of the ratio of the instantaneous and initial diameters $d/D=\lambda_\theta$.

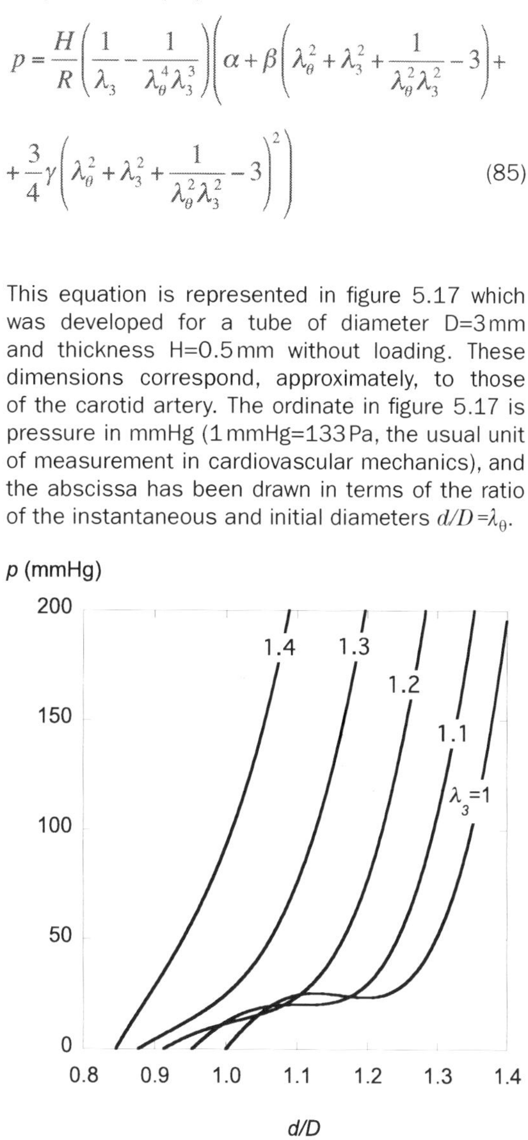

Fig. 5.17. *Pressure-diameter test for a tube subjected to different axial elongations. Initial diameter 3 mm, initial thickness 0.5 mm. Yeoh's incompressible material (α = 40 kPa, β = -142 kPa and γ = 1200 kPa).*

The parameters of the model are α = 40 kPa, β = −142 kPa and γ=1200 kPa, reasonable values for an elastomer, and which simulate the behaviour of the artery wall. In figure 5.18 the uniaxial tension curve of the material is represented. This was obtained following the same process as with the tube, making use of the results of section 5.2.3.

As we can see in figure 5.17, for the indicated geometry and material, the pressure-diameter curve has a relative minimum when the test is performed in the absence of elongation (λ_3=1), so that on reaching a pressure of ~26 mmHg, an instability will appear giving rise to two sections with different diameters. When the tube is subjected to an axial elongation, the relative minimum tends to disappear, which happens in the figure, for elongations of 1.2 and greater. In this case the strain in the tube during the whole inflation process is uniform.

Blood vessels and other biological ducts avoid the occurrence of instabilities such as those mentioned by having constitutive equations which don't give rise to pressure-diameter curves with a relative minimum. Furthermore, blood vessels are naturally stretched, with elongations λ_3 of about 1.2, which increases their stability.

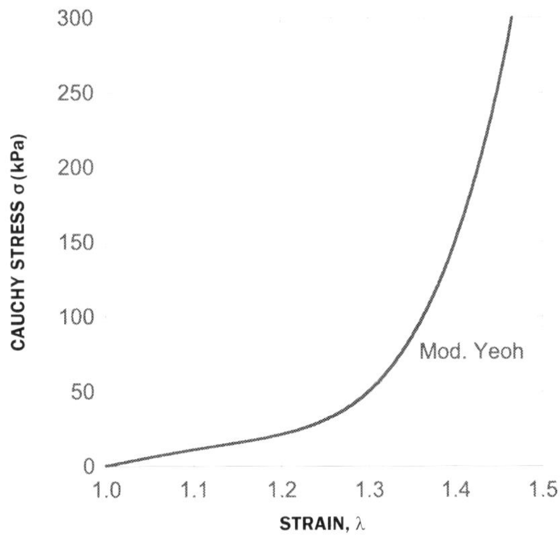

Fig. 5.18. *Uniaxial stress-strain curve for Yeoh's incompressible material (α = 40 kPa, β = -142 kPa and γ = 1200 kPa).*

5.3.3. Rupture by tearing test

Figure 5.19a illustrates, in a schematic form, how a thin membrane can be torn. This type of test–called the Elmendorf tear test, in the paper industry– is a rupture test. During the test, the force F is recorded as a function of the displacement of the point of application of the force and, from this plot, it is possible to determine the specific fracture strength R, that is, the energy expended per unit area of the tear.

The simplest model of the Elmendorf test is based upon the assumption that the material behaviour is *linear elastic* (characterised by an elastic modulus E) up to rupture, and that the elastic energy due to flexure is negligible compared to that of traction. It is also postulated that the work done by the external forces is converted exclusively into elastic deformation of the material and in tearing (consuming a constant energy per unit tear surface R).

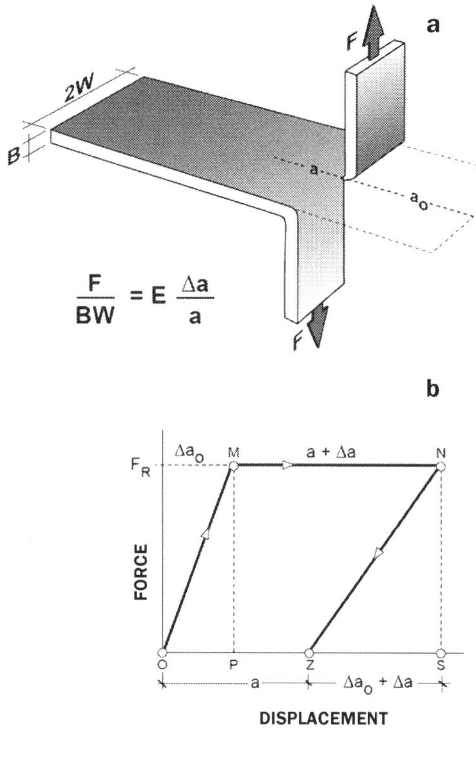

Fig. 5.19. *Schematic of the Elmendorf tear test on a lamina, along with the corresponding force-displacement diagram. The displacement corresponds to one of the forces, considering that the plane of the untorn membrane remains immobile.*

Bearing in mind this hypothesis, the diagram in figure 5.19 represents an elementary process of loading, rupture and unloading: along the section *OM* the band, of length a_0 and the cross-section *BW*, deforms elastically until the force reaches the value F_R, when rupture is initiated by tearing the membrane. (In these types of tests, tearing starts at a force F_R, but almost always continues to progress under the action of the same force.) If the force is held constant, failure progresses in a stable manner and the crack grows to a length a. On reaching point *N*, at which the elastic displacement is $\Delta a_0 + \Delta a$, the membrane is unloaded and the path *NZ* is followed. The membrane is unstressed and has a residual displacement $OZ = a$.

The work done in the process *OMNZ* will be the energy expended in tearing, that is to say:

$$2F_R\left(a + \frac{\Delta a}{2}\right) = R \cdot Ba \qquad (86)$$

in which the left-hand side represents the work done by the two forces (the double of the area of the trapezium OMNZ) and the term on the right is the tear resistance; the specific strength, R, per unit area of tear Ba.

The value of Δa can be calculated from the hypothesis made at the beginning; that the deformation is linear and elastic, assuming that each one of the torn pieces is subjected to a state of uniaxial tension under the force F, from which:

$$\Delta a = \frac{F}{EBW} a \qquad (87)$$

Introducing this result into equation (86) yields the specific tearing strength R, given by:

$$R = \frac{2F_R}{B} + \frac{F_R^2}{B^2 WE} \qquad (88)$$

Often, when the elastic modulus is high, the second term in this expression is negligible compared to the first, producing a very simple expression for estimating the tear strength of a membrane from the force obtained in an Elmendorf test:

$$R = \frac{2F_R}{B} \qquad (89)$$

This procedure remains valid for *nonlinear elastic* materials –elastomers and biological materials in particular– provided that the work done by the external forces is distributed only between the tearing of the material and elastic deformation. Measuring the area, *A*, enclosed by the *force-displacement* diagram results in:

$$R = \frac{2A}{Ba} \tag{90}$$

Note that the number 2 appears because there are two forces doing work.

For Elmendorf tear tests on *elastomers*, which exhibit a nonlinear *force-displacement* curve F-, Rivlin and Thomas (1953) derived an expression similar to equation (88), (see also Atkins and Mai (1985), section 5.10):

$$R = \frac{2F_R\lambda_R}{B} - 2W W_R \tag{91}$$

where $\lambda = l/l_0 = 1 + e$, and e is the engineering strain remote from the tear and W the elastic energy density referred to the volume in the undeformed configuration, also remote from the torn region.

Equation (91) can be written in a more compact form as

$$R = 2W\int_0^{s_R} \lambda\, ds = \frac{2}{B}\int_0^{F_R} \lambda\, dF \tag{91bis}$$

since $s = F/(BW)$.

These results are proved in the following exercises.

Values of the specific fracture resistance R obtained from tear tests on the epithelia of several animals are presented in Table 5.1.

TABLE 5.1. TENACITY OF ANIMAL EPITHELIA

ANIMAL	LOCATION OF EPITHELIA	R (KJ/M²)	REFERENCE
Rat	back (longitudinal)	25	Purslow (1983)
	back (transversal)	20	Purslow (1983)
	belly (longitudinal)	17	Purslow (1983)
	belly (transversal)	11	Purslow (1983)
Rabbit		20	Atkins, Mai (1985)
Rhinoceros	back (deep)	43	Shadwick *et al.* (1992)
	back (shallow)	78	Shadwick *et al.* (1992)

For the tearing of a small crack subjected to remote traction, either in the centre of the membrane or on one of the sides, as shown in figure 5.20, the following expression, valid for *elastic* materials (linear and nonlinear), can be used for the tear resistance R:

$$R = 2kaW_R \qquad (92)$$

where a is the crack length, W_R the elastic energy density (as discussed) remote from the crack and k a parameter which depends upon λ and which can be approximated by $\pi/\lambda^{1/2}$ (Atkins and Mai, 1985).

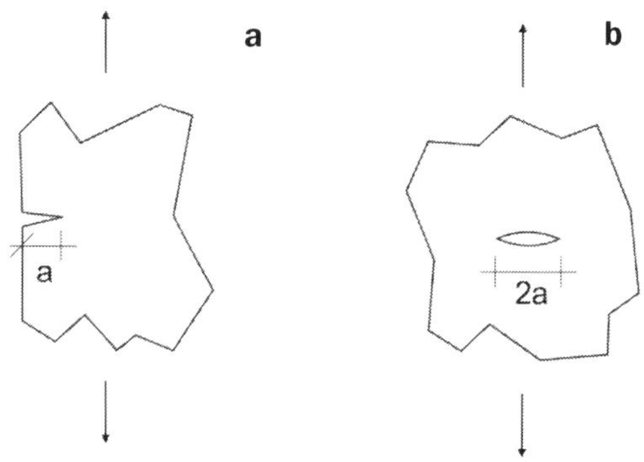

Fig. 5.20. *Tearing of a large membrane subjected to remote traction.*

Equation (92) can be derived from a nondimensional analysis of the fracture process (see exercise 43). The value of the parameter $k(\lambda)$ can be approximated using the empirical relationship $k = \pi/\lambda$, due to Lakes (see Linley 1972). For a linear-elastic material, where $W_R = \sigma^2/2E$, and for small deformations when $\lambda \approx 1$, the value of the parameter k becomes equal to π.

An interesting aspect of the tear resistance of epithelia is the shape of the stress-strain curve; materials with a J-form curve are less sensitive to cracks than those with a linear shape (that is to say, whose stress-strain curve is a continuation of the tangent at the origin).

The *critical value* of the crack length can be deduced from equation (92), as a function of the strain, as shown in figure 5.21, in which the behaviour of the epithelium of a cat in a test like that shown in figure 5.20 is presented. It is assumed that the stress-strain curve is that shown in figure 5.8, and the tear strength is 25 kJ/m^2, a value similar to that of the rat (see table 5.1).

Note that if the behaviour of the epithelium were linear elastic (with an elastic modulus equal to the tangent at the origin), the corresponding curve would be that indicated by the dashed line in figure 5.21. This would imply that the epithelium would be much more sensitive to defects; a tiny fissure under small strains would be enough to tear the skin. The epithelium of cats can withstand large strains –up to 80%– with little risk. In general, epithelia can support large deformations without storing much elastic energy, which would have induced fracture, and for this reason they are difficult to tear. Another consideration is why it is difficult to initiate crack propagation. This is due to the deformability of epithelia at the tip of the crack which blunts very quickly and, as a result, the stress concentration is not very high.

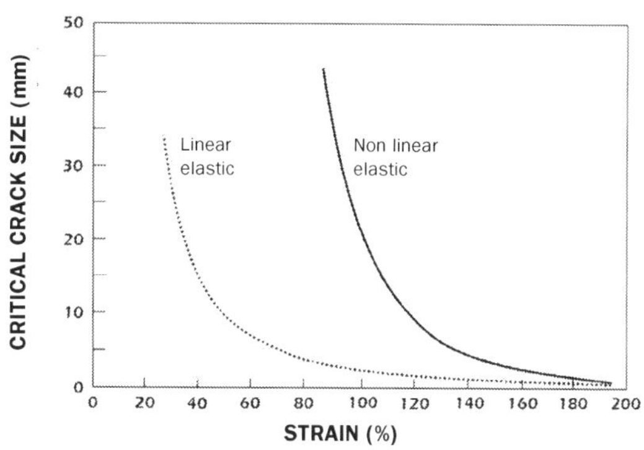

Fig. 5.21. *Critical crack size (in a tear test using the configuration shown in figure 5.20) as a function of the strain in the lamina for the epithelium of a cat. The solid stress-strain curve was obtained from a uniaxial tensile test. The dashed curve was obtained by assuming the curve is linear elastic, with the modulus of elasticity equal to the tangent at the origin of the previous curve.*

EXERCISE 41

Derive equation (91) and show that this coincides with equation (89) if the material is assumed rigid, or if the material is linear elastic. Ignore the elastic energy outside the two tear zones and assume that these are subjected to uniaxial tension under a force F.

SOLUTION:

The Elmendorf tear test is shown in the figure. Let x be the length of the tear, and B and W the dimensions of the unloaded specimen. Under the action of the force F, the length of each piece becomes λx, where λ is the elastic elongation.

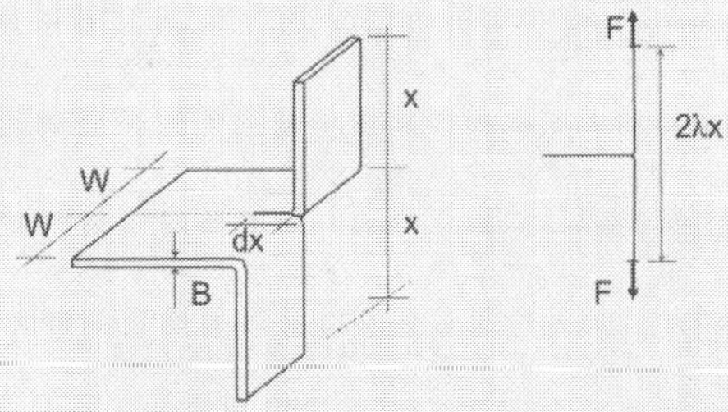

If the crack grows by dx, the work done by the applied forces will be $2F\, d(\lambda x) = 2F\, \lambda dx + 2F\, x\, d\lambda$

in which we see that the length of the tear zones in the specimen can grow either because they deform by ($d\lambda$), or they tear by (dx). Since the body is elastic (linear or otherwise) work is only done in changing the elastic energy or in the fracture process, since no energy has been consumed through plastic deformation or other dissipative mechanisms.

Assuming that the two torn pieces are under a state of uniaxial tension, their elastic energy will be the product of their initial volume $2BWx$ and the energy density $\mathcal{W}(\lambda)$, which will be the same at every point and will not depend upon the level of strain reached, which we characterise by the elongation λ. The change in energy will be

$$d(2BWx\, \mathcal{W}) = 2BW\, \mathcal{W} dx + 2BW\, (d\mathcal{W}/d\lambda)\, x\, d\lambda$$

We will then have $2F\, d(\lambda x) = 2F\, \lambda dx + 2F\, x\, d\lambda = 2BW\, \mathcal{W} dx + 2BW\, x\, (d\mathcal{W}/d\lambda)\, d\lambda + RB\, dx$
where $RB\, dx$ is the energy expended in tearing the material. This expression can be written as

$$(2F\, \lambda - 2BW\, \mathcal{W} - RB)\, dx = (2BW\, x\, (d\mathcal{W}/d\lambda) - 2F\, x)\, d\lambda$$

which must be true for any conceivable process $(dx,\, d\lambda)$, since x and λ are independent variables. This condition requires that the coefficients of dx, $d\lambda$ are identically zero, and therefore that

$$R = 2F\, \lambda\, /\, B\ - 2W\, \mathcal{W}$$

which is equation (91), and $F = BW\, (d\mathcal{W}/d\lambda)$ or $\sigma = d\mathcal{W}/d\lambda$

which is none other than equation (67) applied to a uniaxial tension test (see exercise 40).

For a *rigid material* the strains are very small (≈ 0), and the value of λ will be close to unity. Furthermore, for the same reason, the energy density W is also very small, and the term $2W\mathcal{W}_R$ in equation (91) can be neglected. With these approximations, equation (89) is recovered.

In the case of a nonlinear elastic material with engineering stress and strains s and e (see the exercise in section 5.2.5)

$$\mathcal{W} = \frac{1}{2} se = \frac{F_R}{2BW} e_R$$

whereby equation (91) becomes $R = \dfrac{2F_R(1+e_R)}{B} - \dfrac{F_R}{B} e_R = \dfrac{2F_R}{B} + \dfrac{F_R}{B} e_R$

since $\lambda = 1 + e$, and then $e_R = \dfrac{s_R}{E} = \dfrac{F_R}{BWE}$

substituting in the expression for R, we recover equation (88). Note that linear-elastic materials are only defined in the small deformation range where $s \approx \sigma$ and $e \approx \varepsilon$.

EXERCISE 42

Show that equation (91) can never be negative and that it can be rewritten as equation (91bis).

SOLUTION:

The strain energy is equal to (see the exercise in section 5.2.5)

$$\mathcal{W} = \int_0^e s\,de$$

which can be written as $R = \frac{2F_R\lambda_R}{B} - 2W\mathcal{W}_R = 2W\left[s_R\lambda_R - \int_0^{e_R} s\,de\right]$ since $s_R = F_R/(BW)$.

The value of the integral can be bounded, since the stress will always be $s < s_R$

$$\int_0^{e_R} s\,de < s_R\int_0^{e_R} de = s_R e_R = s_R(\lambda_R - 1)$$

and we conclude that $s_R\lambda_R - \int_0^{e_R} s\,de > s_R\lambda_R - s_R(\lambda_R - 1) = s_R > 0$

so that $R > 0$ always. On the other hand, since $de = d(\lambda - 1) = d\lambda$

$$R = 2W\left[s_R\lambda_R - \int_0^{e_R} s\,de\right] = 2W\left[s_R\lambda_R - \int_0^{\lambda_R} s\,d\lambda\right]$$

and integrating by parts we recover equation (91bis).

$$R = 2W\left[s_R\lambda_R - \int_0^{\lambda_R} s\,d\lambda\right] = 2W\left[s_R\lambda_R - s_R\lambda_R + \int_0^{s_R} \lambda\,ds\right] = 2W\int_0^{s_R} \lambda\,ds$$

EXERCISE 43

Derive equation (92) $R = 2k\,a\,W_R$.

SOLUTION:

If we examine either of the two cases shown in figure 5.20, we can see that the only relevant geometrical variable is the size of the crack a, since the dimensions of the plates are infinite in both cases.

Therefore, for the problem posed, each equation for R must be able to be written as a function of:

- R
- the geometric parameter a
- the stress-strain level, which we can characterise by the strain energy per unit volume W_R, at points remote from the crack
- the response of the material, characterised by the elongation λ in the direction of loading, evaluated at points remote from the crack

Considering the possible nondimensional groups that can be assembled from these parameters, we see that the required equation must have the form:

$$f\left(\frac{R}{aW_R}, \lambda\right) = 0$$

since $R/(a\,W_R)$ and λ are the two monomials that can be formed. The previous equation can be rewritten to express one monomial as a function of the other

$$R/aW_R = g(\lambda)$$

which we can easily recognise as a form of equation (92) by making $g(\lambda) = 2k(\lambda)$.

EXERCISE 44

Examine how the membrane material influences the value of the critical crack length when it is subjected to a traction as illustrated in figure 5.20. Use equation (92) to obtain the critical crack length. For this, consider the following forms of the uniaxial stress-strain curves for the material:

- linear: $s = e\,E$
- concave: $s = eE + e^m E\,;\; m>1$
- convex: $e = s/E + s^m/E^m;\; m>1$

SOLUTION:

The three stress-strain curves are shown in figure (a) below for $m=3$.

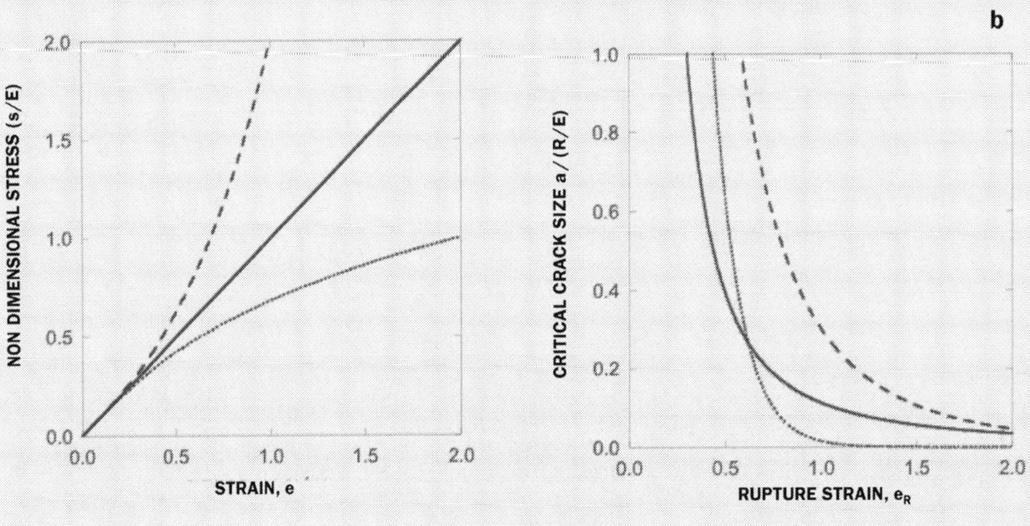

In order to calculate the critical crack size, isolate a in equation (92) and write it in the nondimensional form

$$\frac{a}{R/E} = \frac{\sqrt{\lambda}}{2\pi}\frac{1}{\mathcal{W}_R/E} = \frac{\sqrt{1+e}}{2\pi}\frac{1}{\mathcal{W}_R/E}$$ and we calculate $\mathcal{W}_R/E$ by integration.

In the case of the **concave** curve s-e, this will be $\mathcal{W}_R/E = \int_0^{e_R}(s/E)\,de = \int_0^{e_R}\left(e+e^m\right)de = \frac{e_R^2}{2} + \frac{e_R^{m+1}}{m+1}$

For **linear** behaviour $\mathcal{W}_R/E = \int_0^{e_R}(s/E)\,de = \int_0^{e_R} e\,de = \frac{e_R^2}{2}$

and for the **convex** case, we integrate by parts to obtain

$$\mathcal{W}_R/E = \frac{1}{E}\int_0^{e_R} s\,de = \frac{1}{E}\left(s_R e_R - \int_0^{S_R} e\,ds\right) = \left(s_R/E\right)^2\frac{1}{2} + \left(s_R/E\right)^{m+1}\frac{m}{m+1}$$

in which we have used the equation of the curve $e = s/E + s^m/E^m$. From these, we can relate each value of the strain e_R with the stress s_R, and with the last one we obtain $\mathcal{W}_R/E$.

The values of the critical crack size are shown in figure (b) above as a function of the failure strain. In this, it is noted that the most beneficial form (the largest critical value for a given rupture strain) is always the concave curve, as exemplified in the text with the cat epithelium.

Observe also that there is a certain value of the strain at rupture below which a convex curve is more favourable than that of linear behaviour.

EXAMPLE: RAT EPITHELIUM

During the preparation of his doctoral theses at the end of the 1970s, Purslow experimented with soft tissue and measured the specific tear strength of a rat epithelium using the tear test. A summary of these results is discussed below. Details can be found in Purslow (1983).

The epithelium, of approximately 1 mm thickness, is represented schematically in figure 5.22. Sixteen specimens were cut from the epithelium: half were tested in the longitudinal and the other half in the transverse direction, in order to investigate possible anisotropy in the tissue, as shown in figure 5.22a.

The force displacement curves, representative of the tear tests, are reproduced in figure 5.22b. The specific fracture energy, R, was calculated, initially, from equation (91), but Purslow argued that, in this case, the following simplifications could be made:

Since the applied forces were small, about 5 N, when the skin tears, the deformation remote from the crack could also be considered small and the value of λ approximated to 1. For this reason, Purslow neglected the elastic energy density $\mathcal{W}$ far from the region of the crack. (Paradoxically, it is as if the material were rigid, see exercise 41).

After these simplifications, the specific strength is given by

$$R \backsimeq \frac{2F_R}{B}$$

The average values of R (from 8 tests) obtained by Purslow in the longitudinal and transverse directions were, respectively, 17 ± 1 kJ/m^2 and 11 ± 2 kJ/m^2, results which showed an appreciable anisotropy in the specific fracture energy.

Purslow also studied the influence of the load rate and temperature on the values of R. Increasing the deformation rate (from 5 to 500 mm/minute) doubled the value of R. As far as temperature was concerned, experiments were performed between 7 and 23 °C, but no clear trend was observed, perhaps only a small increase with temperature.

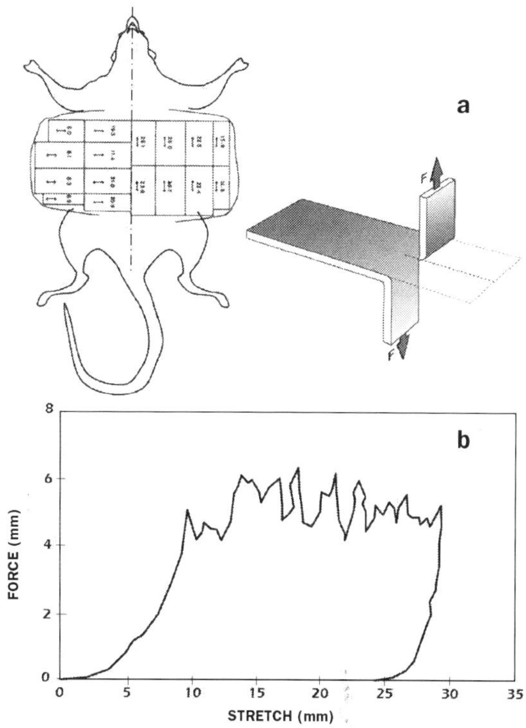

Fig. 5.22. *Tear test on the epithelium of a rat (Purslow, 1983).*
a. Schematic illustration of the extraction of the test pieces.
b. Typical result of the tear test.

5.3.4. Separation tests

Simple peeling test

The peeling of a thin membrane from a substrate is shown in figure 5.23a. This type of test –attributed to Dannenberg (1961) by some authors– is a rupture test which provides knowledge of the specific adhesion energy between a lamina and substrate. During such a test, the force F is recorded as a function of the displacement of the point of application of the force. The specific adhesion strength can be calculated from a plot of this data following the same steps outlined in the previous section.

The simplest way to model this test is to assume that the membrane behaves as a *linear-elastic* material (small deformations, and elastic modulus E), and that the elastic energy due to flexure is negligible compared to that of traction. Furthermore, it is postulated that the work done by the external force is converted only into elastic deformation of the membrane and in separating it (while consuming constant specific energy R). In this simple model, any viscoelastic effects are ignored. Again, as happened in the case of tearing, separation is usually produced under the action of a constant force F_R.

Incorporating these assumptions, the diagram in figure 5.23b outlines the basic loading, separation, and unloading process. In the process represented by the line *OM* the membrane, of length a_0 and cross-section *BW*, deforms elastically until the force reaches a value F_R, at which point separation starts. If the force is held constant, separation progresses in a stable manner, increasing up to a length a. On reaching point *N*, the membrane is unloaded, and the path *NZ* is followed. The membrane is left stress-free but with a residual displacement $OZ = a$.

An energy balance performed for the OMNZ process, in a similar manner to that carried out in the previous section, results in the expression:

$$F_R \left[a(1 - \cos\theta) + \frac{\Delta a}{2} \right] = R \cdot Wa \tag{93}$$

which the term on the left-hand side represents the work done by the force F (the area under the trapezium *OMNZ*) and the term on the right the separation strength; the separated area, Wa, multiplied by the specific strength, R. The diagram in figure 5.23c is drawn to help explain the calculation.

The value of Δa is calculated bearing in mind the hypothesis that the membrane behaves linear elastically, so that:

$$\Delta a = \frac{F}{EBW} a \tag{94}$$

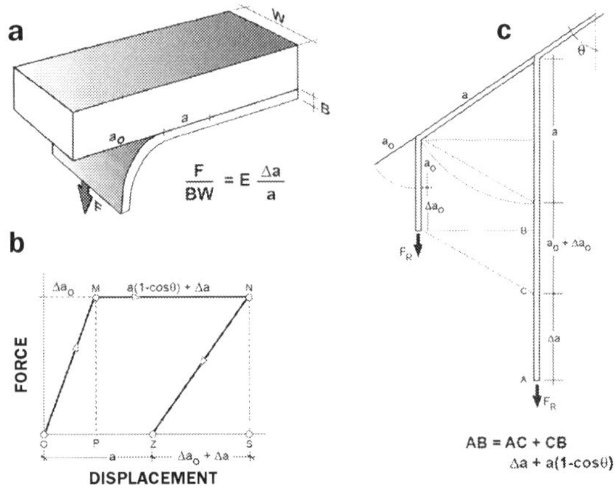

Fig. 5.23. *Schematic diagram of a simple separation test on a lamina, together with the corresponding force-displacement plot.*

Substituting this result into equation (93), the specific separation (or adhesion) strength R is obtained:

$$R = \frac{F_R}{W}\left(1 - \cos\theta + \frac{F_R}{2EBW}\right) =$$

$$= \frac{F_R}{W}\left(1 - \cos\theta + \frac{\varepsilon}{2}\right) \tag{95}$$

Note that $F_R/2EBW$ is $\varepsilon/2$, half of the elastic strain in the membrane.

When the modulus of elasticity is high, the last term inside the bracket is usually negligible compared to the others (analogous to the previous case of rupture by tearing). In these circumstances, a very simple expression is obtained for estimating the specific adhesion strength between the membrane and substrate. If the membrane separation is perpendicular to the substrate, the $R \approx F_R/W$.

This procedure is also valid if the membrane exhibits *nonlinear elastic* behaviour. The specific strength is obtained by dividing the area under the F-u diagram by the separated area: aW.

There are two relatively simple arrangements which create a separation zone in a membrane, one like that just discussed –subjected to a state of uniaxial tension– and the other a separation in a tympanum object, described in the following section, which is a membrane subjected to a state of biaxial stresses.

The first arrangement is shown in figure 5.24a, in which the lamina is separated using a bracket. The same hypotheses are made as before; linear elasticity, small deformations, and ignoring the elastic energy due to flexure compared to that stored in traction. It is also assumed that initially an area a_0W is separated and the initial strain is zero.

The problem can be solved following the procedures described above; from the area under the F-u diagram. The result can also be written directly, if it is observed this problem is analogous to the previous one, but in this case the force F is replaced by $F/2$ $\sin\theta$. The final result is then:

$$R = \frac{F_R}{2W\sin\theta}\left(1 - \cos\theta + \frac{\varepsilon}{2}\right) = \frac{F_R}{2W\sin\theta}\frac{1 + \cos\theta - 2\cos^2\theta}{2\cos\theta} \quad (96)$$

in which the strain ε has been written as:

$$\varepsilon = \frac{1-\cos\theta}{\cos\theta} \quad (97)$$

It is sometimes convenient to write the specific strength as a function of the height H and the pressure that causes separation p. (In this case we can define an average pressure equivalent to that obtained from the quotient of the force F and the area 2aW.) That is to say:

$$R = \frac{FH}{2aW}\frac{1 + \cos\theta - 2\cos^2\theta}{2\sin^2\theta} \equiv f_1(\theta)\cdot Hp \quad (98)$$

For small values of $\theta, f_1(\theta)$ tends to ¾ and the separation strength can be approximated by:

$$R \simeq \frac{3}{4}Hp \quad (99)$$

A second configuration, referred to previously and shown in figure 5.24b, involves a lamina which is separated under the action of an internal pressure p. The same hypotheses are made as before; linear elasticity, small deformations and treat the lamina as a membrane. Under these conditions, it can be assumed that the deformation is a cylindrical surface, of radius ρ, and the separating force becomes, $F = p \cdot 2\rho\, sin\,\theta \cdot W$.

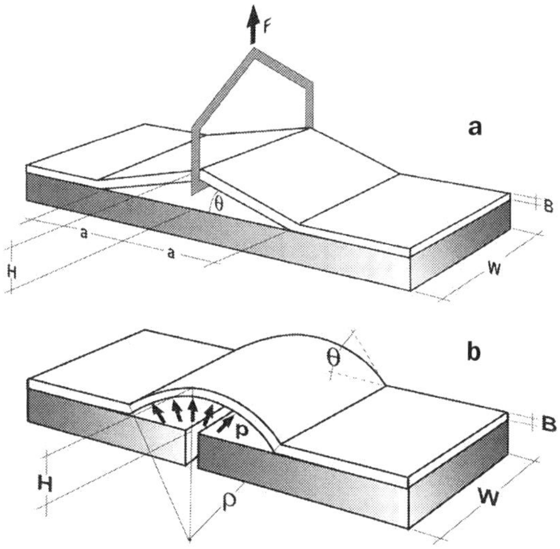

Fig. 5.24. *Schematic representations of two lamina separation tests.*
a. Using a bracket.
b. Using internal pressure.

Using similar reasoning as in the previous example, the specific adhesion will be given by equation (95)

$$R = \rho p\left(1 - \cos\theta + \frac{\varepsilon}{2}\right) = \rho p\left(1 - \cos\theta + \frac{\theta - \sin\theta}{2\sin\theta}\right) \quad (100)$$

where, now, the strain ε can be approximated by

$\varepsilon = (\theta - \sin\theta)/\sin\theta$

This result can also be written as a function of H and p, in the following form

$$R = Hp\left[1 + \frac{\theta - \sin\theta}{2\sin\theta\,(1 - \cos\theta)}\right] \equiv f_2(\theta)\cdot Hp \quad (101)$$

which, for small values of θ, reduces to

$$R \simeq \frac{7}{6}Hp \quad (102)$$

Once again, if the lamina exhibits *nonlinear elastic* behaviour, the specific separation strength can be found by measuring the work done and dividing it by the area of the separated surface.

Separation test on a tympanum

In order to measure the adhesion of thin laminas (or films) it is sometimes helpful to use a technique based on the formation of a blister or vesicle from a tympanum. This test, proposed by Dannenberg (1961) and modified by Wan and Mai (1995), has the advantage of not requiring jaws or grips which could cause plastic deformation, and the necessary measurements –pressure and height of the blister– are not difficult to obtain.

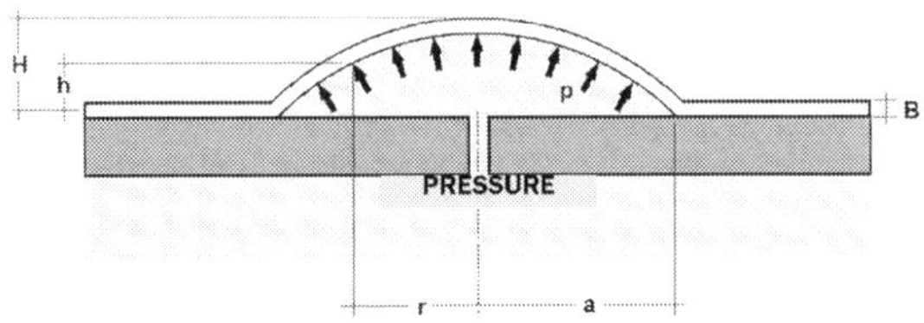

Fig. 5.25. *Schematic representation of the biaxial separation of a film (tympanum or blister test).*

The test is shown schematically in figure 5.25. When a pressure p is introduced through a hole in the substrate, the film (of thickness B) can separate and form a blister (with a cross-section which is assumed to be circular, of radius a and height H). To model this test, several distinct hypotheses can be made about the behaviour of the film, which range from considering it as a thin plate, with a certain flexural stiffness, to considering its behaviour to be like a membrane with no flexural rigidity. Later, both situations will be analysed in order to obtain bounds to the solution of the problem.

If it is supposed that the film behaves like a **plate**, the displacements of the tympanum due to the pressure p can be calculated by assuming that the deformations are small and the behaviour *linear elastic* (see, for example, Timoshenko and Woinowsky, 1959). The result is:

$$h(r) = \frac{p}{64D}\,(a^2 - r^2)^2 \qquad (103)$$

where a, h and r are shown in figure 5.25, D is the flexural rigidity of the plate, given by $D=EB^3/12\,(1-\nu^2)$, E is the modulus of elasticity and ν is the Poisson ratio. The maximum displacement, H, is produced at the centre $(r = 0)$ and is:

$$H = \frac{pa^4}{64D} \qquad (104)$$

The specific adhesive strength can be calculated from the elastic energy (see, for example, Elices 1998, page 90) yielding:

$$R = \frac{p^2a^4}{128D} = 0.5Hp \qquad (105)$$

This is a very simple result that depends only upon the pressure p and the blister height H.

If the hypothesis is that the film behaves like a **membrane**, the displacements can also be calculated by assuming small deformations and *linear-elastic* behaviour. The maximum displacement is produced at $r=0$ and has a value:

$$H = c_1\left(\frac{pa^4}{EB}\right)^{1/3} \qquad (106)$$

where c_1 is approximately 0.6 and varies weakly with the Poisson ratio ν (Williams 1997).

The specific adhesion strength calculated from the elastic energy (Elices 1998) becomes:

$$R = \frac{5}{4}\frac{c_1c_2}{(EB)^{1/3}}(pa)^{4/3} = \frac{5}{4}c_2Hp \qquad (107)$$

where c_2 is approximately 0.52 (Elices, 1998).

This simple result is very similar to that obtained with the hypothesis that the film behaves like a thin plate, that is to say $R=0.5Hp$. Assuming that it behaves like a membrane $R \approx 0.65\,Hp$ (again, this last result changes very little with the value of ν).

These two hypotheses represent two extreme cases and they suggest that a lamina tested in this way will exhibit an intermediate behaviour. In reality, the two solutions are similar, and the error in adopting one or the other is not great. The specific adhesive strength can then be written as:

$$R = \lambda\,Hp \qquad (108)$$

where p is the pressure that produces the separation, H is the height of the blister formed and λ is a nondimensional coefficient which lies between 0.5 and 0.65, depending upon the hypothesis made.

It might be imagined that if the film is very flexible, the small deformation approximation, on which the previous formulae were based, is not valid. This was examined by Williams (1997) who concluded that, even in such cases, the approximation was reasonably good. To be more precise, he obtained

$$R = Hp\left[\frac{5}{8} + \frac{(H/a)^2}{8\,[(H/a)^4 + 3\,(H/a)^2 + 1]}\right] \qquad (109)$$

This result indicates that R is almost insensitive to (H/a); for $H/a=0$,

$$R = \frac{5}{8}Hp = 0.625\,Hp \qquad (110)$$

and the maximum value of $R\,/\,Hp$ is 0.644 when $H/a = 3^{0.25} \approx 1.31$.

EXERCISE 45

The figure below shows data obtained by Kita and co-workers (1990) who used an ocular micro-injection device to determine the bonding of the retina to the back of the eye. The investigators measured simultaneously the radius of the separation zone *a* and the subretinal pressure Δp which induces detachment of the retina.

From the data provided, evaluate which separation model of the tympanum –plate or membrane– is more appropriate for analysing the experiment, and estimate the value of the specific adhesion strength of the retina. Assume that the displacements and deformations are very small and that the retina behaves linear elastically with $E = 0.43\,\text{MPa}$, $\nu = 0.42$ and thickness $B = 0.3\,\text{mm}$. For the membrane analysis, model the separated area as a spherical shell.

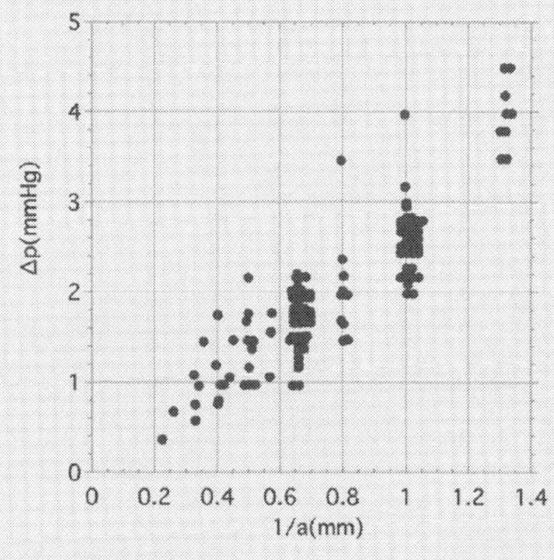

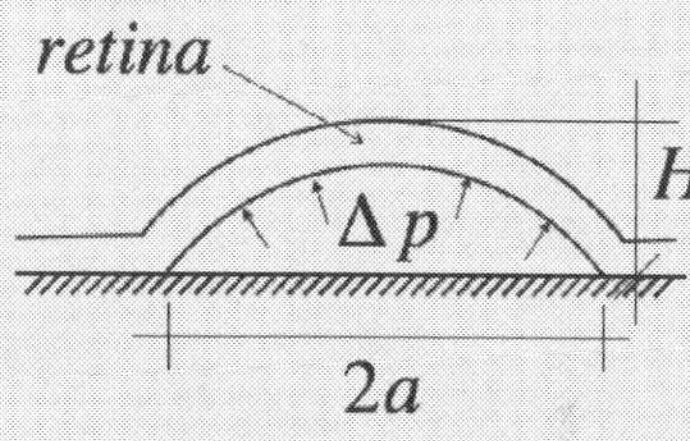

SOLUTION:

For the **plate** model, equation (105) provides the relationship between the pressure and the radius of the separated zone. This Δp - a relationship can then be written as

$$\Delta p = \sqrt{\frac{32EB^3R}{3(1-\nu^2)}}\left(\frac{1}{a}\right)^2$$

which indicates that the experimental $\Delta p - 1/a$ curve should be parabolic.

The corresponding equation for the **membrane** model is deduced from equation (108), but it is also necessary to know the relationship between the height of the separation zone *H*, its radius *a* and the pressure Δp. For that we assume that the shape of the tympanum is spherical, as shown in figure a on the next page. Then

$$H = \frac{1-\cos\theta}{\sin\theta}a \approx \frac{\theta a}{2}$$

The value of θ is obtained by assuming that the deformation occurs under a stress σ, which we obtain from equilibrium of forces on the membrane in the vertical direction ($2\pi a B\sigma\sin\theta = \Delta p\pi a^2$):

$$\sigma = \frac{\Delta pa}{2B\sin\theta} \approx \frac{\Delta pa}{2B\theta}$$

By symmetry of the problem, the stress state in the membrane is equiaxial, and the stress which acts in two mutually perpendicular directions is equal to σ, giving rise to a strain (applying Hooke's Law)

$$\varepsilon = \frac{\sigma}{E} - \frac{\nu}{E}(\sigma + 0) = \frac{1-\nu}{E}\sigma = \frac{r\theta - a}{a} \approx \frac{\theta^2}{6}$$

since $a = r\ \sin\theta \approx r(\theta - \theta^3/3! + \ldots)$. From the previous two expressions, we can eliminate σ to obtain

$$\theta = \left(\frac{3(1-\nu)\Delta pa}{EB}\right)^{1/3}$$

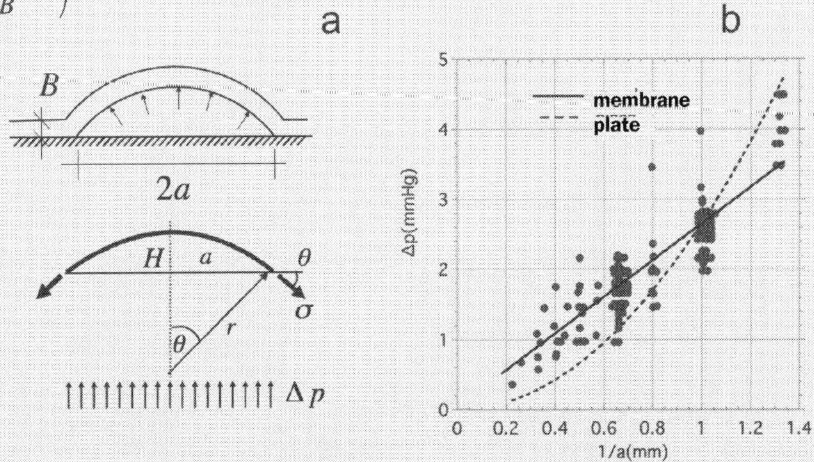

and then $H = \frac{\theta a}{2} = \left(\frac{3(1-\nu)\Delta pa^4}{8EB}\right)^{1/3}$

with which, and using equation (108), we finally obtain

$$R = \lambda pH = \lambda\left(\frac{3(1-\nu)(\Delta p)^4 a^4}{8EB}\right)^{1/3}$$

The Δp - a relationship in this case becomes:

$$\Delta p = \left(\frac{8EBR^3}{3\lambda^3(1-\nu)}\right)^{1/4}\frac{1}{a}$$

which represents a linear relationship between Δp and $1/a$.

The Δp - $1/a$ fits for both models are shown above in figure b. Despite the large scatter, it seems that the membrane model provides a better fit to the experimental data (the membrane data correlation coefficient is 0.89 as opposed to 0.67 for the plate model). The values of R can be obtained from B, E and v data, and the constants from the curve fit:

- Plate model: Δp = 2.65 mmHg mm² $(1/a)^2$
- Membrane model: Δp = 2.63 mmHg mm $(1/a)$

from which we deduce

- Plate model: R=8.29 10^{-4} J/m²
- Membrane model: R=185 10^{-4} J/m²

6 CHAPTER Examples of Biological Membranes

6.1. INTRODUCTION

The objective of this chapter is to provide some applications of the concepts developed above –the mechanical properties of biological membranes. When dealing with practically two-dimensional materials, the mathematical treatment is rather more complex than that of the fibres considered in the Section 1 of this book.

In the following examples, attempts are made to characterise the mechanical behaviour of membranes using certain parameters which will need to be measured. For this reason relevant experiments are also detailed. From this information, attempts will be made to predict the mechanical response of the membrane under particular external loads using constitutive models. In practice, these models often have a limited scope due to the complexity of biological structures.

The first example concerns the *pericardium*, the membrane which surrounds and protects the heart. The mechanical behaviour of the pericardium has been studied extensively because of the interest in its use as a biomaterial in bio-prosthetics for vascular grafts, repair patches on the abdominal wall and in heart valves.

The second example is the *wall of the arterial aorta*. The objective is to describe and model its biomechanical behaviour in order to understand its physiological characteristics, the ageing process and the development of its pathologies. This information is useful in applications of particular medical interest, such as the rupture of aneurisms or the effects of arterial hypertension.

The third example will consider *elasmoid scales* of fish, a type of rigid membrane, unlike the previous ones. Interest in the study of the mechanical properties of these scales lies in understanding their tenacity, flexibility and resistance to penetration. These properties have been, and are, a source of inspiration for the design of armours.

In all of these examples, when attempting to model the mechanical behaviour, we encounter materials which possess complex structures because they are very hierarchical. In these materials there is no clear separation between *material* and *structure* because the different components are joined together without interruption across several scales of observation, from the molecular to macroscopic levels.

In order to develop models using continuum mechanics, it is assumed that the material is composed of homogeneous parts, a distinction which is arbitrary and made exclusively as a function of the available data and the type of information needed for the model. This implies that the stresses and strains obtained from the models will have no physical sense below a certain level.

6.2. PERICARDIUM

6.2.1. Description

The pericardium is a membrane that covers and protects the heart, separating it from the neighbouring structures while allowing its freedom of movement. It covers and extends to the roots of the great vessels.

It is an anisotropic membrane composed essentially of collagen and elastin fibres in layers of different orientation, embedded in a matrix of glycol proteins and glycosaminoglycans. Thanks to the collagen content, the pericardium exhibits good mechanical strength and prevents the excessive dilation of the chambers of the heart, limiting hypertrophy under conditions of maximum activity.

The pericardial sac has a conical form (figure 6.1a). In this figure, two preferential anatomical directions are identified: the apex-caudal or *radial*, from the upper part of the heart to the apex, and the *circumferential*, in a direction orthogonal to the previous one. The pericardium contains collagen fibres grouped in bundles of micrometric thickness which, together with its small thickness, can be observed through the membrane by means of a beam of light. Low-angle light scattering experiments (Sacks *et al.* 1997) revealed that the collagen fibres are arranged preferentially in the radial and circumferential directions (figure 6.1b).

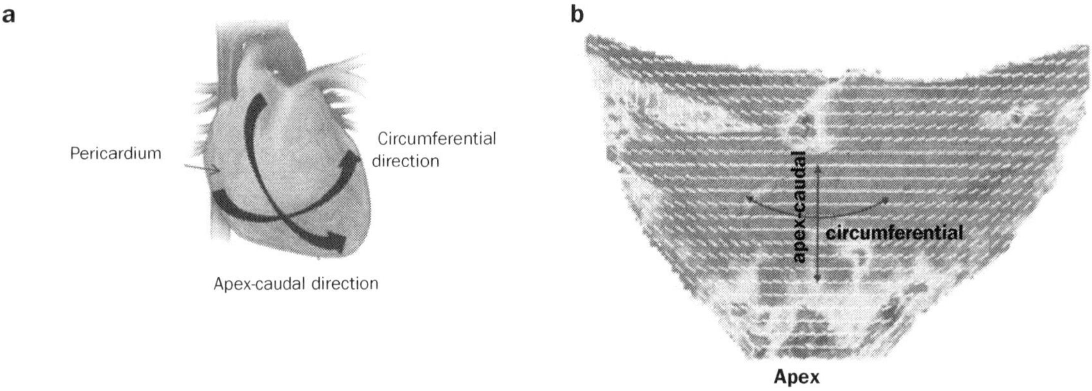

Fig. 6.1. *a) Anatomical arrangement of the pericardium and preferential directions. b) Collagen fibre orientation obtained by light diffraction (Sacks et al. 1997).*

The pericardium has two layers (figure 6.2): the fibrous pericardium and the serous pericardium. The fibrous pericardium is the outer layer and is attached to the adjacent structures (diaphragm and sternum) by ligaments. It is a strong stiff film with a high collagen content (figure 6.3). Its inner surface is intimately connected to the parietal layer of the serous pericardium, which is thinner and formed of mesothelial cells. The innermost layer of the serous pericardium, also called the epicardium, is attached to the heart, forming its external coating. Between the two layers of the serous pericardium is the pericardial cavity which contains a small quantity of fluid, similar to blood plasma, and which minimises the friction produced by the beating heart.

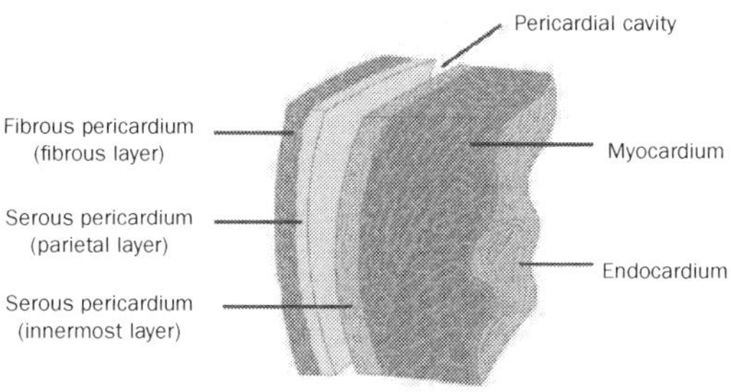

Fig. 6.2. *Layers of the pericardium.*

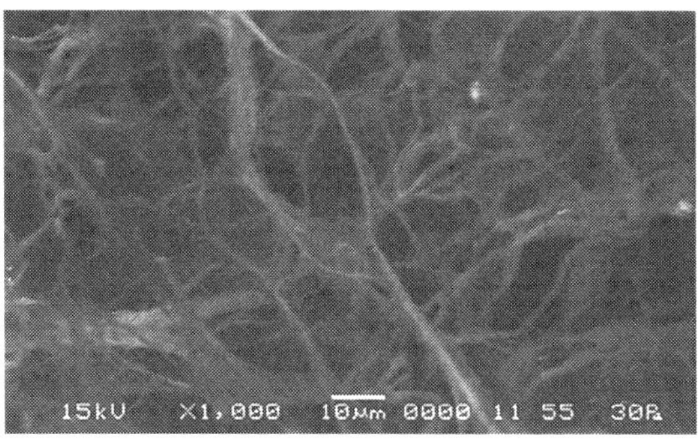

Fig. 6.3. *Scanning electron microscope image of the fibrous layer (Mendoza-Novelo et al., 2013).*

Due to the discontinuity produced by the pericardial cavity, the combined fibrous and serous/parietal layers are often improperly called pericardium because they form a quite distinct set that separates easily from the heart.

Animal, typically bovine, pericardium is a material often used in medicine for the repair of all sorts of structures; in cardiovascular, abdominal, ophthalmic or genitourinary surgery. It is also used as a bioprosthetic heart valve sheath (figure 6.4). In these cases, the fibrous layer is used –and the parietal zone of the serous layer attached to it– since, as well as being easy to acquire, it is the strongest layer.

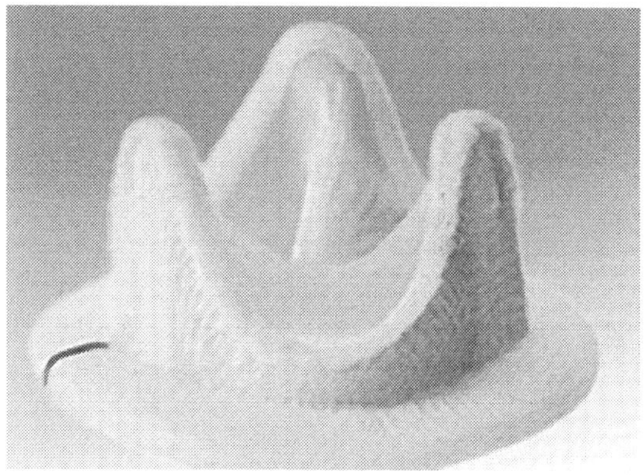

Fig. 6.4. *Bioprosthesis for replacing a heart valve. The veils are made of three crosslinked membranes of pericardium.*

In order to increase the stability of the tissue, avoid its decomposition in the interior of the body and to reduce its antigenicity, the pericardium (that is the fibrous layer) is treated by physical (heat or UV radiation) or chemical means to form irreversible crosslinks of the functional groups of the collagen molecules.

Among the chemical treatments, the most common uses glutaraldehyde at a concentration of 0.625%. This treatment provides greater strength, although it induces calcium deposits on the membrane.

6.2.2 Mechanical behaviour

The mechanical behaviour of the pericardium has been studied extensively because of its interest as a structural biomaterial, conferring mechanical strenth to biological structures. The material analysed is usually pericardium (its fibrous and serous-parietal layers) with crosslinks, generally treated with glutaraldehyde. In the following, and unless indicated to the contrary, these will be the conditions of the pericardial membrane to which we refer.

Uniaxial tensile tests

Figure 6.5 shows representative curves of the tensile behaviour of bovine pericardium under physiological conditions along the two preferential anatomical directions. In this figure, the true stress (or Cauchy stress, as we saw above) σ, obtained by dividing the load by the instantaneous cross-sectional area, is plotted against the ratio of the instantaneous and initial lengths (the elongation λ). We can observe a clear dependency on the direction (anisotropy). Furthermore, the shape of the curves reflects the process of regrouping and progressive alignment of the collagen fibres with increasing load, producing the typical J-shape.

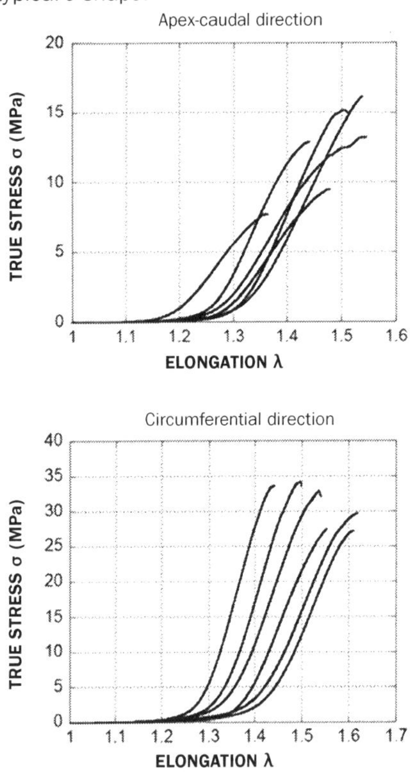

Fig. 6.5. *Uniaxial tensile tests on bovine pericardium specimens (18 x 2 mm) treated with glutaraldehyde. Average thickness 0.52 mm. Loading rate 0.2%/s. Specimens immersed in physiological serum at 37°C (Tobaruela, 2016).*

The influence of the reorientation of the fibres is more apparent if the pericardium specimens are subjected to some tensile fatigue cycles. Figure 6.6 shows such tests, in which 100 loading and unloading cycles were applied, up to a maximum value of 3 MPa which corresponds to approximately 10% of the strength of the material in the circumferential direction and 25% of its strength in the apex-caudal direction. As we can see, almost after the first cycle of loading and unloading the response stabilises, which suggests that the structure of the membrane has essentially reached its final configuration.

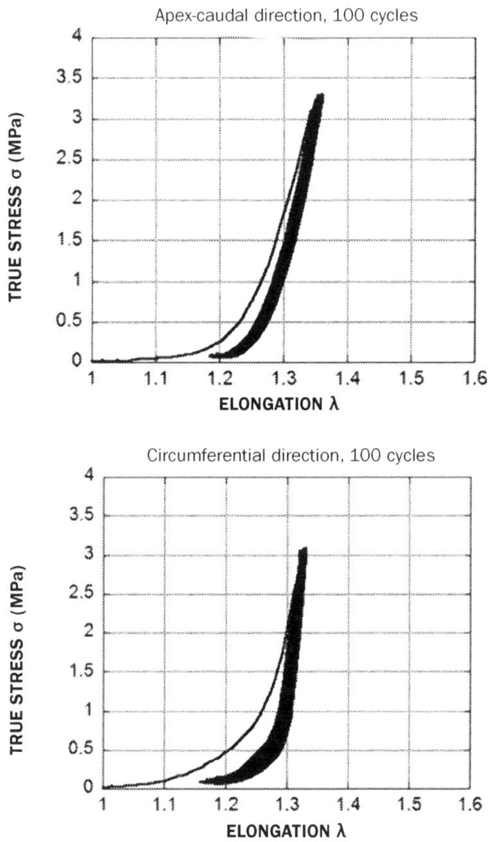

Fig. 6.6. *Uniaxial tensile fatigue tests on bovine pericardium specimens (18 x 2 mm) treated with glutaraldehyde. Average thickness 0.52 mm. Strain rate 2%/s. Specimens immersed in physiological serum at 37 °C. A = apex-caudal direction, Ci = circumferential direction (Tobaruela, 2016).*

In figures 6.7 and 6.8 the tensile strength is shown as a function of the number of cycles previously applied to the specimen, which, as we have seen, indicates a rearrangement of the internal structure of the material. While the tensile strength remains substantially constant (at about 10MPa for the apex-caudal direction and 30MPa in the circumferential), the value of the elongation drops significantly with increasing cyclic treatment (from ~1.5 to ~1.3). In both cases the anisotropy of the material remains unchanged.

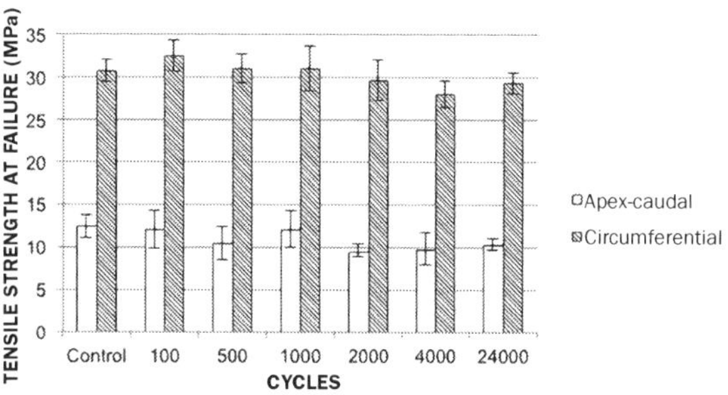

Fig. 6.7. *Tensile strength values (mean value ± standard error, N=33) of bovine pericardium after various uniaxial tensile fatigue cycles (maximum stress = 3MPa). Dimensions of the specimens are 18 x 2 mm. Average thickness 0.52 mm. Specimens treated with glutaraldehyde. Fatigue strain rate 2%/s. Strain rate in the tensile strength tests 0.2%/s. Specimens immersed in physiological serum at 37°C (Tobaruela, 2016).*

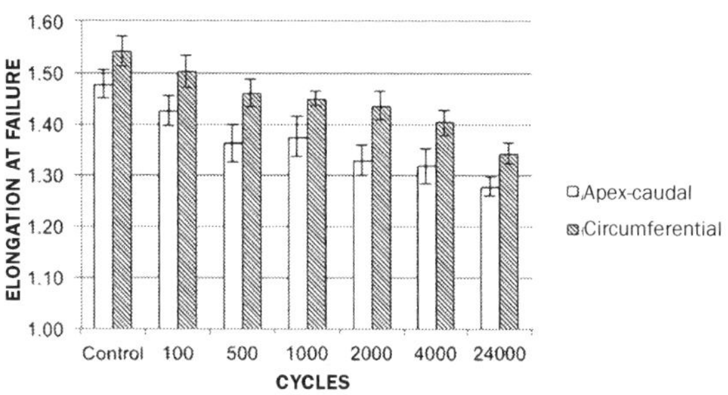

Fig. 6.8. *Values of the elongation at failure (mean value ± standard error, N=33) of bovine pericardium after various uniaxial tensile fatigue cycles (maximum stress = 3MPa). Dimensions of the specimens: 18 x 2 mm. Average thickness 0.52 mm. Specimens treated with glutaraldehyde. Fatigue strain rate 2%/s. Strain rate in the tensile strength tests 0.2%/s. Specimens immersed in physiological serum at 37°C (Tobaruela, 2016).*

It is interesting to determine the slope of the linear part (tangent modulus) of the $\sigma-\lambda$ curves before rupture of the uniaxial tensile specimens, since these can provide an idea of the stiffness of the membrane when all of the fibres have attained maximum alignment. In figure 6.9, these values are shown both for the specimens without prior fatigue (control) and those which have been subjected to a pre-determined number of accommodation cycles. It can be seen that the stiffness remains essentially constant, which confirms that the final state, with the fibres aligned, is reached very quickly and in every case within the first 100 fatigue cycles. If it is assumed that there is a direct relationship between the fibre content and the specimen stiffness, we can estimate that the relative amount of collagen in the two anatomical directions should be about 8/3, which is the ratio of the corresponding stiffnesses (approximately 200MPa in the circumferential and 75MPa in the apex-caudal direction).

In 2001, Carrera *et al.* tested specimens which were 20 mm wide and 120 mm long, treated with 0.625% glutaraldehyde and loaded in the apex-caudal direction. Specimens were subjected to uniaxial tension in load steps of 2, 3, and 8 MPa. The values of the rupture stresses lay between 26.6 and 32.6 MPa.

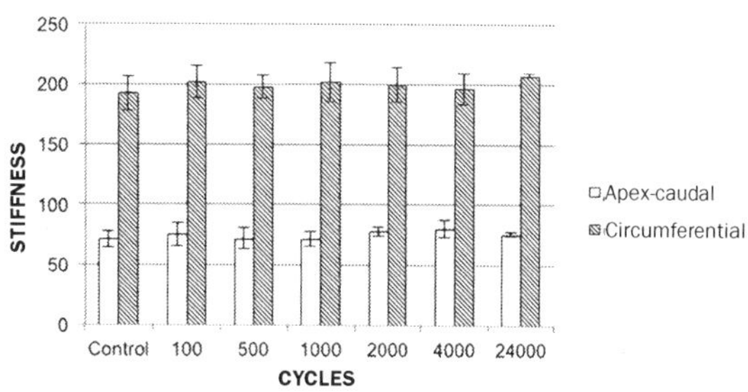

Fig. 6.9. *Stiffness values (slopes of the $\sigma-\lambda$ curves, or tangent modulus) (mean value ± standard error, N=33) of bovine pericardium after various numbers of fatigue cycles (maximum stress = 3MPa). Specimen dimensions are 18 x 2 mm. Average thickness 0.52mm. Specimens treated with glutaraldehyde. Strain rate in fatigue 2%/s. Strain rate in the rupture tests 0.2%/s. Specimens immersed in physiological serum at 37°C (Tobaruela, 2016).*

Biaxial tensile tests

Biaxial tensile tests have been performed in order to obtain a constitutive equation for the pericardium. Figure 6.10 shows the arrangement used by Sacks and co-workers (Sacks and Chuong 1998) in which the pericardium membrane is secured using sutures on its edges, taking advantage of the good tear resistance demonstrated by this material.

The approximation, which is frequently applied to the modelling of other soft materials such as skin, tendons or blood vessels, is that the behaviour of the pericardium, once stabilised after a few fatigue loading and unloading cycles (pre-conditioning), can be represented by a hyperelastic constitutive equation, in which the relationship between the stress and strain energy is given by equation (67):

$$S_{ij} = \frac{\partial \mathcal{W}}{\partial E_{ij}}$$

where $\mathcal{W}=\rho_0 \mathcal{W}_m$ is the strain energy per unit volume, $\mathcal{W}_m$ is the energy per unit mass and ρ_0 the density of the body in the reference configuration.

The fact that the behaviour may be different between loading and unloading is resolved by adopting a different strain energy function for each situation. This is the *pseudo-elastic* model proposed by Fung.

Several strain energy functions specific to the pericardium have been proposed, such as that used by Choi and Vito (1990) and also employed by Sacks and Chuong (1998):

$$\mathcal{W} = b_0[\exp(\frac{1}{2}b_1E_{11}^2) + \exp(\frac{1}{2}b_2E_{22}^2) + \\ + \exp(b_3E_{11}E_{22}) - 3] \qquad (1)$$

or the simplified expression of Billiar and Sacks (2000)

$$\mathcal{W} = \frac{c}{2}[\exp(a_1E_{11}^2 + a_2E_{22}^2) - 1] \qquad (2)$$

in which a_1, a_2, b_0, b_1 , b_2 , b_3 and c are constants. In neither of these is account taken of the effect of shear strains (E_{12}), given their small experimental relevance.

Other more elaborate expressions separate the contributions to the strain energy due to collagen fibres, $\mathcal{W}^c$, elastin fibres $\mathcal{W}^e$ and the matrix $\mathcal{W}^m$

$$\mathcal{W} = \mathcal{W}^c + \mathcal{W}^e + \mathcal{W}^m$$

In this case, the values of $\mathcal{W}^e$ and $\mathcal{W}^c$ depend only on the elongation produced in the preferential direction of the fibres. More details can be obtained by consulting Humphrey (2002).

As an illustration of the process of obtaining the constitutive equation from the results of biaxial tensile tests, it is interesting to review the work of Sacks and Chuong (1998) on bovine pericardium membranes crosslinked with glutaraldehyde. The experimental arrangement is shown in figure 6.10.

Square, 25x25mm, specimens were cut from the pericardium membrane so that the preferential orientation of the collagen fibres is one of the axes of the test (1-axis). In this way, and by symmetry of the specimen, biaxial tension does not produce angular deformation and the situation is equivalent to that studied in section 5.2.3. The specimens were pre-conditioned with biaxial loading and unloading with equal strains along both axes. As is often the case when dealing with soft biological materials it was assumed that the volume remains constant.

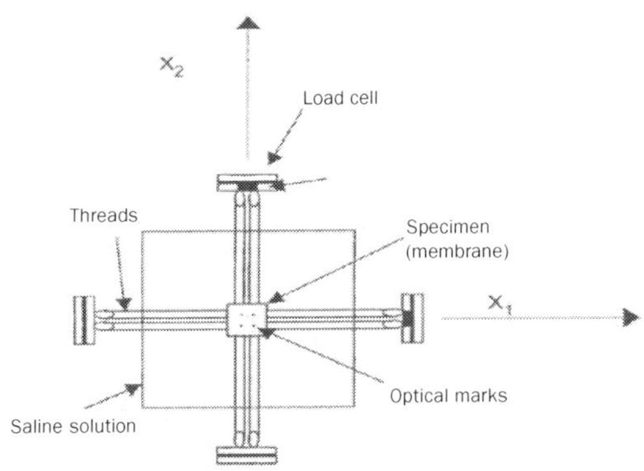

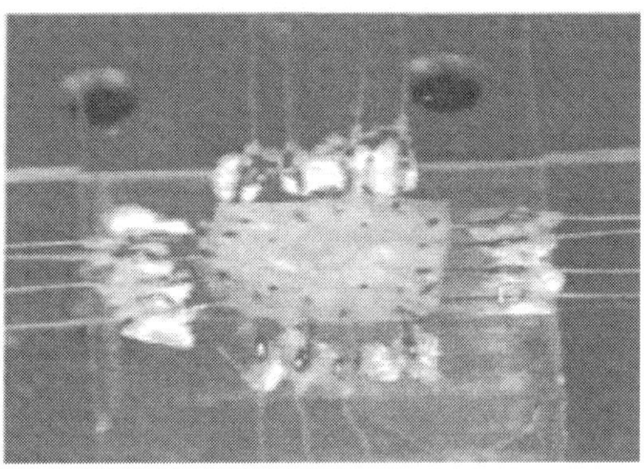

Fig. 6.10. *Biaxial tensile tests. The load is applied using suture threads and the strain measured from the displacement of marks drawn on the surface of the membrane (Sacks and Chuong 1998).*

The value of the strain was measured from the positions of four reference points on the specimen (figure 6.10), using equation (41), and the second Piola-Kirchhoff stress tensor was obtained from the forces exerted on the sides of the membrane, by means of equations (47).

Biaxial tests were performed with the ratios E_{11}/E_{22}= 1/3, 1/2, 1, 2 and 3, up to a maximum strin value of 0.16, equal to that used in the pre-conditioning cycles. This is equivalent, according to equation (41), to applying an elongation of λ_2 varying with λ_1 as shown in figure 6.11. The maximum level of strain was chosen so that the stresses did not approach 0.16MPa, which is the upper limit of stress estimated that can be reached during the normal function of the valve veils. Note that these values are far from the rupture values of the pericardium (stresses approximately between 10 and 30MPa).

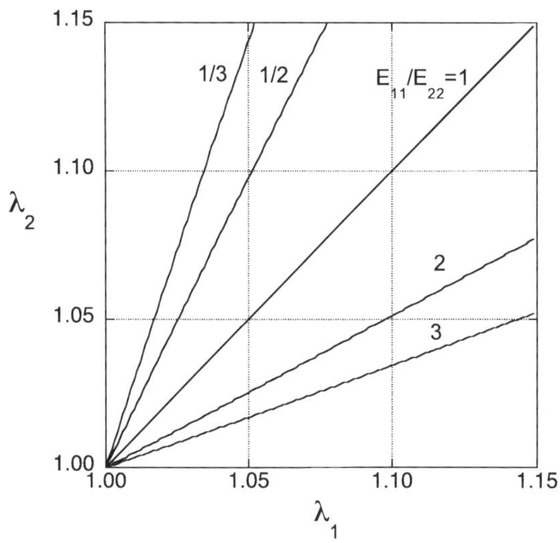

Fig. 6.11. *Values of the elongation in Sacks and Chuong's (1998) biaxial tensile tests.*

Figure 6.12 contains the results of the biaxial tensile tests performed and the stresses obtained from a fit to the constitutive equation (1):

$$W(\text{kPa}) = 1.873\ [\exp(53.46E_{11}^2) + \exp(35.7E_{22}^2) + \exp(45.52E_{11}E_{22}) - 3] \quad (3)$$

which provides the stresses

$$S_{11}(\text{kPa}) = 200.3E_{11}\exp(53.46E_{11}^2) + 85.26E_{22}\exp(45.52E_{11}E_{22})$$

$$S_{22}(\text{kPa}) = 133.7E_{22}\exp(53.46E_{22}^2) + 85.26E_{11}\exp(45.52E_{11}E_{22})$$

$$S_{12} = 0 \quad (4)$$

which agree reasonably well with the experimental results (r^2=0.97).

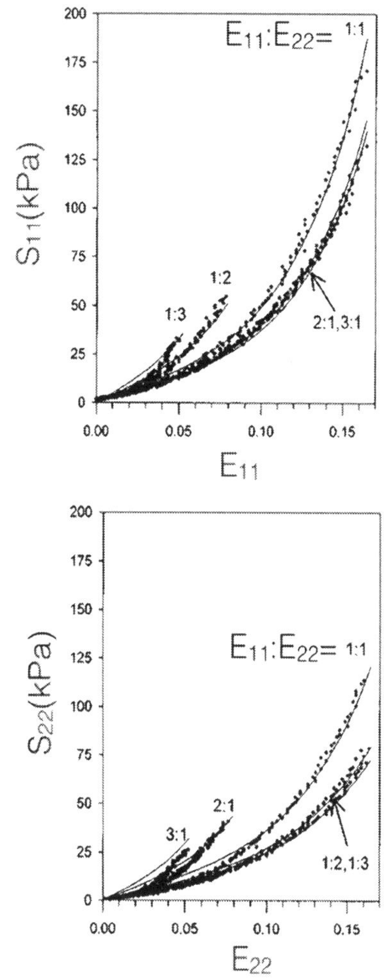

Fig. 6.12. *Biaxial tensile tests on bovine pericardium. Second Piola-Kirchhoff stress, S, as a function of the Green-Lagrange strain, E, and fit of the model given by equation 111. Sacks and Chuong (1998).*

Sacks and Chuong (1998) also showed that the predictions from equations (117) are reasonable for simulating an experiment in which the deformation is restrained along the 1-axis (that of the collagen orientation). Figure 6.13 shows the experimental results and the predictions from equations (4).

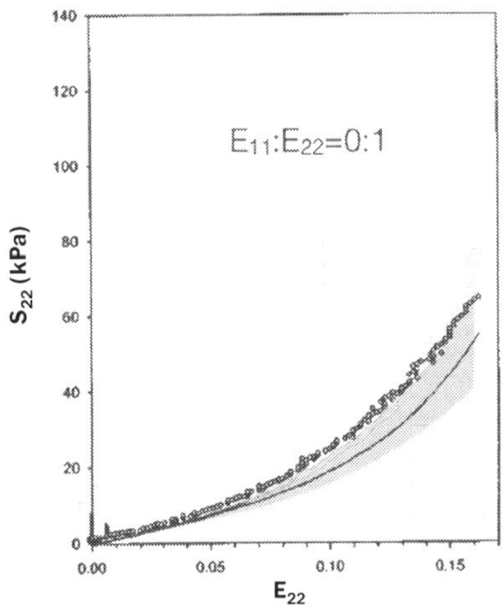

Fig. 6.13. *Biaxial tensile test on bovine pericardium with restrained deformation E_{11} Experimental results and predictions of the model from equation (117). The margin of error is shaded. S_{22}=second Piola-Kirchhoff stress, E_{22}=Green-Lagrange strain. Sacks and Chuong (1998).*

It is also possible, with the aid of equations (4), to predict the results of a uniaxial tensile test, in which stress is applied in only one direction. Figure 6.14 shows the stress-elongation curve deduced from equations (4) for the cases S_{11}=0 (uniaxial tension along the 2-axis, perpendicular to the collagen fibres) and S_{22}=0 (uniaxial tension along the 1-axis, parallel to the collagen fibres). The second Piola-Kirchhoff stresses (S_{11} or S_{22}) have been represented together with the Cauchy stresses (σ_{11} or σ_{22}) obtained by using equations (44) and (46). The elongations λ_1 and λ_2 were obtained from the Green-Lagrange strains using equation (41).

Claramunt *et al.* (2011) performed biaxial fatigue tests by pressurising membranes, adopting the arrangement shown in figure 5.12. This process simulates the loading of the pericardium when it is used as a veil in valve prostheses. With these it was shown that the behaviour in the first few fatigue cycles is a good indicator of the fatigue strength of the bovine pericardium.

Also using pressurisation tests, García Páez *et al.* (2001) showed that ostrich pericardium has a greater static strength than the bovine pericardium (both crosslinked with glutaraldehyde) even though its resistance to fatigue is much lower (Claramunt *et al.* 2012).

By means of biaxial fatigue tests by pressurisation at different frequencies (from 1 to 30 Hz) and average maximum stresses between 0.5 MPa and 1 MPa –which are estimated to be the maximum produced in a valve veil– it has been demonstrated that biaxial fatigue induces residual strains which can significantly alter both the geometry of the specimen (Sun *et al.* 2004) and its mechanical properties (Martin and Sun, 2014), increasing the rupture load and stiffening the pericardium membrane.

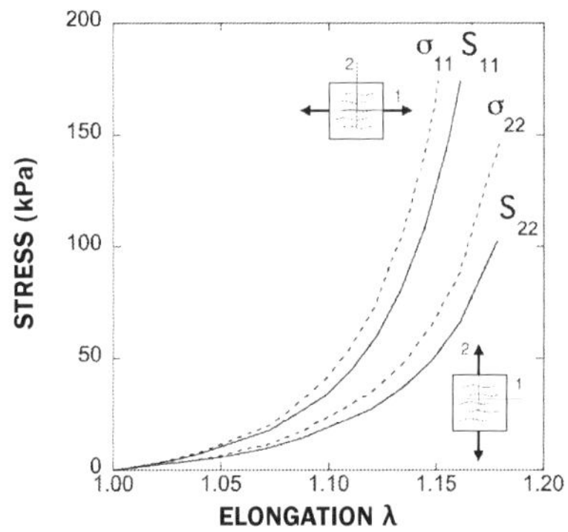

Fig. 6.14. *Uniaxial tensile curves deduced from the constitutive equation (4).*

In these experiments it has also been demonstrated that–just as in uniaxial fatigue– the first fatigue cycle is different from the rest, being focussed primarily in permanent deformation of the membrane by alignment of the fibres in the loading direction (Sellaro *et al.* 2007). This fact plays a critical role in the mechanical response of the material, since various studies have shown that the energy dissipated in the first fatigue cycle is an indicator of damage in the material, which can be used to estimate its durability and strength; see figure 6.15 (García Páez *et al.*, 2009; Rojo *et al.*, 2010; and Claramunt *et al.*, 2011).

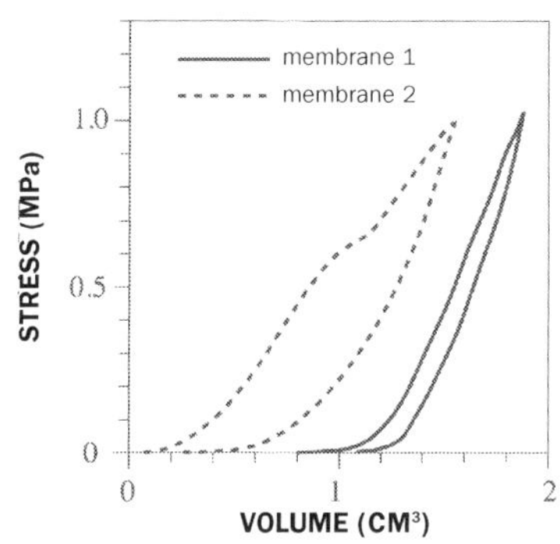

Fig. 6.15. *First loading-unloading cycle in pressurisation tests. The hysteresis curve of greater area indicates a lower specimen durability to fatigue (Rojo et al. 2010).*
Membrane 1: Hysteresis energy in the first cycle = 0.09J; failure after 600 cycles. Membrane 2: Hysteresis energy in the first cycle = 0.30 J; failure after 13 cycles.

Tear tests

The tear strength of the pericardium has been investigated by several authors, given its use as a structural biomaterial, and the need to know its strength when subjected to loading through sutures. Unfortunately, the variability of the results obtained is very large, and depends not only on the type of pericardium but also on the particular crosslinking treatment and the animal from which it was obtained, so that it is only possible to establish trends and general values of the rupture parameters.

In 1989 Trowbridge studied the sensitivity of bovine pericardium membranes to the presence of triangular notches such as that shown in figure 6.16. As can be seen in figure 6.17, the pericardium membranes are quite tolerant to the presence of this type of defect, which do not significantly affect the rupture stress or strain. In this work, it was observed that the tearing of the pericardium was always accompanied by a process of delamination between the fibrous and serous-parietal layers (see figure 6.2).

Using Elmendorf's tear test, the specific rupture energies of bovine and ostrich pericardium have been measured. It was observed that there is a significant variation with the length of the tear. Bovine pericardium specimens of smaller dimensions (tear 2.5mm) had an average energy of 33 N/mm compared to the 270 N/mm recorded for a tear length ten times bigger (25mm) (García Páez, *et al.*, 2006). In this work no significant differences were observed between the circumferential and apex-caudal directions.

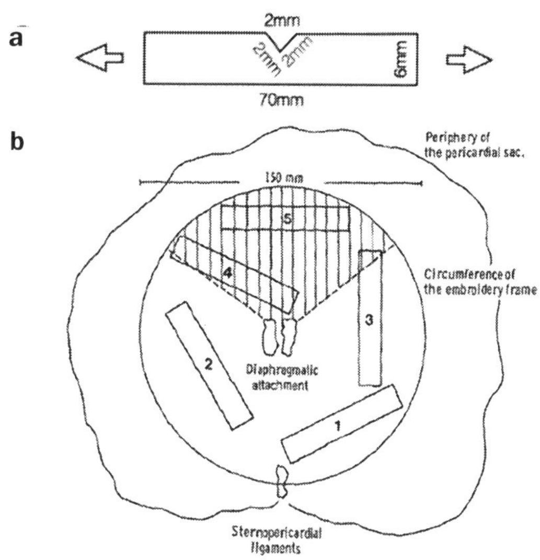

Fig. 6.16. *Tear tests on notched specimens. a. specimen geometry; b. anatomical position (Trowbridge 1989).*

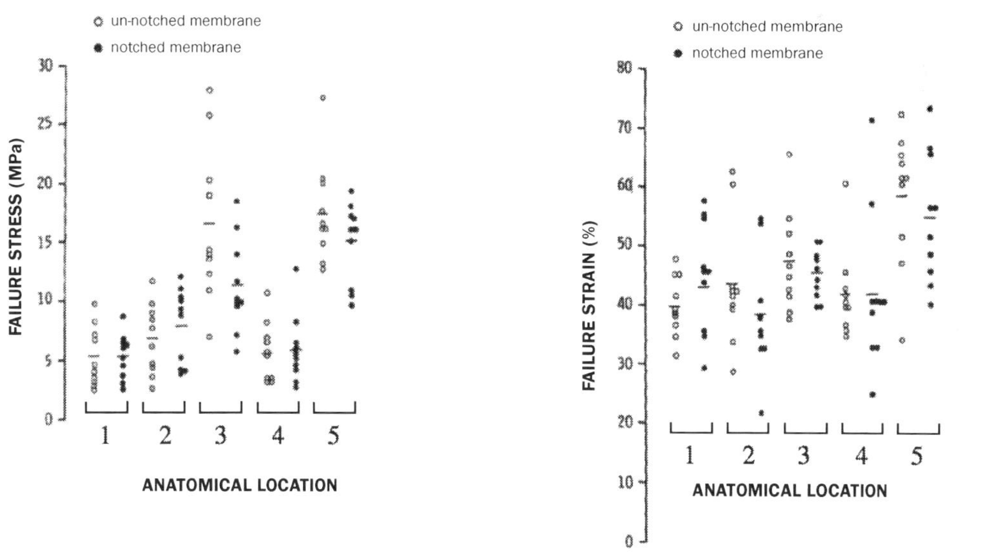

Fig. 6.17. *Tear stress and strain in specimens with triangular notches. There are no significant differences between the notched and un-notched specimens, nor between the different locations and orientations of the specimens (Trowbridge 1989)*

García Páez *et al.,* in 2004, also measured the load which produces the tearing of sutures in bovine and ostrich pericardium (figure 6.18). The specimen, in the form of trousers, had dimensions of 20 mm wide and 120 mm long. The load to initiate tearing in sutured specimens was similar to, or somewhat larger than, that in the pericardium without suturing.

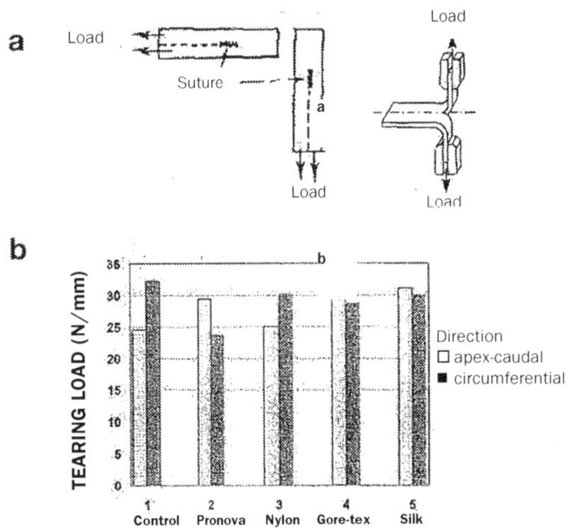

Fig. 6.18. *a. Specimens used to evaluate the strength of sutures in pericardium.*
b. Values of the tearing load (García-Páez et al. 2004).

However, the energy absorbed in tearing was greater in the material without the suture, as shown in some Elmendorf tests. In these, the average tear strength measured in the apex-caudal direction was 120N/mm in specimens without sutures compared to a value of 47 N/mm in 10mm sutured specimens. (García Páez *et al.,* 2010). In these cases, it was observed that tearing of the membrane produced delamination between the layers.

A detailed inspection of pericardium membrane ruptured by the tearing process shows that delamination always occurs between its layers, in the manner indicated in figure 6.19.

The phenomenon accounts for the increase in rupture energy as the tear length increases (and, therefore, the delimitation area). In their work, Tobaruela *et al.* (2016) estimated that the "pure" tear resistance (that is to say, tearing of the pericardium layers, without producing delamination) is of the order of 30 N/mm, whether the test is performed on the apex-caudal or circumferential direction, while the delamination process –inseparable from the former– consumes about 0.5 N/mm, also nearly independent of the test direction.

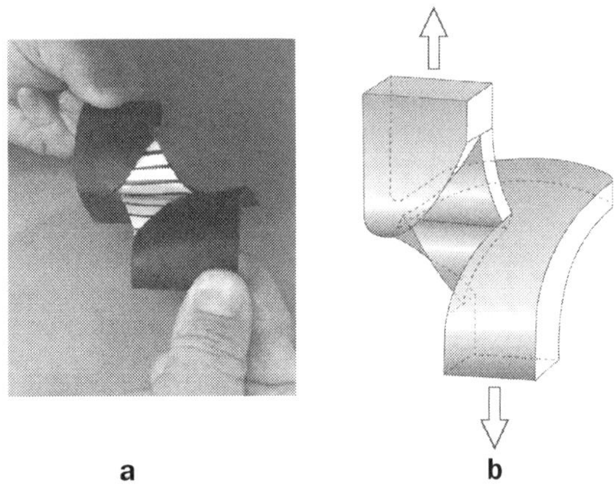

Fig. 6.19. *Delamination during the tearing process (Tobaruela et al. 2016).*

Effect of crosslinking

For use in bioprostheses it is essential, as we have noted, that the pericardium is treated and crosslinked, in order to avoid it degrading in the interior of the body and to reduce its immunogenicity potential. Hence, the vast majority of studies performed have been carried out on treated membranes, as noted in previous sections.

There are several pieces of work on natural membranes which show the effect of crosslinking treatment on the mechanical properties of the pericardium. In general, the treatment with glutaraldehyde substantially increases the stiffness of the material and the elongation at rupture, but there are no conclusive results as to its effect on the tensile strength (Trowbridge *et al.*, 1989; and Mirnajafi *et al.*, 2005).

It has been demonstrated that some decellularisation processes with ionic detergents such as SDS (sodium dodecyl sulphate or sodium laurylsulphate, commonly used in biology and personal hygiene products) can produce irreversible denaturation and swelling of the collagen, which reduces the tensile strength (Courtman *et al.*, 1994; García-Paéz *et al.*, 2000; and Mendoza-Novelo *et al.*, 2009). Therefore, it is preferred, if possible, to avoid non-ionic detergents (Mendoza-Novelo *et al.*, 2011).

6.2.3. Applications

The pericardium used as a biomaterial is commonly bovine or porcine, although on occasions other sources such as equine, canine, ostrich or kangaroo have been employed. Table 6.1 shows the most important biomedical applications found in the market.

TABLE 6.1 APPLICATIONS OF PERICARDIUM MEMBRANES (RÉMI *et al.*, 2011)

SOURCE OF PERICARDIUM	SURGICAL APPLICATIONS	COMMERCIAL PRODUCT	COMPANY
Bovine or Porcine	- Repair of soft tissue - Hernias - Defects in the abdominal wall and thorax	- Implantable surgical tissue Peripatch® - TutoMesh®	Neovasc, Maverick Biosciences PTY Ltd, Tutogen Medical GmbH, RTI Biologics,Med&Care, Biovascular Inc, Novomedics
	- Suture reinforcement	- Veritas Peristrips® Dry	Synovis Life Tech
	- Orbital reinforcement	- Tutopatch® - Ocugard®	Tutogen Medical GmbH, RTI Biologics, Med&Care, Biovascular Inc, Novomedics
	- Dural repair	- Lyolem® All BP	National Tissue Bank Malaysia
	- Perivascular repair	- Parche vascular Peripatch®	Neovasc
	- Cardiac repair and reconstruction	- Implantable surgical tissue Peripatch®	Neovasc, Maverick Biosciences PTY Ltd
	- Heart valve replacement	- PercevalS® - Mitroflow® - Freedom solo® - Carpentier-Edwards PERIMOUNT® Magna EaseAortic Heart Valve - Carpentier-Edwards PERIMOUNT® Magna Mitral Ease Heart Valve - Carpentier-Edwards PERIMOUNT® Theon Aortic Heart Valve - Carpentier-Edwards PERIMOUNT® Theon Mitral Replacement System	Grupo Sorín Edwards Life Sciences
Equine	- Tendon repair	- OrthADAPT®	Synovis Life Tech
Human	- Valvuloplasty - Heart valve	- Autologous, homologous or heterologous grafts	

Heart valves

The most important use of the pericardium as a biomaterial, as much as for mechanical as economic reasons is found in the field of biological heart valves (figure 6.4). The pericardium is used as a valve veil and must support the continuous functioning of these prostheses for millions of cycles. Currently, even though the means of selecting the biomaterial with a certified durability is not available, which would allow biological prostheses to compete for reliability with mechanical ones, the latter have the inconvenience of requiring the prescription of an anticoagulant treatment for life.

The most common cause of premature rupture of the pericardium valve veils is calcification, which stiffens the membrane and promotes the propagation of cracks. This phenomenon is known to be associated with the collagen crosslinking treatments (glutaraldehyde induces calcium deposits) and, moreover, this is sensitive to the stress-strain state in the membrane. Currently, this is one of the most active areas of scientific study.

On the other hand, the introduction of, and the advances in, the techniques of valve implantation by catheter (percutaneous implantation) is driving the miniaturisation of bioprostheses in a form that can be folded and transported through the interior of a catheter as far as its end point, where it is unfolded and attached the walls of the duct (figure 6.20).

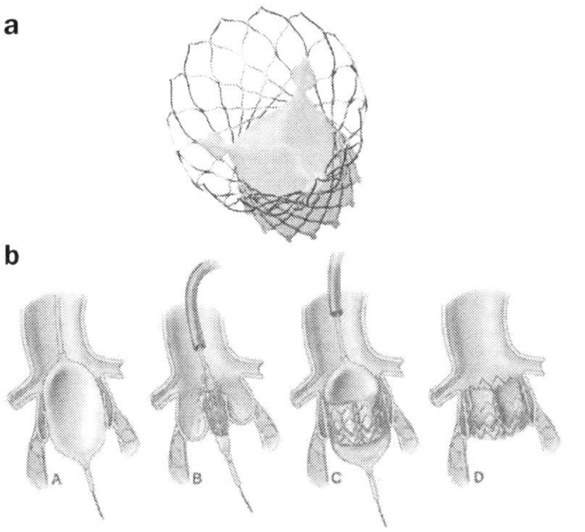

Fig. 6.20. *a. Valve for percutaneous implantation.*
b. Surgical procedure

This type of application is very demanding in terms of the thickness and uniformity of the veils, which cannot exceed about 0.3 mm thickness. Neonatal bovine pericardium is a good candidate for these applications, being finer and more elastic than that in the adult. However, the expected duration of this type of valve, less than 10 years, imposes severe restrictions on its use, which is limited to patients whose condition is highly deteriorated, or at high risk in invasive surgery.

Repair of soft tissue

The traditional use of pericardium membranes has been in the repair of ruptured membranes of the abdominal cavities and blood vessels. I. these cases, a piece of pericardium is cut into the desired shape and stitched to the adjacent tissue, in order to return the structural integrity (figure 6.21).

Pericardium membranes are indicated for patients with severe obesity or diabetes, where biological grafts, rather than synthetic meshes, enhance revascularisation and reduce the risk of postoperative infections.

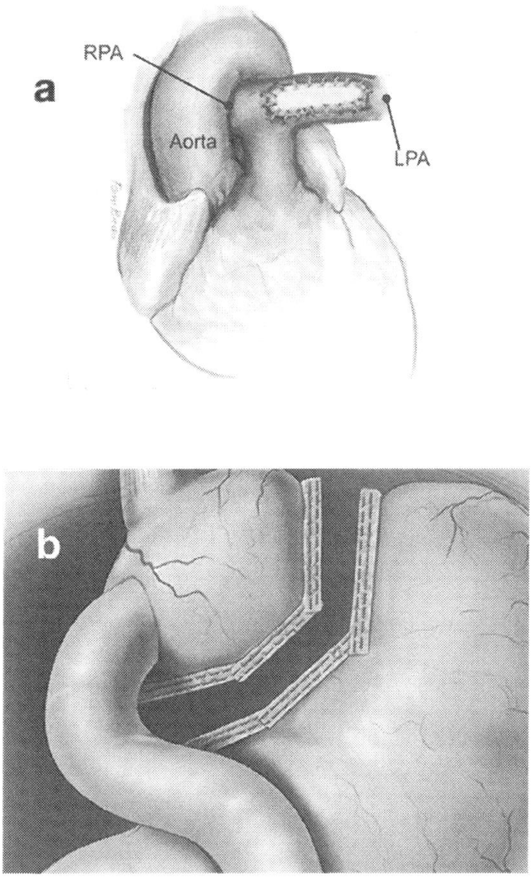

Fig. 6.21. *a. Pulmonary artery repair using a pericardium patch.*
b. Reinforcement of the suture line in a stomach reduction procedure with pericardium

Conclusion

The pericardium, primarily that of bovine or porcine origin, is a biological tissue that is widely used as a biomaterial in bioprostheses for vascular grafts, patches for the repair of the abdominal or vaginal wall and, more frequently, heart valves. However, despite the advances achieved over the last few years, pericardium membranes still present some important drawbacks such as the deterioration of their biological matrix and degeneration in the body, often associated with the appearance of calcification.

There are a large number of decellularisation and crosslinking protocols aimed at providing stability and durability to the pericardium to eliminate cell antigens. Of these the most common is the treatment with glutaraldehyde, which despite its undeniable advantages, changes the mechanical properties, the integrity of the extracellular matrix and increases the risk of calcification.

The problem of calcification, together with the difficulties in finding material selection procedures which guarantee uniform properties, are the principal drawbacks which limit the use and development of biological heart valves. A better understanding of the structure of the pericardium and improved selection and processing methods will improve the conception and design of pericardium bio-prostheses and increase the safety of the patients.

6.3. AORTIC ARTERIAL WALL

6.3.1. Introduction

The aorta is the human artery which distributes blood from the heart to the rest of the arteries which form the major circulation system, responsible for carrying oxygenated blood throughout the body. It is also the largest artery, with an average diameter and thickness of 25 mm and 2 mm, respectively (Burton, 1954). By means of its elastic deformation, the aortic artery smooths out the pulsating blood flow coming from the heart, facilitating the passage of blood to the other organs and tissues.

The aortic artery is prone to injury induced and/or aggravated by the blood pressure, the most common being the formation of plates, the development of aneurysms and dissections (or tears) in the wall. It has been found that such injuries, like other histopathological changes in the aortic wall, are spatially correlated with the mechanical stresses present (Bäck *et al.* 2013). The aortic artery is also sensitive to ageing, which produces important changes in its thickness, diameter and stiffness (Lanne *et al.* 1992).

There is great interest in describing and modelling the biomechanical behaviour of the aortic wall in order to better understand its physiological behaviour and ageing process, and to improve and optimise the diagnostic processes and treatment of aortic illnesses, and the development of its pathologies. There have, therefore, been a large number of expositions and constitutive equations proposed for the biomechanical study of this biological material. The task is not easy, and a satisfactory knowledge is still a long way off, since it is difficult to incorporate the complex architecture and composition of the arterial wall into biomechanical models which are robust and readily transferable to clinical practice.

In this section, a brief summary of the behaviour of the aortic arterial wall is presented, together with the current state of biomechanical modelling, beginning with a short description of its composition and structure. Its mechanical behaviour and the most important models used for its study will then be reviewed. The section is concluded with some applications of medical interest, such as rupture of aneurysms and the effect of arterial hypertension.

6.3.2. Description of the aortic wall

The aortic wall, like other blood vessels, is composed of an extracellular matrix of proteins and proteoglycans which maintain its integrity, and a set of cells which, as well as generating and preserving that matrix, have a direct mechanical action on the geometry of the vessel in order to control its physiological function.

The principal components of the extracellular matrix are *elastin* and *collagen* proteins, which determine the passive mechanical behaviour –without muscular activation– of the artery. The smooth muscle cells present in the aortic wall are responsible for its active behaviour and, together with the fibroblasts, produce and maintain the extracellular matrix (McDonald, 1974).

The structure of the aortic wall is also similar to that of other blood vessels, being organised in three layers: *intima, media* and *adventitia* (figure 6.22).

The **intima** layer is the innermost and is formed of a monolayer of endothelial cells. These cells develop and grow on a very thin basal plate, about 80nm thick, formed of type IV collagen and adhesive molecules, fundamentally laminin 1 and fibronectin. The basal plate is supported, in turn, by a sub-endothelial lamina composed of connective tissue and smooth muscle cells aligned with the axial direction of the vessel. The endothelial cells are plane and elongated in the axial direction, the direction of the blood flow. Their dimensions are approximately 0.3μm x 12μm x 40μm (thickness x width x length) (Humphrey, 2002).

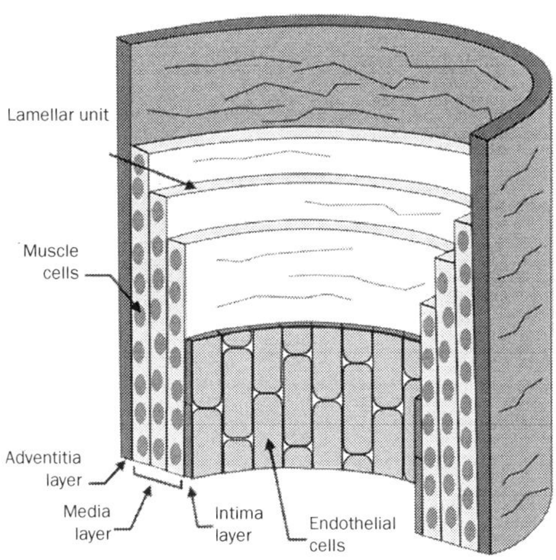

Fig. 6.22. *Structure of the aortic wall. For simplicity, only three lamellar units are shown.*

The endothelial cells are in constant contact with the blood flow and provide an anti-thrombogenic, low-friction substrate, bearing the frictional shear stress with the blood whose physiological value lies between 1 and 2 Pa. These cells communicate the variations in the blood flow to the other cells in the aortic wall, so that they can adapt to the new conditions (Humphrey, 2002; Chen *et al.*, 2003).

The **media** layer is primarily responsible for the mechanical behaviour of the aorta, as much because of its dimensions as its components and configuration. It is composed of sublayers called *lamellar units* (figure 6.23a. which are about 14μm thick (Clark and Glagov, 1985). Each lamellar unit consists of concentric sheets of fine highly crosslinked elastin fibres (0.1 to 0.5μm) between which are inserted smooth muscle cells and some fibroblasts, all preferentially arranged in the circumferential direction (O'Connell *et al.*, 2008). The number of lamellar units varies as a function of the thickness of the aortic wall so that the circumferential force which each unit supports is, essentially, constant at about 2 ± 0.4 N/m, figure 6.23b (Wollensky and Glagov, 1967).

Bundles of collagen fibres lie between the lamellar units, running parallel to the muscle cells, surrounding them. These fibres are wavy and, at physiological pressures, only about 7% of them carry any stress (Armentano *et al.*, 1991).

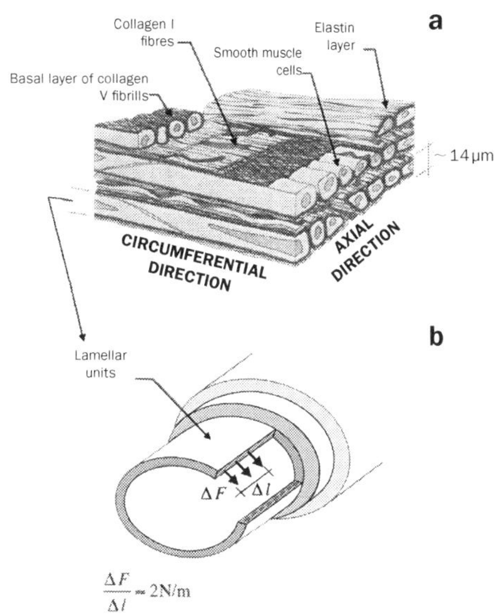

Fig. 6.23. *a) Schematic representation of lamellar units (Clark and Glagov 1985)*
b. Circumferential force carried by each lamellar unit.

The **adventitia** is the outermost layer (figure 6.22) and is composed principally of bundles of micrometric-sized collagen fibres and loose connective tissue which anchors the artery to its surrounding tissue. Some soft muscle cells and, mainly, fibroblasts are responsible for its maintenance, being able to adjust the expression and synthesis of collagen in response to mechanical signals (Bishop and Lindahl, 1999).

The structure and composition of the aortic artery are not uniform and change with position, as shown in figure 6.24. The elastin content, and in consequence the elastic character of the vessel, reduces significantly from the diaphragm, into the abdominal region of the aorta. The diameter and thickness also diminish monotonically away from the exit from the heart (Sokolis, 2007).

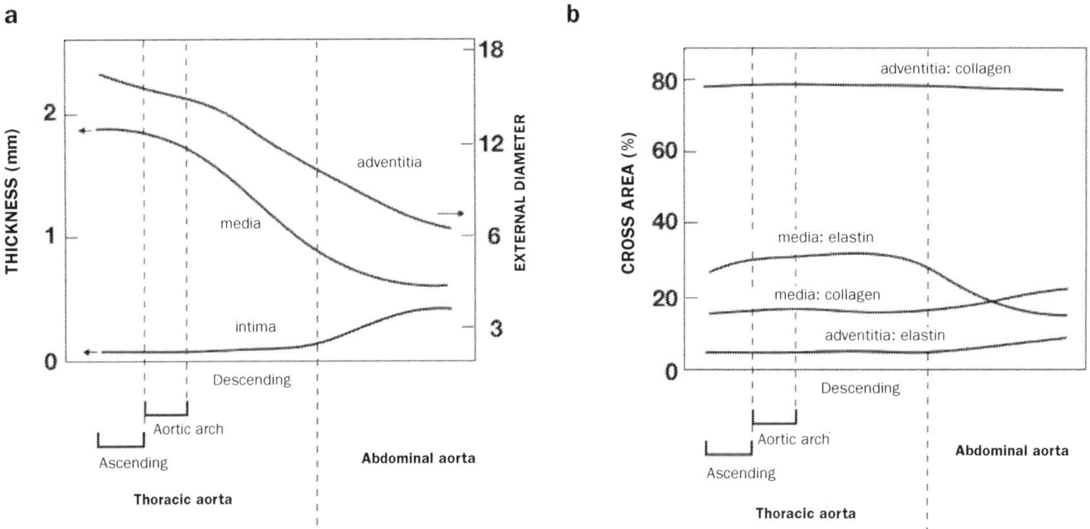

Fig. 6.24. *Structure of the porcine aortic wall as a function of its position.*
a. External diameter of the vessel and thickness of each layer.
b. Percentage of cross-section occupied by each component.
Note that the intima layer occupies less than 1% of the total thickness (Sokolis, 2007).

6.3.3. Mechanical behaviour

Owing to essentially mechanical functions –blood flow and pressure regulation (by deformation and elastic recovery)– and also by the number of mechanical defects which it can suffer (wall ruptures, formation of aneurysms), the aortic artery has always been at the forefront of the study of biomechanics, since the pioneering works of C. S. Roy in 1881.

In a study of the behaviour of the aorta, we must distinguish between its passive and active states. The former corresponds to the situation in which the smooth muscle cells are not contracted, while in the latter there is muscular contraction. Although, ideally, the passive state is reached when the cell population dies, some authors regard the passive as corresponding to the basal cellular state, in which there is some degree of contraction (basal muscle tone).

Figure 6.25 shows pressure-diameter tests carried out on rat aorta under the maximum muscle activation and in the passive state. The curves were obtained for two different values of axial elongation in the aorta, since in their *in vivo* physiological state the arteries are always stretched. In the human aorta the average elongation *in vivo* reaches a value of $\lambda = 1.4$ in young persons, decreasing with age (Horný *et al.*, 2014). Muscular activation changes the behaviour of the vessel making it stiffer, especially in low pressure zones. Note that the arterial pressure of rats is similar to that of humans, having an average value of around 100 mmHg.

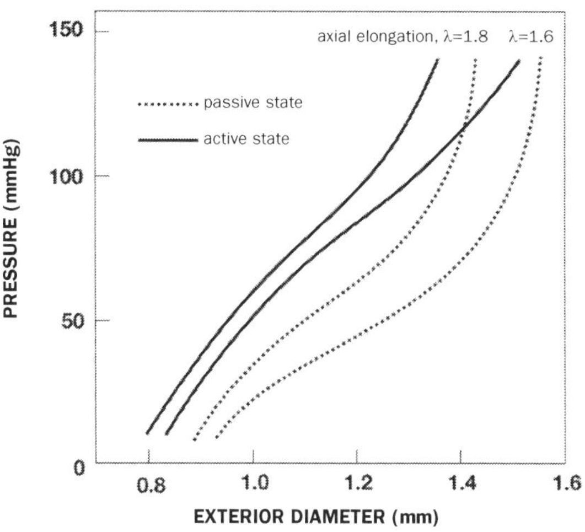

Fig. 6.25. *Pressure-diameter curves of rat aorta in the passive and active states. The rat aorta is subjected in vivo to an axial elongation. Adapted from Agianniotis et al. (2012).*

The results in figure 6.25 indicate that, if it is required to simulate correctly the *in vivo* behaviour of the artery, muscular contraction cannot be ignored. The same is true of the other arteries, since the effect of the smooth muscle contraction on the artery wall increases as we move away from the heart and becomes very important in the farthest arteries– called muscular– such as those originating in the celiac trunk, renal or radial arteries. The presence of the smooth muscle and its activation not only change the elastic response of the aorta but also significantly affect its viscoelastic behaviour.

Examination of passive behaviour shows that the pressure-diameter curves of the aorta have two zones of very different stiffnesses, figure 6.26. The first of these zones corresponds to the region with pressures equal to, or less than, the physiological, while the second is where they are higher than the normal physiological pressure of 120 mmHg. When the vessel exceeds this value of pressure, its stiffness increase greatly–by almost an order of magnitude– due to the collagen fibres coming into play, which were initially wavy and stress-free.

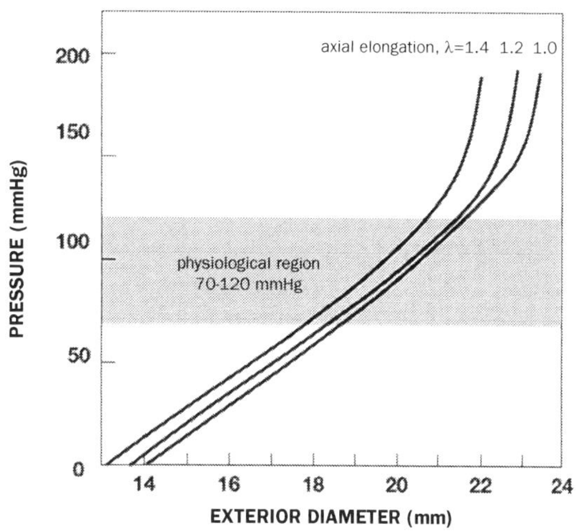

Fig. 6.26. *Pressure-diameter curves for human aorta (20-year old female) in the passive state (García-Herrera, 2008).*

The functioning physiological zone, of pressures between 70 and 120 mmHg, is situated in the most flexible region of the pressure-diameter curve, immediately before the transition. In this interval, the mechanical behaviour of the aorta is almost exclusively controlled by the elastin fibres, which are very elastic (Roach and Burton, 1957). Therefore, when working in the physiological range, the aorta exercises its entire capacity to deform elastically to accommodate the volume of blood pumped by the heart, and to smooth out its delivery to the rest of the arterial system, in a phenomenon known as the Windkessel[27] effect. When the pressure rises above the physiological level, the collagen fibres are activated, as a reinforcement which prevents excessive deformations which could lead to rupture of the vessel.

The uniaxial tensile test on the arterial wall, although somewhat removed from the real mechanical situation of the pressure-diameter test on the vessel, is more frequently used due to its ease of execution and data interpretation. Figure 6.27 contains data from uniaxial tensile tests to rupture, performed on healthy adult (ages between 16 and 36 years) human aortas.

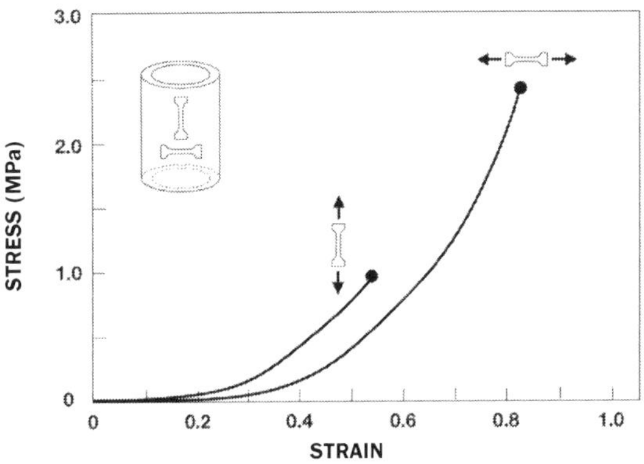

Fig. 6.27. *True stress-strain curves for specimens from the human aorta of healthy young adults (16 and 36 years), in the passive state (García-Herrera, 2008).*

These experiments provide a graphic illustration of the anisotropy in the aortic wall, which is much stiffer and weaker in the axial than in the circumferential direction, since, as has been seen in the previous section, the collagen fibres, like those of elastin, are distributed preferentially in the circumferential direction.

Uniaxial tensile tests are also very useful in characterising the behaviour of aortic material as a function of time. Figure 6.28 shows loading and unloading cyclic tests performed on specimens orientated in the longitudinal and circumferential directions. The material initially shows an important hysteresis loop which is significantly reduced after a small number of cycles, attaining a repetitive and stable behaviour[28]. From a physiological point of view, the hysteresis cycle has an important damping effect which helps reduce the flow and pressure fluctuations produced by the waves generated by the heart beat.

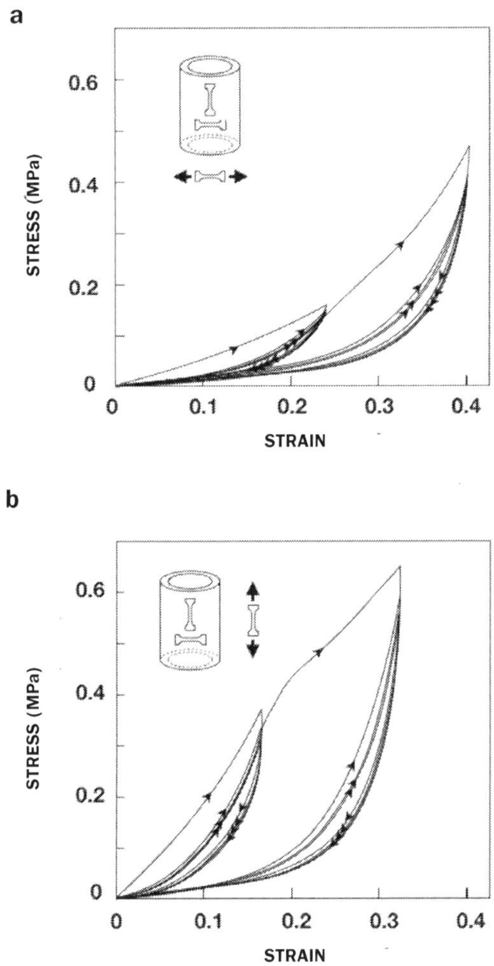

Fig. 6.28. *Cyclic true stress-strain curves of human thoracic aorta specimens. Adapted from Horný et al. 2010.*

[27] In German *windkessel* means *air chamber*, referring to devices used in water tanks by eighteenth century firemen to obtain a continuous flow of water, and eliminate pulsations of the pump piston. The elastic deformation of the arterial aorta and other elastic arteries store some of the blood pressure from the heart systole and return it to the arterial system during the diastole, smoothing out the blood flow. This idea was proposed by the English scientist and cleric Stephen Hales (1677–1761), and mathematically modelled by the German physiologist Otto Frank (1865–1944).

[28] This phenomenon, known as the Mullins effect, was first described for rubbers by the English researcher Leonard Mullins (1918–1997).

Biomechanical models

The first challenge appears when considering the structure of the aortic wall. The aorta is a combination of solid components (protein, proteoglycan and cellulose fibres) and water, the great part of which have little mobility being tied by the proteoglycan and elastin macromolecules. Even so, there is a radial flux generated by pressure differences in the blood, in the interior of the artery, and the interstitial fluid in the exterior which is essential for the diffusion of nutrients and signalling molecules. On the other hand, the disposition of the solid components is very complex as we have seen, and there is not –as occurs in all biological materials– a clear distinction between material and structure, since the different components interact over a wide range of scales of observation, from the molecular to the microscopic level, without interruption.

In order to overcome this difficulty using the mechanics of continuous media it is necessary to assume that the aortic wall is a continuous solid[29] composed of homogeneous parts. Although this assumption is common and has been made for all types of materials, in the case of biological materials, due to its internal organisation, this distinction is totally arbitrary and is made solely on the basis of the available data and/or the type of information required in the model. Therefore, we will have to consider the simplest models of the aortic wall as a single homogeneous material (for example, Hoppmann and Wan, 1970), or a homogeneous material reinforced with fibres, as in the case of Gasser *et al.* 2012. The most complex models consider the aorta as a material composed of several phases, each one with its own laws of evolution. Clearly, an increasing level of detail requires more sophisticated techniques and more refined experimental data which allow a reliable determination of the parameters of the models, which is not always possible.

It is not the aim of this text to go deeply into the biomechanical models of the aorta, so we will only provide a few simple examples, concentrating on the passive behaviour. For a deeper understanding of the topic, the interested reader can consult the bibliography at the end of the chapter.

TABLE 6.2 VALUES OF SOME THREE-DIMENSIONAL STRAIN ENERGY FUNCTIONS USED TO MODEL THE BEHAVIOUR OF THE AORTIC WALL IN THE PASSIVE STATE

MATERIAL	STRAIN ENERGY $\mathcal{W}$	REFERENCE
Isotropic	$\mathcal{W} = c_0 \exp\left(c_1 I_{1E} - 1\right)$	Deimirai (1981) Polzer *et al.* (2013) Riveros *et al.* (2013)
	$\mathcal{W} = \sum_{i=1}^{N} c_i I_{1E}^i$	Yeoh (1993) N=2: Raghavan and Vorp (2000) N=5: Polzer *et al.* (2013)
Orthotropic	$\mathcal{W} = c_0\left(e^Q - 1\right)$ $Q = c_1 E_{\theta\theta}^2 + c_2 E_{33}^2 + c_3 E_{rr}^2 + c_4 E_{\theta\theta} E_{33} + c_5 E_{rr} E_{33} + c_6 E_{rr} E_{\theta\theta} + c_7 E_{\theta 3} E_{3\theta} + c_8 E_{r3} E_{3r} + c_9 E_{r\theta} E_{\theta r}$	Chong and Fung (1986)
	$\mathcal{W} = c_0\left[\exp\left(c_1 E_{\theta\theta}^2\right) + \exp\left(c_2 E_{33}^2\right) + \exp\left(c_3 E_{\theta\theta} E_{33}\right) - 3\right]$	Choi and Vito (1990) Van der Geest *et al* (2006)
	$\mathcal{W} = -c \ln\left(1 - Q\right)$ $Q = \frac{1}{2} c_1 E_{\theta\theta}^2 + \frac{1}{2} c_2 E_{33}^2 + c_3 E_{\theta\theta} E_{33}$	Matsumoto and Hayashi (1996)
	$\mathcal{W} = c_0 I_{1E} + \sum_{i=1}^{N} c_{1i}\left\{\exp\left[c_{2i}\left(I_{4i} - 1\right)^2\right] - 1\right\}$	Holzapfel and Gasser (2001)

$\mathcal{W}$ is the strain energy per unit volume in the reference configuration.
c_i are the appropriate constants for each models.
I_{1E} is the first invariant of the Green-Lagrange strain tensor.
I_{4i} is a parameter which takes into account the strain according to the orientation of the i-th family of fibres.

[29] We understand a continuous solid to be one which completely occupies the space over which it extends, so that it leaves no "gaps". Therefore, a continuous solid can be subdivided infinitely, with every one of its parts keeping the properties of the original solid.

The simplest models represent the aortic wall as a continuous, elastic, incompressible material. This hypothesis is reasonable as a first approximation, since the materials can be considered continuous at the macroscopic scale (in the region of millimetres), which is the level at which surgical procedures are carried out (sutures, prosthetic implants and grafts, *etc.*) and also the level of precision of the common experimental measurements, both purely mechanical (forces and displacements), and flow and blood pressure. This implies that the stresses and strains obtained from the models lose their physical meaning below the millimetric scale, similar to that which occurs with homogeneous solids such as concrete, in which it is meaningless to specify stresses and strains obtained from continuous material models in regions smaller than the size of the aggregate.

On the other hand, as we have seen in figure 6.28, if the loading is cyclic the arterial tissue is stabilised and presents a response which does not vary with time. In this way, under the action of cyclic pressures originating in the heart, the arterial wall is found to be *preconditioned* in the sense indicated in section 5.2, so that one-to-one relationships can be established between stresses and strains, for the loading and unloading branches. This situation allows a hyperelastic representation of the constitutive equation[30] valid (and only valid) for each branch. This is the pseudoelastic process proposed by Fung in 1979 (Fung *et al.* 1979) and which is still used today.

Finally, as a result of its high water content, the aortic wall –like the majority of biological tissues– can be considered to be an incompressible material. This does not mean that there are no volumetric strains, but rather that these are several orders of magnitude less than the others, so they can be neglected without loss of accuracy[31].

As was observed in section 5.2.5, hyperelastic models are defined in terms of the strain energy per unit volume in the reference configuration $\mathcal{W}=f(\mathbf{E})$. From this, employing equations (66) the second Piola-Kirchhoff stress tensor $\mathbf{S}$ can be obtained, while using equation (67) (for the two-dimensional problem) or (69) (for the three-dimensional problem) yields the Cauchy stress tensor σ.

When the material is isotropic, the strain energy function $\mathcal{W}=f(\mathbf{E})$ can be written as $\mathcal{W}=f(I_{1E}, I_{2E}, I_{3E})$, depending only upon the invariants[32] of the strain tensor, $\mathbf{E}$. Table 6.2 shows some of the more popular isotropic functions for aortic tissue, in which, normally, $\mathcal{W}$ only depends upon the first invariant. Even though the aortic wall is not isotropic, these expressions are useful for one-dimensional representations in which it is required to model a single relationship, such as the force-displacement curve in a uniaxial test, or the pressure-diameter behaviour under axial elongation *in vivo*, without concern for the values of the other strains or forces acting.

In order to model the anisotropic behaviour of the aortic wall the anisotropy is introduced directly into the hyperelastic constitutive equation, either by writing it as a function of the individual components of the strain tensor, or through the particular invariants that take into account the structure of the material.

From the configuration of the constituents, described in section 6.3.2, the aortic wall is considered as an orthotropic[33] material whose principal directions are the radial, circumferential and axial axes, which were defined in figure 5.6c. For this reason, some models define the strain energy function $\mathcal{W}$ directly in terms of the components of the Green-Lagrange strain tensor in these axes, as shown in table 6.2.

Alternatively, some authors model the anisotropic behaviour using an *isotropic* strain energy function to which the dependence on some specific parameters I_m is added, which account for the directions of anisotropy $\mathcal{W}=f(I_{1E}, I_{2E}, I_{3E}, I_4, I_5, I_6 \ldots)$. In general, each parameter I_m takes into account the effect of a family of reinforcing fibres, according to their orientation in the aortic wall. One of these models is also shown in table 6.2. Details can be found in the references.

[30] As the reader will have already observed, in order to avoid confusion with the elastic behaviour under small deformations, it is usual to adopt the word hyperelastic in order to refer to elastic behaviour in the presence of large deformations.

[31] An incompressible material does not change its volume no matter what load is applied. From a practical point of view, we can consider a material to be incompressible when its bulk modulus is much greater than its elastic modulus.

[32] In an isotropic material the strain energy does not vary with the spatial orientation of the strain tensor and therefore depends only upon its invariants, defined as $I_1 = tr\mathbf{E}$, $I_2 = (tr^2\mathbf{E} - tr\mathbf{E}^2)/2$ and $I_2 = (tr^3\mathbf{E} - 3\, tr\mathbf{E}\, tr\mathbf{E}^2 + 2tr\mathbf{E}^3)/6$. The number of independent invariants is equal to the number of dimensions of the tensor: (I_1, I_2) for two dimensions and (I_1, I_2, I_3) for three.

[33] An orthotropic material has three planes of material symmetry.

6.3.4. Application: Arterial hypertension

Sustaining a high pressure in the arteries (typically more than 100 mmHg diastolic and 140 mmHg systolic pressures) causes remodelling and a permanent change in their geometry and mechanical properties. Later we will present, as an illustration, the study performed by Matsumoto and Hayashi (1994, 1996) who investigated the behaviour of rat aortas into which hypertension was induced over 8 weeks by mechanically reducing the cross-section of the renal artery.

After dispatching the animals and extracting their thoracic artery, the authors found that the thickness of the aortic wall increased considerably owing to the increased thickness of the lamellar units, while the internal radius of the artery was unchanged, or slightly increased. As a result of this change in geometry, the circumferential stress in aortas subjected to the hypertension remained essentially constant at the physiologically normal value of $\sigma_{\theta\theta}=300\,\text{kPa}$. The thickening of the wall also increased the value of the residual stresses, whose angle of aperture angle α (see section 4.3, figure 4.10) appeared to increase linearly with the arterial pressure. The remodelling of the vessel also produced a reduction in the *in vivo* elongation, λ_3, of 15%.

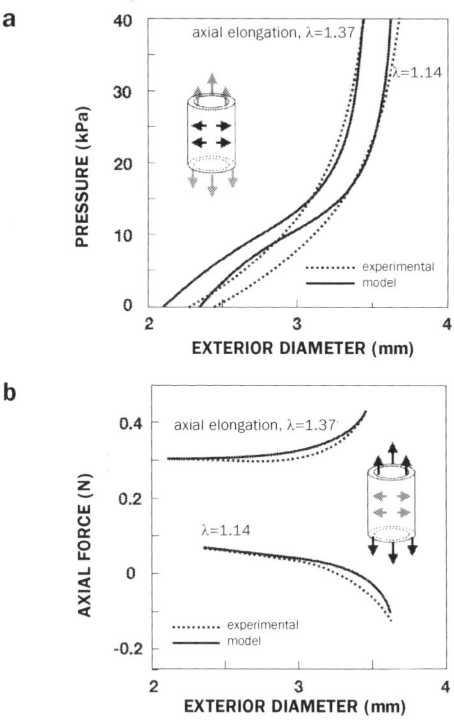

Fig. 6.29. *Pressure tests under axial elongation of arteries subjected to chronic hypertension. a. Pressure-diameter curves and fit of the biomechanical model. b. Value of the axial force and fit of the biomechanical model. The in vivo elongation of the normotensive vessels is about 1.37, and that in hypertension is about 1.14. Adapted from Matsumoto and Hayashi (1996).*

Figure 6.29 shows a comparison between Matsumoto and Hayashi's constitutive equation proposed in table 6.2 and experimental data from the pressure-diameter test. In this case a calculation was performed in which the vessel's internal residual stresses (measured by the angle of aperture) were incorporated.

As can be observed, the model agrees reasonably well with multiaxial loading produced by the pressure and simultaneous elongation of the vessel. The values of the parameters obtained were:

$c = 69.1\,\text{kPa}$, $c_1=1.317$, $c_2=2.001$, $c_3=0.355$ for the aorta which had suffered arterial hypertension (hypertensive aorta), and

$c=94.2\,\text{kPa}$, $c_1=0.773$, $c_2=1.031$, $c_3=0.237$ for the aorta under normal pressure conditions (normotensive aorta).

Figure 6.30 contains the predictions of the model for uniaxial tests in the circumferential and axial directions. The model adequately reproduces the large stiffness of the vessels in the axial direction and the general stiffening produced by the hypertension.

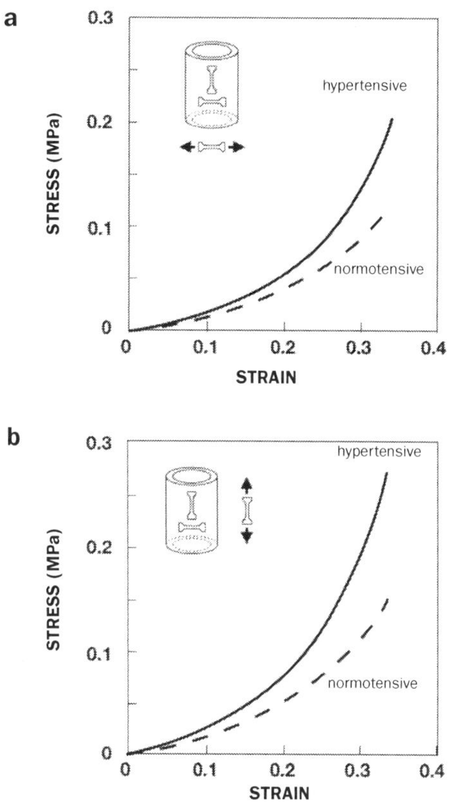

Fig. 6.30. *Modelling of uniaxial tensile tests on normotensive arteries and those subjected to chronic hypertension. a. Tested in the circumferential direction. b. Tested in the axial direction. Adapted from Matsumoto and Hayashi (1996).*

Several models developed to predict the evolution of the geometry and properties of hypertensive arterial walls have been developed. Perhaps the first and simplest –despite its complexity– was proposed by Rachev and co-workers (Rachev *et al.* 1998) which used the strain energy function suggested by Chong and Fung (1986).

The model proposes phenomenological approaches for the temporal development of the variables which control the geometry and the strain energy function $\mathcal{W}$.

Predicted values of the temporal variation of the thickness and angle of aperture (as a measure of the residual stresses) are shown in figure 6.31 for the rabbit thoracic aorta, using the model.

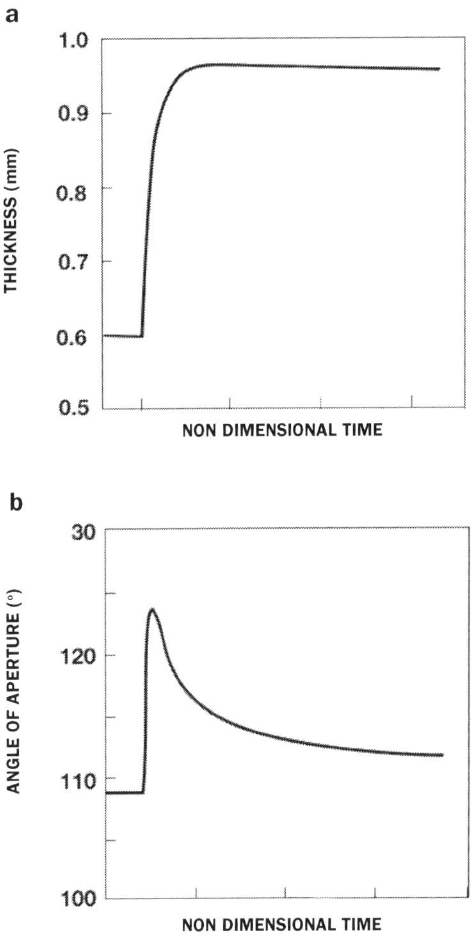

Fig. 6.31. *Model of the temporal variation of the hypertensive rabbit thoracic aorta. a. Variation of the aortic wall thickness. b. Variation of the angle of aperture. Adapted from Rachev et al. 1998.*

The model correctly predicts a monotonic increase in the aortic wall, reaching an equilibrium state when the thickness is approximately 50% greater than that of the normotensive vessel. The temporal evolution of the angle of aperture presents a very rapid increase followed by a drop to an asymptotic equilibrium value which can be above or below the control value, depending upon the location of the vessel, an observation also made by several authors (Fung and Liu, 1989). A similar trend (not shown in the figure) is observed for the internal radius, which tends rapidly to its initial value.

This model also correctly predicts the evolution of the mechanical behaviour of the vessel, which reproduces the experimental observations of Matsumoto and Hayashi (1994), as shown in figure 6.32.

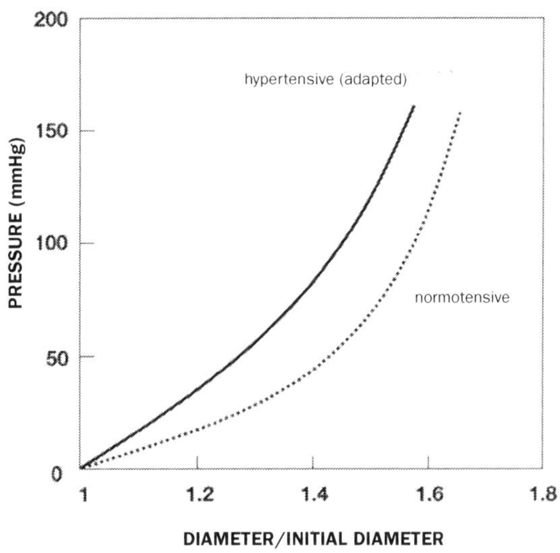

Fig. 6.32. *Pressure-diameter curves for the normotensive aorta, and that after reaching the equilibrium state under hypertension. Adapted from Rachev et al. 1998.*

Conclusion

Aortic tissue –and biological tissues in general– manifests a huge capacity to adapt to mechanical changes in its surroundings, so it is increasingly important to have accurate, experimentally validated, biomechanical models which allow a better understanding of these processes, and are useful as a guide to the design and implementation therapeutic interventions and new medical devices. In this context, the biomechanics concepts illustrated in these pages are useful for establishing a sound and objective foundation upon which to build and develop new theories which, with the aid of powerful computational techniques, can be brought to bear in their biological study and medical practice.

6.4. ELASMOID SCALES

6.4.1. Structure of elasmoid scales

The structure of *elasmoid* scales is, at first glance, simple. In a cross-section of the scales two layers can be distinguished: a hard surface layer, of collagen which is highly mineralised (with apatite), and a soft lower layer, with less-mineralised collagen, formed by packets of superimposed fibres of different orientations. These scales, which are thinner than *ganoid* scales, are anchored to the dermis and partially overlap adjacent ones.

Two types of *elasmoid* scales are shown in figure 6.33: *Cycloids* (figure 6.33a) have smooth exposed parts and their form is oval or almost circular, while *ctenoids* (figure 6.33b) have rough patches or spines on their exposed parts. Rings, corresponding to stages of growth, can be observed on the scales. The spacing and other details of these lines give clues to the life of the fish –growth, spawning, and other development events– which can be interpreted by expert biologists. Radial lines can also be seen, which appear to emanate from a focus, located at the centre of the rings.

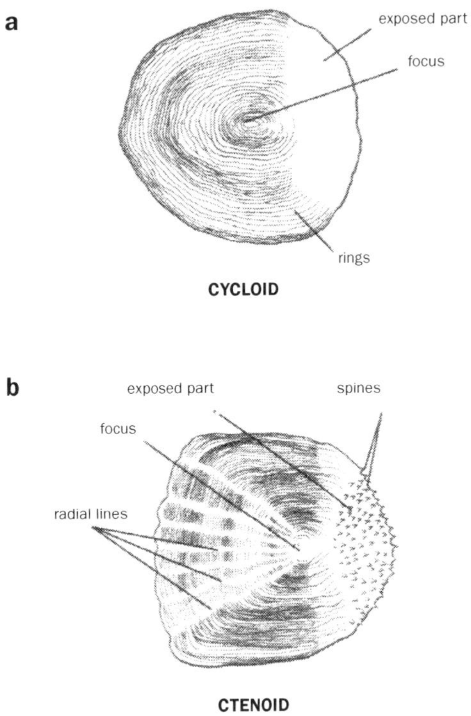

Fig. 6.33. *Types of elasmoid scales.*
a. Cycloid
b. Ctenoid

One of the scales studied in most detail is that of the fish *Arapaima gigas* (Sherman *et al*. 2017), previously mentioned in section 4.5. The scales overlap, so that, on average, each part of the skin is covered with three scales. Part of each scale is exposed and the other part covered by the adjacent scales.

The hierarchical structure of these scales is shown, in a simplified form, in figure 6.34. Note that the structures considered range from centimetres to nanometres, that is, about 7 orders of magnitude. The *exposed* part of the scale is covered with a hard corrugated, highly mineralised, layer (inhabitants of the Amazon used these scales to file their nails) which can reach a thickness of 0.5 mm. Underneath this hard layer there is another, softer, one formed of collagen fibre sheets arranged in parallel packs, with thicknesses of about 50 μm. These sheets are stacked almost orthogonally (figure 6.34). The sheets themselves are formed of cylindrical bundles of fibres, with diameters of approximately 1.0 μm. Each bundle is, in turn, composed of collagen fibrils of about 100 nanometres in diameter (Yang *et al*. 2014).

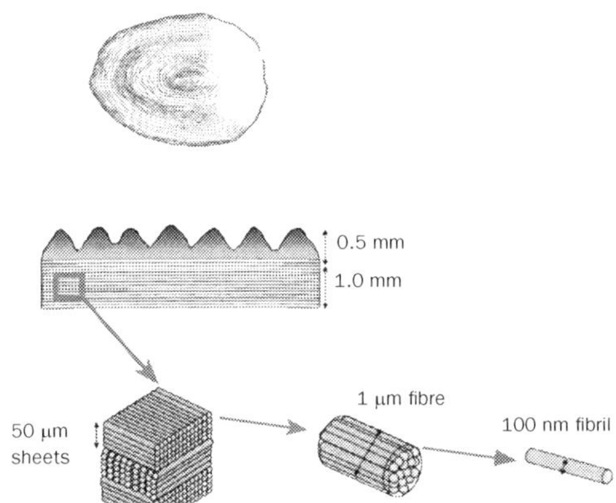

Fig. 6.34. *Hierarchical structure of the scale of* Arapaima gigas.

Another *elasmoid* scale which has been studied in some detail is that of the *Megalops atlanticus*, the shad or tarpon (Murcia et al. 2018, Gil-Duran *et al*. 2016). It is a large silver fish with a compressed fusiform body, which can reach a length of two metres. While it is a sea fish, it can survive the river waters because it can tolerate waters which lack oxygen by inhaling air directly into its swim bladder, which acts as a lung.

A simplified cross-section of these scales is shown in figure 6.35a in which three regions can be identified: A hard external layer formed of highly mineralised collagen, with an apatite mineral/collagen ratio greater than 3, and an internal softer layer, also formed of less-mineralised collagen fibres, known as *elasmoidine*. This second region can, in turn, be subdivided in two (figure 6.35a): an upper layer, with a mineral/collagen ratio which varies between 3.0 and 0.5, and an inner layer of elasmoidine with a mineral/collagen ratio of less than 0.5. The collagen fibres are grouped into helically stacked sheets, whose average thickness is about 20 μm. In this structure, called a *Bouligand structure* (Bouligand, 1972), the angles between the adjacent sheets can vary from 35 to 85°.

a

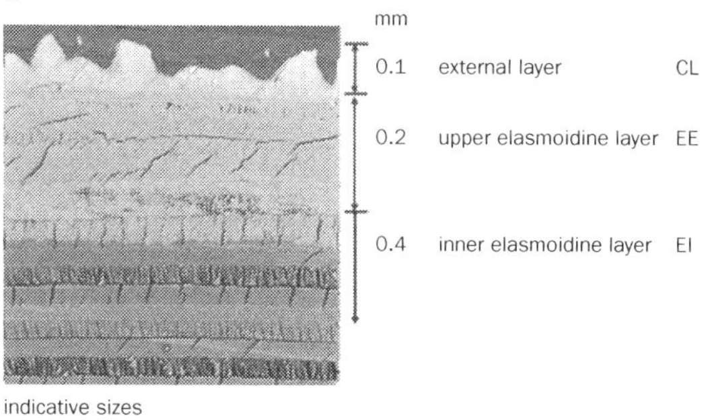

b

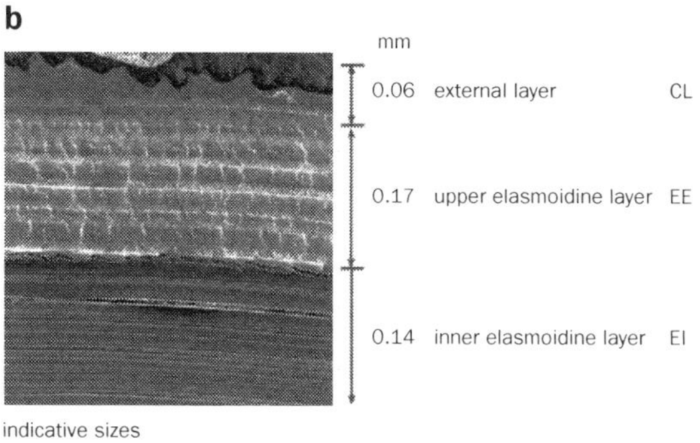

Fig. 6.35. *a. Cross-section of the scale of a shad b. Cross-section of the scale of a carp*

Other types of *elasmoid* scales which are worth examining, in order to compare the structures in different circumstances, are those fish which do not have special needs, both in locomotion and in protection against predators. Carp, *Cyprinus carpio*, is a fresh water fish which serves this purpose.

A simplified cross-section of a carp scale is shown in figure 6.35b. The structure is very similar to that of the shad, but thinner (Murcia *et al.* 2015). Three regions can also be identified: a hard external layer, of highly mineralised collagen, and another inner layer of less-mineralised collagen (elasmoidine). The collagen fibres are also grouped in sheets, whose average thickness is about 10 μm, and the sheets are helically stacked, as in the shad.

The *elasmoid* scales of other fish are similar but differ in size and detail, depending upon their anatomical position. Murcia *et al.* (2017) performed a comparative study between the three scales discussed, with the following results:

- The materials which form *elasmoid* scales are a combination of type I collagen and apatite. Unlike hard mineral tissues, such as bone, dentine or enamel which are highly mineralised, *elasmoid* scales have lower mineralisation, which allows the construction of an armour which may be flexible to facilitate the motion of the fish.
- In all three examples, the microstructure is similar. Three distinct zones are apparent, although the thickness of the scales varies; about 1.5 mm for the arapaima, less than 1 mm for the shad, and less than 0.5 mm for the carp.
- The thickness of the surface layer, the hard layer, is much larger in the arapaima than in the other two fish; between 500 and 300 μm for the arapaima, and less than 100 μm for the shad and carp. Perhaps this is due to the fact that the arapaima must withstand the teeth of piranhas. A curious fact is that, in both the shad and carp, this thickness is the same in the scales of the head, as those of the body and those of the caudal peduncle. This is not the case in the arapaima, in which the thickness decreases with distance from head to tail.
- The thickness of the sheets formed of collagen fibres also varies; about 50 μm for the arapaima, more than 20 μm for the shad and 10 μm for the carp. The most important difference is found in the stacking of these sheets; in the shad and the carp the stacking sequence repeats itself every five lamina, while in the arapaima it is every two.

6.4.2. Mechanical properties

The objective in studying the mechanical properties of *elasmoid* scales is to try to understand their tenacity, flexibility and resistance to penetration, for two reasons: first, in order to determine the role each component plays in such a hierarchical structure and, second, to derive ideas which would allow the design of armours with the same or greater benefits than those offered by nature. To this end, conventional tensile, tearing and puncturing tests have been performed.

Tensile tests

Experiments were carried out with prismatic specimens, such as those shown in figure 6.36, and tested in tension, recording the force as a function of the displacement. These data reflect the response of a complex structure –this does not relate to a homogeneous isotropic material but a heterogeneous and hierarchical one– and, in consequence, represent the behaviour of the whole structure and not that of some component. In order to evaluate the contribution of the different elements, when it is not possible to test them separately, models have been developed based on the available information and, to test their validity, they are compared with the experimental results. Because of the characteristic structure of the scales it should not be surprising that the tensile tests show that the material is anisotropic and that, moreover, the results have a large dispersion introduced by the heterogeneity of the samples.

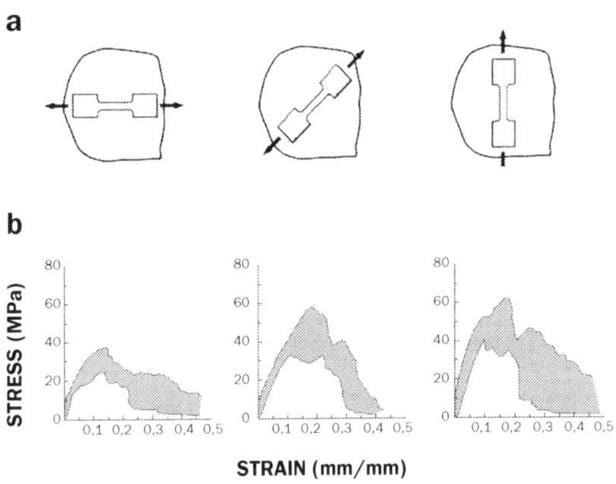

Fig. 6.36. *Tensile tests on scales of* Morone saxatilis. *Influence of anisotropy (Zhu et al. 2012).*

Anisotropy has been investigated in a series of tests performed by Zhu *et al.* (2012) on scales of the striped bass (*Morone saxatilis*), an anadromous fish found along the Atlantic coast of North America, and which, though living in the sea, enters freshwater to reproduce. This fish can reach a length of 150 cm and a weight of 30 kg. The shape and disposition of the specimens, which were tested in tension, are shown in figure 6.36a and the results of these tests presented in figure 6.36b, in which a large dispersion of the results is apparent (shaded zones). The average values of the initial elastic modulus ranged from 600 MPa, for specimens with an orientation of 45°, up to 850 MPa, for specimens orientated in the longitudinal direction of the fish, 0°. Specimens with a 90° orientation yielded intermediate values. As for the tensile strength, the values obtained were: 30 MPa for 0° and 50 MPa for 90°, with intermediate values when the orientation was 45°.

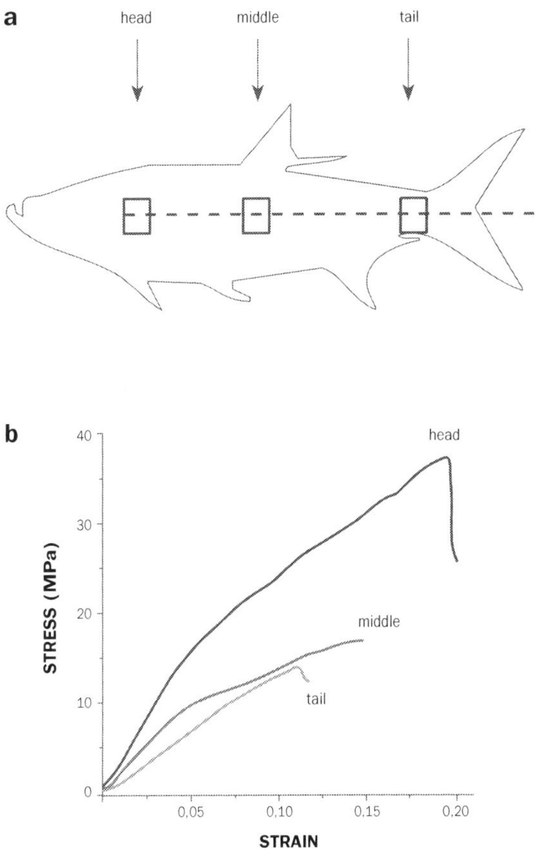

Fig. 6.37. *Tensile tests on scales of* Megalops atlanticus. *Influence of position (Murcia et al. 2018).*

The *heterogeneity* of the scales varies according to their position, that is to say, whether they are in the head, middle of the body or tail, as is also apparent. For example, in the tests performed by Murcia *et al.* (2018) on shad scales (*Megalops atlanticus*), mentioned above, the heterogeneity can be appreciated. The locations from which the specimens were taken are identified in figure 6.37a and in this experiment the orientation was always the same, in the longitudinal direction of the fish (0°, following the previous notation). Engineering stress-strain curves which represent the average values of the results are shown in figure 6.37b. In summary, the values obtained, which confirm the heterogeneity of the material, are as follows: The initial elastic modulus diminishes from head to tail; from 300 MPa for scales taken from the head to 150 MPa for those from the tail. The tensile strength data have a large spread in values, the average strength of specimens taken from the head was greater than those of specimens taken from the other locations, reaching 35 MPa. The average value for specimens from the middle was about 18 MPa, while that for those taken from the tail was about 15 MPa.

In order to substantiate these results, Murcia *et al.* (2018) proposed a semi-empirical model –valid only for elastic behaviour– in which account is taken of the contribution of the three regions of the *elasmoid* scales (figure 6.38a), discussed in the previous section; the hard external layer, surface layer (SL) and the internal and external layers of the elasmoidine layer, (IE) and (EE), respectively. (Later, a brief summary of how to develop this type of model will be provided, only for information purposes. The reader interested in the detail of the derivation should consult the cited reference).

The model assumes that each layer behaves as a Hookean spring and that the three-layer act in parallel, so that the relationship between the applied force F, (in a direction parallel to the layers) and the displacement, Δ is given by

$$F = (\Delta/L)\ \Sigma\ (A_i\ E_i)$$

where E_i and A_i (i = 1, 2, 3) are the area and elastic modulus of each layer, so that

$$\sigma = F/A = (\Delta/L)\ \Sigma\ (A_i\ E_i\ /\ A)$$

in which $A = A_1 + A_2 + A_3$ is the total area.

Since the width of the three layers is the same, the ratios of the areas are the same as those of the thicknesses (t_1, t_2, t_3 and the total thickness t) which are known values (from microscopic observation of the cross-sections). The above expression can then be written as

$$\sigma = E\ \varepsilon = (t_1E_1/t + t_2E_2/t + t_3E_3/t)\ \varepsilon =$$
$$= (\alpha_1E_1 + \alpha_2E_2 + \alpha_3E_3)\ \varepsilon$$

where $\varepsilon = \Delta/L$ is the total strain. Approximately, α_1 = 0.12, α_2 = 0.30 and α_3 = 0.58. It then remains to evaluate E_1, E_2, and E_3.

The elastic modulus of the hard layer, E_1, is calculated by discretising the layer into six slices, as shown in figure 6.38c, bearing in mind that the percentage of mineralisation of the layer varies from 70% to 17%, as shown in figure 6.38b. The average value of each slice is calculated from values of the moduli of the matrix and collagen fibres, using data from the literature, and employing Nielsen and Chen's (1968) model for composite materials with short, randomly distributed, fibres.

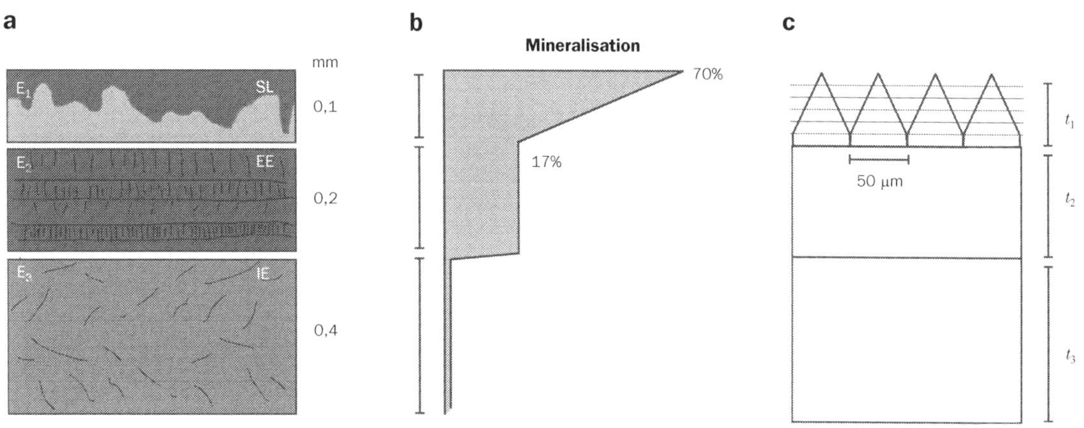

Fig. 6.38. *a. Cross-section of a carp scale. b. Distribution of the mineralisation across the three layers. c. Scheme used for the calculation. Area A_i of each layer is the product of its thickness t_i and the width of each scale (in the direction perpendicular to the plane of the drawing).*

Calculation of the elastic modulus of the external elasmoidine layer, E_2, is more complicated, because this region has a laminar structure, with the sheets helically stacked. The sheets are formed of mineralised collagen fibres. Several, quite sophisticated, models were suggested for the calculation of the elastic modulus of these structures, but the authors (Nielsen and Chen) proposed an average value, given by

$$E_2 = \Sigma\ (t_i / t_T)\ E(\theta_i)$$

where t_i is, now, the thickness of each lamina, t_T the thickness of the layer, and θ_i the angle the two sheets form between each other. $E(\theta_i)$ is calculated using the above cited Nielsen and Chen model.

Finally, since the internal elasmoidine layer is hardly mineralised (figure 6.38b) it is assumed that the elastic modulus, E_3, will be of the order of that of hydrated collagen fibres, at about 0.15 GPa.

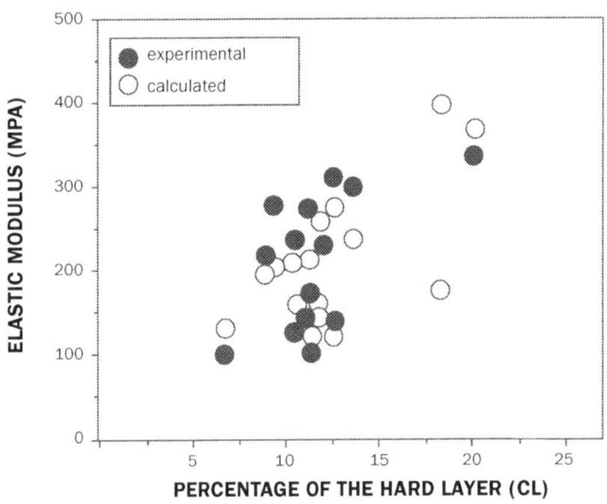

Fig. 6.39. *Elastic modulus values as a function of the percentage of the hard layer (CL). Comparison between experimental and calculated results.*

Experimental values of the initial elastic modulus are compared in figure 6.39 to those calculated from this model. The average error between the predictions and experiment is about 7%, a reasonable value bearing in mind the nature of this type of material. From the results obtained it can be seen that the elastic modulus depends greatly upon the contribution of the hard layer, CL, which is preponderant in the head region of the fish. The predictions of this model are largely based upon the mineralisation of its components, the model could be improved if account were taken of the interactions between the collagen fibres in the elasmoidine regions, EE and IE. Anyway, the models put into relief the various roles –percentage and distribution– of some of the components which allow the design of better armours.

A further step in modelling is to resort to smaller levels; to model the collagen fibres, their distribution in the sheets, and the effectiveness of different forms of stacking. For such a proposition, numerical analysis is a very useful tool and, in particular, in models based on molecular dynamics.

Yang *et al.* (2018) have used these techniques in order to understand how the collagen fibres behave in tensile test on arapaima scales. The structure of the elasmoidine was modelled as a superposition of *bundles* of collagen fibres. Each bundle was composed of three superposed *laminas* which form angles of 42, 78, and 18° between each other. And each lamina was composed of parallel collagen fibres. Each fibre was, in turn, modelled like a string of rosary beads in which the beads are connected by springs (here, lower levels were not considered, and the fact that the collagen fibres were formed of fibrils was ignored). This model allowed an investigation of how the angle between two sheets, the interaction between fibres, and the stiffness of the fibres influence the mechanical response of the scales.

In the tensile tests, calculation of the stress as a function of the strain was performed assuming a hyperelastic material, with a potential function which took into account the interaction between two beads, between three beads, and that between fibres. Not all of the parameters needed for this calculation were known, so it was necessary to pick the unknown parameters which gave the best fit to the experimental results, such as in the case of the interaction energy between the fibres (this value can vary between 0.05 and 3.0 J/m^2, and the best fit in this case was about 1.0 J/m^2). If it is assumed that failure of the scale is initiated with the first separation of the collagen sheets, the calculated value was about 26.4 MPa, which fits quite well with the experimental data, between 30 and 40 MPa. Once again, the model can be refined to provide a further step toward evaluation of the contribution of each element, and of the microstructure in detail, which would permit a better understanding of the mechanical behaviour of this very hierarchical material.

Rupture by tearing tests

The greatest threat met by a fish is the teeth of an adversary. An attack can cause tearing of the scales. From this perspective, an analysis of the tear resistance of the scales it is useful to know how the distinct components of this hierarchical structure stand up in against the cited threat.

Murca *et al*. (2015) investigated the tear resistance of carp (*Cyprinus carpio*) scales and demonstrated that the specific tear strength R (see section 5.3.3) depends, to a large extent, on the mineralised external layer and the external elasmoidine layer, such that the tear resistance increases with the thicknesses of these layers.

In these tests the *anisotropy* and heterogeneity of the scales were taken into account, depending upon their position. Figure 6.40a outlines how the anisotropy was studied; by testing specimens prepared with different orientations: 0, 45, and 90°. The *heterogeneity* was also examined by testing scales near the head, from the middle region of the body and the tail, as indicated in figure 6.40b.

The average specific tearing resistance, R, was determined following the process outlined in section 5.3.3. The dimensions of the specimens are shown in figure 6.40c, together with a sketch of the test configuration.

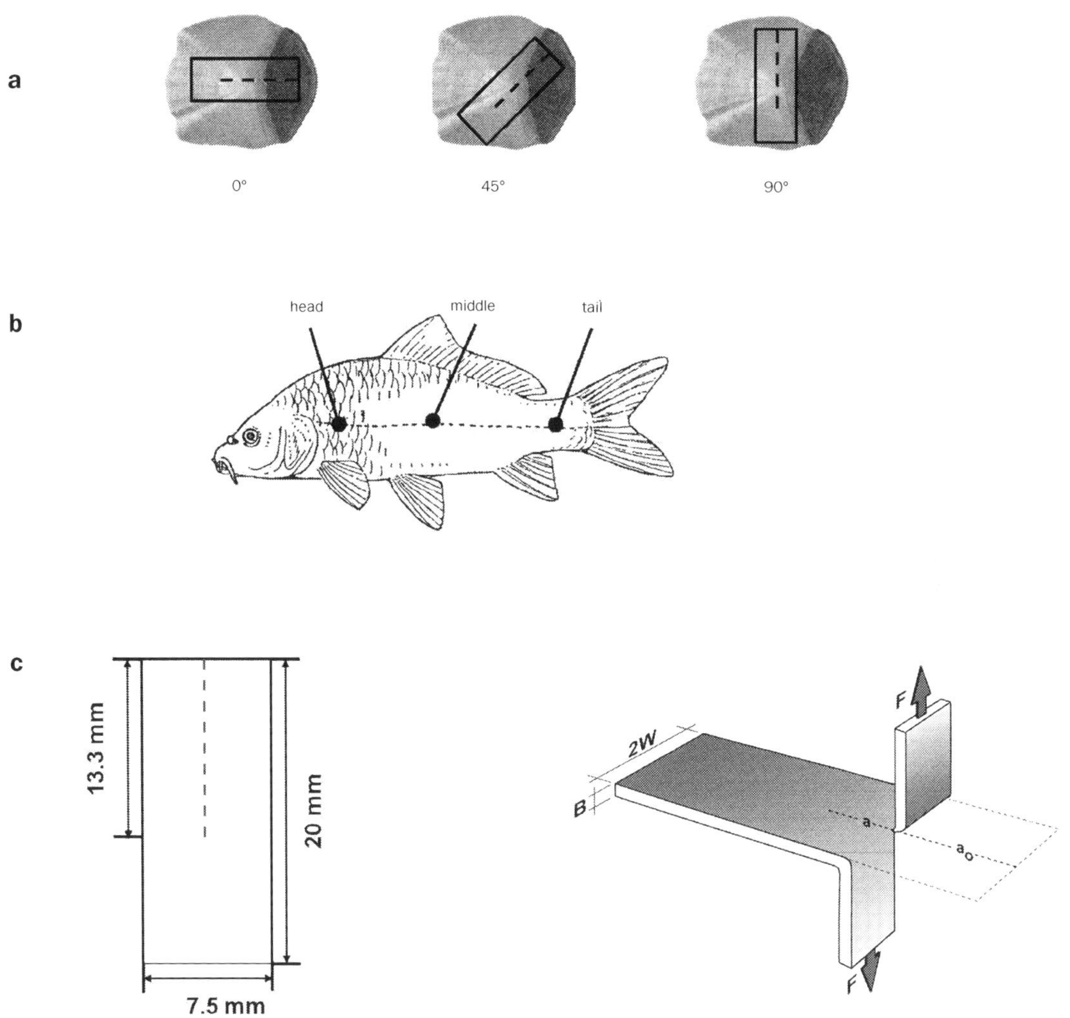

Fig. 6.40. *Tear test on the scales of* Cyprinus carpio. *a. Position of the samples to investigate the anisotropy. b. Position of the samples to investigate the heterogeneity. c. Dimensions of the samples and schematic arrangement of the test.*

Since the scales have a nonlinear behaviour in the tensile tests, it was assumed that they could be treated as an *elastomeric* material, and that equation (91) could be used to calculate R, that is to say

$$R = \frac{2F_R\lambda_R}{B} - 2W\mathcal{W}_R$$

where F, B and W are defined in figure 6.40c, $\lambda = l/l_0 = 1 + \varepsilon$, and ε is the engineering strain remote from the tear, and $\mathcal{W}$ is the elastic strain energy density referred to the undeformed volume of the configuration, also far from the tear.

The cited authors argue that, in this case, the strain, ε, far from the tear is negligible compared to 1, and that the elastic strain energy density $\mathcal{W}$ is also small. Consequently, the previous expression can be approximated by

$$R = 2\,F_R/B$$

The authors performed tear tests at ambient temperature (21 °C) and at low temperatures; 0°C, -30°C and -150°C. The results of an experiment at ambient temperature, on a scale near the head and with an orientation of 0°, are presented in figure 6.41.

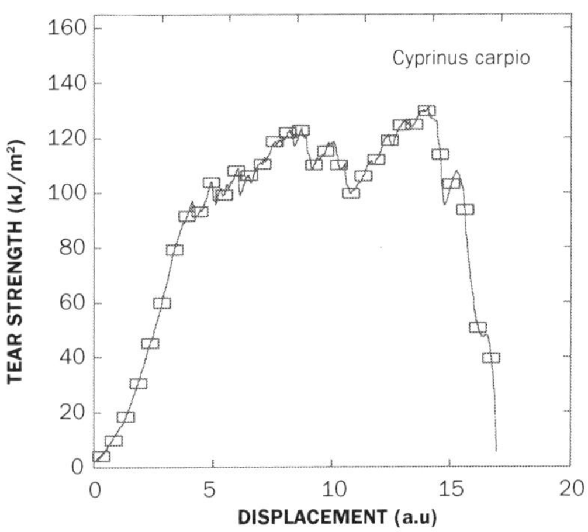

Fig. 6.41. *Tear test on scales from* Cyprinus carpio, *at ambient temperature, with a scale taken from a region near the head, and an orientation of 0°.*

The scales near the head showed a marked *anisotropy* in the tests at ambient temperature. The average values were: 120 kJ/m^2 with scales orientated at 0°, 100 kJ/m^2 for a 45° orientation, and 90 kJ/m^2 for those at 90°. For scales taken from the middle region of the body and tail anisotropy was not so apparent and, in both cases, the average value, for the different orientations, was about 90 kJ/m^2.

Heterogeneity was also observed in tests performed on scales taken from different locations on the body. At ambient temperature, scales located near the head gave higher values than those from the middle and tail; these last two gave similar values.

The authors of this study concluded that the specific tear resistance depended primarily on the thickness of the mineralised external layer, and the number of laminas in the external elasmoidine layer. These two microstructural features are functions of the anatomical position of the scales and are, possibly, related to the functional needs of the fish.

Puncture tests

The principal function of scales is in the mechanical protection against aggressors, particularly from the teeth of predators. The tensile strength and tear resistance, although important properties, are not sufficient for the evaluation of the resistance to sharp puncture. Indentation tests (see section 8.2.3) provides more information on the ability of scales to resist the action of pointed objects, capable of perforating the skin and damaging the softer, weaker internal tissue.

D. Zhu *et al.* (2012, 2013) studied the resistance of the elasmoid scales of the Atlantic sea bass (*Morone saxatilis*), whose characteristics were outlined above, in section 6.4.2. For this, they first investigated the behaviour of a single scale under shear punching and then, the more representative, case of three overlapping scales.

The penetration resistance of a scale was investigated using a device –shown in simplified form in figure 6.42a– which was designed to simulate the tooth of a predator. The indenter terminates in a rounded point of radius 25 µm. The scale rests on a silicon substrate which is meant to simulate the skin protected by the scale. Typical output data from these tests is the force as a function of the displacement of the indenter, which are shown in figure 6.42b. In all cases, a small step was detected, at about 2 N, and the maximum values of the force lay between 3.0 and 3.5 N.

Theses authors distinguished three stages in the tests (figure 6.42b): The first, associated with the initial linear part of the force-displacement curve, corresponds to the bending of the scale. At the end of this stage –at a load of about 2 N– cracking of the hard layer of the scale, and sliding and separation of the most mineralised layers of the elasmoidine, are initiated. Finally, the indenter perforates the scale and causes a sudden drop in the load, giving rise to the third stage.

In a more detailed analysis of this process, D. Zhu *et al.* (2012) showed that during sharp puncture, the scale –in the second stage– goes through two phases: First, the hard external layer cracks, following the pattern shown in figure 6.42c, and creates four plates of hard material. In a second phase, the deformation of these plates is impeded by the fibrous elasmoidine layer which acts as a resisting membrane due to the elasticity of the collagen fibres. Thanks to these two processes, the strength of the scale increases by about 50% rising from 2 to 3 N. The combination of the hard layer, which can crack in a controlled manner, and the soft elastic substrate, results in an effective design to defend it against puncture.

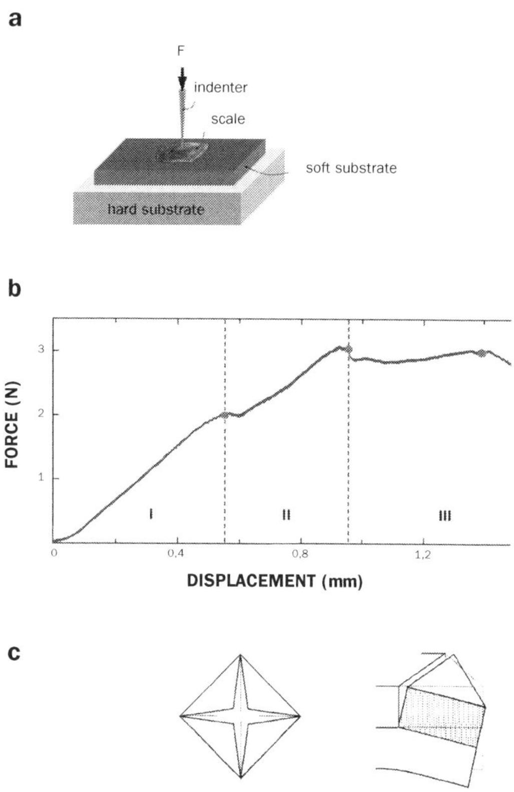

Fig. 6.42. *a) Schematic representation of a device for the puncture test.*
b) Distinct stages in the test.
c) Schematic representation of the rupture of the hard layer.

In reality, a sharp puncture attack is not resisted by just one scale, the scales overlap and share the load among themselves. D. Zhu *et al.* (2013) investigated this phenomenon in striped bass, in which the scales are arranged in overlapping rows so that each scale overlaps with another six –three on the front and another three on the back– in this arrangement, every point on the surface is covered with three scales. For these test, the authors used a device similar to that described above, and studied the influence of the stiffness of the substrate, the number of scales, the friction between them, and the way they overlap.

The stiffness of the *overlap* influences the results. Three substrates were tried; a rigid one, an elastomer, and the other simulating the natural substrate of the fish. The force-displacement curves had similar values for the maximum force, but very different values of the displacement under the maximum load; small values for the rigid substrate, large values for the elastomeric and even larger for the natural.

The experiments on different superposed scales showed, as might be expected, that the puncture force increased with the number of scales. The most striking result was that the perforation mechanism was similar to that observed in experiments on a single scale, which was replicated in passing from one scale to another.

The influence of friction between the scales was investigated by means of a puncture experiment with two scales and numerical simulation of two extreme situations; two scales bonded together, and two in contact, without friction between them. The experimental results showed that the influence of friction was, at least in this test, negligible.

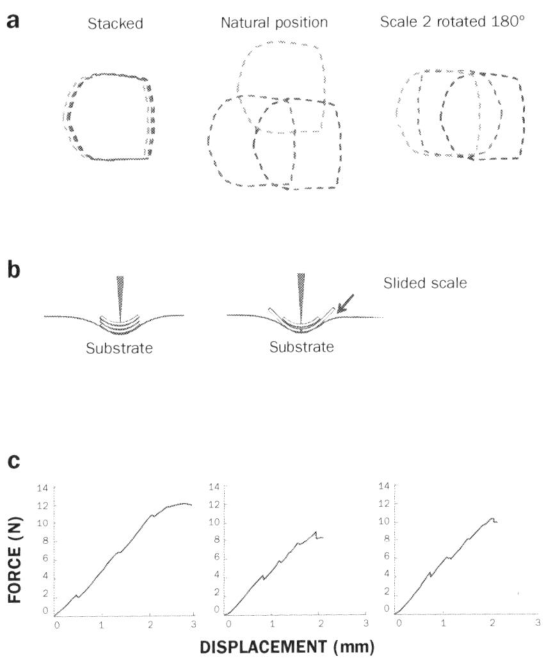

Fig. 6.43. *a) Different arrangements of the three overlapping scales.*
b) Different behaviours of the scales and substrate.
c) Force-displacement curves.

Finally, the effect of the arrangement of the overlap between the scales was investigated using the three configurations shown in figure 6.43a; in one of these the three scales are stacked on top of each other, in the second they are in the natural position and in the third, the second of the three scale has been rotated through 180° (as shown in the figure). The force displacement curves indicated that the initial stiffness is practically the same in all three cases, and that maximum load was greater when the scales were stacked on top of each other and somewhat less in the other two cases (figure 6.43c). The most interesting result lay in the difference between the deformation mechanisms of the scales and substrate, illustrated, in a simplified form, in figure 6.43b. When the scales overlapped in the natural position, they slid over each other, the area of contact with the substrate increased, and the load was spread over a larger area and the damage decreased.

Summarising; the stiffness of the substrate does not affect the value of the puncture force for the range of stiffnesses studied, which were representative of the stiffnesses of fish substrate and of materials used in the artificial armours inspired by biomimetics. Stacking three scale tripled the protection against puncture, compared to a single scale. Possible friction between the scales did not appear to increase the puncture resistance. Finally, the experiments demonstrated that, even though the scales could resist puncture, sliding between the scales and soft substrate redistributed the stresses caused by the indenter over a larger volume, and soothed the damage.

Similar studies have been carried out by other scientists. M Meyers and co-workers (W Yang *et al.* 2014) performed detailed investigations into the strength of the elasmoid scales of the *Arapaima gigas* using indentation tests, X-ray diffraction analysis, and molecular dynamics calculations, and achieved similar results to those already discussed: When a predator's teeth bite into the scales, the hard external layer breaks first and the soft elasmoidine layer is left exposed. If the teeth continue penetrating, the mineralised collagen fibres dispersed in sheets with different orientations must be separated, reorienting the collagen fibres and stressing them. All this consumes energy, facilitates crack branching and, ultimately, the damage is less.

P Y Chen *et al.* (2012) also studied the behaviour of arapaima scales under the action of piranha and shark teeth, using nanoindentation techniques, and obtained similar results.

6.4.3. Biomimetics: Flexible artificial armour

Elasmoid scales are a source of inspiration for the design of armours; the *hardness* of the scales protecting the fish against the teeth of its predators, and being *imbricated* and anchored to a soft substrate, to provide sufficient flexibility to the skin so as not to impede movement.

A very simple design, based upon these two concepts, is to fabricate armour which protects the front against penetration while being flexible, has been suggested by several authors (S Rudykh *et al*. 2015). The architecture of the armour shown, schematically, in figure 6.44a consists of two layers bonded together: The upper layer is formed of thin hard acrylic polymer sheets (0.5 mm thick) and the lower layer is a soft elastomeric material.

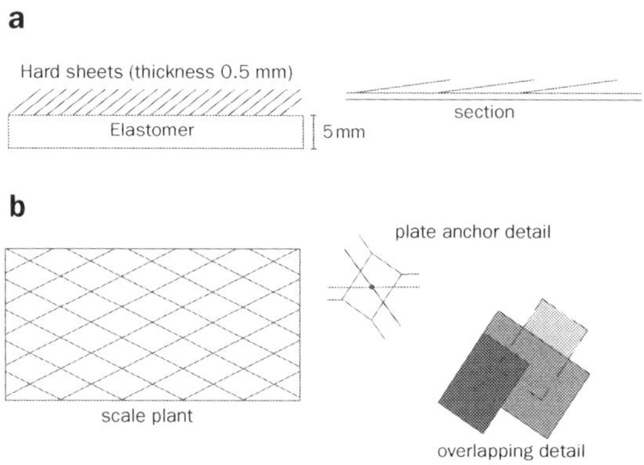

Fig. 6.44. *a) Schematic representation of the armour proposed by S. Rudykh et al. (2015).*
b) Schematic representation of the armour proposed by N. Funk et al. (2015).

The resistance to penetration was measured by three-point flexure tests. The resistive capacity can be controlled by altering the percentage of the hard layer, and the flexibility by means of the angle, which the sheets form with the soft layer (figure 6.44a). Increasing the resistive capacity reduces the flexibility, and increasing the inclination of the sheets (between 10° and 45°) increases the flexibility. With a judicious combination of the percentage of the hard layer and the angle, it is possible to optimise the design of the armour. The authors suggest that the resistance can be increased by a factor of 40, at the cost of reducing the flexibility by a factor of 5. Playing with these two factors, an armour can be designed with *locally* different properties; providing more resistance to the more delicate zones and greater flexibility where needed.

Another simple design, taking into account the hierarchical structures that nature has to offer, is also based upon covering a soft substrate with hard scales. In this case the soft base is a mesh of flexible polypropylene fibres (rather than an elastomer) and the scales are, now, thin acetate sheets (N Funk *et al*. 2015).

The mesh, primarily, provides a flexible support to the *scales* –the average value of its elastic modulus is seven orders of magnitude less than that of the scales– and, moreover, it serves to anchor the plates at the nodes (figure 6.44b). The anchor is made such that the plates can overlap and rotate a little, very much like the way elasmoid scales behave. The cited authors performed several tests – on what they called *synthetic skin*– tensile, flexure and puncture, with specimens measuring 180 mm by 150 mm, and which contained 690 *scales*, with satisfactory results. The overlapping of the scales (between two and three) allowed a better redistribution of the forces and a reduction in the damage produced by the punching. They also developed a simple model of the architecture –mesh and plates– which was able to reproduce their experiments. The advantage of a model which replicates the experimental results, is that it allows modification of the parameters (shape and properties of the mesh, and dimensions and properties of the plates) in order to optimise the performance of the armour; perhaps by using stronger fibres to resist rupture of the mesh, modifying the shape of the scales or increasing their flexural stiffness.

Based on these two ideas –hard overlapping plate and a soft substrate– several examples of light, flexible armour have been designed. Biomimetics has attempted to provide a further step towards integrating both properties in the same material, by means of the concept of *flexible ceramics* (M A Meyers *et al*. 2012). The concept, was inspired by the elasmoid scales of the *Arapaima gigas*, whose corrugations on the hard external layer play an important structural role. If the thickness of this layer were homogeneous (always the same) the flexibility of the scale would be more restricted and more likely to crack on bending because the material is brittle. However, if it were corrugated it becomes more flexible. In the referenced article, the authors confirm this advantage by means of an elastic calculation. Consequently, cracking is localised at the bottom of the valleys of the corrugations and is less serious. S Rudykh *et al*. (2015) have demonstrated that it is possible to fabricate very flexible armours with these concepts.

SECTION 3
BIOLOGICAL MATERIALS

7 CHAPTER General Concepts

7.1. INTRODUCTION

The focus of this section is the class of biological materials in which it is necessary to consider the three spatial dimensions (3D) for an adequate description and study. Previous sections reviewed materials in which only one or two dimensions were relevant –fibres and membranes. However, in the case of bulk materials, different techniques and approaches must be used to test and model their mechanical behaviour.

The structure of these materials is complex and, almost always, extremely hierachical. Biological materials are usually soft and when they must perform structurally, the required stiffness and strength are achieved by introducing minerals, producing composite materials and sophisticated cellular structures. Cancellous (spongy) bone is a classic example of this strategy; showing great strength combined with low density thanks to a cellular structure and, in turn, to the cell walls which are true composite materials in which the reinforcements are particulate minerals, in this case hydroxiapatite.

Mechanical tests on 3D biological materials are, to some degree, a generalisation of the simple experiments mentioned in the previous volumes: tensile, compression and torsion tests. To these is added the flexure test, almost exclusive to 3D materials, since very thin fibres and membranes are usually very flexible and do not lend themselves to this type of characterisation.

Modelling the mechanical behaviour of 3D biological materials is also, to some extent, a generalisation of the methods introduced earlier, but with some caveats.

Even though the basic concepts are essentially the same, the mathematical tools are more complicated, but they can be greatly simplified when the materials are rigid –which occurs in many cases– since the deformations are usually small and, often, can be performed in an analysis in a linear-elastic regime. Powerful computers are commonly used in numerical calculations where complex geometries are modelled; for example, for cellular materials it is not necessary to use regular models in which a unit cell is repeated periodically, it is possible to construct Voroni cells, which have different geometries and positions at random. Defects and other imperfections can also be modelled with finite element methods. The difficulty in this approach lies not in the modelling but in the material constitutive equations; in practice, for small deformations, equations which provide acceptable results are available, but when the material's response depends upon time and the load history, the problem becomes very complicated.

At the end of this volume some 3D biological materials of great interest are described in some detail: Mollusc shell, an example of a composite material; wood, an example of a bidimensional cellular material; spongy bone, an example of a 3D cellular material, and cartilage, an example of a multiphase material.

7.2. COMPOSITE AND CELLULAR MATERIALS

In the majority of man-made materials, and particulalry in metals and ceramics, strength and toughness are conflicting properties: the greater the strength, the more brittle (less tough). The design of very strong and simultaneously tough materials is a challenging objective for the materials engineer. Nature has achieved this in both the plant and animal kingdoms, fabricating composite materials with polymers and minerals. In some materials, the matrix is a polymer –such as bone, in which the matrix is collagen– and in others, the polymer is the reinforcement –such as wood, in which the reinforcement consists of cellulose fibres.

When the increase in stiffness implies an increase in weight, a drawback for the living organism –such as in the case of birds– nature overcomes the problem by achieving maximum stiffness for a minimum weight by means of cellular structures.

Strength and toughness are optimised in many biological materials through hierarchical structures, with a skillful combination of composite materials and cellular structures which provide stiffness and energy dissipation mechanisms during crack propagation. We can find examples –in both animal and plant tissue– of how nature has engineered them by combining composite materials and cellular structures to achieve both strength and toughness.

Plant tissues have a cellular structure. The cell walls are composite materials in which the *reinforcement* consists of cellulose fibres –a glucose polymer which forms very strong and tough fibres– and the *matrix* is an amorphous polymer (lignin) and a semi-crystalline glucose polymer (hemicellulose). The cellulose content in the walls can reach 50%. The walls are not homogeneous but formed of various layers in which the orientation of the fibres varies from one to another, conferring stiffness and toughness. Details of the cellular structure and cell arrangement in tree trunks are discussed below in section 9.2, but it is worth mentioning that evolution has endowed trees with sufficient strength and toughness to withstand storms; when a tree falls during a gale generally this is because it has been uprooted, rarely because the trunk has broken; some branches may break, but the tree will survive.

In some plants the flexural strength of the stem is optimised using all of the above strategies; composite material in the cells and cellular hierarchical structure.

The stem of the horsetail –genus *Equisetum* provides a simple example of the side of the pteridophytes; ferns and allies. Figure 7.1a shows, in a simplified form, the stem section in which a compact cellular structure in the outer and inner edges of the ring can be observed, together with a more open cellular structure with large pores on the inside. The sandwich construction strategy is repeated in many biological materials and the accumulation of material in a ring rather than distributing it homogeneously optimises the bending resistance of the stem.

Animal tissues can be classified into two large families: *organic tissue,* such as skin, composed almost entirely of organic materials and *bioceramic tissue,* such as bone, with a polymeric base and a high mineral content. The latter tissues have a structural role, and in order to increase the stiffness they need to include large amounts of mineral which is a brittle component. Nature has solved the problem by creating bioceramic materials which are stiff, strong and tough. This time, a different process from that in plants is followed; the reinforcements are now crystals (rather than cellulose fibres), and the matrix is a polymer. The toughness is achieved thanks to the hierarchical structure of the composite which provides diverse energy dissipation mechanisms when a crack is propagated, as will be discussed later.

The mineral component of mollusc shells is calcium carbonate, a brittle but abundant material in the sea. From this, apparently brittle, material molluscs fabricate very strong and tough shells capable of inspiring armour for military vehicles (Yao *et al.*, 2010). The calcium carbonate, almost always in the form of aragonite, occupies about 95% of the volume of the composite material. In some cases the matrix which represents only 1% of the volume, is a complex biopolymer with chitin and a protein similar to the fibroin present in silk fibres. The shells are formed of densely packed microscopic ceramic bricks bonded by a fine glue which constitutes the matrix. To create some patterns, in pearlescent shells such as mother-of-pearl, the aragonite bricks have a thickness of some tenths of a micron and an average length of several microns, while the thickness of the cement which binds them is a few tens of nanometres. The value of the rupture stress in a flexural test can reach 200 MPa and the specific fracture energy about 500 J/m^2. Other shells with a crossed lamellar structure are tougher, achieving fracture energies of an order of magnitude greater. The structure and mechanical properties of shells are discussed in detail in section 9.1.

The skeleton of a living fossil, the *armoured crab* (also known as the pan or bayonet crab), provides an example of the combination of a composite material with a hierarchical cellular structure, similar to that of the stem of the horsetail. Figure 7.1b shows the cross-section of the shell in which the sandwich structure can be observed, together with its stiff walls and the interior with a progressive cellular structure, including large holes. Once again, this structural arrangement optimises the mechanical behaviour of the shell.

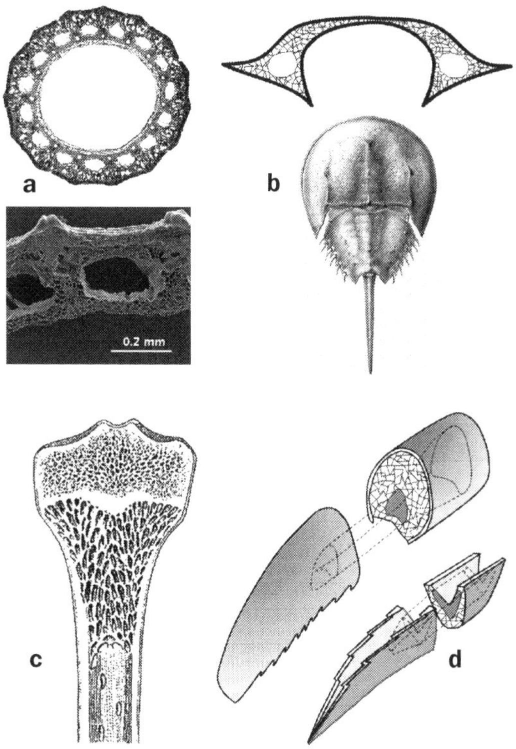

Fig. 7.1. *Hierarchical cellular structures.*
a. Stem of the horsetail.
b. Section of the shell of the armoured crab.
c. Section of a long bone.
d. Toucan beak.

The mineral component of bones is calcium phosphate (hydroxyapatite), and the matrix consists of collagen fibres. Both components have similar proportion by volume. There is a certain parallel between the cellular structure and hierarchy of bone and that of wood, despite the great differences in the chemistry of the constituents. In bone the collagen fibres are the matrix, while in wood cellulose fibres are the reinforcement. Young bones, with abundant collagen, are enriched by the disposition of hydroxyapatite nanoplates. With the passage of time the mineral content increases and osteons are formed –cylindrical laminas concentrated around a neurovascular canal– which have the strength and stiffness to withstand the loads that will act upon the mature organism. At the macroscopic scale, it is possible to identify a dense type of bone, called compact (or cortical), which covers the bone and a form of bone with a cellular structure, called **cancellous** (or trabecular) bone, that usually fills the vertebrae, the extremities of long bones (figure 7.1c) and the interior of some bony plates such as in the skull. This combination of dense structures and pores serves to increase the stiffness for a given amount of material, or in other words, to optimise the specific stiffness. Bones can be very tough, the specific resistance values ranging from 500 to 5000 J/m^2, considerable values compared to some advanced ceramics (about 500 J/m^2). Bones have properties that are very attractive to engineers; they are materials which are strong, tough, multifunctional and capable of self-repair and also of adapting themselves in response to mechanical loads. The structure and mechanical properties of bones are considered in detail in section 9.3.

The beaks of some birds offer another example of the combination of composite materials and cellular structures. Bird beaks are often short and thick or long and thin. The beak of the toucan (a climbing American bird) is an exception; the beak is long –a third of the length of the body– and very thick. It probably needs a long beak to get fruit, but one that is not too heavy because its habitat is in very tall trees. For this, evolution has furnished it with a light, stiff beak which has a structure formed of a hard keratin cover and an interior with a cellular structure (figure 7.1d). Keratin can be considered as a composite material reinforced with keratin fibres embedded in a protein matrix. Mineralisation with calcium and other elements contributes to give the keratin rigidity. The cellular structure is not open, such as in cancellous bone, but closed by fine membranes. Most of the loads borne by a toucan's beak make it work in flexure, and the structure is especially adapted for that; the flexural strength is much greater than that which would be achieved if all the mass of the beak were concentrated in an outer layer –as if it were a cylindrical bone– thanks to the interior cellular structure, as demonstrated by Meyers and co-workers (Seki *et al.* 2005).

7.3. BIOMINERALISATION

The previous section described how some biological materials acquire the required stiffness by incorporating minerals into their structure, a process known as *biomineralisation,* and which is worth acomment in its relation to Materials Science.

Organic molecules in solution can influence the shape and orientation of inorganic crystals if there is some affinity between them. An understanding of the processes in which living organisms control the growth and structure of inorganic materials can aid important advances in Materials Science and Engineering and open the door to many novel techniques for synthesising components from the nanometric scale.

Biomineralisation implies the selective extraction and incorporation of elements that are in the environment and their integration into functional structures, all of which under biological control. Unlike other biological transformations that leave no trace in the environment, the formation of hard minerals by biomineralisation –such as the shells of molluscs or bones– leave a fossil footprint that provides valuable information for paleoentology, sedimentology, geochemistry, marine science and many other fields related to chemistry, biology and geology.

Furthermore, biomineralisation, as well as providing strength and stiffness to organisms, can fulfil other biological functions, such as protection, locomotive systems, defensive mechanisms, and gravity, optical or magnetic sensors.

Among the minerals associated with the biomineralisation process, **calcium** plays an essential role, not just because it is very abundant but because it is the component in very familiar structures such as bones and mollusc shells. Note, as has already been mentioned, that the calcium in bones is in the form of calcium phosphate, while in shells it is calcium carbonate; the reason for these differences is, as yet, unknown. Both calcium carbonate and calcium phosphate have a high bond energy and have low solubilities and, in consequence, are thermodynamically stable in biological environments. This is not the case with the hydrates, such as calcium oxalate or calcium sulphate, which are more soluble and therefore less common.

Silicon is another element which takes part in biomineralisation but with a lesser role than that of calcium. It seems that the use of the salts of calcium is metabolically more economical, which justifies its abundance. The motive as to why some organisms prefer to use amorphous silica over calcium carbonate is also unclear. According to some authors, it could be because there are advantages when it comes to making very fine but strong spicules, because amorphous silica contains no defects that would initiate cracks, like the steps in calcite crystals. In some plants, the presence of silica spicules in cell walls serves to make them unappetising to predators. Equisetales, such as horsetail mentioned above, have high silica content, up to 25% by weight when dry, and due to their abrasive qualities have been used for brushes. Similar quantities of silica are found in rice husks which have come to be used as a raw material for the production of silicon nitride. More delicate silica structures are found in many unicell organisms, such as diatoms and radiolaria capable of displaying beautiful filigree.

The mineralised silica skeletal system of the hexactinellid sponge, *Euplectella aspergillum*, commonly known as the Venus' flower basket, has received considerable attention from the engineering and materials science communities for its remarkable hierarchical architecture and mechanical robustness across multiple length scales. The sponge's diagonal reinforcement strategy achieves the highest buckling resistance for a given amount of material (Fernandes *et al.* 2020).

8 CHAPTER

Mechanical Properties

8.1. INTRODUCTION AND OBJECTIVES

The three-dimensionality of the material adds further complexity to the study and interpretation of experiments on biological materials in which none of the three dimensions can be ignored.

This chapter first addresses the peculiarities of the most common mechanical tests (tensile, compression and torsion) which require consideration of the three dimensions of the material, in particular the associated extensometry (displacements of the movements of the specimen) and the possible anisotropy. The flexural test, frequently used to measure the **modulus of rupture,** is also presented. Special attention is paid to fracture tests due to an increasing interest in determining the toughness of biological materials.

Previously, in the study of membranes and laminas, tear and peel tests were mentioned, which strictly speaking are also three-dimensional tests, but ones in which the transverse dimension could, to a first approximation, be ignored. Indentation tests are also used when standard specimens cannot be fabricated or when it is not possible to obtain a sample of the material, and the whole element must be tested. It is also used to obtain local properties of the material. The information provided by these tests is not very precise, but they are cited in scientific articles, so their scope and limitations should be recognised.

Secondly, the previously discussed concepts of stress, strain and the relationships between them for one and two dimensions will be amplified. The commentary is practically limited to materials that behave elastically and experience small deformations. Even so, the results obtained are very useful in the study of anisotropic materials, composite materials, such as mollusc shells, or cellular materials, such as wood or bone.

8.2. TEST AND CHARACTERISATION TECHNIQUES

8.2.1. Tensile, compression, torsion and flexure tests

Tensile tests

Tensile tests (simple or uniaxial tension, and biaxial tension) were analysed in detail in the sections corresponding to fibres (2.2.1) and membranes (5.3.1). The tests are preferably used on homogeneous isotropic materials. They are also applied to anistropic materials where it is possible to find the traction directions which coincide with the axes of symmetry of the material, since in this way the stress and strain tensors are diagonal in the test directions and do not produce shear (angular) deformations.

Tensile tests are useful for determining the properties of fibres and membranes since they are easy to interpret because these materials are, in general, very flexible, and it is easy to obtain a relatively large zone in which the stress and strain are uniform, that is to say, independent of position (figure 8.1a). Measuring the extension between two points within this zone allows simple determinations of the strain throughout the zone, in order to relate it to the applied stress, which is also easy to calculate merely by averaging the force applied through the grips, as shown in the corresponding sections.

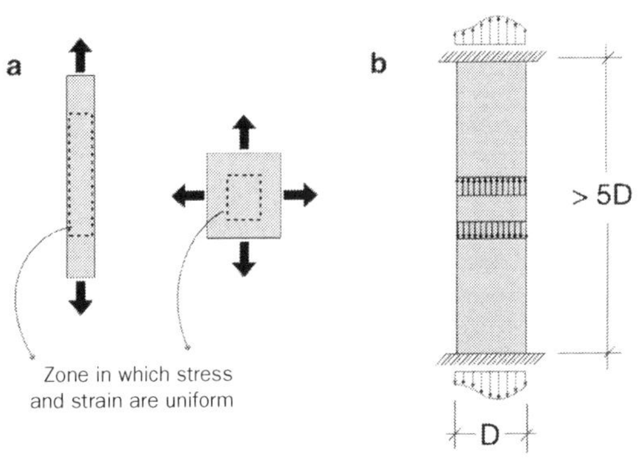

Fig. 8.1. *Tensile test*
a. Uniform deformation zone
b. Specimen dimensions

The tensile test on three-dimensional geometries which due to their proportions cannot be treated as linear bodies is particularly complicated since it is very difficult, if not impossible, to obtain a zone of uniform deformation. This is due the difficulty of gripping the specimen in a way that the load is transmitted uniformly. A practical rule for establishing the specimen dimensions is usually to take the free length between grips to be greater than five times the mean diameter. In this way, the inhomogeneity produced by the grips is attenuated sufficiently[34] so that in the central section the stress-strain state can be considered uniform (figure 8.1b).

As we saw in the analysis of the tensile test the values of the stress and strain which can be achieved during the test are limited by the appearance of a necking instability, originating from a local deformation. The heterogeneity of the material and the use of specimens which are too short generated stress concentrations which accelerated the appearance of necking.

The above issues restrict the tensile test to situations in which sufficiently slender specimens can be obtained. This is the case with Evans' tests on specimens taken from the proximal third of the human femur, shown in figure 8.2. The speciments are cut in the longitudinal direction in the compact (cortical) area of the bone, using the dimensions recommended in the ASTM E8 standard.

The results show that the human compact (cortical) bone is quite strong and stiff. The fracture stresses lie at about 80MPa and the bone has a marked initially linear behaviour, with an elastic modulus of about 18GPa (dry) and 14GPa (wet). The values of the rupture strains are small; they are about 0.4% (dry) and 1.2% (wet).

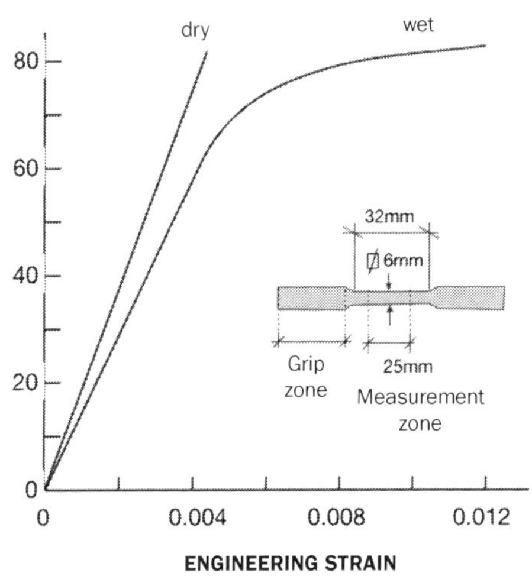

Fig. 8.2. *Tensile tests on the human femur. Adapted from Evans 1969.*

34 According to Saint Venant's principle, the difference between the effects of two statically equivalent loading systems becomes arbitrarily small at sufficiently large distances for the points of application of the loads.

EXERCISE 46

Calculate the relative error in estimating the strains in the femoral bone when true and engineering stresses are taken to be equivalent. What will the relative error in stresses and the initial elastic modulus be? Use the data from figure 8.2.

SOLUTION:

In exercise 1 in section 2.2.1, the relative error considering engineering strain is calculated. Its value depends upon the level of strain reached and is given by

$$\text{relative error} = \frac{e-\varepsilon}{\varepsilon} = \frac{1}{\varepsilon}\left(\exp(\varepsilon)-1-\varepsilon\right) = \frac{e}{\ln(1+e)} - 1$$

where the engineering strain $e = (L–L_0) / L_0$ and the true strain $\varepsilon = ln\ (L / L_0)$. L is the deformed length of the guage section, whose initial length is L_0.

If the maximum possible strain is e_{max} we obtain

$$\text{maximum relative error} = \frac{e_{\max}}{\ln(1+e_{\max})} - 1 = \frac{0.004}{\ln(1.004)} - 1 = 0.002 \quad \text{(dry)}$$

$$\frac{0.012}{\ln(1.012)} - 1 = 0.006 \quad \text{(wet)}$$

We see that the maximum error is very small and lies between 0.2 and 0.6%, so that $e \approx \varepsilon$ and it will be fully justified to use the theory of small deformations with this material.

In section 2.2.1 we saw that, assuming conservation of volume, the relationship between the engineering stress (s) and the true stress (σ) was (equation 7)

$$\sigma = s(1+e)$$

the maximum relative error will, therefore, be

$$\text{relative error} = \frac{s-\sigma}{\sigma} = \frac{1}{1+e} - 1 = \frac{-e}{1+e}$$

which yields values of –0.4% (dry) and –1.2% (wet). The actual value will lie between these values and zero, since bone is not strictly incompressible its volume increases with deformation and true stresses will be less than those given by equation 7. In any case, just as with the strain, we see that the values are very small and that we can consider that $s \approx \sigma$.

The initial elastic modulus is calculated by

$$\mathrm{E} = \left.\frac{d\sigma}{d\varepsilon}\right|_{\varepsilon=0} = \left.\frac{d\sigma}{de}\right|_{e=0}$$

since $\varepsilon = ln\ (1+e)$ and $d\varepsilon = de / (1+e)$, and at the origin $(e = \varepsilon = 0)$, then $de = d\varepsilon$.

On the other hand, from the definition of stress, $\sigma A = s A_0$ so that

$$d\sigma = \frac{A_0}{A} ds - \frac{A_0}{A^2} s dA$$

at the origin, when $A = A_0$ and $s = 0$, then $d\sigma = ds$. Using this result we see that

$$\mathrm{E} = \left.\frac{d\sigma}{d\varepsilon}\right|_{\varepsilon=0} = \left.\frac{ds}{de}\right|_{e=0}$$

and the initial elastic modulus will be independent of which stress or strain is used for its calulation –true or engineering– since both representations coincide at the origin. We conclude then that the error will always be zero.

For practical purposes, when handling stress-strain curves determined by points, there will always be a small error due to the impossibility of calculating the tangent at the origin with precision.

Compression test

A compression test consists of applying a compressive load through two opposing faces of a specimen while measuring its shortening (figure 8.3a). This test is suitable for characterising tissues which, during their physiological function, are usually subjected to compressive stresses, such as bone and cartilage. Furthermore, the compression test has certain advantages over the tensile test since it requires smaller, generally cubic or cylindrical, specimens which are easier to produce, and do not need complex gripping systems to transmit the loads. Also, as noted above, the compression test does not introduce intrinsic instabilities, as can happen in the tensile test when necking occurs.

Compressive tests, like those in tension, are preferably perfomed on homogeneous isotropic materials, or on anistropic materials whose axes of symmetry are parallel or perpendicular to the axis of compression. In this way complex states of stress, which make the analysis of the results difficult, are avoided. The interpretation of the results is easier and more direct if there is a region of the specimen in which the stress state can be taken as uniform (figure 8.3b).

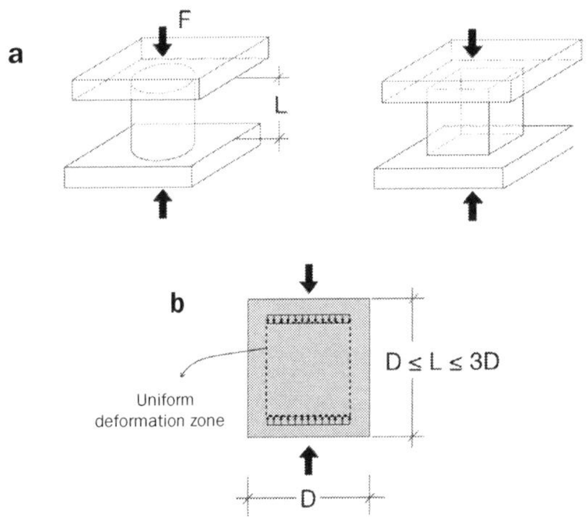

Fig. 8.3. *Compression test*
a. Cylindrical and prismatic specimens
b. Uniform deformation zone

Despite its apparent simplicity, performing and interpreting a compression test is not without its difficulties which have to be taken into account. In the first place, although the test –as has been said– is intrinsically stable, an instability can be caused by buckling if the specimen is too slender. In order to avoid this it is recommended that the height of the specimen be between one and three times the width ($D \leq L \leq 3D$).

Regarding these difficulties of an experimental nature, first it is necessary to ensure sufficient parallelism of the surfaces on which the compressive load is applied, in order to avoid stress concentrations which produce local damage and a non-uniform stress-strain state (figure 8.4a). A practical criterion is to establish a maximum difference between the strains produced on the load surface which is less than 1/10 of the strain corresponding to the elastic limit. Since this reqirement is especially demanding in rigid materials (such as cortical bone) it is common to recourse to placing a ball joint on one of the load plates with the aim of achieving a uniform contact (figure 8.4b). In order to avoid instabilities during the test it is important that the ball joint is adjusted at the start of the experiment, with the specimen unloaded or with a very small value of load, and preventing its movement during the rest of the test.

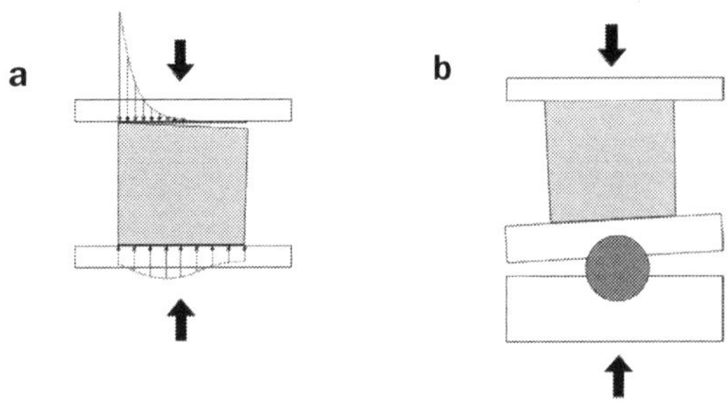

Fig. 8.4. *Lack of parallelism in the specimen*
a. Local stress concentration
b. Pivoting support

Another relevant experimental problem is the occurrence of friction between the loading surfaces and the faces of the specimen due to the differential lateral expansion. The friction increases with the test load and can become very important. There is no effective solution and it is usual to try to reduce the coefficient of friction as much as possible by using very flat surfaces (such as glass when testing cartilage) and lubricants such as low viscosity oils (Linde *et al.*, 1989).

The edge effect, generated by the abrupt interruption of the microstructure caused by machining the surface of the specimen (figure 8.5), produces a local deformation increase in the zones near the surfaces, which leads to an overestimate of the ductility and flexibility of the specimen. Its effect is greater as the size of the microstructure approaches that of the specimen. The zone affected by the machining can be estimated at about five times the size of the microstructure[35] so the solution to the problem is to measure the deformation of the specimen in regions sufficiently remote from the surfaces, in which there is a uniform state of stress and strain. In materials with large-scale microstructure (such as cancellous or trabecular bone, with millimetric microstructure), the edge effect can greatly influence the minimum size of specimen required for testing.

Compression tests on biological materials are usually performed using, as a guide, the existing standards for artificial materials (ASTM C39, ASTM E-9, ASTM D-695), adapted for the conditions of the tissue, especially the amount of available material. Like the other mechanical tests, the compression test is very sensitive to the heterogeneity and anisotropy of the material, and to the boundary conditions, especially the load rate and the moisture content of the sample, as well as the movement of tissue fluid into and out of it. Compression tests are usually carried out on saturated samples, submerged in physiological serum or in phosphate buffered saline.[36]

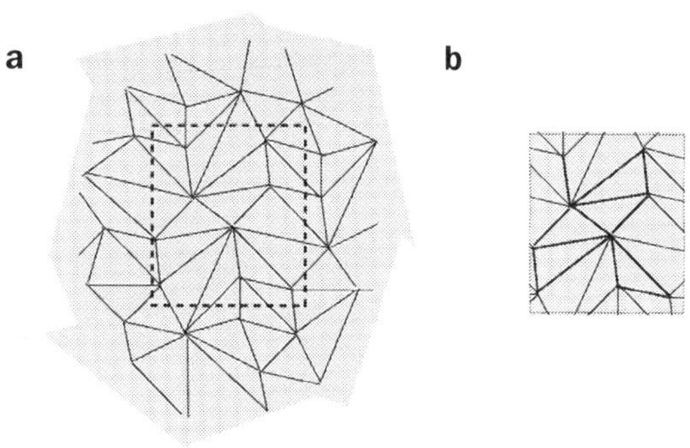

Fig. 8.5. *Edge effect in a trabecular material*
a. Unmachined material. The dotted lines represent the surfaces of the cut.
b. Machined specimen. Only the darkest trabeculae retain their bonds.

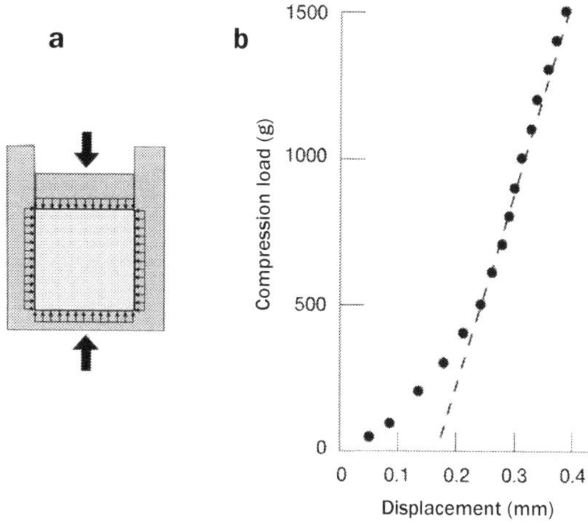

Fig. 8.6. *a. Constrained compression test*
b. Load-displacement curves for a constrained compression test with cylindrical specimens of bovine cartilage, short term (1s). Specimen diameter 6.4 mm; thickness 1.8 mm. (Adapted from Mow et al. 1980)

The **constrained compression** test (figure 8.6a) is a special case of the compression test, which is useful for soft tissue (including cartilage) in which it is difficult to maintain the sample geometry during the test, and where there may be interstitial fluid movement. In this test the sample is constrained laterally adjusting its size, with the greatest possible precision, to the more rigid side walls. Generally the walls are impermeable while the upper plate through which the load is applied is usually porous to allow drainage of the sample.

By way of example, figure 8.6b contains the load-displacement curve for a constrained compression test on cartilage from the bovine femoral condyle, obtained by Mow and co-workers (Mow *et al.* 1980). In this case drainage of the specimen was allowed although the data were obtained in the short term, applying each load for one second. The results indicate an initial nonlinear region while the sample adjusts to the diameter of chamber which was a little wider. From ~500g, true specimen constraint is produced. From the figure we can obtain the value for the aggregate modulus H_A of the material (defined as the ratio of the compressive stress to the strain in the sample) which comes out at 3.6MPa, five times greater than that obtained (~0.7MPa) if each load step allowed to stabilise until it reached equilibrium.

[35] Using a criterion similar to that set forth for tensile tests, based on Saint Venant's principle.

[36] Phosphate buffered saline is an aqueous solution of sodium chloride, sodium phosphate, potassium chloride and potassium phosphate whose osmolarity and ion concentration are very similar to mammal tissue extracellular liquid.

EXERCISE 47
Analyse the uniaxial compression test and determine the relationship between the engineering stress and strain (s, e) and the true stress and strain (σ, ε). Apply the results to the case of a material whose true stress-strain relationship is linear $\sigma = E\,\varepsilon$.

Could an instability be produced during the test?

Hint: To solve the problem assume that there is no friction at the load surfaces and use the principle of conservation of volume.

SOLUTION:
In section 2.2.1 the uniaxial tensile test was analysed. Everything stated there is valid for the uniaxial compression test with no more than taking into account that now $F < 0$ and therefore:

$$\lambda = \frac{L}{L_0} < 1 \quad ; \quad e = \frac{L - L_0}{L_0} = \lambda - 1 < 0 \quad ; \quad \varepsilon = \ln\lambda = \ln(1+e) < 0 \quad ;$$

$$s = \frac{F}{A_0} < 0 \quad ; \quad \sigma = \frac{F}{A} < 0 \quad \text{and} \quad \sigma = s\lambda = s(1+e) \quad \text{due to volume conservation} \quad AL = A_0 L_0$$

Applying the previous expressions we obtain the s-e relationship:

$$s = \frac{\sigma}{1+e} = \frac{\ln(1+e)}{1+e}E$$

which is shown in the Figure along with $\sigma = E\,\varepsilon$

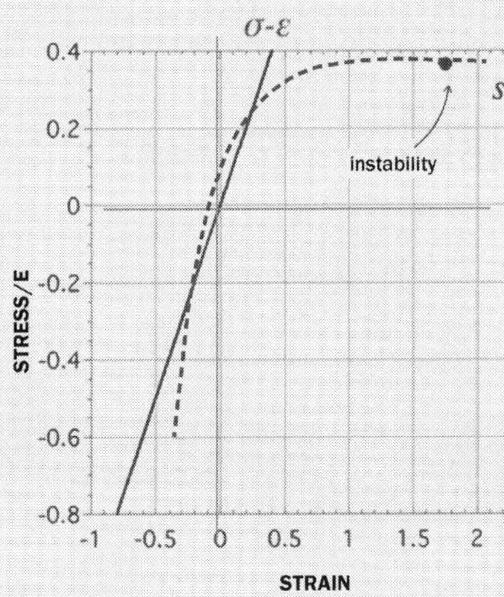

As stated in section 2.2.1, equation (8), a loading instability occurs when $dF = 0$, which –using conservation of volume– leads to the condition $\sigma = d\sigma/de$.

In a compression test it is impossible to fulfil the previous relationship since the true stress-strain curve of real material is always monotonically increasing and $d\sigma/d\varepsilon$ 0, while in the compression test we have $\sigma < 0$.

This is illustrated by the curve $\sigma = E\,\varepsilon$ in the figure, in which we observe that the engineering stress-strain curve s-e (or s/E - e) has a maximum (instability) in tension but there is no similar point in compression.

The instability point in a tensile test will occur when

$$\sigma = \frac{d\sigma}{d\varepsilon}$$

In this example

$$\sigma = \frac{d\sigma}{d\varepsilon} = E \quad ; \quad \varepsilon = \frac{\sigma}{E} = 1$$

and,therefore, in the s-e diagram the instability values will be

$$e = \mathrm{e}^{\varepsilon} - 1 = \mathrm{e} - 1 = 1.72 \quad ; \quad s = \frac{\sigma}{1+e} = \frac{E}{\mathrm{e}} = 0.37E$$

EXERCISE 48

Analyse the constrained compression test for a linear-elastic, homogeneous, isotropic material (small deformations) with modulus of Elasticity E and Poisson's ratio ν. Calculate the aggregate modulus, H_A, defined as the ratio of the compressive stress and strain in the sample. Assume that there is no friction at the loading surfaces.

SOLUTION:

Given the linear relationship between stress and strain, we will solve the problem using the principle of superposition. Later in this chapter (section 8.3.2) we will draw up the equations that relate the stresses and strains for a linear-elastic, homogeneous, isotropic body. With these it is possible to obtain a much more direct and rapid solution to the problem, which we leave, as an exercise, to the reader.

The load state in which we find the specimen is the sum of three uniaxial compressions, one along the axis of compression (σ_{11}) and two along the perpendicular axes (σ_{22}, σ_{33}).

The compressive stress-strain state along the 1-axis will be (see figure 8a of the "Biological Fibres" section)

$$\sigma_{11} \quad \sigma_{22}=0 \quad \sigma_{33}=0 \quad \varepsilon_{11}=\frac{\sigma_{11}}{E} \quad \varepsilon_{22}=-\nu\frac{\sigma_{11}}{E} \quad \varepsilon_{33}=-\nu\frac{\sigma_{11}}{E}$$

where σ_{11} is the applied compressive stress. A compressive stress will appear along the 2 and 3 axes due to the constraint effect. For the 2-axis we have

$$\sigma_{11}=0 \quad \sigma_{22} \quad \sigma_{33}=0 \quad \varepsilon_{11}=-\nu\frac{\sigma_{22}}{E} \quad \varepsilon_{22}=\frac{\sigma_{22}}{E} \quad \varepsilon_{33}=-\nu\frac{\sigma_{22}}{E}$$

and for the 3-axis

$$\sigma_{11}=0 \quad \sigma_{22}=0 \quad \sigma_{33} \quad \varepsilon_{11}=-\nu\frac{\sigma_{33}}{E} \quad \varepsilon_{22}=-\nu\frac{\sigma_{33}}{E} \quad \varepsilon_{33}=\frac{\sigma_{33}}{E}$$

The resulting strain state, the sum of the three previous, will be equal to

$$\sigma_{11}\ ;\ \sigma_{22}\ ;\ \sigma_{33}\ ;\ \varepsilon_{11}=\frac{\sigma_{11}}{E}-\frac{\nu}{E}(\sigma_{22}+\sigma_{33})\ ;\ \varepsilon_{22}=\frac{\sigma_{22}}{E}-\frac{\nu}{E}(\sigma_{11}+\sigma_{33})\ ;\ \varepsilon_{33}=\frac{\sigma_{33}}{E}-\frac{\nu}{E}(\sigma_{11}+\sigma_{22})$$

The resultant state must be such that the strain in the 2 and 3 axes is zero, which leads to

$$\varepsilon_{22}=\frac{\sigma_{22}}{E}-\frac{\nu}{E}(\sigma_{11}+\sigma_{33})=0 \quad\rightarrow\quad \sigma_{22}=\nu(\sigma_{11}+\sigma_{33})$$

$$\varepsilon_{33}=\frac{\sigma_{33}}{E}-\frac{\nu}{E}(\sigma_{11}+\sigma_{22})=0 \quad\rightarrow\quad \sigma_{33}=\nu(\sigma_{11}+\sigma_{22})$$

which we solve to obtain

$$\sigma_{22}=\sigma_{33}=\frac{\nu}{1-\nu}\sigma_{11}$$

Substituting the values of σ_{22} and σ_{33} in the strain expression ε_{11} along the 1-axis

$$\varepsilon_{11}=\frac{\sigma_{11}}{E}-\frac{\nu}{E}(\sigma_{22}+\sigma_{33})=\frac{\sigma_{11}}{E}\frac{(1+\nu)(1-2\nu)}{(1-\nu)}$$

with which we obtain

$$H_A=\frac{\sigma_{11}}{\varepsilon_{11}}=\frac{(1-\nu)E}{(1+\nu)(1-2\nu)}$$

Torsion test

The torsion test was introduced in section 2.2.2. In it, a circular cylinder was subjected to two opposing torsional moments (torques) T at its ends, while measuring the rotation ϕ with respect to the axis (figure 8.7).

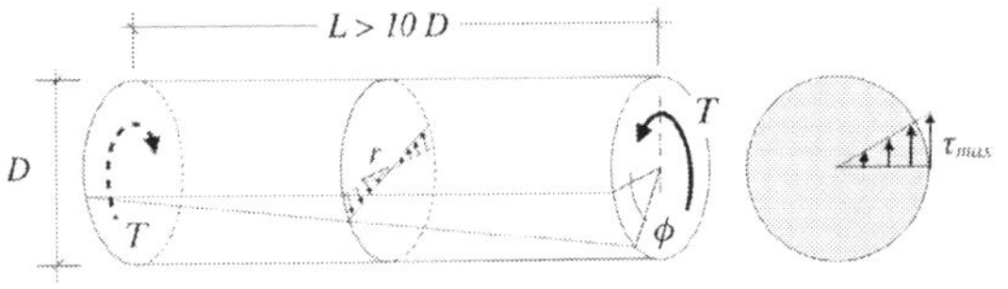

Fig. 8.7. *Torsion test on a circular section bar*

As in the tensile test, the objective of the torsion test is to produce a state of pure shear, which easy to characterise from the external actions, in this case the torsional moment T applied at the ends, and the rotation ϕ of one end of the test piece with respect to the other. The principal advantage of the torsion test is that it does not cause necking, and larger values of strain can be attained.

In order to avoid the occurrence of complex stress states in the torsion test it is only performed on homogeneous and isotropic materials or transversely isotropic materials whose axis of symmetry is the axis of the cylinder (the torsional axis). Furthermore, the specimens must be sufficiently slender, where it is normal to take as a practical rule a separation between the load points of at least ten times the diameter of the cylinder. In the following we shall assume that both conditions are satisfied.

The torsion test produces, as expained in section 2.2.2, a state of pure shear whose shear strain is the same in every section of the cylinder, and only varies linearly with the distance r from the torsional axis (equation 11, section 2.2.2).

$$\gamma(r) = r\,\phi / L \qquad (1)$$

If the material behaviour is **linear elastic,** the shear stress and strain are proportional through the shear modulus ($\tau = G\,\gamma$). In this situation it is easy to calculate the maximum shear stress (see section 2.2.2).

For a **circular** section of diameter D (figure 8.7) the maximum shear stress is reached at the periphery and its value is:

$$\tau_{\max} = \frac{16T}{\pi D^3} \qquad (2)$$

When the section is **elliptical** with semi-axes a and b (figure 8.8a) the calculation is much more complex. At any point in the elliptical section the shear stress is proportional to the distance from the centre of the ellipse and its direction is parallel to the tangent at the point of intersection of the radius with the surface, as shown in the figure. The shear stress at point (x, y) is equal to (Timoshenko, 1975):

$$\tau_{xz} = -\frac{2T}{\pi a b^3}\,y \qquad \tau_{yz} = \frac{2T}{\pi a^3 b}\,x \qquad (3)$$

The maximum value of the shear stress occurs at the ends of the minor sem-axes, and its value is (Timoshenko, 1975):

$$\tau_{\max} = \frac{2T}{\pi a b^2} \qquad (4)$$

The rotation per unit length ϕ / L for this problem is:

$$\phi / L = \frac{T}{G}\,\frac{a^2 + b^2}{\pi a^3 b^3} \qquad (5)$$

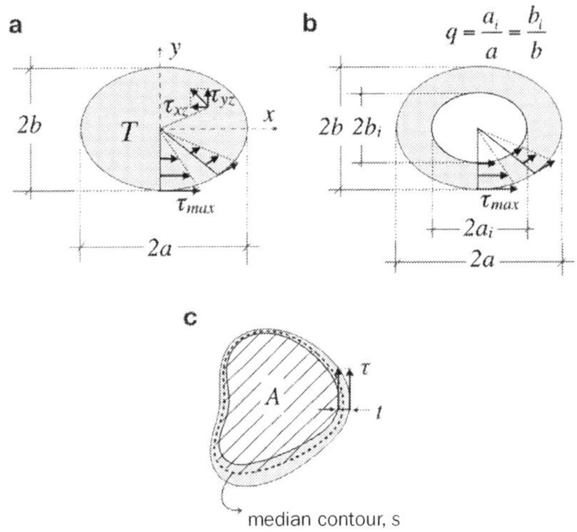

Fig. 8.8. *Cross-sections in torsion tests*
a. Elliptical
b. Hollow elliptical
c. Thin-walled tube

A frequent use of the torsion test is the characterisation of long bones whose geometry resembles that of a tube. Using the solution for an elliptical section it is easy to obtain the solution for a tube with elliptical walls whose semi-axes are proportional, as shown in figure 8.8b. In this case the stresses at (x, y) are:

$$\tau_{xz} = -\frac{2T}{\pi ab^3}\frac{y}{1-q^4} \qquad \tau_{yz} = \frac{2T}{\pi a^3 b}\frac{x}{1-q^4} \tag{6}$$

where $q = a_i/a = b_i/b$ the ratios of the semi-axes. The maximum shear stress occurs, as before, at the ends of the minor semi-axes:

$$\tau_{max} = \frac{2T}{\pi ab^2(1-q^4)} \tag{7}$$

and the rotation per unit length ϕ/L is:

$$\phi/L = \frac{T}{G}\frac{a^2+b^2}{\pi a^3 b^3(1-q^4)} \tag{8}$$

A very practical solution for thin-walled tubes of linear-elastic material, variable thickness and arbitrary geometry (figure 8.8c) is Timoshenko (1975):

$$\tau = \frac{T}{2At} \quad , \quad \frac{\phi}{L} = \frac{T}{4A^2G}\oint\frac{ds}{t} \tag{9}$$

which provide the shear stress at a point in the section and the value of the rotation per unit length. The wall thickness t (always small) can vary. In the expressions above A is the value of the area enclosed by the median contour of the section and s is the arc measured on the median contour (figure 8.8c). When the section thickness is constant:

$$\tau = constant = \frac{T}{2At} \quad , \quad \frac{\phi}{L} = \frac{Ts}{4A^2Gt} \tag{10}$$

where s is the length of the median contour of the tube.

For a thin-walled circular tube, the stress and strain are determined directly from the boundary conditions:

$$\tau = \frac{2T}{\pi tD^2} \quad , \quad \gamma = \frac{\phi D}{2L} \tag{11}$$

where it is not necessary to assume any relationship between them (linear elastic or otherwise), since the value is fixed by the equations of equilibrium and compatibility.

EXERCISE 49
From the solution for an elliptical cross-section, derive expressions 6, 7 and 8 for tubes with an elliptical boundary.

SOLUTION:
Consider an elliptical cylinder subjected to a torsional moment T as represented in the figure. If we imagine a cylinder inside this with proportionate semi-axes, we can see that the stresses on the lateral surface of this inner cylinder must be zero, as indicated by the expressions 3, since the shear stress is tangent to the elliptical boundary.

We can then consider that the internal cylinder function as if it were decoupled from the outermost part–it alone– supporting a torque T^*. The stresses on the boundary will be (equation 3)

$$\tau_{xz} = -\frac{2T^*}{\pi a_i b_i^3} y \qquad \tau_{yz} = \frac{2T^*}{\pi a_i^3 b_i} x$$

where

$$\frac{x^2}{a_i^2} + \frac{y^2}{b_i^2} = 1$$

The stresses on the boundary of the inner cylinder, now calculated from the stress state of the solid cylinder, will be

$$\tau_{xz} = -\frac{2T}{\pi a b^3} y \qquad \tau_{yz} = \frac{2T}{\pi a^3 b} x$$

since these must be coincident, then $T^* = Tq^4$, where $q = a_i/a = b_i/b$ the ratio of the semi-axes.

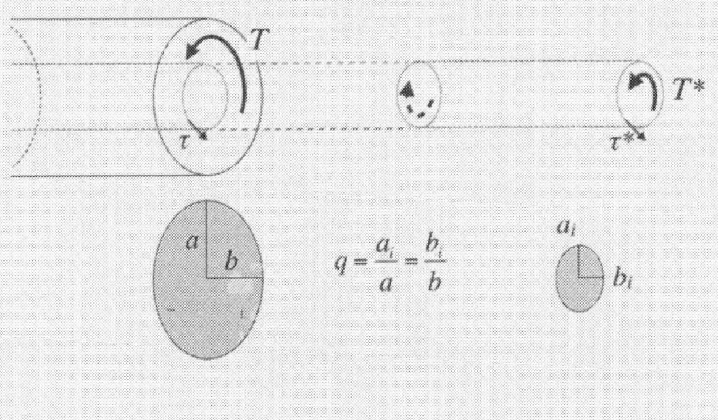

From the above analysis we see that the inner cylinder supports the torque T^* while the external zone, which is an elliptical tube with external semi-axes (a, b) and interior (a_i, b_i) carries the remaining torsional moment $T - T^* = T(1-q^4)$.

If we now consider an elliptical tube on its own, we see that stresses induced are caused only by the torque which acts upon it: $T_{tube} = T - T^* = T(1-q^4)$ so that the stress solution for the tube problem will be the same equations 3 but with the substitution $T = T_{tube}/(1-q^4)$, that is:

$$\tau_{xz} = -\frac{2\boldsymbol{T_{tube}}}{\pi a b^3}\frac{y}{1-q^4} \qquad \tau_{yz} = \frac{2\boldsymbol{T_{tube}}}{\pi a^3 b}\frac{x}{1-q^4}$$

which are equations 6. In a similar manner we arrive at equations 7 and 8.

EXERCISE 50

The figure provides the results of a torsion test (torque T – twist angle ϕ) in a rat femur, performed by Ekeland and co-workers (Ekeland *et al.* 1981, 1982). The figure also shows the central cross-section and length of the femur.

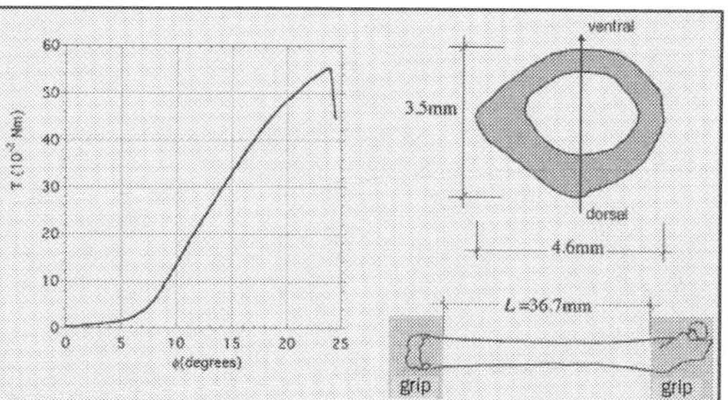

From these data, calculate the shear stress at rupture, and estimate the shear modulus G of the bone.

SOLUTION:

The initially concave part of the torsion curve is due –as in many other tests– to the initial articulation of the experimental apparatus and provides no information on the material. To correct for this, a straight line is fitted to the rectilinear part of the response which follows it, representative of the linear-elastic behaviour of the material, and its intersection with $T = 0$ is taken as zero on the abscissa.

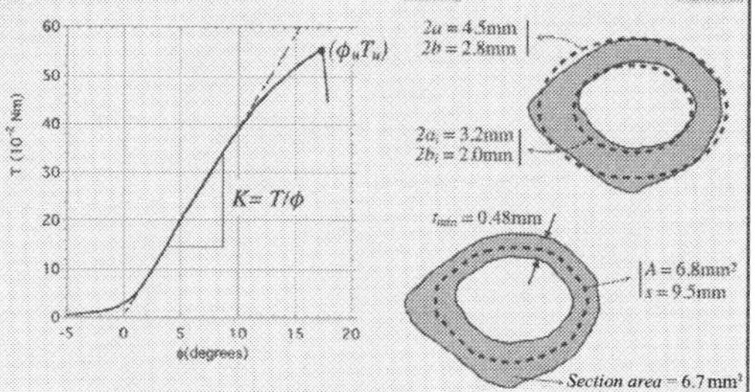

From the corrected T - ϕ curve, we measure the rupture data which are T_u = 0.55 Nm; ϕ_u = 17° = 0.297 rad and the elastic slope: K=0.0387 Nm/°=2.215 Nm/rad

To obtain the stresses and strains we approximate the bone section to an elliptical tube, as shown in the figure, obtaining the parameters:

$2a$ = 4.5 mm; $2b$ = 2.8 mm

$2a_i$ = 3.2 mm; $2b_i$ = 2.0 mm

where $q = a_i/a = b_i/b = 0.71$.

From equation 7 we obtain: $\tau_u = \dfrac{2T_u}{\pi a b^2 (1-q^4)} = \mathbf{53\ MPa}$

From equation 8 we can determine the shear modulus in which we have substituted $T/\phi = K$, obtained from the fitted line.

$$G = \frac{T}{\phi}\frac{L\left(a^2+b^2\right)}{\pi a^3 b^3 (1-q^4)} = \mathbf{487\ MPa}$$

If we use the *thin-walled* assumption, we can calculate the maximum stress with the aid of equation 9, which yields

$$\tau_u = \frac{T_u}{2At} = \mathbf{84\ MPa}$$

In this case we have taken the area A=6.8 mm² enclosed by the median boundary (see the figure) and the minimum wall thickness t = 0.48 mm.

If we calculate G from equation 10, taking a mean thickness calculated from the median contour and the area of the cross-section

$$\langle t\rangle = \frac{\text{Area enclosed}}{s} = \frac{6.7}{9.5} = 0.7\text{mm}$$

we obtain $G = \dfrac{T}{\phi}\dfrac{sL}{4A^2\langle t\rangle} = \mathbf{5960\ MPa}$

We see that these values are very different from those assuming an elliptical tube.

In any case, the poor uniformity of the section generates large discrepancies between the values obtained, so most of the time these tests are used to make comparisons between bones with different treatments rather than to obtain the intrinsic properties of materials.

Flexure test

Introduction

The flexure test is a simple method of characterising a material and consists of measuring the load and deflection produced in a bar of constant and symmetric cross-section, usually rectangular, subjected to loads which are perpendicular to its axis (figure 8.9). Generally the bar is supported at its ends and loaded in its central region by one (three-point bending, figure 8.9a) or two forces (four point bending method, figure 8.9b).

The flexure test provides a relatively straightforward method of measuring the strength and stiffness of materials. It is also useful as a technological test for evaluating and comparing the suitability of elements that must work in flexure; for example, for long bones where it is very common to use the test to study the strength after fracture and repair of the bone.

The flexure test is used preferentially on homogeneous and isotropic materials, and also on orthotropic materials whose axis of symmetry is perpendicular to the axis of the bar and contained in the plane of the load. In all of these cases, the resulting stress state can be simplified by assimilating it to a combination of uniaxial tensile and compressive states which act parallel to the axis of the specimen (figure 8.9c). If the bar were ideally divided into layers parallel to its axis, the upper parts would be compressed while the lower portions would carry tensile stresses, and would be stretched. The line (in projection) parallel to the axis which separates the compressed and stretched regions is called the *neutral axis*. All points on the neutral axis maintain their original length, suffering neither stretching nor shortening.

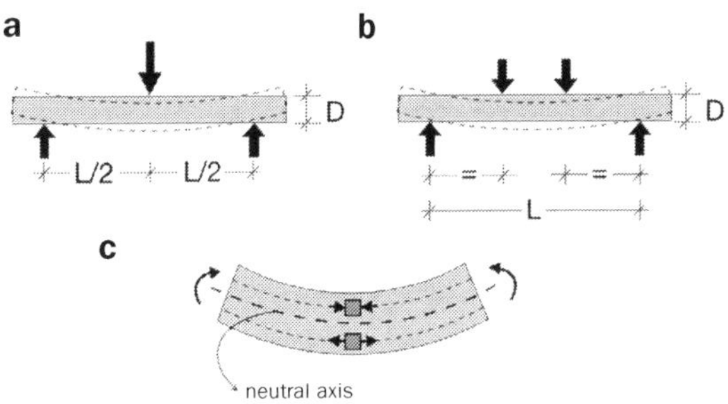

Fig. 8.9. *Flexure test*
a. Three-point bending
b. Four-point bending
c. Stress state

In contrast to tensile and compression tests, the flexure test generates a stress state which is not homogeneous, making it difficult, if not impossible, to obtain the stress-strain curve for the material directly[37]. Thus, in almost all cases it is usual to assume a particular constitutive equation –generally linearelastic– whose parameters are determined from the results of the test.

Experimental configuration

With regard to its experimental configuration –as in all tests on biological materials– it is necessary to take into account the effect of load rate and the environmental variables (humidity and temperature). The flexure test is mechanically robust and relatively insensitive to specimen surface imperfections or the load arrangement. Even so, it is advisable to use rollers to even out possible geometric defects as shown in figure 8.10a. The rollers must be coupled together and locked before the start of the test to avoid instabilities during test execution.

Loads in the test are applied through rigid cylindrical, usually steel, fixtures whose radius of curvature must be large enough to avoid indentation and contact damage, and possible local compression fracture (figure 8.10b). For stiff, strong, biological materials such as bone it is customary to take as a guide the ASTM standards for testing plastics (ASTM D790 and ASTM D6272) which recommend a minimum radius of curvature of 5 mm.

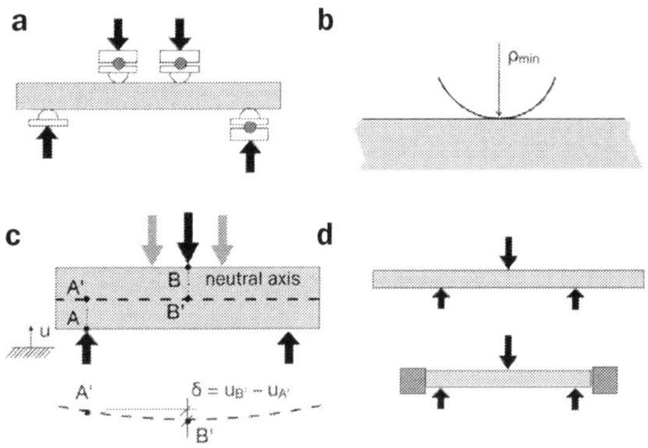

Fig. 8.10. *Experimental configuration in flexure tests*
a. Pivoting supports
b. Curvature in the loading zone
c. Measurement of the vertical displacement δ of the central section
d. Compensation for self-weight

[37] Deriving the constitutive relationship from the results of a flexure test (finding the best fit to the experimental data by means of a numerical model which simulates the test) turns out to be a mathematically ill-conditioned problem which lacks a unique solution even in the ideal case of having accurate experimental results.

The curvature of the supports also has an important effect on the measurement of the *displacement in the central section of the specimen*, δ, which is often used–as will be seen later– to determine the elastic modulus of the material. If the displacements were evaluated by taking as a reference the displacement of the load points (points A and B in figure 8.10c) it must be borne in mind that these values depend critically upon the size of the contact zone, which is, in turn, a function of the radius of the cylindrical load surface (Planas *et al.* 1994). In the absence of a customised analysis that accounts for the extent of the true support region, it is advisable to determine the displacement of the central section of the specimen from the displacements measured on the neutral axis of the specimen, ($\delta = u_{B'} - u_{A'}$), as illustrated in figure 8.10c.

With regard to the slenderness of the specimen, it may be taken that, in both three-point and four-point tests, above a length/depth ratio (L/D) of 4 a state of flexure is fully developed. Nonetheless, the ASTM D790 and ASTM 6272 standards usually recommend that the tests are performed on specimens which are much more slender ($L/D = 16$) in order to reduce the effect of the shear deformation produced by the vertical forces. Some authors, for tests on bone, increase the recommended value to as much as 20 or even 25 (Spatz *et al.* 1996). With high slenderness ratios, it is essential, on the other hand, to take care not to induce nonlinear effects due to excessive changes in the position of the loads during the test.

Even though the self-weight of the specimen is generally very small, its effect can be easily eliminated using double length specimens, although this will depend upon the availability of the material. An alternative is to add counterweights to balance the test piece (figure 8.10d).

Analysis of the mechanics

The analysis of the mechanics of the flexure test on a homogeneous isotropic material is shown in figure 8.11. It is assumed that the strains in the fibres of the bar are very small (infinitesimal strain theory) and also Navier's hypothesis[38] that plane cross sections of the bar remain plane after deformation. This hypothesis is satisfied reasonably well provided that the specimen is sufficiently slender and the effects of the shear stresses can be ignored.

Consider the coordinate system shown in figure 8.11a, whose axes (y, z) are perpendicular to the axis of the bar while the x-axis runs along the length of the bar. The x-y plane lies in the plane of symmetry of the bar, in which the flexure is produced. The location of the neutral axis $y = y_0$, is also shown in the figure, which lies in the plane[39].

The deformation of a small section of the bar $\Delta x = \Delta l_0$ is illustrated in figure 8.11b. Its extremities rotate through an angle $\Delta\theta$ remaining plane (according to Navier's hypothesis) and the fibres of the section are curved. If ρ_0 is the radius of curvature of the neutral axis, located in y_0, then:

$$\Delta l_0 = \rho_0 \Delta\theta \quad (12)$$

The radius of curvature $\rho(y)$ of the section of fibre located at position y will be:

$$\rho(y) = \rho_0 - (y - y_0) \quad (13)$$

and the change in length of this section will be equal to:

$$\rho(y)\Delta\theta - \Delta l_0 = [\rho_0 - (y - y_0)]\Delta\theta - \Delta l_0 \quad (14)$$

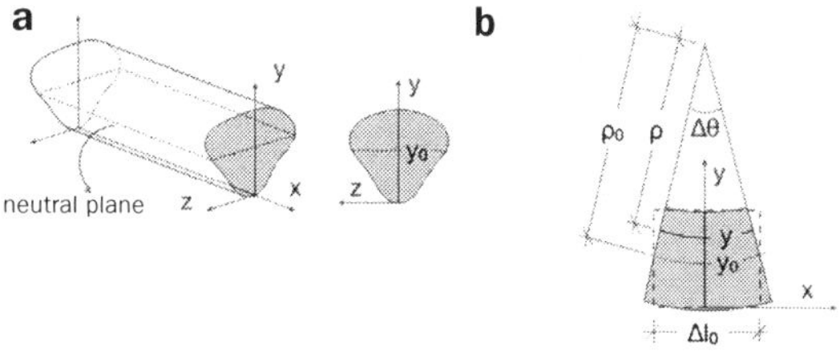

Fig. 8.11. *Analysis of the flexure test*
a. Coordinate system
b. Deformation between two adjacent cross-sections

[38] As a result of this hypothesis and his contribution to the study of beam bending, the French engineer Claude-Louis Marie Henri Navier (1785-1836) is considered the father of the theory of Strength of Materials and structural analysis.

[39] Note that the term "neutral axis" refers to the projection on the x-y plane.

The **strain** in the fibre can be calculated from this value, by simply dividing by Δl_0 and using equation (12) for $\Delta\theta$:

$$\varepsilon(y) = -\frac{1}{\rho_0}(y - y_0) \qquad (15)$$

Every point in the cross-section of the bar will be subjected to a normal tensile or compressive stress whose value will be determined by the strain (equation 15) and the constitutive equation $\sigma=\sigma(\varepsilon)$ of the material. Points where $y>y_0$, above the neutral axis (neutral plane), will be under a state of compression producing a shortening of the fibres. The opposite happens for points where $y<y_0$, located below the neutral axis (figure 8.11c).

c

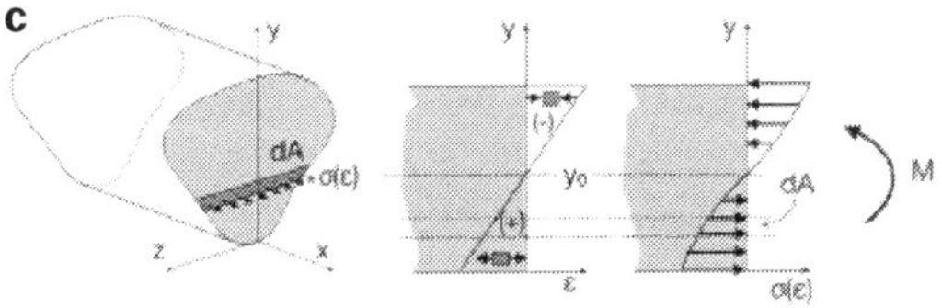

Fig. 8.11. *c. Stress-strain state*

In order to evaluate the strains at a given point in the cross-section of the bar it will be necessary to know the value of the radius of curvature ρ_0 and the position of the neutral axis y_0, as deduced from equation (15). Both values are obtained by setting up the equations of equilibrium of forces and moments on the cross-section.

Figure 8.12 shows the forces acting on a cross section of the bar, caused by the external loads. The axial force on any cross-section is zero while the moment of the external forces varies with position along the bar. For a three-point bending test (see figure 8.12a) the bending moment is smallest at the supports and reaches its maximum at the central section ($M_{max,3P}=FL/4$). In the case of the four-point bending specimen (see figure 8.12b), we see that the region between the two central loads is subjected to a maximum, constant, value of $M_{max,4P}=Fs/2$, where L and s are shown in figure 8.12.

The first equation of equilibrium relates to the forces normal to the cross-section of the bar. Since there are no forces in the direction of the axis of the bar (x-axis), the resultant of the stresses on the cross section of the bar (of area A) must be zero:

$$\int_A \sigma(\varepsilon(y))dA = 0 = \int_A \sigma\left(-(y-y_0)/\rho_0\right)dA \qquad (16)$$

The second equation, establishes that the moment M on the cross-section produced by the external loads (whose value depends upon the section analysed, as can be seen in figure 8.12) must be equal to the internal moment produced by the stresses (figure 8.11c). Taking moments about the neutral axis[40] we obtain:

$$M = -\int_A (y-y_0)\sigma\, dA \qquad (17)$$

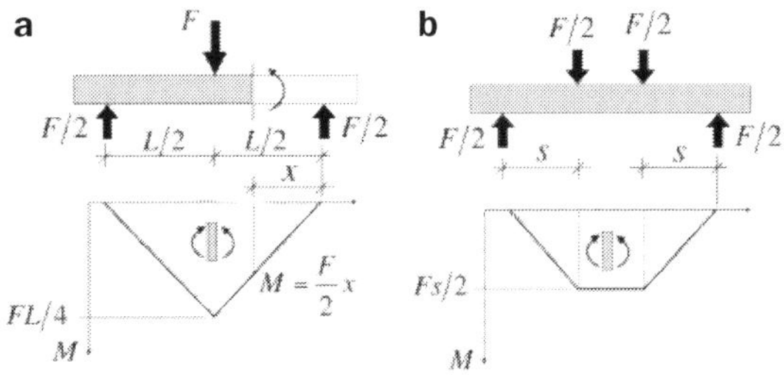

Fig. 8.12. *Moment of the external forces*
a. Three-point bending
b. Four-point bending

Equations (16) and (17), together with the function $\sigma(\varepsilon)$ which relates the stresses and strains in the material, allow calculation of the values of (ρ_0, y_0) at any cross-section along the bar. From these, the **strain** and **stress** at any point are immediately obtained by means of equation (15) and the constitutive equation $\sigma(\varepsilon)$.

When the function $\sigma(\varepsilon)$ is odd (that is to say $\sigma(-\varepsilon)=-\sigma(\varepsilon)$) and the cross-section of the bar has a centre of symmetry CS, then the position of the neutral axis coincides with the centre of symmetry (see Exercise 51).

$$y_0 = y_{CS} \qquad (18)$$

In the most common case of assuming a linear-elastic relationship between stresses and strains ($\sigma=\varepsilon E$), the position of the fibre coincides with the centroid C of the cross-section (see Exercise 51).

$$y_0 = y_C = \frac{1}{A}\int_A y\, dA \qquad (19)$$

In this case, and taking into account equation (15), equation (17) becomes:

$$M = -\int_A (y-y_0)E\varepsilon\, dA = \frac{E}{\rho_0}\int_A (y-y_C)^2\, dA = \frac{EI}{\rho_0} \qquad (20)$$

in which I is the second moment of area of the section about an axis parallel to the z-axis and which passes through the centroid of the section.

[40] Since the sum of the stresses is zero, the moment is always the same and the point about which moments are taken is irrelevant.

The stresses on the cross-section of the bar follow a linear law given by:

$$\sigma(y) = -\frac{E}{\rho_0}(y - y_C) = -\frac{M}{I}(y - y_C) \quad (21)$$

and the maximum tensile stress (positive) occurs in the lowest fibre (figure 8.11c), when $(y-y_C)$ is minimum (negative):

$$\sigma_{max} = \sigma_f = -\frac{M}{I}(y_{min} - y_C) \quad (22)$$

The value of the maximum tensile stress produced in the bar at the instant of rupture is called the *rupture modulus* (σ_f) or *flexural strength* of the material. Even though theoretically its value should coincide with the *tensile strength,* in reality it is always greater[41].

The calculation of the central displacement of the bar is somewhat more laborious. The interested reader may refer to Timoshenko (1980). The result, for the case of a linear-elastic material ($\sigma = \varepsilon E$) is

$$\delta_{3P} = \frac{FL^3}{48EI} \quad (23)$$

for three-point bending, and

$$\delta_{4P} = \frac{Fs}{48EI}(3L^2 - 4s^2) \quad (24)$$

for a specimen in four-point bending.

EXERCISE 51

Derive equations (18) and (19).

SOLUTION:

Proof of equation (18) follows directly, since if we take $y_0 = y_{CS}$ equation (16) is satisfied. To see this it is sufficient to consider the coordinate $v = y - y_{CS}$, with origin at the centre of symmetry of the cross-section, and separate the integral (16) into two parts depending upon whether v is greater than or less than zero:

$$\int_A \sigma\left(-\frac{y-y_0}{\rho_0}\right) dA = \int_A \sigma(-v/\rho_0)\,dA = \int_{v>0} \sigma(-v/\rho_0)\,dA + \int_{v<0} \sigma(-v/\rho_0)\,dA$$

By symmetry, for each one of the elemental surfaces dA located in $v<0$ there is an identical corresponding one in $v>0$. Therefore, the integral over the region $v<0$ will be identical to the integral over $v>0$ with only the sign of v changed

$$\int_{v>0} \sigma(-v/\rho_0)\,dA + \int_{v<0} \sigma(-v/\rho_0)\,dA = \int_{v>0} \{\sigma(-v/\rho_0) + \sigma(+v/\rho_0)\}\,dA$$

and given that σ changes sign with its argument we see that the integral is zero.

When $\sigma = \varepsilon E$ in equation (16) the result is: $\int_A \sigma\left(-\frac{y-y_0}{\rho_0}\right) dA = 0 = \int_A -\frac{E}{\rho_0}(y-y_0)\,dA = -\frac{E}{\rho_0}\int_A (y-y_0)\,dA$

and recalling the definition of the centroid of a plane section: $y_C = \frac{1}{A}\int_A y\,dA$

we see that we can write equation (16) as: $-\frac{E}{\rho_0}\int_A (y-y_0)\,dA = -\frac{E}{\rho_0}(y_C - y_0)A = 0$

from which equation (19) follows immediately.

41 In ceramic materials it is about 30% greater. A much simplified explanation is that in the flexure test the volume of material subjected to the elevated stresses is much smaller than in a conventional tensile test and, therefore, it is more difficult to find a large crack which causes rupture in the volume (of elevated stress).

EXERCISE 52

Ekeland and co-workers (Ekeland *et al.* 1981) used a cantilever flexure test to measure the rupture strength of rat femurs. The figure gives a sketch of the experimental set-up. In the test a force is applied to the end of a disc which rotates about a shaft O and induces a bending moment $M = Fa$ on the cross-section of the bone located in O. If rupture occurs when $F = 24N$ calculate the flexural strength (rupture modulus) of the bone.

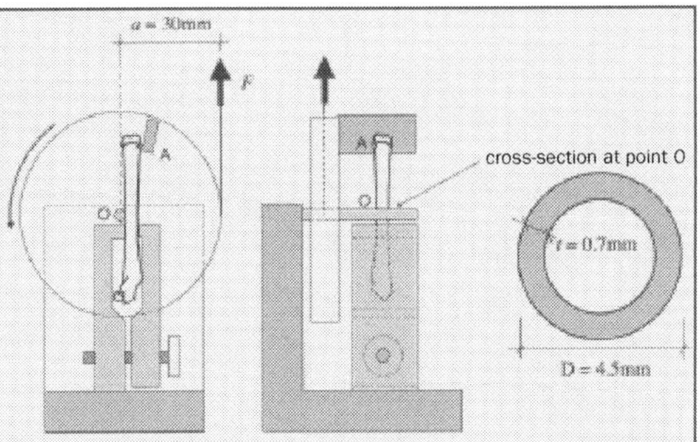

Assume that the material is linear elastic and that the cross-section of the bone in the area of contact with the shaft O can be approximated to a circular tube of 4.5 mm external diameter and thickness 0.7 mm.

SOLUTION:

The experimental rig is called a "cantilever" since the test piece is rigidly clamped at one end while the free end is loaded, as shown schematically in the adjoining figure.

In these circumstances, the maximum bending moment is reached at the end of the embedded region (grip), which corresponds to the point O. The cross-section of the bone at this point is subjected to a bending moment

$$M = Fa = 24\,0.03 = 0.72\,\text{Nm}$$

The bending moment will produce a linear distribution of stresses, given by equation (21). The maximum value of the stress is reached at the surface of the bone opposite the contact with the shaft O, and its value will be given by equation (22)

$$\sigma_{\max} = -\frac{M}{I}(y_{\min} - y_C) = +\frac{Fa}{I}\frac{D}{2}$$

The second moment of area I of the tube cross-section about its centroidal axis (neutral axis) will be:

$$I = \frac{\pi}{4}\left[(D/2)^4 - (D/2 - t/2)^4\right] = 9.8935\ 10^{-12}\ \text{m}^4$$

with which the resulting value of the maximum stress is:

$$\sigma_{\max} = \sigma_f = \frac{Fa}{I}\frac{D}{2} = 163.7\ \text{MPa}$$

which will be the value of the rupture modulus of the bone.

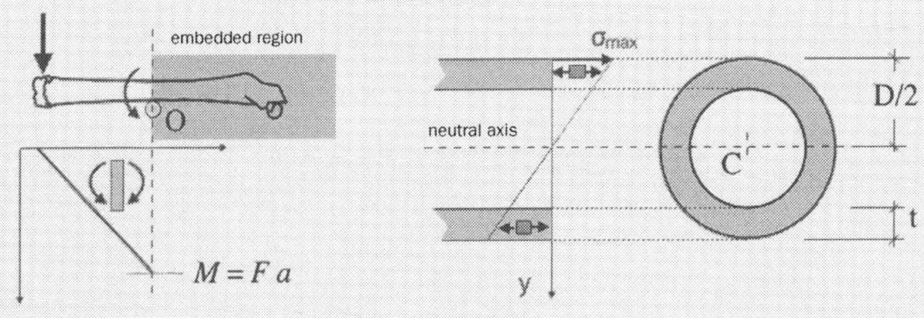

EXERCISE 53

The figure shows the force-displacement curve for a three-point bending test carried out by Libonati and Vergani (2016) on a bovine cortical bone specimen whose length was aligned with the direction of the osteons. The specimen dimensions are shown in the figure.

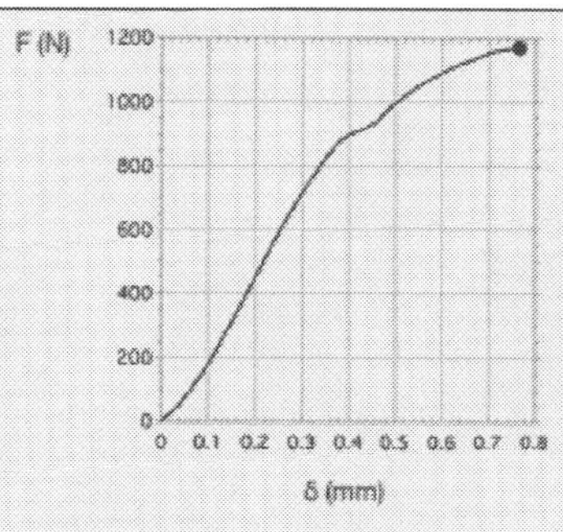

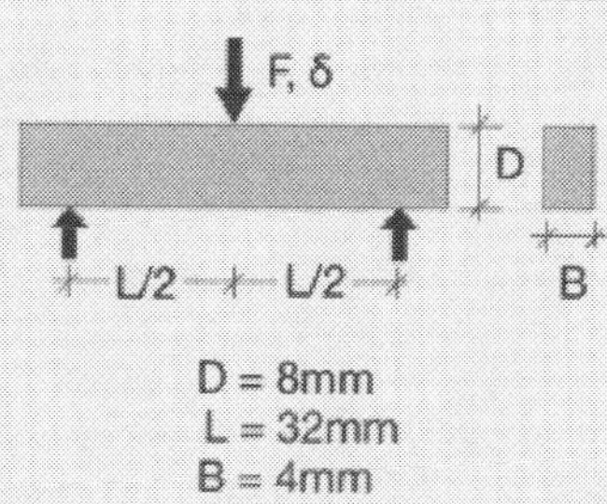

Calculate the rupture modulus of the bone and its elastic modulus assuming linear-elastic behaviour.

SOLUTION:

In order to calculate the modulus of rupture, σ_f, we turn to equation (22) taking into account that the second moment of area of the rectangular cross-section about its centroidal axis is:

$$I = \frac{1}{12} BD^3$$

with which the expression for the rupture modulus becomes: $\sigma_{max} = \sigma_f = \frac{M}{I}\frac{D}{2} = \frac{3FL}{2BD^2}$

Examining the F–δ curve we see that rupture occurs when $F = 1170$ N, which, when substituted in the above equation, yields:

$$\sigma_f = 219.4\text{MPa}$$

In order to obtain the elastic modulus, E, we need to determine the slope of the initial linear-elastic section of the curve.

From the F–δ curve we observe that, for very low loads at the start of the test, the curve is concave which is usually due to the adjustment (settling in) of the loading system. For that reason we make the fit to the subsequent linear zone, which gives:

$$\left(\frac{F}{\delta}\right)_{linear} = 2703\text{N/mm}$$

and, with the aid of equation (22), we calculate the value of the elastic modulus:

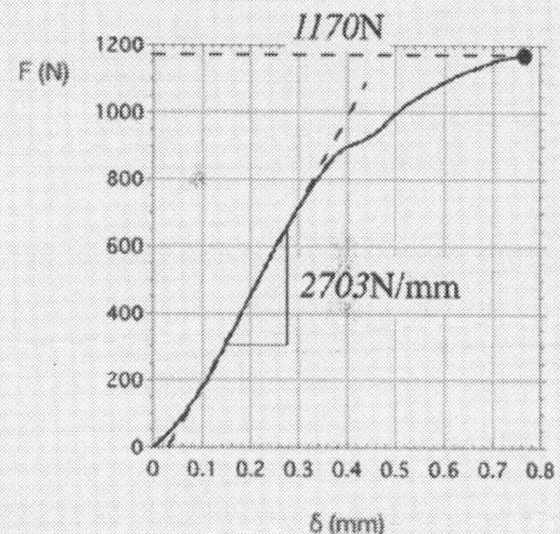

$$E = \frac{FL^3}{48I\delta} = \left(\frac{F}{\delta}\right)\frac{L^3}{4BD^3} = 10.8\text{GPa}$$

In their series of experiments the authors obtained an average value of the flexural strength (modulus of rupture) of $\sigma_f = 220 \pm 10$ MPa, and an elastic modulus $E = 12 \pm 1$ GPa, values that fit in well with our measurements.

In their series of experiments the authors obtained an average value of the tensile strength of $\sigma_u = 111 \pm 30$ MPa and $E = 20 \pm 5$ GPa, rather different from the flexure test results which gave, approximately, double the strength and half the stiffness. In this case, in addition to the differences between the two types of test (observed on isotropic material), there is also the fact that the cortical bone is orthotropic.

8.2.2. Rupture tests

General Fracture Mechanics concepts

Rupture of a material begins with the propagation of a preexisting defect –nearly always, a crack– or one which is formed during the loading process. *Fracture Mechanics* aims to determine the load which a cracked material can support, or up to what size a crack might grow without producing catastrophic failure. This branch of mechanics was developed and consolidated in the second half of the twentieth century.

Rupture criteria are well established for elastic materials –linear or otherwise– but this is not the case when there is plasticity or viscoplasticity (behaviour commonly found in biological materials). For the moment, our observations on fracture analysis will focus on elastic materials because almost all of the standard fracture tests are based on this type of behaviour and they are the most frequently used to characterise biological materials.

The treatment of failure criteria is based on two approaches, **global** and **local.**

The *global* approach is based on an energy balance between energy available (supplied from the outside and the elastic stored energy) and the energy expended in fracture. The energy approach has several advantages; it is simple (being based on the principle of conservation of energy) and it is general (allowing application to any elastic material, even if nonlinear).

The *local* approach is based on the fact that for elastic materials it is possible to describe the fracture criterion locally, as a function of the stress and displacement near the crack tip. These (stress and displacement) fields –in linear-elastic materials– are fully characterised if the stress intensity factor (K) is known and failure is produced when K reaches a critical value.

It is, then, equally possible to express the failure criterion (for a linear-elastic material) as a function of the available energy or the stress intensity factor: a crack becomes unstable when either the strain energy, or the stress intensity factor, reaches a critical value.

The advantage of the local criterion over the global is that it is usually much easier to calculate the stress intensity factor than the available energy (because for most problems these factors are tabulated in handbooks and because, when dealing with linear elastic materials, the principle of superposition can be applied). The drawback is that its application is restricted to linear-elastic materials.

Later, brief comment will be made on some issues and definitions that are needed in order to apply the procedures, described above, in the fracture tests. The concepts and the theory behind *Fracture Mechanics* have no place here but the interested reader can refer to a book written by the present authors (Elices and Guinea, 2021), and the references contained in it.

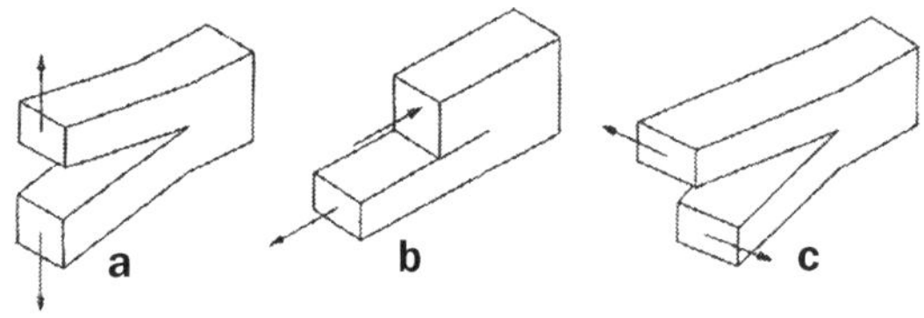

Fig. 8.13. *Basic modes of fracture.*
a. Mode I: Opening (tensile) mode.
b. Mode II: (In-plane) Shearing mode.
c. Mode III: (Out-of-plane shear) Tearing mode.

The analysis of the propagation of a crack in a solid subjected to any distribution of forces and displacements is complex. In order to facilitate its study, the general propagation of a crack is decomposed into three basic modes, represented in figure 8.13; simple tensile opening, or Mode I (figure 8.13a), sliding, by in-plane shear, or Mode II (figure 8.13b), and tearing (out-of-plane shear), or Mode III (figure 8.13c). Most fractures occur in Mode I which, furthermore, is the most studied and that on which the tests discussed below are based. Mode III is characteristic of tearing fractures which originate in laminas and membranes, a subject discussed previously in section 5.3.3.

The **local approach** is written as:

$$K = K_C \tag{25a}$$

The first term, $\boldsymbol{K}$, called the *stress intensity factor,* is a function which must be **calculated** in every case and which characterises the state of stress near the tip of a crack tip. This function, K, is defined for, and only makes sense in, *linear-elastic materials.*

The second term, $\boldsymbol{K_C}$, called the *fracture toughness* (or critical stress intensity factor), is a material property and its **measurement** is governed by standards (for example, in the ASTM E399, ASTM E1304 and ASTM E1820 standards). Strictly speaking K_C is not a material property, it varies with thickness until a state of plane strain is achieved.

The **global approach** can be written as:

$$G = G_C \qquad (25b)$$

The first term is the *strain energy,* per unit area (or strain energy release rate) to break the material – which is represented by G– and the second is the specific fracture energy, which is assumed to be a characteristic of the material.

G is a quantity that is **calculated** and is a function of the applied load, crack length and the geometry of the body. G_C is a property which is **measured** and depends upon the temperature and the loading rate, in a state of plane strain.

The values of K and G, are not independent. For a crack propagating in Mode I

$$K_I^2 = E'G \qquad (26)$$

where E' is the generalised elastic modulus of the material, which is equal to the elastic modulus E in a state of plane stress (thin specimens) and to $E/(1\text{-}v^2)$ in plane strain (thick specimens). v is the Poisson ratio of the material. When the crack propagates the relationship above leads to:

$$K_{IC}^2 = E'G_C$$

Measurement of G_C can be performed directly, measuring the area enclosed in a process of loading and unloading after a small amount of stable crack propagation, or by indirect measurements from a force-displacement record in a cracked specimen of a given geometry. The latter process is specified in the ASTM Standard E1820.

Standard tests

Measurement of the fracture toughness, K_C

When measuring the value of the toughness which initiates the propagation of a crack, in pre-cracked specimens, it is found that this value usually decreases as the specimen thickness increases, until it reaches a practically constant value beyond a certain thickness. This limiting value is a material property called the fracture toughness (or more properly the plane strain fracture toughness) and which is denoted K_{IC}. The fracture toughness does not always depend upon the thickness provided the dimensions of the specimen are large enough compared to the size of the plastic zone, and it can be assumed that around the crack tip there is a state of plane strain.

It has also been established that the fracture toughness (K_{IC}), varies with temperature and the loading rate, as happens with other material parameters; such as the yield stress or the modulus of elasticity.

Measurement of the plane strain fracture toughness, K_{IC}, is standardised –like the measurement of G_{IC}– in the ASTM Standards (E399, E1304 and E1820). All of these contain details of the specimens, test fixtures and the mechanical testing machines.

The test consists of loading a cracked specimen in Mode I and measuring the critical load which initiates crack growth, taking care to ensure that the plastic zone at the crack tip is small and that the test takes place under plane strain conditions. For the measurement of K_{IC} ASTM Standard E399 proposes the specimens shown in figures 8.14a and 8.14b.

The first step towards the determination of K_{IC},once the type of specimen has been chosen, consists of specifying the specimen thickness. The limitations regarding the specimen thickness and the crack length, which are practically the same, are determined by the conditions set out above in order to ensure that the value of the fracture toughness obtained from the test is a characteristic of the material. These conditions are:

- That the stress state is plane strain.
- That the size of the plastic zone at the tip of the crack is small compared to the specimen dimensions.

Once the specimen thickness has been selected the specimen can be fabricated. It is recommended that the cracks are large, with $0.5 \leq a/W \leq 0.70$. Machining of the notch and the generation of the starter crack at the end of the notch are delicate processes. These starter cracks are produced by notching the specimen and inducing a fatigue crack at the tip of the notch, so that the resulting crack is close to the ideal crack (zero crack-tip radius). The test standard provides details as to how the specimen should be fatigued in order to avoid creating an excessively large plastic zone at the tip of the crack because the residual stresses resulting from excessive plasticity would invalidate the test results.

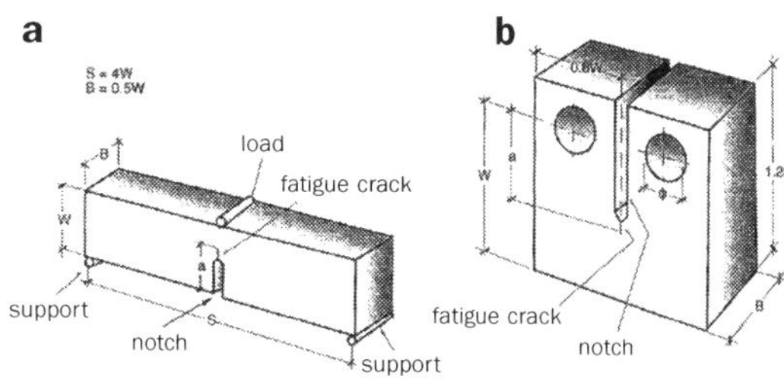

Fig. 8.14. *ASTM Standard E399 for fracture tests*
a. Recommended geometry for a beam type specimen
b. Recommended geometry for a compact tension specimen

Once the specimen has been fabricated the K_{IC} test is performed, subjecting the specimen to an increasing load until rupture, while recording the applied force F and the relative displacement w of the lips of the crack.

After rupture, the length of the crack is determined from the average of its length at the centre of the specimen and those at two points midway between the centre and the two lateral faces of the specimen.

Later, with the length of the crack, a, calculated in this way, the critical stress intensity factor, K_Q, is determined (see figure 8.14c). For this, in the graph (F, w), the secant line which passes through the origin of coordinates and which has a slope of 95% of the gradient of the tangent at the origin of the curve is drawn. The point at which the secant cuts the curve defines a load P_5. If the load at all points in the data record before P_5 is less than this value, then $P_Q=P_5$ is taken. If, on the other hand, there is a maximum in the load prior to P_5 greater than this value, P_Q is taken to be that maximum value.

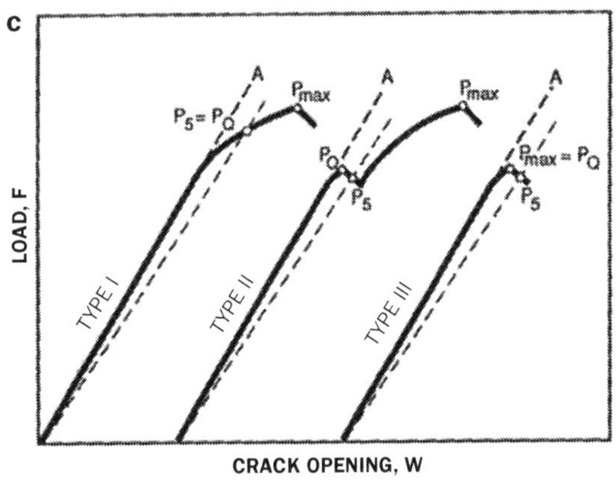

Fig. 8.14. *c. Recommendations for calculating the value of* K_{IC}

Subsequently, P_{max}/P_Q is calculated, where P_{max} is the maximum load borne by the specimen during the test. If that ratio is less than 1.10, P_Q can be used to calculate K_Q; if that is not the case, then this indicates that there is too much plasticity in the test to allow definition of a K_{IC} and the test is invalid.

When there is a valid value of P_Q it is possible to calculate K_Q with the following expressions, depending upon the type of chosen specimen:

– for the beam specimen

$$K_Q = \frac{6P_Q}{BW^{1/2}}\sqrt{\alpha}\,\frac{1.99-\alpha(1-\alpha)(2.15-3.93\alpha+2.7\alpha^2)}{(1+2\alpha)(1-\alpha)^{3/2}} \qquad (27)$$

– for the compact tension specimen

$$K_Q = \frac{P_Q}{BW^{1/2}}\frac{(2+\alpha)(0.886+4.64\alpha-13.32\alpha^2+14.72\alpha^3-5.6\alpha^4)}{(1-\alpha)^{3/2}} \qquad (28)$$

where $\alpha = a/W$, and a, W, B, S, are defined in figures 8.14a and 8.14b.

The final step, before declaring the test as valid, is to check that the condition

$$B > 2.5\left(\frac{K_Q}{\sigma_c}\right)^2 \qquad (29)$$

is satisfied where σ_c is yield stress of the material (or 0.2% off-set stress). The value K_Q is introduced into this expression in order to determine if the thickness of the specimen tested is sufficient to satisfy the inequality. In the affirmative case, it can be taken that the fracture toughness K_{IC} is equal to the determined value K_Q. In the opposite case, it is necessary to repeat the test with a thicker specimen. For ductile materials the thickness can be so large that it is impossible to measure K_{IC} from the available thicknesses.

Another specimen, which has also been proposed for measuring K_{IC}, is that shown in figure 8.15, which is commonly called the short cylinder or short bar specimen. One advantage, together with its compact dimensions, is that this specimen does not have to be pre-fatigue-cracked, which is a major benefit when testing very brittle materials, such as ceramics. A detailed description of the characteristics of the specimen and the interpretation of the test results can be found in the ASTM Standard E1304.

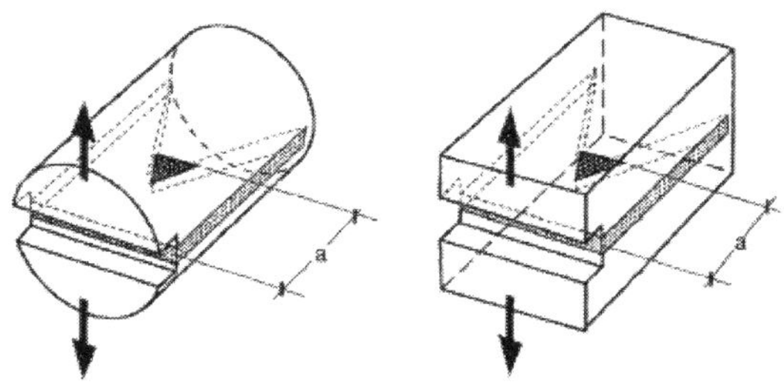

Fig. 8.15. *Specimen geometries for measurement of fracture toughness using the ASTM Standard E130.*

EXERCISE 54

Cook and Zioupos studied the fracture toughness of trabecular bone under pathological conditions. The figure shows the force-crack opening displacement (COD) curve from an experiment performed on a standard specimen in the form of a disc of trabecular bone taken from the head of the femur of a person with osteoporosis. The specimen dimensions are also shown. The plane of the crack lies perpendicular to the predominant direction of the trabeculae.

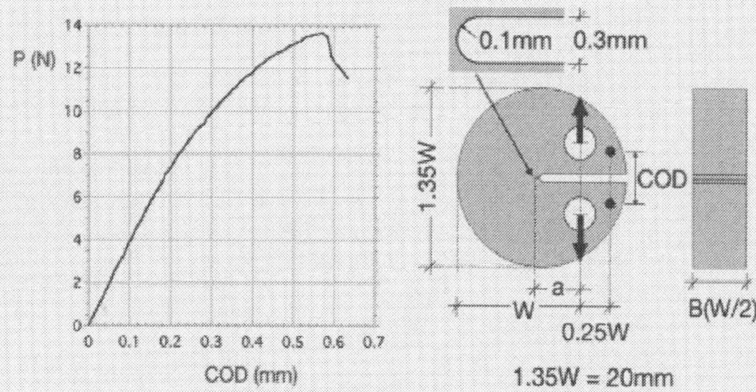

Calculate the value of K_{IC} according to the ASTM Sandard E1820.

SOLUTION:

The compact disc-shaped specimen is one of the standard geometries proposed in the ASTM Standard E1820. The following formula is used for the calculation of K_I:

$$K_Q = \frac{P_Q}{BW^{1/2}} \frac{(2+\alpha)(0.76+4.8\alpha-11.58\alpha^2+11.43\alpha^3-4.08\alpha^4)}{(1-\alpha)^{3/2}}$$

which is very similar to equation (28).

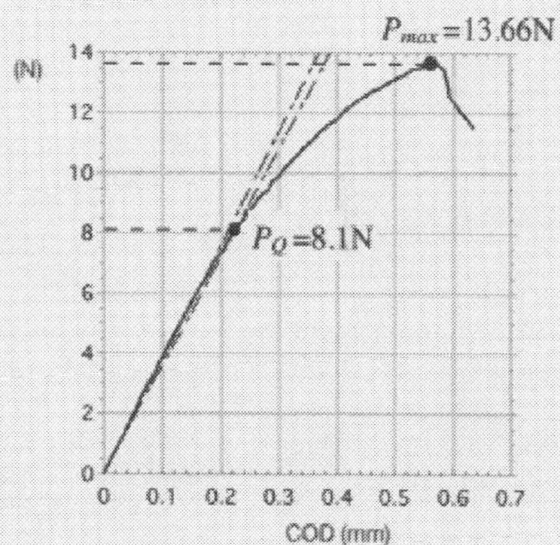

To determine the crack initiation load P_Q a straight line must be drawn from the origin at 95% of the initial slope and the point of intersection with the experimental curve found. This process is better performed numerically, with a digitised curve, since the errors in a hand construction can be quite high.

In our case we obtained the initial tangent by fitting it to the section (0, 7N) which is reasonably linear. The resulting line is: $P_{100}(N)$= 37.9 COD(mm) and the corresponding 95% line $P_{95}(N)$= 36.0COD(mm), which gives a load P_Q = 8.1 N.

Since it turns out that P_{max}/P_Q = 1.7 > 1.1 the specimen has undergone too much plastic deformation so it is not possible to obtain a valid value of K_{IC} and here the test would end.

However, and for the sole purpose of illustrating the procedure, if we suppose that the valid value of P_Q were $P_Q = P_{max}$, from the formula for K_I we would obtain K_Q = 0.1074 MPam$^{1/2}$. To perform a final validation of K_Q we should check equation (29). Taking a conservative value of 10 MPa for the yield stress of the trabecular bone used (with an apparent density of 0.33 g/cm^3) we obtain:

$$B = 7.4mm,\ b = W - a = 9.19mm > 2.5\left(\frac{K_Q}{\sigma_c}\right)^2 = 0.29mm$$

so the validity conditions are met, and $K_{IC} = K_Q$.

An interesting aspect of the test is the realisation of the crack. We see in the figure in the exercise that the specimen had a machined notch (rather than a crack) with a tip radius of curvature of 100 μm. Bearing in mind that the scale of the microstructure (the size of the trabecula) is greater than 1 mm it seems reasonable to regard this notch as a crack, the notch tip radius being an order of magnitude less than the material microstructure.

Measurement of the specific fracture energy, G_C

In this test it is assumed that the work done by the external forces is converted, solely, into cracking the specimen. Therefore, the specific fracture energy is obtained by dividing the work done by the area of the crack.

ASTM Standard E1820 provides a method of measuring G_C for materials which are not very brittle and which, once cracked, are capable of supporting some degree of stable crack propagation before sudden failure. This Standard combines the processes of measuring G_C and K_{IC} and allows the determination of the values of the specific fracture energy under conditions for which ASTM Standard E399 is not applicable, as happens when there are small amounts of plastic deformation which cannot be ignored.

The major difficulty lies in determining the precise instant of crack initiation. The Standard allows an estimation of this value by extrapolation, from specimens in which the crack has propagated a certain distance and, in addition, allows verification that the experiment was performed under elastic conditions. The ideas upon which the test procedure is based are outlined below.

Since one of the options in the ASTM Standard E1820 is to measure the fracture toughness K_{IC} the specimen fabrication processes and the testing are the same as those already described for the E399 Standard. After completing the test, K_Q is calculated and it is checked to see if this value is valid for K_{IC}.

If it is valid, G_C is determined from the expression:

$$G = \frac{K_{IC}^2\left(1-\nu^2\right)}{E} \quad (30)$$

where, E and ν are the modulus of elasticity and Poisson's ratio of the material, respectively.

If it is not valid, the specific fracture energy G_C can be estimated from

$$G = \frac{K_Q^2\left(1-\nu^2\right)}{E} + J_{pl} \quad (31)$$

The specimens must be tested under controlled displacement while recording the force F as a function of the displacement u of the load point. The values of J_{pl} are calculated from this record, applying the following expressions

$$J_{pl} = \frac{1.9A_{pl}}{Bb} \text{ for beam specimens} \quad (32)$$

$$J_{pl} = \frac{\eta A_{pl}}{Bb} \;;\; \eta = 2+0.522\left(1-a/W\right)$$

for compact tension specimens (33)

The area enclosed by the F-u curve, represented in figure 8.16b, by A_{pl}, $b = W-a$, and B is the specimen thickness, indicated in figure 8.14. a includes the length of the notch and fatigue crack, but not the stable crack propagation produced during the test.

In the first version of the ASTM Standard a single equation that integrated the elastic and plastic components was used:

$$G = \frac{2A}{Bb} \text{ for beam specimens} \quad (34)$$

$$G = \frac{\eta A}{Bb} \;;\; \eta = 2+0.522\left(1-a/W\right)$$

for compact tension specimens (35)

where A is the total area enclosed by the F-u curve, without elastic unloading. This procedure has the advantage that it does not depend upon the values of the elastic parameters E and υ in equation 31.

In principle, according to the above, a single test should be enough to obtain the critical value of G, but in reality this is not done since it is difficult to detect the on-set of propagation. In practice, a series of tests are performed in which the crack is allowed to grow different amounts Δa. In each case a value of G is calculated, G is represented as a function of Δa and extrapolated to that when Δa tends to zero. Experience has shown that this extrapolation can be improved if, instead of considering the intersection with the vertical axis ($\Delta a = 0$), the intersection with a line parallel to the line drawn through $\Delta a = 0.2$ mm is used.

$$G = 2\sigma_0 \Delta a \quad (36)$$

The line represents, in a conventional sense, the line of blunting of the crack. The stress σ_0 can be any value between the yield stress $\sigma_{0.2}$ and the fracture stress, σ_R. The Standard recommends that the value $(\sigma_{0.2}+\sigma_R)/2$ is taken.

Essentially, and according to the above Standard, the process for measuring G_C consists of performing a series of tests, representing the corresponding values of G as a function of Δa and performing a least squares fit of the potential curve, $G = C\Delta a^m$ to these values after discarding the extreme values (which are those which fall outside the zone indicated in figure 8.16a, where $G_{max} = b\sigma_0/7.5$) and as long as there are four or more remaining. The value of G_Q is obtained from the intersection of this curve with the parallel to the line of blunting plotted by $\Delta a = 0.2$ mm, as shown in the figure.

Once G_Q is calculated, it is necessary to verify that the experiment has been performed within the hypotheses used for deducing equations (34) and (35) and that this value corresponds to G_C. Following an in-depth experimental study (see the references cited in the Standard), it has been established that G_Q can be taken as a reliable measure of G_C if the following conditions are satisfied:

$$B > \frac{25G_Q}{\sigma_0} \qquad b > \frac{25G_Q}{\sigma_0} \qquad \left.\frac{dG}{da}\right|_Q < \sigma_0 \tag{37}$$

where $(dG/da)_Q$ is the slope of the regression curve at the ordinate point G_Q. As has already been indicated, this summary is only intended to give an idea of how the process works, the quoted Standard contains many more details about how to perform the test and validate the results.

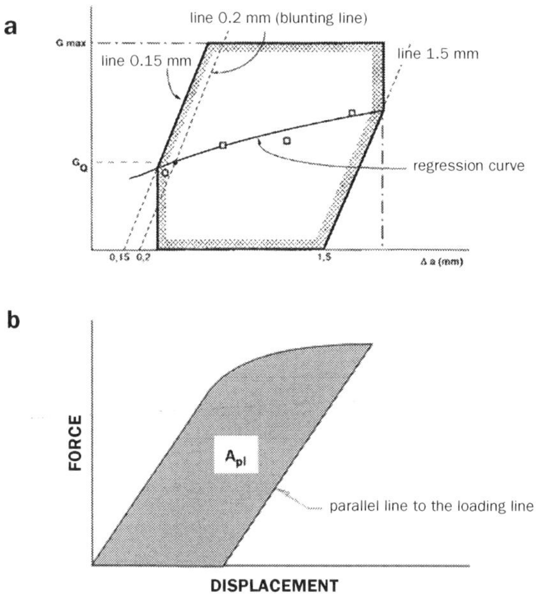

Fig. 8.16. *ASTM Standard E1820 for fracture tests*
a. Recommendations for calculating G_C *and valid zone tests.*
b. Recommendations for calculating A_{pl}

EXERCISE 55
Determine if it is possible to obtain an estimation of the value G_C from the data in Exercise 54.

SOLUTION:
For specimens in which it is not possible to measure K_{IC}, ASTM Standard E1820 prescribes the taking of several measurements with different amounts of stable crack propagation with the aim of estimating the value G_C at the initiation of fracture. In our case, Cook and Zioupos show in their work that the nonlinearity of the test is due to the collapse of the trabecula (plastic deformation) near the crack tip but not to its propagation, which starts when P_{max} is reached. For this reason we will estimate G_C from the maximum load data assuming that at this instant $\Delta a = 0$.

Since there is no direct record of the load line displacement, we are going to estimate its value from the force-COD curve, which takes, as a reference, points very close to the load, as shown in the figure. On the other hand, in order to avoid having to estimate the elastic parameters, E and υ in equation 31, we use equation 35 valid for both elastic and plastic situations.

In this way we arrive at:

$$G_C = \{2 + 0.522(1 - a/W)\}\frac{A}{Bb} = 159.7\,\mathrm{J/m^2}$$

a value which is in agreement with those of other authors.

Taking –as in Exercise 54– a value of the yield stress σ_0=10 MPa, we can verify that the estimated value of G_C, satisfies equations 37:

$$B = 7.4mm, \quad b = W - a = 9.19mm > \frac{25G_Q}{\sigma_0} = 0.4mm$$

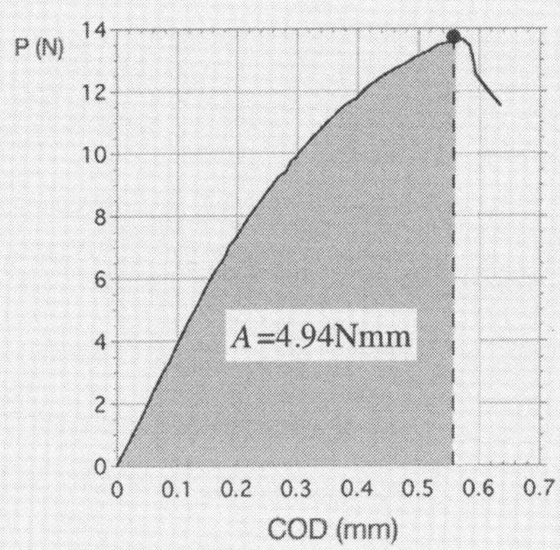

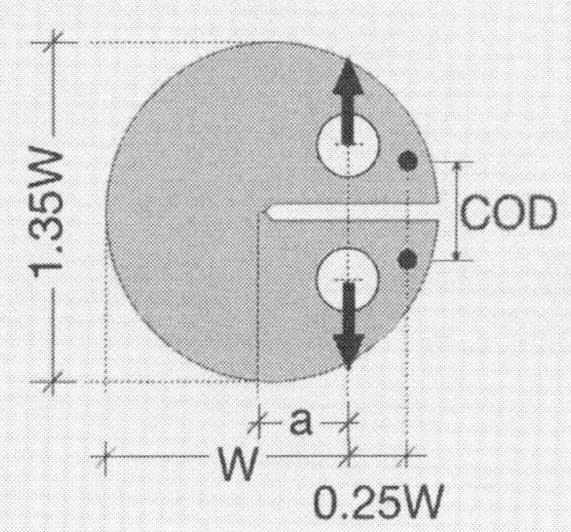

8.2.3. Indentation tests

Introduction

In order to perform conventional tensile, flexure or fracture tests it is necessary to start with a specimen of a particular geometry (cylinders, plates, beams, etc) as has been discussed in the corresponding sections. Sometimes it is not possible to prepare the specimens with the desired dimensions because the precursor material does not allow it, because there is not enough material or because the stiffness of the material makes it impracticable; all of these circumstances may arise with biological materials. In these cases, tests based on *indenting* (pressing with a suitable punch) a small region of the material can, sometimes, give information on the elastic modulus or yield stress of the material.

Mollusc shells, when large, allow specimens to be cut to carry out traditional mechanical tests, as mentioned in section 3.3.1. But when the shell thickness is small, less than 1 mm, then it is necessary to resort to nanoindentation techniques in order to estimate the mechanical properties of its various components (see, for example, Yao *et al.*, 2010).

Thickening of the arterial walls and loss of elasticity are the principal characteristics of a group of illnesses called *arteriosclerosis*, characterised by the formation of lesions called *atherosclerotic plaques* in the walls of the small and medium arteries. The rupture of these plaques can provoke damage to the heart and infarcts.Therefore it is interesting to know the mechanical behaviour of these plaques; in particular their stiffness and fragility. Again, it is not possible to fabricate conventional specimens from these plaques so indentation tests may be used in order to extract some mechanical parameters (see, for example, Ebenstein *et al.*, 2009).

Another situation, in which it is not possible to machine conventional specimens, is when soft materials need to be characterised; a very interesting case, in particular, occurs in the study of the mechanical properties of cells. The atomic force microscope (AFM) is a powerful tool which permits not only a study of the shape of the cells but also to interact with them under physiological conditions through the tip of a cantilever, if used as an indenter (see, for example, Ikai, 2009).

The idea of indenting a material and extracting information of the imprint is old, dating at least from the experiments proposed by Brinell[42] in 1900. The procedure consists of using as an indenter, in the first instance, a hard steel ball and, later, a tungsten carbide ball, and pressing it with 3000 kgf force (approximately 30 kN) onto the material. The load is held in place for 30 seconds and the diameter of the (projection of) the imprint left in the surface is measured. The Brinell hardness (HB), measured in kg/mm^2, is the ratio of the applied, F, and the contact area.

The next step was to correlate the hardness with some mechanical property. For steels there is an empirical correlation between the hardness and the yield stress, σ_c, given by

$$HB = 3.2\sigma_c \qquad (38)$$

In the middle of the last century, with the development of electronics, equipment became available to measure simultaneously the force F and the imprint h, which facilitated the test. If, moreover, the device was attached to a microscope it was possible to extend the range of indentations to micron-sized regions of the material.

At the beginning of this century, with the appearance of atomic force microscopes, these techniques have now entered the nanometre domain.

Hardness test are standardised, for example, in ISO 14577. This Standard defines three areas in which hardness is measured:

- Macroscopic domain: for values of F between 2 N and 30000 N
- Microscopic domain: for values of F less than 2 N and h greater than 0.02 μm
- Nanoscopic domain: for values of h, less than 0.02 μm

Below, a brief introduction to each domain is presented, although for biological materials the domains of interest are the microscopic and nanoscopic.

[42] Johan August Brinell (1849-1925) was a Swedish metallurgical engineer. He proposed his method at the 1900 Paris Exposition.

Indentation in the macroscopic domain

The advantage of macroscopic tests is that they are relatively simple and non-destructive. However, it is not easy to obtain reliable absolute values of the mechanical properties so the tests are, almost always, used to compare materials or different batches of the same material.

The hardness of a material is not a well-defined property; moreover, there are several different modes of indentation for measuring it (Brinell, Rockwell or Vickers). For a detailed description of the different methods see, for example, ASM Handbook Vol. 8. What these tests measure, in reality, is a combination of the elastic, plastic and, sometimes, fracture responses of the material. Deducing the modulus of elasticity or the yield stress of a material from these test data is not always a rigorous exercise.

The Brinell hardness (HB) is defined in the Standard BS EN ISO 6506. The indenter is a tungsten carbide ball, between 5 and 10 mm in diameter, through which a force of between 500 and 3000 kgf is applied for a certain time, between 10 and 30 seconds. After removing the load the diameter, d, of the imprint projection on the initial surface is measured and the value of the hardness calculated from:

$$HB \simeq \frac{2F}{\pi D\left(D-\sqrt{D^2-d^2}\right)} \qquad (39)$$

where F is the applied force and D the diameter of the ball (see figure 8.17). Details of this test, its limitations and the necessary precautions can be found in ASM Handbook Vol. 8.

The Rockwell hardness (HR) is defined in the Standard, ISO 3738. This test is used for hard materials. The HR is different from the HB because the indentation is applied in two successive stages; one with a small force and the other with a greater force, and the difference in the depths is used to better estimate the size of the imprint, in this way initial errors are eliminated and the measurement is more precise. The duration of the test is from 5 to 10 seconds. The indenters are tungsten carbide spheres or diamond in the shape of a cone with a rounded tip, see figure 8.17 (ASM Handbook Vol. 8 contains more details).

The Vickers hardness (HV) is defined in the Standard EN 23878. The indenter is a diamond pyramid (see figure 8.17) and the method is similar to that used for measuring the HB, because the indenter is pressed into the material and then withdrawn. The diagonals of the imprint are measured, the projected area calculated and the hardness obtained by dividing the load by the area. This test is also used in the micro-indentation domain when the applied forces are less than 1 kgf (ASM Handbook Vol. 8 provides more details). For materials in which the values of E or σ_c depend on the time that they bear the load –a common feature of biological materials– it can be expected that the size of the imprint will also vary with time and, consequently, so will the value of HV.

HARDNESS TEST				
Test	Indentor	Indentation shape: Side view	Indentation shape: Plan view	Hardness
Brinell	WC ball			$HB = \frac{2P}{\pi D\left(D-\sqrt{D^2-d^2}\right)}$
Rockwell	Diamond cone			HR
Vickers (Microhardness)	Diamond pyramid			$HV = 1.854\, P/d_1^2$
Knoop (Microhardness)	Diamond pyramid			$HK = 14.2\, P/l^2$
Berkovich (Microhardness)	Pyramid	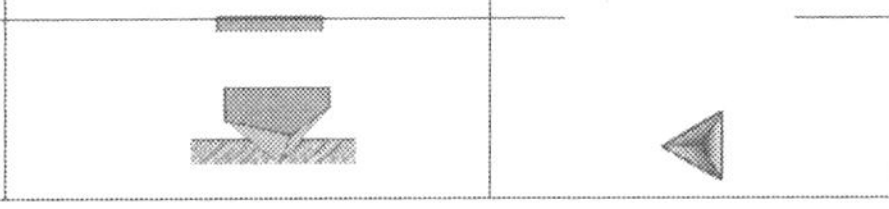		HBk

Fig. 8.17. *Hardness tests: Geometry of the indenters and the notation used.*

Another aspect to bear in mind is that it is always assumed that any material *support* does not affect the results, or in other words, the support material is considered to be a semi-infinite medium. This assumption is used in the following sections. However, at the end of this chapter some comments will be made on how to tackle the indentation of laminas, where it is essential to take account of their thickness in indentation measurements.

EXERCISE 56

Derive the empirical correlation between the hardness and the yield stress for steel equation (38).

Consider a two-dimensional case (a three-dimensional calculation would produce the same results, approximately) in which the material does not strain-harden and yields along the planes of maximum shear stress (at 45°) as shown in the figure, where the value of the maximum shear stress is k. Assume, also, that the force applied to the indenter is F, the contact area A and the vertical displacement u.

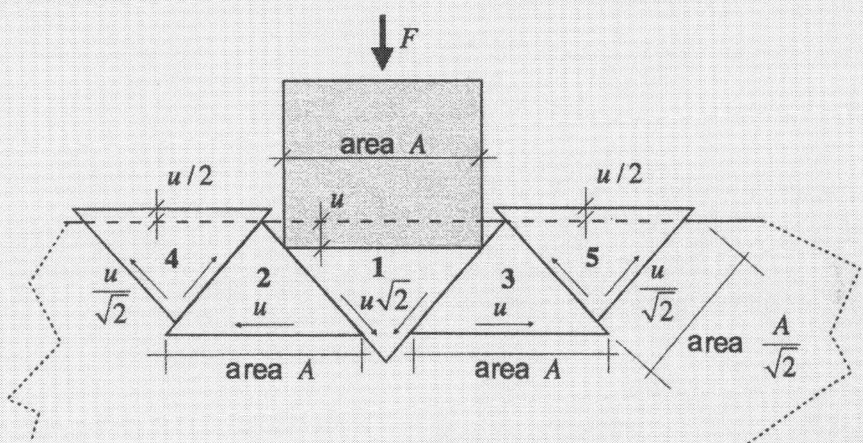

SOLUTION:

Equating the work done by the indenter, Fu, to the work of the shear stress on the sliding planes we obtain:

$$Fu = 2 \cdot \frac{Ak}{\sqrt{2}} u\sqrt{2} + 2 \cdot Ak \cdot u + 4 \cdot \frac{Ak}{\sqrt{2}} \frac{u}{\sqrt{2}} = 6Ak$$

The work done by k on block 1 is the first term, the work on blocks 2 and 3 is the second term and the work on blocks 4 and 5 is the third.

If the relationship between the yield stress σ_c and maximum value of the shear stress k, is $\sigma_c = 2k$ and the hardness H, is expressed as $H = F/A$, resulting in:

$$H = 3\sigma_c$$

In reality, sliding is not confined to the planes drawn in the figure, but along many 45° planes near the indenter. Even if the real geometry were different, the result obtained is an upper bound (this is a consequence of a theorem in the *theory of plasticity*).

Indentation in the microscopic domain

To measure the hardness in microscopic domains – for example, in polycrystalline materials to examine the interior of grains or grain boundaries– very small forces are employed; the ASTM Standard E384 suggests forces between 1 gf and 1000 gf, and to achieve this special indenters have been developed. The hardness is measured in the conventional way by dividing the force, F, by the area of contact, A.

The values obtained from microhardness tests cannot be extrapolated, in general, to those obtained from macroscopic tests because the areas investigated may not be analogous; the data obtained for a nanocrystalline grain does not have to coincide with that measured on a polycrystal, for example.

The most used micro-indentation tests are the Vickers and Knoop tests.

The Vickers hardness (HV) is measured as described in the previous section. The force is applied gradually and contact held for 10 or 15 seconds. After unloading, the diagonals of the imprint are measured and their average value d calculated and the hardness evaluated using the following expression:

$$HV = \frac{2\sin 68^\circ F}{d^2} \approx \frac{1.8544F}{d^2} \qquad (40)$$

where F is measured in kgf and d in mm (for details see, for example the ASTM Standard E384).

The Knoop harness (HK) is measured with a diamond indenter in the form of a stretched pyramid (see figure 8.17) which has a width-to-height ratio of 7:1 and face angles of 172 degrees for the long edge and 130 degrees for the short edge. The hardness is given by the formula:

$$HK = \frac{14229F}{L^2} \qquad (41)$$

where F is measured in gf and L (the length of the major diagonal) in microns. The depth of the imprint is small in these tests, so they are suitable for thin materials, in which the effect of the support could influence the measurements.

For soft materials, such as elastomers or many biological materials, other means of indentation are needed, such as the Shore durometers, in which the indenter is steel stylus which terminates in a truncated cone (Shore A or C) or in a rounded shape (Shore D). Hardness is measured on a scale which goes from 0 to 100 (ASTM D2240). One variant of these durometers is the indenter proposed by the International Rubber Association, in which the hardness scale used, the IRHD (International Rubber Hardness Degrees), is very similar to Shore A.

EXERCISE 57

The figure shows Vickers indentation tests performed by Deliormandi and Guden (2006) on human dentin and enamel. If the indentation load was 1 kgf determine the value of the Vickers hardness for both materials.

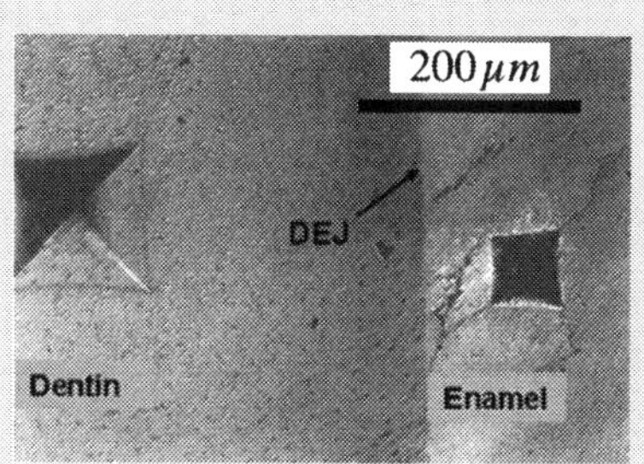

SOLUTION:

We shall use equation 40, which has as its parameters the applied force F and the diagonal d of the imprint.

Measuring from the micrograph, we obtain:

$$d_{dentin} = \frac{1}{2}(172.6 + 182.7) = 177.7\,\mu m$$
$$d_{enamel} = \frac{1}{2}(90.8 + 88.2) = 89.5\,\mu m$$

which are the average values of the two diagonals. With these values and noting that $F = 1$ kgf we obtain

$$HV_{dentin} = \frac{1.8544 \cdot 1}{0.1777^2} = 58.7\,\text{Kgf/mm}^2 = 575.8\,\text{MPa}$$
$$HV_{enamel} = \frac{1.8544 \cdot 1}{0\ 0895^2} = 231.5\,\text{Kgf/mm}^2 = 2.271\,\text{GPa}$$

Note that the hardness has dimensions of stress; therefore it is more appropriate to write it in S.I. units.

However, it is customary to write the hardness without units, as it comes from the formulas that define it, without decimals but referring to the load used in kgf. For example, in our case the results are written in the form:

$$HV_{dentin} \rightarrow 59HV1$$
$$HV_{enamel} \rightarrow 232HV1$$

Generally, the duration of the load application is understood to be between 10 and 15 seconds. If a value outside this range were used, it would also be included with the resulting measurement, for example:

$$HV_{dentin} \rightarrow 59HV1/20$$
$$HV_{enamel} \rightarrow 232HV1/20$$

if the load were applied for 20 seconds.

From the result obtained we demonstrate that enamel is almost four times harder than dentin, which is the substrate supporting it. The hardness of enamel is superior to mild steel (which is about 150 HV) but below ceramics (HV greater than 1000).

Indentation in the nanoscopic domain

The nanometric domain is investigated with more sophisticated indenters which began to evolve in the final quarter of the last century with the development of electronics and the start of this century with the appearance of the atomic force microscope.

First generation nano-indenters were, essentially, computer-controlled indenters which simultaneously measured the force and the depth of the imprint. A schematic representation of these devices is shown in figure 8.18a. The indenter is, almost always, a Berkovich pyramid (figure 8.17), which produces imprints such as that of figure 8.18b. The resolution of the applied loads can be less than 50 nN and the displacement less than 0.02 nm.

During the test, loading-unloading curves, such as those shown in figure 8.18c, are obtained and the hardness measured following the usual process by $H=F_{max}/A$, where F_{max} is maximum load and A is the projection on the plane of the contact area corresponding to the maximum load.

The area A, as a function of the contact depth h_c (see figure 8.18d), is obtained by assuming that the indenter is rigid and hardly deforms, and it is only a function of h_c. The following expression is recommended.

$$A = a + bh_c^{1/2} + ch_c + dh_c^{3/2} + 24.56h_c^2 \qquad (42)$$

where a, b, c and d are coefficients to be adjusted (see, for example, Oliver and Pharr, 1992). For an ideal Berkovic indenter, the coefficients a, b, c and d are zero. In practice, however, they are not zero because the tip of the indenter becomes blunt.

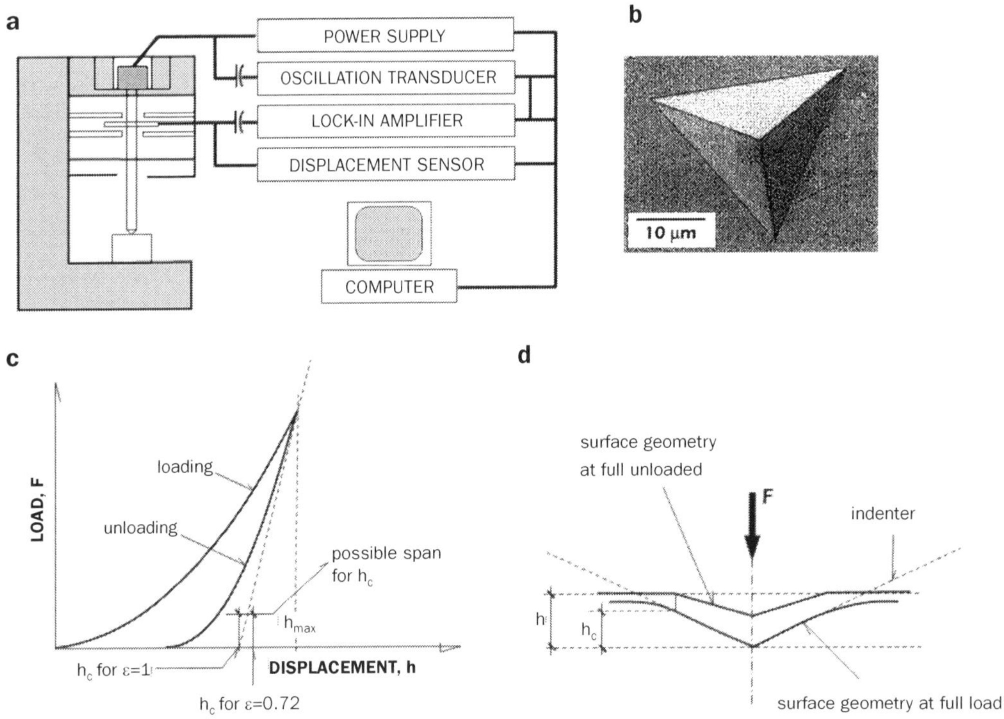

Fig. 8.18. *Nano-indentation tests.*
a. Sketch of a nano-indentation device.
b. Nano-indentation imprint.
c. Load-displacement curve during loading and unloading.
d. Indentation profiles under load and after load removal.

The stiffness S (see figure 8.18c) can be calculated from the following expression (Oliver and Pharr, 1992):

$$S = \frac{dF}{dh} = \frac{2}{\sqrt{\pi}} E_r A^{1/2} \quad (43)$$

where E_r is the reduced modulus of the specimen-indenter combination, which can be approximated by (Dorener and Nix, 1986):

$$\frac{1}{E_r} = \frac{1-\nu_i^2}{E_i} + \frac{1-\nu_m^2}{E_m} \quad (44)$$

where v and E are the Poisson ratio and the elastic modulus and the subindices i and m correspond to the indenter and specimen, respectively.

More accurate results can be obtained by means of dynamic indentation tests in which a small harmonically oscillating force is superimposed and the displacement of the indenter and the phase angle between the force and displacement are recorded continuously (see, for example, Oliver and Pharr, 1992).

The mechanical properties of biological materials are investigated, more and more, using nano-indentation techniques due to the difficulty in producing conventional test pieces. However, biological materials are far removed from exhibiting the elasto-plastic characteristics of the materials that underpinned the development of first-generation indentation tests, so it is not possible to extrapolate those procedures to these new materials whose responses to indentation depend upon time and other variables; temperature, load rate, humidity, or load and deformation history. The article by Oyen and Cook (2009) contains a critical review of the use of nano-indenters on biological materials.

The next stage in the development of nano-indenters was the use of the tip of the cantilever in the atomic force microscope. The device was developed to obtain representations of the topological surface of samples but, if operated in the axial mode (perpendicular to the surface) instead of the tangential mode (or sweep, as initially), it could be used to nano-indent the surface of the specimen and to investigate the mechanical performance at the nanometric scale (Tang and Ngan, 2011). In these tests, the arrow on the cantilever tip and the displacement of the support (as indicated in figure 8.19) are recorded, which allows the force F exerted by the point to be determined and a relationship between it and the penetration, d, into the sample to be calculated. In order to interpret these results the Hertzian elastic contact model which relates F and d to the (plane strain) elastic modulus of the sample is usually employed, that is:

$$F = a.Ed^2 \quad (45)$$

where a is a constant which depends upon the geometry of the indenter (Daza *et al.* 2015). Since the majority of biological materials exhibit visco-elasto-plastic behaviour, the values of E are only indicative but can be used to compare diverse materials or for the control of the same material.

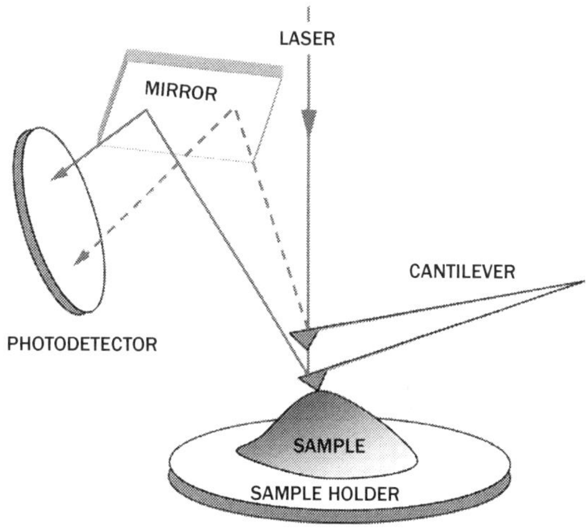

Fig. 8.19. *Schematic representation of a nano-indentation device based on the atomic force microscope (AFM).*

Some of the difficulties which emerge when trying to measure the stiffness of visco-elasto-plastic materials, such as the variation of the modulus with load-rate, can be overcome by measuring the instantaneous modulus during the application of a sudden load (ideally, a step load). This is the rate-jump procedure proposed by Tang and Ngan (2011). Essentially, instead of loading the specimen under a constant rate of displacement, it is applied rapidly so that the load increases rapidly as well. The aforementioned authors claim that by adopting this procedure, stiffness values of the material are obtained which are more representative of the material according to Chan and Ngan (2010) (see, for example, a study performed on cells in Zhou *et al.* 2012).

EXERCISE 58

The figure shows the indentation curve for a murine leukocyte, using an AFM (Daza *et al.*, 2015). The recorded data correspond to the displacement of the cantilever tip as a function of the sample movement produced by a piezoelectric actuator.

Calculate the stiffness (or the elastic modulus) of the leukocyte assuming the simplified linear-elastic Hertzian contact model.

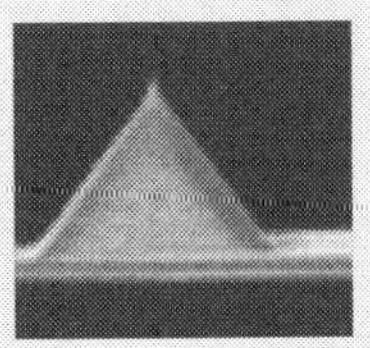

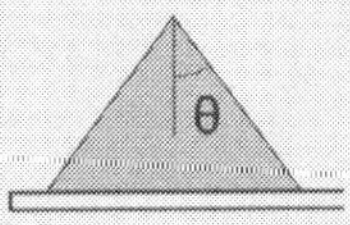

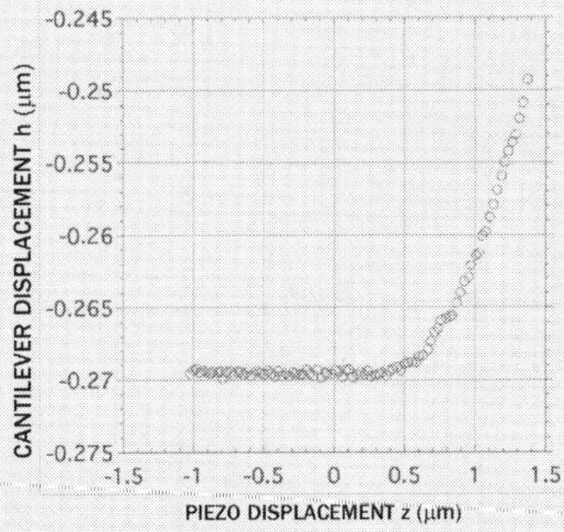

DATA:

Four-sided silicon nitride pyramid indenter; semi-angle $\theta = 39°$
Cantilever 200 μm (long) × 20 μm (wide) × 0.4 μm (thick); elastic constant $k_C = 0.05 \text{N/m}$

SOLUTION:

Sketches of the operation of the atomic force microscope are shown in figures a) and b). The sample support is displaced by means of a piezoelectric actuator, by an amount z in the vertical direction. The interaction between the sample and the indenter flexes the cantilever, producing a tip deflection h. Observe that the reference points for the measurements of h and z are different.

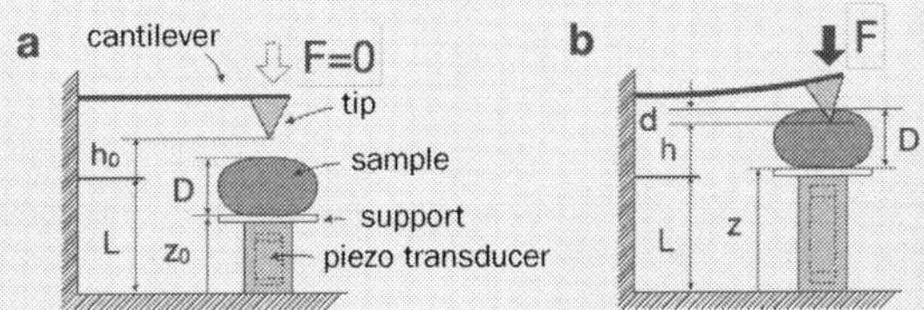

Since the sample is deformable, the following relationship between the displacements of the sample (z), cantilever tip (h) and the resulting indentation (d) is obtained:

$$d = z + D - h - L$$

where D is the sample size and L the difference between the references for z and h.

The relationship between the force F exerted by the point and its displacement h is linear and governed by the flexure of the cantilever, which can be written as:

$$F = \left| \begin{array}{ll} k_C(h - h_0) & \text{if } h \geq h_0 \\ 0 & \text{if } h < h_0 \end{array} \right.$$

where k_C is the elastic constant of the cantilever.

If the pyramidal indenter is much more rigid than the sample –which is almost always the case with biological specimens– the penetration d is small and the sample can be modelled as a linear elastic half-space, the relationship between the applied force F and the penetration d can be approximated by the expression (Rico *et al.*, 2005):

$$F = \frac{1}{\sqrt{2}} \frac{E \tan\theta}{\left(1 - \nu^2\right)} d^2 = k_S d^2$$

where θ is the semi-angle of the indenter, and E and v are the elastic modulus and Poisson's ratio of the sample, respectively.

Combining the three previous equations to eliminate F and d we obtain a quadratic equation in h:

$$k_C(h-h_0)=k_S[z+D-h-L]^2$$

which can be solved to yield:

$$h=z+D-L+\frac{k_C}{2k_S}-\sqrt{(z+D-L-h_0)\frac{k_C}{k_S}+\frac{1}{4}\left(\frac{k_C}{k_S}\right)^2}$$

The sign (of the square root) in the above equation has been determined by taking into account that when $z+D-L=h_0$ then $h=h_0$. In this equation $k_C=0.05N/m$, and the term h_0 is deduced from the constant horizontal section of the experimental graph, when the cantilever has not yet come into contact with the sample (h_0=-0.2696μm) (see figure c).

The values $D-L$ and k_s are unknown, but can be obtained by fitting with the experimental data (note that it is not possible the determine D and L separately).

Figure c) shows the curve fit, which yields the parameters $D-L$ = -0.6297μm and k_s = 1015N/m^2. With this last value, and assuming that the Poisson ratio v = 0.5 for lymphocytes (since they contain high concentrations of water, it may be assumed that deformation occurs under constant volume) we finally arrive at the Rico *et al.* expression for the elastic modulus:

$$E=\frac{\sqrt{2}(1-v^2)}{\tan\theta}k_S=1329Pa$$

Figure d) shows the force-penetration curve obtained with the above values

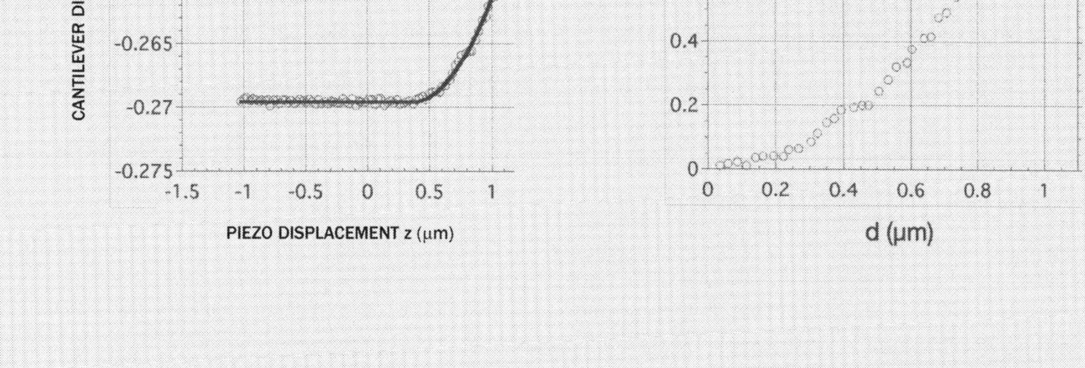

Indentation techniques for measuring fracture toughness

Techniques for measuring fracture toughness based on indentation were developed possibly because of the difficulties encountered with brittle materials, such as ceramics. Measurement of the fracture toughness of these materials, following the traditional procedures, is not easy because, in many cases, it is difficult to introduce the starter crack without breaking the specimen, and if notched specimens (no crack) are used they are not reliable. For these reasons, techniques based on measuring cracks induced by a diamond-tipped indenter are attractive because they are easy and do not need elaborate equipment. The success achieved with ceramic materials has led to its application with hard biological materials but its reliability, for the moment, is limited for reasons which will be set out below.

Indentation techniques for measuring the fracture toughness of ceramic materials are described in depth in Chapter 8 of Lawn's book. In essence, it is assumed that the material behaviour is ideal brittle-elastic and that cracks are created due the stresses produced during unloading of the indenter. The expression obtained for the fracture toughness, K_C, is:

$$K_C = \alpha\sqrt{\frac{E}{H}}\frac{F}{c^{3/2}} \tag{46}$$

where F is the applied force, E the elastic modulus, H the hardness (Vickers, is the most common), and c the trace of the semi-circular crack (see figure 8.20). α is an empirical constant which lies between 0.016 ± 0.004 (Lawn, 1993).

Kruzic and co-workers (Kruzic *et al.* 2009) have published a critical review of indentation techniques for measuring fracture toughness, in particular when applied to hard biological materials, and conclude that they are not reliable for obtaining absolute values but can be useful for comparative purposes. Their arguments are based on the fact that biological materials are composites with soft organic components which can undermine the hypothesis upon which the techniques that were developed for ceramic materials were based. They also drew attention to the possible formation of spurious cracks which could originate during the drying of samples, particularly if they are examined in an electron microscope. Another issue to bear in mind is that some biological materials exhibit an increase in fracture toughness during crack growth (an R-curve), again, this effect is not detectable in an indentation test.

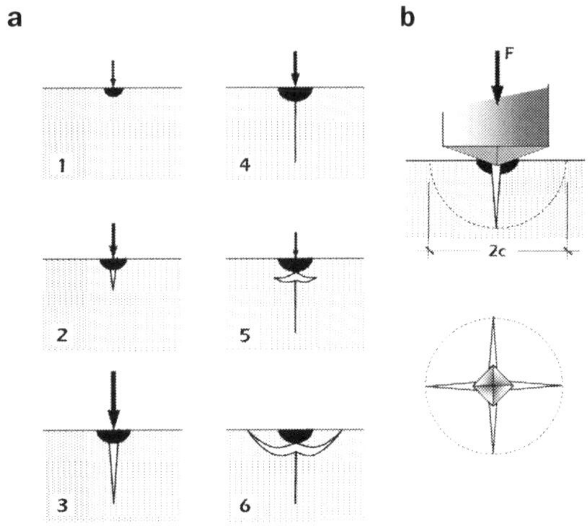

Fig. 8.20. *Cracks created by indentation.*
a. Evolution during a cycle of loading (+) and unloading (-). The dark zones denote regions in which the deformation is irreversible.
b. Trace of the semi-circular crack.

Indentation of laminas and thin films; stiffness and hardness

The main difficulty with the indentation of thin laminas is to avoid contamination of the results, due to the properties of the substrate. The most common approach to addressing this problem is to restrict the depth of the indentation to less than one tenth of the lamina thickness, though this treatment has no firm physical basis (Hay *et al.* 1998).

For laminas of a certain stiffness it is possible to justify an analogous expression to that cited above by considering the stiffness S in the nanoscopic domain, that is:

$$\frac{dF}{dh} = \frac{2}{\sqrt{\pi}}E_{ef}A^{1/2} \tag{47}$$

where F, h and A have the same meaning as before and E_{ef} is an effective modulus.

King 1987 has proposed the following expression for the effective modulus:

$$\frac{1}{E_{ef}} = \frac{1-\exp\left(-\alpha t A^{-1/2}\right)}{E_l^*} + \frac{\exp\left(-\alpha t A^{-1/2}\right)}{E_s^*} + \frac{1}{E_i^*} \tag{48}$$

where t is the thickness of the lamina, α an empirical constant, and E_l^*, E_s^* and E_i^* are the plane strain elastic moduli, for the lamina, the substrate and the indenter (see figure 8.21a).

Another expression for the effective modulus is that proposed by Gao *et al.* (1992),

$$E_{ef} = E_s + (E_l - E_s)\, I(a/t, \nu) \qquad (49)$$

where I is a function of the lamina thickness, t, the radius of contact a and the Poisson ratio ν of the lamina. This function tends to zero as t becomes vanishingly small and tends to 1 as t tends to infinity.

a

b

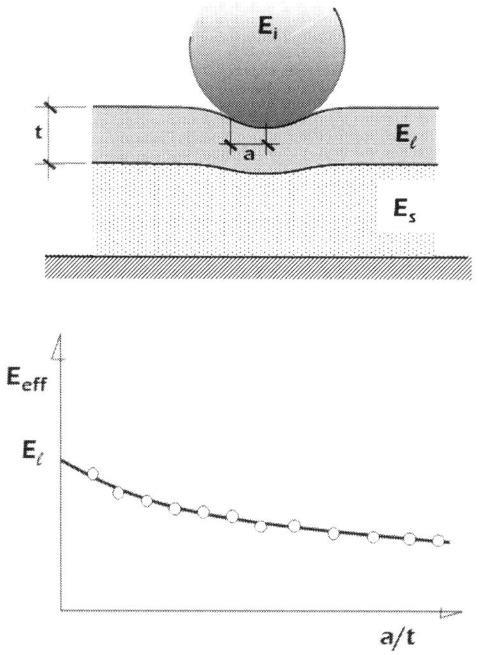

Fig. 8.21. *a. Schematic representation of the indentation of a lamina on a substrate.*
b. Obtaining the elastic modulus of a lamina by extrapolation (for geometrically similar indenters)

A practical means of obtaining the elastic modulus of the lamina E_l, consists of carrying out several measurements for different values of penetration (h or a) and extrapolating the results back to the origin, as indicated in figure 8.21b.

Measurement of the hardness of a lamina presents the same difficulties and uncertainties as mentioned above. The most commonly used expression is analogous to that of the effective modulus, that is to say, an effective hardness, H_{ef} is defined by means of the following expression:

$$H_{ef} = H_s + (H_l - H_s)\, B(h/t, E_s/E_l, \sigma_{c,s}/\sigma_{c,l}, \nu) \qquad (50)$$

where B is now a function of h/t, the relative depth of the indentation, and $\sigma_{c,s}/\sigma_{c,l}$ is the ratio of the yield stresses of the substrate and the lamina, respectively.

For soft laminas on a rigid substrate, Bhattacharya and Nix (1988) have proposed the following expression for B:

$$B = \exp\left[-\frac{\sigma_{c,l}}{\sigma_{c,s}}\frac{E_s}{E_l}\left(\frac{h}{t}\right)^2\right] \qquad (51)$$

As yet there are no sufficiently general expressions for soft or hard laminas on hard or soft substrates. The rule of one-tenth of the thickness is the most used, although finite element modelling suggests that for soft laminas on hard substrates the maximum value of h/t can be relaxed to about 0.3.

EXERCISE 59
Estimate the fracture toughness of human enamel for the data in Exercise 57.

SOLUTION:
We see that in the micrograph indentation of the enamel has produced a diagonal crack. Even though its size cannot be determined with sufficient precision, we can try to bound the length of the crack 2c (see figure 8.19) as

$$2c \geq 270\,\mu\text{m}$$

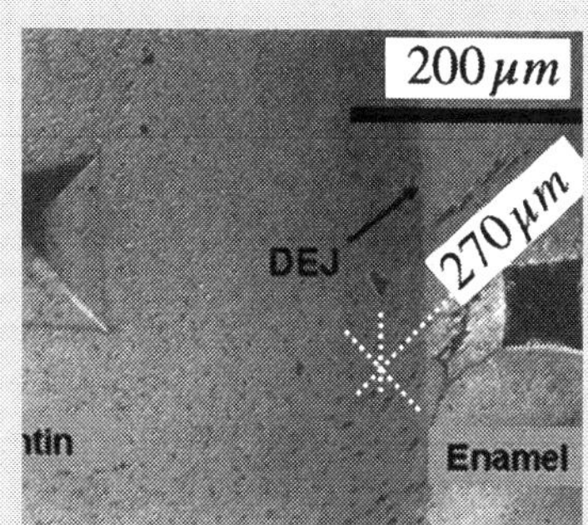

Using equation 46 we obtain

$$K_C = \alpha\sqrt{\frac{E}{H}}\frac{F}{c^{3/2}} \leq 0.016\sqrt{\frac{50}{2.27}}\frac{9.81\text{N}}{(135\,\mu\text{m})^{3/2}} = 0.47\text{MPa}\sqrt{\text{m}}$$

This value is slightly below the range quoted by other authors who tested this material (0.7 to 1.5 $\text{MPam}^{1/2}$).

8.3. MECHANICAL BEHAVIOUR

8.3.1. Introduction

The information obtained from tensile, compression, torsion, flexure, fracture or indentation tests is valuable for characterising the mechanical behaviour of bulk biological materials (those in which all three dimensions are significant). In practice, the number of such materials that are amenable to modelling today is quite small and restricted to biological materials that deform little and behave essentially elastically; mollusc shells, bones or wood, for example. However, this does not prevent a small incursion being made, in this text, into the visco-elasto-plastic regime when, later, cartilage is considered.

Even though it is possible to formulate the constitutive equations for materials undergoing large deformations with inelastic behaviour and which, also, depend upon the load history, the reason that the study is limited to the cited materials is that the associate mathematical complexity and the large number of parameters necessary to fit the models make them impractical and, often, of dubious predictive capacity.

Another issue to consider in the modelling of these materials is the hierarchical structure. Predicting the mechanical behaviour of biological materials is like predicting the behaviour of a large structure –a bridge or a skyscraper– from the properties of their structural components –steel and concrete– and their structural geometry. For this reason emphasis has been placed, in this book, on the behaviour of basic biological materials, since predictions of the mechanical behaviour of hierarchical biological materials is increasingly a structural analysis problem, which is beyond the scope of this book.

For the reasons mentioned above, only the essential basic three-dimensional elasticity concepts are summarised in this section, as a complement to the ideas introduced in the earlier studies of fibres (one-dimensional situations) and membranes and laminas (two-dimensional treatments).

Within the field of homogeneous materials it is worth dwelling on those which have a cellular structure due to their appearance in several biological materials.

In order to account for the heterogeneity of some materials, a brief incursion is also made into the field of composite materials.

¿Material or structure? (With permission of G. Jeronimidis in Structural Biological Materials, *M. Elices Ed.)*

8.3.2. Homogeneous materials[43]

The aim of this section is to summarise the most common expressions that relate the stresses and strains in a three-dimensional homogenous material which experiences small strains and displacements[44] and which behaves elastically. The detailed development of these concepts is beyond the scope of this book; for this the interested reader may find it helpful to consult, for example, the Landau and Lifshitz (1986) book on elasticity.

The infinitesimal strain tensor in three dimensions is a generalisation of the two-dimensional concept introduced in section 5.2.2 (Equation 23), where, now $i = 1, 2, 3$. That is

$$\varepsilon = \frac{1}{2}\left[\frac{\partial \mathbf{u}}{\partial \mathbf{X}} + \left(\frac{\partial \mathbf{u}}{\partial \mathbf{X}}\right)^T\right] \quad \text{or} \quad \varepsilon_{ij} = \frac{1}{2}\left[\frac{\partial u_i}{\partial X_j} + \frac{\partial u_j}{\partial X_i}\right] \tag{52}$$

where $\mathbf{u}$ is the displacement of a material point initially located at reference position $\mathbf{X}$ (see section 5.2.2 and Figure 5.2).

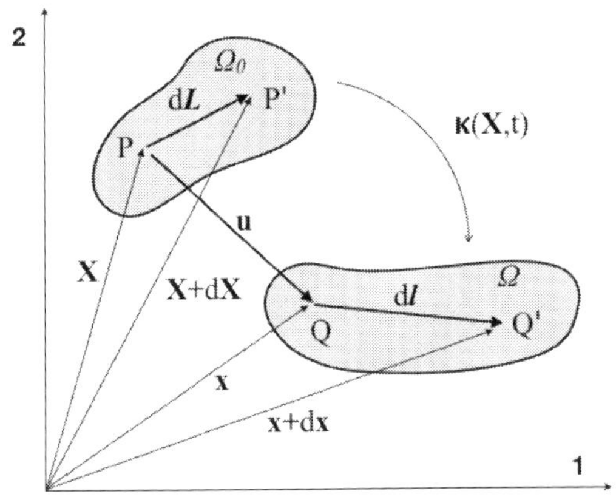

Fig. 5.2. *Notation used in the reference Ω_0, and deformed Ω configurations.*

In order to simplify the notation in what follows we shall use Einstein's summation convention in which a repeated index indicates the sum of the terms over all possible values of the index[45].

Note that the **strain tensor** is symmetrical and, therefore, has six independent components. Note also that since the displacements and strains are very small (ideally infinitesimal) the previous equation can be equally written using the values of the deformed position $\mathbf{x}$, or the reference position $\mathbf{X}$.

Figure 8.22a shows the notation used for the strains

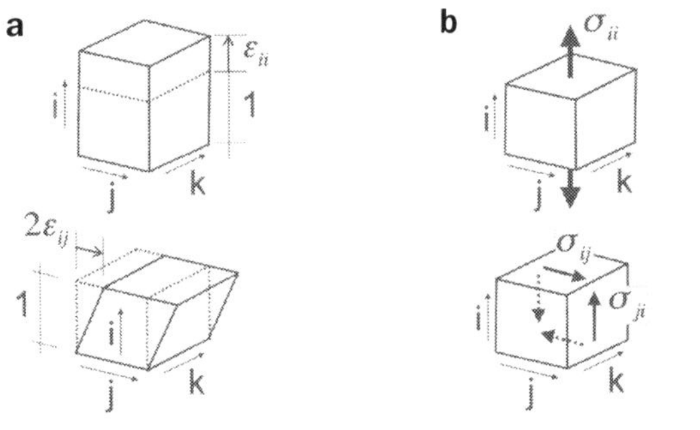

Fig. 8.22 *a. Notation employed for strains. b. Notation employed for stresses.*

The **stress tensor** σ_{ij} is also a generalisation of that which was defined in section 5.2.2, Equation 25. The tensor is symmetrical and, therefore, has only six independent components. Figure 8.22b shows the notation used.

Equilibrium of an elemental cube now obeys the following equations

$$\frac{\partial \sigma_{11}}{\partial x_1} + \frac{\partial \sigma_{21}}{\partial x_2} + \frac{\partial \sigma_{31}}{\partial x_3} + \rho g_1 = 0$$
$$\frac{\partial \sigma_{12}}{\partial x_1} + \frac{\partial \sigma_{22}}{\partial x_2} + \frac{\partial \sigma_{32}}{\partial x_3} + \rho g_2 = 0$$
$$\frac{\partial \sigma_{13}}{\partial x_1} + \frac{\partial \sigma_{23}}{\partial x_2} + \frac{\partial \sigma_{33}}{\partial x_3} + \rho g_3 = 0 \tag{53}$$

which –using the Einstein convention– can be written in the compact form

$$\frac{\partial \sigma_{ij}}{\partial x_i} + \rho g_j = 0 \tag{54}$$

with $i = 1, 2, 3$. g_i are the components of the gravitational acceleration and ρ is the density.

[43] A material is homogeneous if the material properties are the same at every point.

[44] Even if a body undergoes small strains the displacements may not always be small. An example is the bending of a very long cantilever beam. Although the strains in each section of the beam are small, the accumulated movement at the end of the beam can be so large that the change in shape of the beam must be incorporated into the analysis. This case is known as geometrically nonlinear and leads to nonlinear behaviour of the structure even when the material is perfectly linear elastic.

[45] For example, $e_{ii} = e_{11}+e_{22}+e_{33}$ if $i = 1, 2, 3$.

The constitutive equations for the situation under consideration –small strains and elastic behaviour– are relatively simple. Using the first law of thermodynamics we can establish that the variation of the internal energy per unit volume will be equal to the heat gained less the work done by the internal stresses. For a reversible process the amount of heat is TdS, where T is the absolute temperature and S is the entropy per unit volume. The work per unit volume is given by $-\sigma.d\varepsilon = -\sigma_{ij}\, d\varepsilon_{ij}$. Therefore:

$$dU = TdS + \sigma_{ij} d\varepsilon_{ij} \tag{55}$$

If a material is subjected only to a uniform pressure p and the effects of gravity can be ignored, the equilibrium forces on the boundary leads to internal hydrostatic stresses $\sigma = -p\,1$ (or $\sigma_{ij} = -p\delta_{ij}$). Under these conditions the work per unit volume is $-\sigma_{ij} d\varepsilon_{ij} = -p\, d\varepsilon_{ii} = -p\ d(tr\varepsilon)$. Observing that $d(tr\varepsilon) = dV$ dV is the elemental change suffered by a unit volume (see exercise 62), Equation (55) takes on the more familiar form of the first law of thermodynamics for hydrostatic systems (p, T, V):

$$dU = TdS - pdV \tag{56}$$

Taking as independent variables the entropy and the strains (S, ε_{ij}), we can write dU as

$$dU = \left(\frac{\partial U}{\partial S}\right)_{\varepsilon_{ij}} dS + \left(\frac{\partial U}{\partial \varepsilon_{ij}}\right)_S d\varepsilon_{ij} \tag{57}$$

which, compared to Equation (55), leads to

$$\sigma_{ij} = \left(\frac{\partial U}{\partial \varepsilon_{ij}}\right)_S \tag{58}$$

an equation which allows calculation of the stress tensor from the internal energy per unit volume of the material, expressed as a function of the entropy and the strains $U(S, \varepsilon_{ij})$.

If, instead of using the internal energy U, the free energy per unit volume, $F = U - TS$, is introduced into Equation (2.3.2-4), the following is obtained

$$dF = -SdT + \sigma_{ij} d\varepsilon_{ij} \tag{59}$$

and now taking F as a function of the temperature and the strains $F(T, \varepsilon_{ij})$, we arrive at

$$\sigma_{ij} = \left(\frac{\partial F}{\partial \varepsilon_{ij}}\right)_T \tag{60}$$

which allows calculation of the stresses if $F(T, \varepsilon_{ij})$ is known.

EXERCISE 60

Calculate the infinitesimal strains in a homogeneous bar which undergoes a uniform extension along its axis and a narrowing, also uniform, of its diameter, preserving its volume.

SOLUTION:

The situation described is shown in the figure below. The bar, with initial dimensions (L, R) is deformed homogenously until it has dimensions $(L+\Delta L, R+\Delta R)$, complying with $L R^2 = (L+\Delta L)(R+\Delta R)^2$.

The movement of a point $\mathbf{X}$ is given by

$$\mathbf{x} = X_1 \frac{R+\Delta R}{R}\hat{\mathbf{e}}_1 + X_2 \frac{R+\Delta R}{R}\hat{\mathbf{e}}_2 + X_3 \frac{L+\Delta L}{L}\hat{\mathbf{e}}_3$$

and its displacement $\mathbf{u}$ will be

$$\mathbf{u} = \mathbf{x} - \mathbf{X} = X_1 \frac{\Delta R}{R}\hat{\mathbf{e}}_1 + X_2 \frac{\Delta R}{R}\hat{\mathbf{e}}_2 + X_3 \frac{\Delta L}{L}\hat{\mathbf{e}}_3$$

Applying Equation (52) we obtain

$$\varepsilon_{11} = \varepsilon_{22} = \frac{\Delta R}{R} \quad , \quad \varepsilon_{33} = \frac{\Delta L}{L}$$

$$\varepsilon_{12} = \varepsilon_{21} = \varepsilon_{13} = \varepsilon_{31} = \varepsilon_{23} = \varepsilon_{32} = 0$$

We see that the only strains are the normal strains ε_{11}, ε_{22}, and ε_{33}, which have the simple physical interpretation of being the ratio increase in length to the initial length, as illustrated in Figure 8.22.

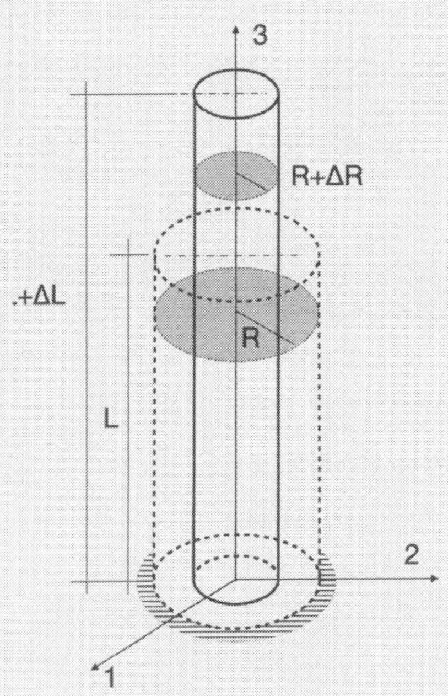

EXERCISE 61
Calculate the infinitesimal strains in a homogenous prism subjected to a displacement of its upper face, which is parallel to the (12) plane, as indicated in the figure.

Assume that all points are displaced only horizontally and uniformly.

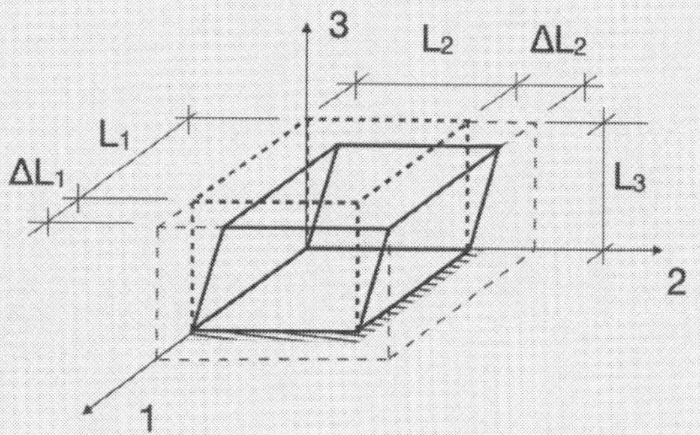

SOLUTION:
The movement of the points in the body can be represented by

$$\mathbf{x} = \left(X_1 + \Delta L_1 \frac{X_3}{L_3} \right) \hat{\mathbf{e}}_1 + \left(X_2 + \Delta L_2 \frac{X_3}{L_3} \right) \hat{\mathbf{e}}_2 + X_3 \hat{\mathbf{e}}_3$$

The displacement $\mathbf{u}$ is

$$\mathbf{u} = \mathbf{x} - \mathbf{X} = \Delta L_1 \frac{X_3}{L_3} \hat{\mathbf{e}}_1 + \Delta L_2 \frac{X_3}{L_3} \hat{\mathbf{e}}_2$$

Applying Equation (52) we obtain

$$\varepsilon_{11} = \varepsilon_{22} = \varepsilon_{33} = 0$$

$$\varepsilon_{12} = \varepsilon_{21} = 0 \quad , \quad \varepsilon_{13} = \varepsilon_{31} = \frac{\Delta L_1}{2L_3} \quad , \quad \varepsilon_{23} = \varepsilon_{32} = \frac{\Delta L_2}{2L_3}$$

in which we show that the shear strains ε_{13} and ε_{23} correspond to half the angle of distortion of the dihedral of the faces, as shown in Figure 8.22.

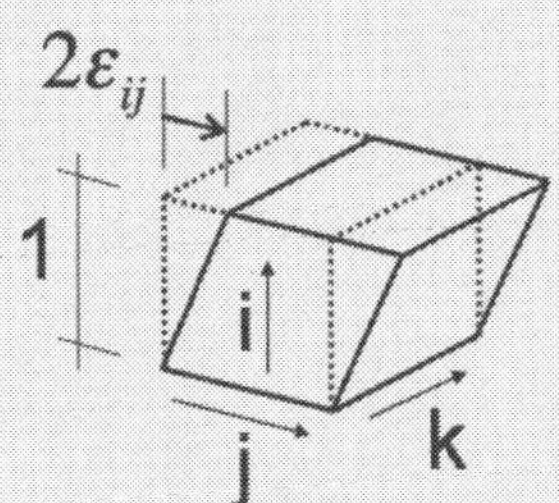

EXERCISE 62

Show that the unit change in volume at a point during the deformation can be calculated by the trace of the strain tensor. Apply the result to Exercises 60 and 61.

SOLUTION:

Given a point in the material, consider, without loss of generality, the special directions in which the strain tensor ε at this point is diagonal (every symmetric tensor such as the strain tensor can always be diagonalised). This means that we take the principal axes (1, 2, 3) as coordinate directions of the strain tensor at that point. Therefore $\varepsilon_{ij} = 0,\ i \neq j$ and also

$$\varepsilon = \frac{1}{2}\left[\frac{\partial \mathbf{u}}{\partial \mathbf{X}} + \left(\frac{\partial \mathbf{u}}{\partial \mathbf{X}}\right)^T\right] = \frac{\partial \mathbf{u}}{\partial \mathbf{X}} \quad \text{or} \quad \varepsilon_{ij} = \frac{1}{2}\left[\frac{\partial u_i}{\partial X_j} + \frac{\partial u_j}{\partial X_i}\right] = \frac{\partial u_i}{\partial X_j}$$

as deduced from Equation (52).

The displacement of a material point $\mathbf{u}$ is given by $\mathbf{u} = \mathbf{x} - \mathbf{X}$ where $\mathbf{x}$ is the deformed position and $\mathbf{X}$ that of the reference. An elemental vector, $d\mathbf{X}$ is transformed into another, $d\mathbf{x}$ given by $d\mathbf{x} = d\mathbf{u} + d\mathbf{X}$ or $dx_i = du_i + dX_i$ which, making use of the strain tensor, we can write as

$$d\mathbf{x} = \frac{\partial \mathbf{u}}{\partial \mathbf{X}} d\mathbf{X} + d\mathbf{X} = (\mathbf{1} + \varepsilon) d\mathbf{X} \quad \text{or} \quad dx_i = \frac{\partial u_i}{\partial X_j} dX_j + dX_i = (\delta_{ij} + \varepsilon_{ij}) dX_j$$

where $\mathbf{1}$ is the three-dimensional unit tensor and δ_{ij} is the Kronecker delta ($\delta_{ij} = 0,\ i \neq j$; $\delta_{ij} = 1,\ i = j$).

We now consider three vectors $d\mathbf{A}$, $d\mathbf{B}$ and $d\mathbf{C}$ parallel, respectively, to the three coordinate axes, which define a cuboid of volume $dV = dA\ dB\ dC$. These vectors are transformed into vectors $d\mathbf{a}$, $d\mathbf{b}$ and $d\mathbf{c}$ according to the above expression:

$$d\mathbf{a} = (1 + \varepsilon_{11}) d\mathbf{A} \ , \ d\mathbf{b} = (1 + \varepsilon_{22}) d\mathbf{B} \ , \ d\mathbf{c} = (1 + \varepsilon_{33}) d\mathbf{C}$$

We see that since ε is a diagonal tensor, each transformed vector will be parallel to the initial vector and, therefore, the deformed figure will also be a cuboid of volume dv

$$dv = da\ db\ dc = (1 + \varepsilon_{11}) dA\ (1 + \varepsilon_{22}) dB\ (1 + \varepsilon_{33}) dC = dV(1 + \varepsilon_{11} + \varepsilon_{22} + \varepsilon_{33} + \varepsilon_{11}\varepsilon_{22} + \varepsilon_{11}\varepsilon_{33} + \varepsilon_{22}\varepsilon_{33} + \varepsilon_{11}\varepsilon_{22}\varepsilon_{33})$$

Neglecting second and higher order terms (by considering small strains) we calculate the unit volumetric strain as:

$$\frac{dv - dV}{dV} = \varepsilon_{11} + \varepsilon_{22} + \varepsilon_{33} = tr\varepsilon$$

which is independent of the axes we take for ε, the trace being a tensor invariant.

In the case of **Exercise 60** we see that the trace of the strain tensor is

$$\varepsilon_{11} + \varepsilon_{22} + \varepsilon_{33} = \frac{2\Delta R}{R} + \frac{\Delta L}{L}$$

For the conditions set out we know that $LR^2 = (L + \Delta L)(R + \Delta R)^2$, an expression that we can expand

$$LR^2 = LR^2 + L(\Delta R)^2 + 2\,LR\Delta R + \Delta L R^2 + \Delta L(\Delta R)^2 + 2\,\Delta L R\ \Delta R \approx LR^2 + 2\,LR\ \Delta R + \Delta L R^2$$

since ΔL and ΔR are small. From this equation we deduce that $2\Delta R/R = -\Delta L/L$ with which we show that $tr(\varepsilon) = 0$.

The strains calculated in **Exercise 61** yield $tr(\varepsilon) = 0$, which indicates that the volume remains constant. The same result is obtained by observing that the deformed figure is an oblique prism of volume $L_1\ L_2\ L_3$. Note that, whatever the material, a movement without normal stresses such as that described does not produce a change in volume.

Isotropic materials[46]

If it is assumed that the *temperature remains constant* and that *initially there are no residual stresses,* the free energy per unit volume, for an isotropic and homogeneous material, can be expressed up to second-order terms in ε_{ij} by (Landau, 1986)

$$F = F_0 + \frac{1}{2}\lambda(\varepsilon_{kk})^2 + G\varepsilon_{ij}\varepsilon_{ij} \quad (61)$$

since F cannot depend upon the orientation of the axes in which we express ε_{ij} and, therefore, must be function only of the invariants of ε. The symbols λ and G are the Lamé coefficients.

In Exercise 62 it was shown that during deformation the unit variation of volume is given by $tr\varepsilon = \varepsilon_{ii}$. When $tr\varepsilon = 0$ there is no change in volume and the shape of the material is only a distortion during deformation. This state is denoted *pure shear.* Equation (61) shows that, in this case, the constant term F_0 corresponds to the strain-free state; furthermore, there is a term due to the distortion. The corresponding coefficient G, is given the name *modulus of rigidity or shear modulus.*

In the contrary case, when there are no shear stresses ($\varepsilon_{ij}=0,\ i \neq j$) the material maintains its shape but changes size. The state is called *hydrostatic tension (or compression).*

It should be observed that the states refer to particular points in the material and can change from one point to another. When the strain is the same at every point in the material it is said that the body is subjected to a *homogeneous* deformation.

Taking into account these concepts, any deformation can be written as the sum of hydrostatic and a pure shear strain components. For this, it is sufficient to consider the identity:

$$\varepsilon_{ij} = \left(\varepsilon_{ij} - \frac{1}{3}\varepsilon_{kk}\delta_{ij}\right) + \frac{1}{3}\varepsilon_{kk}\delta_{ij} \quad \text{or}$$

$$\varepsilon = \left[\varepsilon - \frac{1}{3}(tr\varepsilon)\mathbf{1}\right] + \frac{1}{3}(tr\varepsilon)\mathbf{1} \quad (62)$$

The first of the two addends corresponds to a pure shear strain, or deviator ε'_{ij}

$$\varepsilon'_{ij} = \varepsilon_{ij} - \frac{1}{3}\varepsilon_{kk}\delta_{ij} \quad \text{or} \quad \varepsilon' = \varepsilon - \frac{1}{3}(tr\varepsilon)\mathbf{1} \quad (63)$$

since the sum of their diagonal terms (ii) is zero. The second tensor $\varepsilon_{kk}\delta_{ij}/3$ is purely hydrostatic, with no off-diagonal components and proportional to the unity tensor.

Form these expressions, the free energy (Equation 61) can be written separating both types of strain into different terms

$$F = \frac{1}{2}\lambda(\varepsilon_{kk})^2 + G\varepsilon_{ij}\varepsilon_{ij} = \frac{1}{2}K(\varepsilon_{kk})^2 + G\varepsilon'_{ij}\varepsilon'_{ij} \quad (64)$$

We have eliminated, for simplicity, the constant term F_0 which, as will be seen later, does not affect the relationship between stress and strain. The constant K is called the *bulk, compressibility* or *hydrostatic modulus* and G is, as before, the shear or modulus of rigidity. It can be shown that $K = \lambda + 2G/3$.

The **relationship between stress and strain** is obtained from Equation (60) by differentiating F:

$$\sigma_{ij} = K\varepsilon_{kk}\delta_{ij} + 2G\varepsilon'_{ij} = K\varepsilon_{kk}\delta_{ij} + 2G\left(\varepsilon_{ij} - \frac{1}{3}\varepsilon_{kk}\delta_{ij}\right) \quad (65)$$

or, in vector form

$$\sigma = K(tr\varepsilon)\mathbf{1} + 2G\varepsilon' = K(tr\varepsilon)\mathbf{1} + 2G\left(\varepsilon - \frac{1}{3}(tr\varepsilon)\mathbf{1}\right) \quad (66)$$

We see that the stress tensor is also decomposed into a hydrostatic stress tensor σ_h, responsible for the volumetric strain (relative change in volume), which can be characterised by a pressure $p = -K\varepsilon_{kk}$

$$\sigma_{h,ij} = K\varepsilon_{kk}\delta_{ij} = -p\delta_{ij} \quad \text{or} \quad \sigma_h = K(tr\varepsilon)\mathbf{1} = -p\mathbf{1} \quad (67)$$

to which is added a pure shear tensor, called the deviatoric stress tensor σ', which does not give rise to a change in volume

$$\sigma'_{ij} = 2G\varepsilon'_{ij} = 2G\left(\varepsilon_{ij} - \frac{1}{3}\varepsilon_{kk}\delta_{ij}\right) \quad \text{or}$$

$$\sigma' = 2G\varepsilon' = 2G\left(\varepsilon - \frac{1}{3}(tr\varepsilon)\mathbf{1}\right) \quad (68)$$

Equation (67) shows that K, the bulk modulus, measures the relationship between the applied pressure relative change in volume of the material when it is compressed; therefore

$$K = -\frac{p}{\varepsilon_{kk}} = -\frac{p}{(\Delta V / V_0)} \quad (69)$$

since ε_{kk} is the unitary change in volume, as was shown in Exercise 62. An incompressible material will have an infinite value of K.

46 A material is isotropic when its properties do not depend upon the direction in which it is examined.

The **inverse relationship between the strain and stresses** is

$$\varepsilon_{ij} = \frac{\sigma_{kk}\delta_{ij}}{9K} + \frac{\sigma'_{ij}}{2G} = \frac{\sigma_{kk}\delta_{ij}}{9K} + \frac{1}{2G}\left(\sigma_{ij} - \frac{1}{3}\sigma_{kk}\delta_{ij}\right) \tag{70}$$

These results show that there is a linear relationship between the strain tensors ε_{ij} and the stress tensors σ_{ij}. Or, in other words, that the strain is proportional to the applied forces. This conclusion, valid only for small strains, is given the name *Hooke's Law*.

Another, more common, form of writing Equations (65) and (70) is:

$$\sigma_{ij} = \frac{E}{1+\nu}\left(\varepsilon_{ij} + \frac{\nu}{1-2\nu}\varepsilon_{kk}\delta_{ij}\right) \tag{71}$$

and

$$\varepsilon_{ij} = \frac{1}{E}\left[(1+\nu)\sigma_{ij} - \nu\sigma_{kk}\delta_{ij}\right] \tag{72}$$

where E is the *elastic modulus* and ν is the *Poisson ratio*, parameters related to K and G by means of the expressions

$$E = \frac{9KG}{3K+G} \tag{73}$$

and

$$\nu = \frac{1}{2}\frac{3K-2G}{3K+G} \tag{74}$$

The reader is invited to consult section 2.3.2 of the *Biological Fibres* section for a better interpretation of the physical meaning of the parameters E, G and ν.

EXERCISE 63

From Equations (71) and (72), write expressions for σ_{ij} and ε_{ij} $(i, j = 1,2,3)$. Apply the result to the constrained compression test and the uniaxial tensile test.

SOLUTION:

The stresses are obtained from Equation (71) when the volumetric strain $\varepsilon_{kk}=\varepsilon_{11}+\varepsilon_{22}+\varepsilon_{33}$ and taking into account the properties of the Kronecker delta function δ_{ij}

$$\sigma_{11}=\frac{E}{1+\nu}\left[\varepsilon_{11}+\frac{\nu}{1-2\nu}(\varepsilon_{11}+\varepsilon_{22}+\varepsilon_{33})\right]=\frac{E}{(1+\nu)(1-2\nu)}\left[\varepsilon_{11}(1-\nu)+\nu(\varepsilon_{22}+\varepsilon_{33})\right]$$

$$\sigma_{22}=\frac{E}{(1+\nu)(1-2\nu)}\left[\varepsilon_{22}(1-\nu)+\nu(\varepsilon_{11}+\varepsilon_{33})\right]$$

$$\sigma_{33}=\frac{E}{(1+\nu)(1-2\nu)}\left[\varepsilon_{33}(1-\nu)+\nu(\varepsilon_{11}+\varepsilon_{22})\right]$$

$$\sigma_{12}=\sigma_{21}=\frac{E}{1+\nu}\varepsilon_{12}\ ,\quad \sigma_{13}=\sigma_{31}=\frac{E}{1+\nu}\varepsilon_{13}\ ,\quad \sigma_{23}=\sigma_{32}=\frac{E}{1+\nu}\varepsilon_{23}$$

expressions in which we have made use of the symmetry of the stress and strain tensors.

In order to obtain the strains, we proceed in a similar manner using Equation (72)

$$\varepsilon_{11}=\frac{1}{E}\left[\sigma_{11}-\nu(\sigma_{22}+\sigma_{33})\right]$$

$$\varepsilon_{22}=\frac{1}{E}\left[\sigma_{22}-\nu(\sigma_{11}+\sigma_{33})\right]$$

$$\varepsilon_{33}=\frac{1}{E}\left[\sigma_{33}-\nu(\sigma_{11}+\sigma_{22})\right]$$

$$\varepsilon_{12}=\varepsilon_{21}=\frac{(1+\nu)}{E}\sigma_{12}\ ,\quad \varepsilon_{13}=\varepsilon_{31}=\frac{(1+\nu)}{E}\sigma_{13}\ ,\quad \varepsilon_{23}=\varepsilon_{32}=\frac{(1+\nu)}{E}\sigma_{23}$$

In a **constrained compression** test, all of the strains are zero except for that in the direction of the loading axis. If the compression is applied in the 1-direction, the stress state is obtained from the previous expressions for the stresses by making all of the strains zero except ε_{11}, that is:

$$\sigma_{11}=\frac{E(1-\nu)}{(1+\nu)(1-2\nu)}\varepsilon_{11}$$

$$\sigma_{22}=\sigma_{33}=\frac{\nu E}{(1+\nu)(1-2\nu)}\varepsilon_{11}=\frac{\nu}{1+\nu}\sigma_{11}$$

$$\sigma_{12}=\sigma_{21}=\sigma_{13}=\sigma_{31}=\sigma_{23}=\sigma_{32}=0$$

From these expressions, it is easy to obtain the aggregate modulus H_A for the constrained compression test, already calculated in Exercise 48. The ratio of the stress and strain becomes: $\sigma_{11}/\varepsilon_{11}=H_A=$ $=E(1\text{-}\upsilon)/[(1+\upsilon)\ (1\text{-}2\upsilon)]$.

In the case of a **uniaxial tensile test** in the 1-direction, the only stress acting is σ_{11} ,and we can calculate the strains as:

$$\varepsilon_{11}=\frac{\sigma_{11}}{E}$$

$$\varepsilon_{22}=\varepsilon_{33}=\frac{-\nu}{E}\sigma_{11}=-\nu\varepsilon_{11}$$

$$\varepsilon_{12}=\varepsilon_{21}=\varepsilon_{13}=\varepsilon_{31}=\varepsilon_{23}=\varepsilon_{32}=0$$

We see that the ratio of the applied stress and the strain in the loading direction is $\sigma_{11}/\varepsilon_{11}=E$, the elastic modulus of the material. We also note that the ratio of the strain in the direction perpendicular to the loading direction and that in the loading direction is $-\upsilon$, the value of the Poisson ratio with the sign changed.

EXERCISE 64
Derive the expressions (73) and (74) and discuss the conditions under which it is possible to consider that the deformation of the material occurs without change in volume (an incompressible material).

SOLUTION:
From Equation (71) we obtain

$$\sigma_{ij} = \frac{E}{1+\nu}\varepsilon_{ij} + \frac{E\nu}{(1+\nu)(1-2\nu)}\varepsilon_{kk}\delta_{ij}$$

which compared to Equation (65) yields the identities

$$2G = \frac{E}{1+\nu} \quad \text{and} \quad K - \frac{2G}{3} = \frac{E\nu}{(1+\nu)(1-2\nu)}$$

from which expressions (73) and (74) may be readily deduced

$$E = \frac{9KG}{3K+G} \quad \text{and} \quad \nu = \frac{1}{2}\frac{3K-2G}{3K+G}$$

and with which we can also write the inverse relationships

$$G = \frac{E}{2(1+\nu)} \quad \text{and} \quad K = \frac{E}{3(1-2\nu)}$$

Looking at the above equations it is interesting to reflect upon the concept of incompressibility of a material, and the criteria which allow us to assume that a particular material maintains its volume (that is to say, it is incompressible) during any deformation process. All of this uses a very simple approximation, the hypothesis of linear-elastic behaviour.

Strictly speaking, taking the value of the bulk modulus K as the criterion, no real material is incompressible (K would have to be infinite, according to Equation (69)) and, even less so, biological materials with elevated water content. The bulk modulus of water $K_{agua} \approx 2\,\text{GPa}$ is much less than that of steel (160 GPa) or cortical bone (15 GPa), and similar to that of polymers (2 to 4 GPa).

However, the assumption of incompressibility is common and often taken as a given for many biological materials, including tissues with moderate amounts of water such as bone.

By measuring the Poisson ratio υ –if it is known– we can estimate accurately the degree of incompressibility of a material, since from Equations (72) we see that when $\upsilon = 1/2$

$$\varepsilon_{kk} = \varepsilon_{11} + \varepsilon_{22} + \varepsilon_{33} = \frac{1}{E}\left[\left(1+\frac{1}{2}\right)\sigma_{kk} - \frac{1}{2}3\sigma_{kk}\right] = 0$$

always and the volumetric strain is zero, no matter what the state of stress. This, as we have seen, is a state of pure shear in which only the deviatoric strains ε' exist.

Given that the experimental determination of v is not easy in biological materials, it is more practical to measure parameters like E and G (see sections 2.3.2 and 8.2.1). From equations developed in the Exercise 64 above we can see that $E/G=2(1+v)$.

For $v=1/2$ the previous expression leads to $E=3G$. This is the maximum possible value of the elastic modulus which satisfies the relationship

$$0 < E < 3G$$

since $-1 < v < 1/2$, as was shown in section 8.3.2.

The closer the value of E gets to $3G$ the more incompressible the material and the volume is conserved to a greater degree during deformation.

In the case of soft biological materials –constituted to a large part by water, between 70 and 90%– then G is much reduced. (Ideal fluids –inviscid– offer no resistance to shear stress and in these $G = 0$. Soft materials, as a function of their degree of hydration, have a behaviour similar to viscous fluids, in which G may also be small if the speed of the application of the shear stress is reduced). The average values of different biological materials, even in the case of tissue such as cortical bone (~20% water), have been shown to follow this hypothesis to a reasonable degree (Cowin and Dotv. 2007).

Alternatively, it is possible to assess the incompressibility of a material by means of the ratio G/K, which is directly related to the ratio of the shear and the volumetric strains. In this case Equation (74) provides the relationship between the Poisson ratio v and G/K

$$v=\frac{1}{2}\frac{3K-2G}{3K+G}=\frac{1}{2}\frac{3-2G/K}{3+G/K}$$

In this we see that when $G/K = 0$ then $v=1/2$. From Equation (73) we also see that, in this case, $E=3G$. For a ratio $G/K=1/30$ then $v \approx 0.48$ and $E \approx 2.97G$.

Note that in order to establish the degree of incompressibility of a material in no case is it possible to consider the value of E (or of K) in isolation, since the shear stresses –which do not produce a change in volume, unlike the normal strains– are determined by the shear modulus G of the material, as was shown in Equation (70).

Anisotropic materials

When the material is no longer isotropic, but still linear elastic, the problem is complicated. If the process is isothermal and there are no residual stresses initially the free energy per unit volume F for small strains can be approximated by a quadratic function of the strain tensor, but instead of having two coefficients there are many more. In general, the free energy will be of the form

$$F = \frac{1}{2}\lambda_{ijkl}\varepsilon_{ij}\varepsilon_{kl} \tag{75}$$

According to Equation (60), the stress tensor σ_{ij} will be given by,

$$\sigma_{ij} = \frac{\partial F}{\partial \varepsilon_{ij}} = \frac{1}{2}\lambda_{ijkl}\varepsilon_{kl} + \frac{1}{2}\lambda_{klij}\varepsilon_{kl} = C_{ijkl}\varepsilon_{kl} \tag{76}$$

where $C_{ijkl}=(\lambda_{ijkl}+\lambda_{klij})/2$ is a second-order symmetric tensor, called the **elastic modulus tensor.**

In principle, the tensor C_{ijkl}, which relates the nine components of stress to the nine components of strain, will have 81 constants. The symmetry condition which we have noted reduces the number of independent constants to 45. On the other hand, the symmetry of the stress tensor $(\sigma_{ij}=\sigma_{ji})$ leads to $C_{ijkl}=C_{jikl}$ and that of the strain tensor $(\varepsilon_{ij}=\varepsilon_{ji})$ to $C_{ijkl} = C_{jilk}$, leaving only 21 truly independent constants.

It is often more convenient to represent the six independent tensor components σ_{ij} and ε_{ik} as vectors (Voigt notation). In this form, the elastic modulus tensor is represented as a second-order 6 x 6 tensor C_{ij} which relates the vectors $[\sigma_{11}, \sigma_{22}, \sigma_{33}, \sigma_{23}, \sigma_{31}, \sigma_{12}]$ and $[\varepsilon_{11}, \varepsilon_{22}, \varepsilon_{33}, \varepsilon_{23}, \varepsilon_{31}, \varepsilon_{12}]$. Since it is symmetric, the maximum number of independent components is 21.

These relationships are set out in Figure 8.23 for the case of general anisotropy (21 independent constants), for an orthotropic material (9 independent constants), for a transversely isotropic material (5 independent constants), and for an isotropic material (2 independent constants).

An **orthotropic** material is one which has three orthogonal planes of symmetry; the expression in Figure 8.23 refers to the three axes formed by the intersection of said planes.

Symmetry with respect to the plane $x_1 = 0$ implies that states of symmetric loading with respect to this plane such as σ_{11}, σ_{22}, σ_{33} or σ_{23} cannot provoke unsymmetric strains with respect to the same, like the strains ε_{31} or ε_{12}. Therefore, $C_{15} = C_{16} = C_{25} = C_{26} = C_{35} = C_{36} = C_{45} = C_{46} = 0$. Proceeding in a similar manner with planes of symmetry $x_2 = 0$ and $x_3=0$ we see that also $C_{14}=C_{24}=C_{34}=C_{56}=0$, resulting in the 9-coefficient matrix of Figure 8.23.

A **transversely isotropic** material is one in which every point has a plane in which the behaviour is isotropic, that is to say, one in which the properties are the same in any direction in the plane. In Figure 8.23, the plane $x_3=0$ is taken as the plane of isotropy. Every direction contained within this plane will have identical properties, so that the axes 1 and 2, and consequently the indices 1 and 2 are interchangeable. With that we see that $C_{11}=C_{22}$, $C_{13}=C_{23}$ and $C_{44}=C_{55}$. Furthermore, any state of pure shear strain $\varepsilon_{12}= \gamma/2$ in the plane of isotropy must be equivalent to the sum of a state of traction $\varepsilon_{11}=\gamma/2$ together with another perpendicular compression $\varepsilon_{22}= \gamma/2$ (see the first exercise in section 2.3.2 "Biological Fibres") so that $C_{66}=(C_{11}-C_{12})/2$ and only 5 independent constants remain.

The case of extreme symmetry occurs when the material is **isotropic** and every plane is a plane of symmetry. In this case all of the directional indices are interchangeable so that $C_{13}=C_{12}$, $C_{33}=C_{11}$ and $C_{44} = (C_{11}-C_{12})/2$ resulting in only 2 independent constants.

The inverse relationship between the stresses and strains is $\varepsilon_{ij}=S_{ijkl}\sigma_{kl}$ where $[S_{ijkl}]$ is known as the **elastic flexibility tensor** and which can be obtained by inverting Equation (76). Using the Voigt notation, we can also write $\varepsilon_i=S_{ij}\sigma_j (i=1, 6)$, and the tensor $\mathbf{S}=[S_{ij}]$ is obtained by inverting the tensor $\mathbf{C}=[C_{ij}]$ shown in Figure 8.23. Except in cases of isotropy and transverse isotropy, this operation is not simple. In any case, the symmetry conditions are still valid for $\mathbf{S}$, whose structure will be identical to that shown in Figure 8.23 for C, with the exception that now the diagonal term $S_{66}=S_{11}-S_{12}$, and similarly the terms S_{44}, S_{55} and S_{66} of the isotropic material are equal to $S_{11}-S_{12}$.

GENERAL ANISOTROPY (21 constants)								
σ_{11}		C_{11}	C_{12}	C_{13}	C_{14}	C_{15}	C_{16}	ε_{11}
σ_{22}		C_{12}	C_{22}	C_{23}	C_{24}	C_{25}	C_{26}	ε_{22}
σ_{33}	=	C_{13}	C_{23}	C_{33}	C_{34}	C_{35}	C_{36}	ε_{33}
σ_{23}		C_{14}	C_{24}	C_{34}	C_{44}	C_{45}	C_{46}	ε_{23}
σ_{31}		C_{15}	C_{25}	C_{35}	C_{45}	C_{55}	C_{56}	ε_{31}
σ_{12}		C_{16}	C_{26}	C_{36}	C_{46}	C_{56}	C_{66}	ε_{12}

ORTHOTROPIC (9 constants)								
σ_{11}		C_{11}	C_{12}	C_{13}	0	0	0	ε_{11}
σ_{22}		C_{12}	C_{22}	C_{23}	0	0	0	ε_{22}
σ_{33}	=	C_{13}	C_{23}	C_{33}	0	0	0	ε_{33}
σ_{23}		0	0	0	C_{44}	0	0	ε_{23}
σ_{31}		0	0	0	0	C_{55}	0	ε_{31}
σ_{12}		0	0	0	0	0	C_{66}	ε_{12}

TRANSVERSELY ISOTROPIC (5 constants)								
σ_{11}		C_{11}	C_{12}	C_{13}	0	0	0	ε_{11}
σ_{22}		C_{12}	C_{22}	C_{23}	0	0	0	ε_{22}
σ_{33}	=	C_{13}	C_{23}	C_{33}	0	0	0	ε_{33}
σ_{23}		0	0	0	C_{44}	0	0	ε_{23}
σ_{31}		0	0	0	0	C_{44}	0	ε_{31}
σ_{12}		0	0	0	0	0	$(C_{11}-C_{12})/2$	ε_{12}

ISOTROPIC (2 constants)								
σ_{11}		C_{11}	C_{12}	C_{12}	0	0	0	ε_{11}
σ_{22}		C_{12}	C_{11}	C_{12}	0	0	0	ε_{22}
σ_{33}	=	C_{13}	C_{12}	C_{11}	0	0	0	ε_{33}
σ_{23}		0	0	0	$(C_{11}-C_{12})/2$	0	0	ε_{23}
σ_{31}		0	0	0	0	$(C_{11}-C_{12})/2$	0	ε_{31}
σ_{12}		0	0	0	0	0	$(C_{11}-C_{12})/2$	ε_{12}

Fig. 8.23 *Second-order (6 x6) elastic modulus tensors C_{ij} for various classes of material.*

EXERCISE 65

a) Cortical bone is a material which responds very differently depending upon the direction of the applied loading, as will be discussed in detail later in section 9.3. The figure, taken from that section, contains three curves from uniaxial tensile tests performed on bovine cortical bone specimens, orientated in the radial (1), circumferential (2) and axial (3) directions. Discuss the applicability of the linear-elastic model to this material.

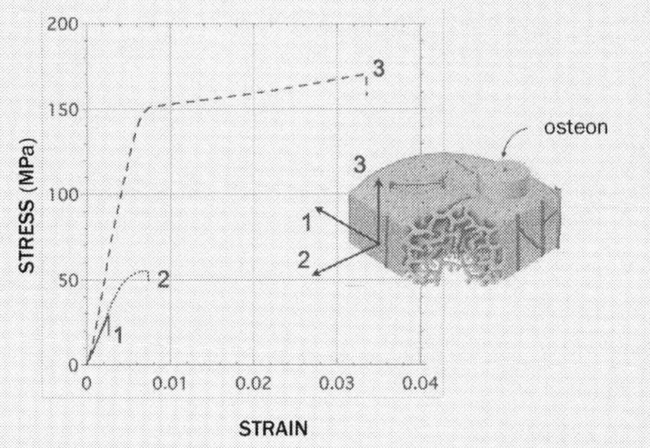

b) Measurements of the elastic modulus and Poisson's ratio performed by Knets and Malmeister (1977) on human cortical bone (tibia) are shown in the table.

Assuming that the linear-elastic model is applicable discuss which type of material (orthotropic, transversely isotropic or isotropic) best fits the experimental data.

E_1	E_2	E_3	G_{12}	G_{31}	G_{23}	ν_{12}	ν_{13}	ν_{23}
6.91	8.51	18.4	2.41	3.56	4.91	0.49	0.12	0.14

E, G (GPa) directions: (1) radial, (2) circumferential, (3) axial

$$E_i = \frac{\sigma_{ii}}{\varepsilon_{ii}} \qquad \nu_{ij} = -\frac{\varepsilon_{jj}}{\varepsilon_{ii}} \qquad G_{ij} = \frac{\sigma_{ij}}{2\varepsilon_{ij}}$$

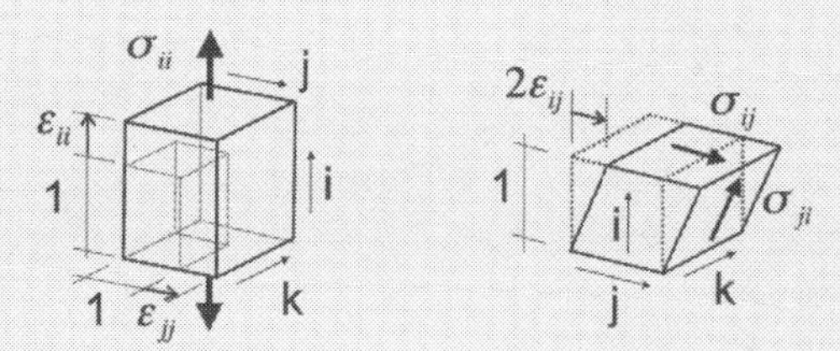

SOLUTION:

a) We observe from the curves in the figure that cortical bone exhibits an initially very linear stress-strain response, practically up to the value of the yield stress in all of them. In this linear zone it is possible to establish a proportionality between the stresses and strains.

On the other hand, the strains at which the yield stress is reached are very low, between 0.0026 and 0.006 depending upon the measurement direction, which implies that mechanical functioning of the bone will always be within the zone of very small strains, as is required by the theory of linear elasticity. In the first Exercise in section 2.2.1 we show that the relative error introduced by using the engineering rather than the true stress () is of the order of half of the maximum strain achieved which, in the worst case, becomes 0.006/2 = 0.3%. Another way of arriving at the same result is to estimate the values of the second-order terms (proportional to the square of the strain), which can be neglected in the linear elastic approximation. In our case we see that $0.006^2 = 0.000036$ which is almost three orders of magnitude less than 0.006 and justifies the approximation of infinitesimal strains.

The characteristics that we have discussed (linearity and small strains) are appropriate for hard biological materials. The linear-elastic model may be adequate for characterising their behaviour, provided the effects of temperature are excluded. For this to occur, generally only those situations in which the mechanical load is either very rapid or very slow are considered.

b) Using the moduli E and G and v values defined in the problem statement, we can write an equation which links the flexibility tensor, the stresses and strains for an **orthotropic** material. When a stress σ_{ii} produces the strains

$$\varepsilon_{ii} = \sigma_{ii}/E_i \quad \text{and} \quad \varepsilon_{jj} = -v_{ij}\,\varepsilon_{ii} = (-v_{ij}/E_i)\sigma_{ii}$$

If the applied stress is σ_{ij} then the strains produced are

$$\varepsilon_{ij} = \sigma_{ij}/(2\,G_{ij})$$

Therefore, bearing in mind that for a linear strain model, when the several stresses act, it is possible to sum the individual strains from each state, we write the complete relationship by means of the flexibility tensor

$$\begin{bmatrix} \varepsilon_{11} \\ \varepsilon_{22} \\ \varepsilon_{33} \\ \varepsilon_{23} \\ \varepsilon_{31} \\ \varepsilon_{12} \end{bmatrix} = \begin{bmatrix} 1/E_1 & -\nu_{21}/E_2 & -\nu_{31}/E_3 & & & \\ -\nu_{12}/E_1 & 1/E_2 & -\nu_{32}/E_3 & & & \\ -\nu_{13}/E_1 & -\nu_{23}/E_2 & 1/E_3 & & & \\ & & & 1/(2G_{23}) & & \\ & & & & 1/(2G_{31}) & \\ & & & & & 1/(2G_{12}) \end{bmatrix} \begin{bmatrix} \sigma_{11} \\ \sigma_{22} \\ \sigma_{33} \\ \sigma_{23} \\ \sigma_{31} \\ \sigma_{12} \end{bmatrix}$$

With the data from the problem statement, all the parameters except υ_{21}, υ_{31} and υ_{32}, can be calculated, noting that the tensor **S** is symmetric, and therefore

$$-\frac{\nu_{21}}{E_2} = -\frac{\nu_{12}}{E_1} \quad -\frac{\nu_{31}}{E_3} = -\frac{\nu_{13}}{E_1} \quad \text{and} \quad -\frac{\nu_{32}}{E_3} = -\frac{\nu_{23}}{E_2}$$

expressions which lead to values $\upsilon_{21} = 0.603$, $\upsilon_{31} = 0.320$ and $\upsilon_{32} = 0.303$. With these data the orthotropic material model is complete

$$\begin{bmatrix} \varepsilon_{11} \\ \varepsilon_{22} \\ \varepsilon_{33} \\ \varepsilon_{23} \\ \varepsilon_{31} \\ \varepsilon_{12} \end{bmatrix} = \begin{bmatrix} 0.144 & -0.0709^* & -0.0174^* & & & \\ -0.0709 & 0.118 & -0.0165^* & & & \\ -0.0174 & -0.0165 & 0.0543 & & & \\ & & & 0.102 & & \\ & & & & 0.140 & \\ & & & & & 0.207 \end{bmatrix} \begin{bmatrix} \sigma_{11} \\ \sigma_{22} \\ \sigma_{33} \\ \sigma_{23} \\ \sigma_{31} \\ \sigma_{12} \end{bmatrix}$$

with all these coefficients in GPa. The values with (*) correspond to those calculated from the experimental data.

As was seen above, for a material to behave in a **transversely isotropic** manner, it must be the case that

$$S_{11} = S_{22} \quad (0.144 \ vs \ 0.118)$$
$$S_{13} = S_{23} \quad (-0.0174 \ vs \ -0.0165)$$
$$S_{44} = S_{44} \quad (0.102 \ vs \ 0.140)$$
$$S_{66} = S_{11} - S_{12} \quad (0.207 \ vs \ 0.214)$$

conditions which, even if not completely satisfied, indicate that the behaviour of bone is close to being that of a transversely isotropic material.

However, it is more difficult to fit the **S** matrix to that of **purely isotropic** behaviour since the requirements are

$$S_{11} = S_{22} = S_{33} \quad (0.144 \ vs \ 0.118 \ vs \ 0.0543)$$
$$S_{13} = S_{23} = S_{12} \quad (-0.0174 \ vs \ -0.0165 \ vs \ -0.0709)$$
$$S_{44} = S_{44} = S_{66} = S_{11} - S_{22} \quad (0.102 \ vs \ 0.140 \ vs \ 0.207 \ vs \ 0.214)$$

conditions that are not satisfied, even approximately.

8.3.3. Heterogeneous materials[47]

All biological materials are heterogeneous if observed with sufficiently close detail but, in practice, many of them are treated as if they were homogeneous when making calculations, by considering an **elemental volume** which is sufficiently large for the average value of properties to be representative of the material as a whole. In other words, the heterogeneous material is replaced with a homogenous one whose average properties on the boundaries are identical when considering volume elements of equal or larger size than the elemental volume. Consequently, in the analysis, the stresses and strains which are obtained by this method at any point only make sense when they are understood to be average values for these volumes, losing all physical meaning if they are ascribed to regions smaller than the elemental volume. Note that this simplification does not prevent the homogenised material from being anisotropic, as happens in the case of wood or bone, for example.

The study of heterogeneous materials, or composite materials, is interesting because it permits an understanding of the behaviour of both from a knowledge of the components and, even more, the design of new materials with superior properties to each of the separate material components. Nature, through biological materials, demonstrates these aspects continuously; for example, the shells of molluscs –a composite material of aragonite and a polymeric cement– has a strength and toughness far greater than that of its components.

The simplest composite materials consist of two phases: the *matrix*, which may be polymeric, metallic or ceramic, and the *reinforcement*, in the form of particles or fibres (which, in turn, may also be polymeric, metallic or ceramic). In some cases, one of the phases may be a gas, as occurs in foams and sponges; these are very interesting materials to which the following chapter is dedicated. The next step in material complexity of composite materials is in their hierarchical structures which, as has been mentioned on many occasions, appear to a greater or lesser extent in all biological materials.

In this chapter the aim is only to summarise some of the useful basic concepts for modelling the mechanical behaviour of the simplest composite materials which occur in the analysis of biological materials. The purpose of these models is to predict the behaviour of the composites in terms of the constituent properties, the proportion of each of them and, in the more sophisticated models, their geometry. There are numerous books on composite materials, with various levels of sophistication and complexity, for example, Chawla (2012), Christensen (1979) or Capaldi (2012).

An accurate prediction of the mechanical behaviour of composite materials is only possible in very few cases. One of these is the prediction of the elastic modulus for a perfectly aligned continuous fibre reinforced material, from the values of the moduli of the matrix and fibre, provided that both have the same Poisson ratio. The problem is more complicated when the fibres are not continuous and have distributions of length and orientation. This is also the case for particle reinforced materials where exact solutions are only found for ordered and special particles. Even so, it is often possible to find very simple models –as will be seen later– which permit estimates of the elastic modulus, strength or other parameters of interest.

47 A material is heterogeneous if its composition and therefore its properties vary from point to point

There are two very simple models for predicting the elastic modulus of a material composed of two alternating phases of linear-elastic material which extend throughout the volume. Both models are shown in Figure 8.24 and, to a certain degree, can be considered as two aspects of the same homogenisation model, according the type of loading (parallel or perpendicular to the laminas).

In the model, the composite material is represented as two joined families of perfectly parallel laminas and made equivalent to a homogeneous linear-elastic material, analysing their combined response in one direction or another. In this case, the elemental volume must be large enough to contain sufficient laminas to allow averaging the forces.

If the loading is **parallel to the laminas,** as shown in Figure 8.24a, the *strain* will be the same in all of them and it is easy to show that the elastic modulus of the composite material E_c, as a function of the modulus of the matrix E_m, and that of the reinforcement E_f is given by,

$$E_c = E_m v_m + E_f v_f \tag{77}$$

where v_m and v_f are the respective volume fractions of the matrix (one of the lamina families) and the reinforcement (the other family). This result is known as the *Voigt model, rule of mixtures, two-phase parallel model* or the *homogeneous strain model.*

If the loading is **perpendicular to the laminas,** as shown in Figure 8.24b, the *stress* in them will be the same and it can be shown that the elastic modulus of the composite material will now be given by:

$$\frac{1}{E_c} = \frac{v_m}{E_m} + \frac{v_f}{E_f} \tag{78}$$

This result is known as the Reuss model, the two phases in series model or the homogenous stress model.

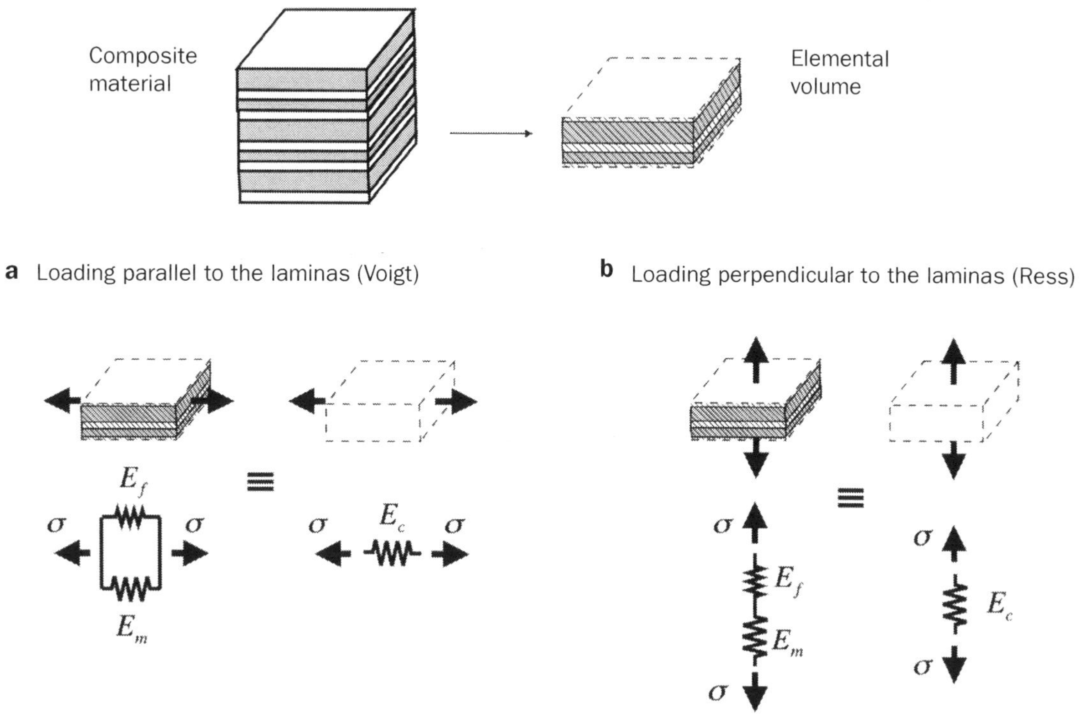

Fig. 8.24 a. *Ideal composite material in which the loading is parallel to the laminas (Voigt or parallel model).* b. *Ideal composite material in which the loading is perpendicular to the laminas (Reuss or series model).*

EXERCISE 66
Derive Equations (77) and (78).

SOLUTION:
Consider, in the first instance, the configuration shown in Figure 8.24a, in which **loading is parallel** to the laminas, and let E_m and E_f be the elastic moduli of the matrix and reinforcement.

The strain in each lamina will be the same, while they are perfectly bonded, having the same length and subjected to the same elongation. Thus we can write $\varepsilon_m = \varepsilon_f$ where ε_m is the strain in any of the matrix laminas and –similarly– ε_f is the strain in the reinforcement laminas.

Treating it as a uniaxial tensile test with linear-elastic behaviour we have that $\varepsilon_m = \dfrac{\sigma_m}{E_m}$ and $\varepsilon_f = \dfrac{\sigma_f}{E_f}$ where σ_m and σ_f are the stresses to which the laminas of each component are subjected.

If ε_c, E_c and σ are, respectively, the strain, elastic modulus of the composite material and the stress, it can be shown that $\sigma = E_c\,\varepsilon_c$

Consequently $\sigma A = \sigma_m A_m + \sigma_f A_f$, where A is the area of the transverse section and A_m and A_f are the total area of those sections corresponding to each phase. Clearly, $A = A_m + A_f$. If we substitute the values of the stresses as functions of the strains in the above equation, observing that for compatibility all of them must be equal, we obtain

$$\varepsilon_c E_c A = \varepsilon_m E_m A_m + \varepsilon_f E_f A_f \quad \rightarrow \quad E_c = E_m \frac{A_m}{A} + E_f \frac{A_f}{A}$$

and then with the configuration in Figure 8.23, the volume fractions of each material can be calculated directly from the ratio of the areas of the transverse sections

$$v_m = \frac{A_m}{A} \quad \text{and} \quad v_f = \frac{A_f}{A}$$

we obtain Equation (77).

In the case of **loading in series,** Figure 2.24b, we must observe that now *the stress in each phase is the same* and, therefore, the strain in each phase and also the combined material will be

$$\varepsilon_m = \frac{\sigma}{E_m} \quad , \quad \varepsilon_m = \frac{\sigma}{E_m} \quad \text{and} \quad \varepsilon_c = \frac{\sigma}{E_c}$$

If each lamina has a thickness L_i and it stretches ΔL_i, its strain ε_i will be $\varepsilon_i = \dfrac{\Delta Li}{Li}$ and the strain in the whole element in the load direction

$$\varepsilon_c = \frac{\Delta L}{L} = \frac{\sigma}{E_c} = \sum \frac{\Delta L_i}{L} = \sum \frac{\Delta L_i}{L_i}\frac{L_i}{L} = \sum \varepsilon_i \frac{L_i}{L} = \sum \frac{\sigma_i}{E_i}\frac{L_i}{L} = \sigma \sum \frac{1}{E_i}\frac{L_i}{L}$$

since all the stresses are equal to σ.

If we group the terms corresponding to each phase in the above equation, we obtain

$$\varepsilon_c = \frac{\sigma}{E_c} = \sigma\left(\frac{1}{E_m}\frac{L_m}{L} + \frac{1}{E_f}\frac{L_m}{L}\right) \quad \rightarrow \quad \frac{1}{E_c} = \frac{1}{E_m}\frac{L_m}{L} + \frac{1}{E_f}\frac{L_m}{L}$$

in which we have called L_m and L_f the thicknesses of each phase.

From this expression, observing that the volume fraction of each phase can be written as $v_m = \dfrac{L_m}{L}$ and $v_f = \dfrac{L_f}{L}$ we recover Equation 78).

a

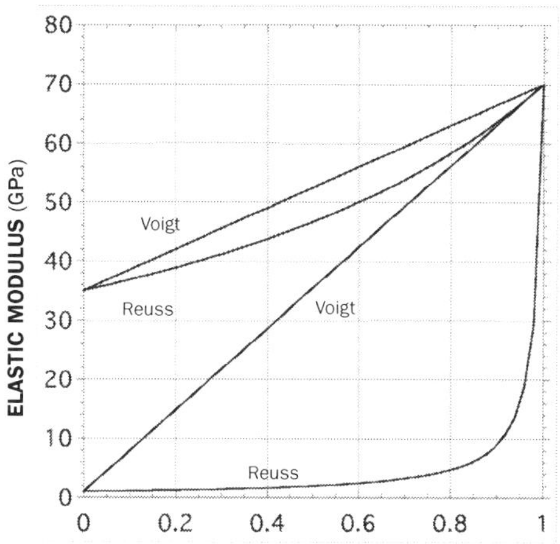

b

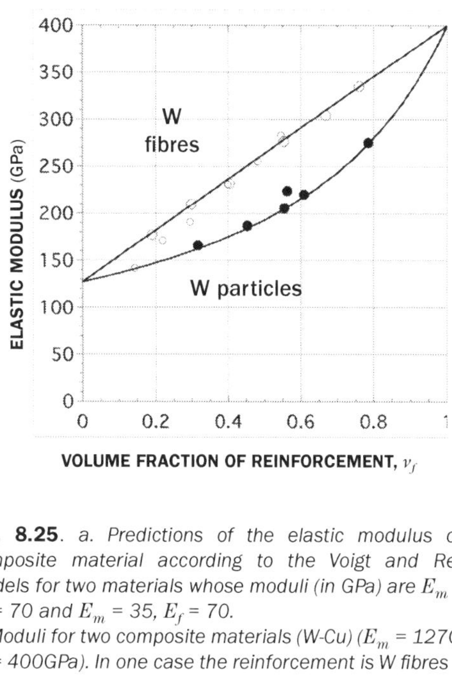

Fig. 8.25. *a. Predictions of the elastic modulus of a composite material according to the Voigt and Reuss models for two materials whose moduli (in GPa) are E_m = 1, E_f = 70 and E_m = 35, E_f = 70.*
b. Moduli for two composite materials (W-Cu) (E_m = 127GPa, E_f = 400GPa). In one case the reinforcement is W fibres and in the other W particles (after Weidmann, 1990).

Figure 8.25a contains values of the elastic modulus of a composite material as a function of the volume fraction of reinforcement, according to the Voigt and Reuss models. Two situations have been illustrated: one in which the modulus of the matrix is 35GPa and the other in which it is 1GPa; in both cases the modulus of the reinforcement is the same and equal to 70GPa.

It may be observed that the Voigt model always predicts values greater than those of Reuss and that this difference is more apparent the greater the difference between the moduli of the two phases. The hypothesis of the Voigt, or homogeneous strain, model appears appropriate for a material reinforced with fibres aligned with the loading direction, while that of Reuss, or homogeneous stress, is better suited to particle reinforced materials. Figure 8.25b illustrates these observations. The results relate to a material composed of a copper matrix reinforced with tungsten fibres, in one case, and tungsten particles in the other. It can be seen that the Voight model gives very good results for fibre reinforcement and that the Reuss model provides acceptable results for particle reinforcement.

It should be borne in mind that both models are extremely simple, that they are based on a minimum information on both phases (their moduli and volume fractions), and that they make no distinction between the phases. Among other things, they ignore the effects of geometry of the phases, how they are distributed, their Poisson ratios, the bonding between the reinforcement and the matrix and, particularly for polymer matrices, viscoelastic effects.

The influence of the **orientation of the reinforcement,** whether it be fibres or particles (if they are not spherical), can be accounted for if an assumption is made about the orientation distribution function. One practical way of tackling this problem is by means of the so-called efficiency factor, B, in the rule of mixtures expression, that is

$$E_c = E_m v_m + B E_f v_f \qquad (79)$$

where B can take the following values:

- $B = 1$ When the fibres are completely aligned and the loading is parallel to them.
- $B = 1/2$ When the fibres are arranged in two orthogonal direction and the loading is in one of these directions.
- $B = 3/8$ When the fibres are arranged randomly in a plane and the loading lies in this plane.
- $B = 1/5$ When the fibres are arranged randomly in three dimensions.

Models of reinforced composite materials are widely used for biological materials since, without loss of generality, it can be said that, at the super-cellular scale, practically any biological material can be considered as a material of this type. So that, for example, wood can be analysed as an amorphous matrix of hemicellulose and lignin reinforced with cellulose fibres, bone as an inorganic hydroxyapatite matrix reinforced with collagen fibres, or soft tissue –blood vessels, cartilage or skin– as glycosaminoglycans matrix reinforced with elastin and collagen fibres.

Unfortunately, except in a few cases such as shells and bones, biomechanical models of fibre-reinforced materials are extremely complex, since it is necessary to take into account the large deformations produced, and this is without including the effect of temperature, time or biological variables which modify the composition and the geometry of the tissue.

In section 6.3.3 comment is made on some models for blood vessels which take account of the effect of the collagen reinforcing fibres present in tissues. As a simplified example of the modelling of material with long fibres, we present below an application of Lanir's (1979) model to tensile tests on ligaments and tendons, which shows –for the case of long fibre reinforcement– the importance of considering the progressive loading of the reinforcing fibres during the deformation process.

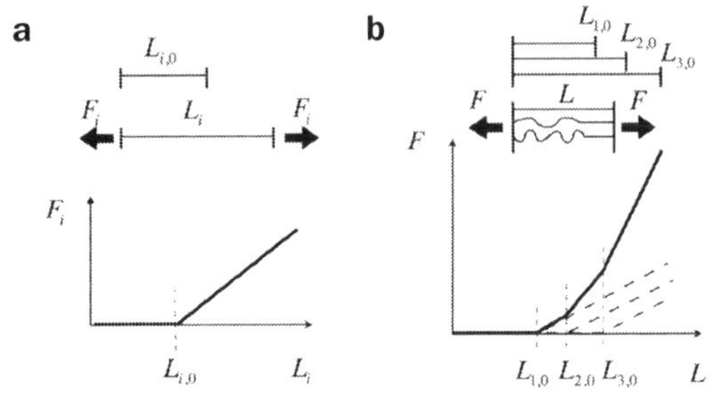

Fig.8.26. *Lanir's (1979) simplified model.*
a. Behaviour of one fibre
b. Behaviour of a group of three fibres with different initial relaxed lengths.

Figure 8.26 shows a very simplified model of tendons and ligaments, in which they are analysed as a mass of collagen fibres subjected to uniaxial tension. The model assumes that all of the fibres are long and not rolled over each other (as in the case of ropes) and reach from one end of the mass to the other. Furthermore, it is assumed that not all of the fibres are orientated and parallel at the start of the deformation, but some possess a degree of laxity, which is progressively lost during the elongation. In the analysis the mechanical contribution of the matrix which bonds the fibres is neglected, considering that its resistance is far lower than these, as well as the interaction between the fibres themselves during deformation.

As can be seen in Figure 8.26, the force required to stretch the mass of fibres depends at any one time upon the number under load. Under these conditions, in order to extend a fibre of relaxed length $L_{i,0}$ (Figure 8.26a) no force is needed until it reaches a length $L_{i,0}$. From this moment the fibre exerts an increasing resistance with the elongation. When, in a similar mass, there are fibres of different relaxed lengths (Figure 8.26b) the effects accumulate, producing a J-shaped force-extension curve, typical of those obtained from tensile tests on soft materials, whose curvature is associated with the process of orientation and regrouping and progressive failure of the fibres which reinforce these materials.

In a more quantitative form, we shall consider the bunch of fibres whose initial relaxed length (without load) is L_0. When the bunch is deformed under tensile loading its elongation is given by the ratio of the lengths

$$\Lambda = \frac{L}{L_0} \tag{80}$$

The elongation of the bunch of fibres Λ should not be confused with the individual fibre elongation λ. However, these two quantities are related by the expression

$$\lambda = \frac{L}{L_{0f}} = \frac{L}{L_0}\frac{L_0}{L_{0f}} = \frac{\Lambda}{\theta} \tag{81}$$

In which the laxity θ, of the fibre, defined as the ratio between the straight relaxed length (without load on the fibre) L_{0f} and the relaxed length of the mass of fibres L_0, that is

$$\theta = \frac{L_{0f}}{L_0} \tag{82}$$

By definition, the smallest value of the laxity is $\theta = 1$, corresponding to those fibres whose relaxed length coincides with the relaxed length of the mass of fibres.

Even though the fibres in the bunch have the same composition and properties, they may have different relaxed lengths and cross-sections. In order to account for these circumstances, a fibre distribution function $\Phi(\theta)$, is introduced, characteristic of the bunch, defined such that $\Phi(\theta)d\theta = dA$ *is the total (relaxed) transverse area of those fibres whose laxity lies between* θ and $\theta + d\theta$. The total cross-sectional area of all of the fibres in the bunch will be given by the integral

$$A = \int_1^{\theta_m} \Phi(\theta)\, d\theta \tag{83}$$

where θ_m is the maximum laxity of the fibres. Note that during the test, beyond an elongation $\Lambda = \theta_m$ all the fibres in the bunch will be stretched.

The function $\Phi(\theta)$ is shown in Figure 8.27 and can take on different shapes depending upon the initial degree of alignment of the fibres. When all of them are initially orientated and parallel, without laxity, the distribution function is a δ-Dirac: $\Phi(\theta) = A\,\delta(\theta - 1)$.

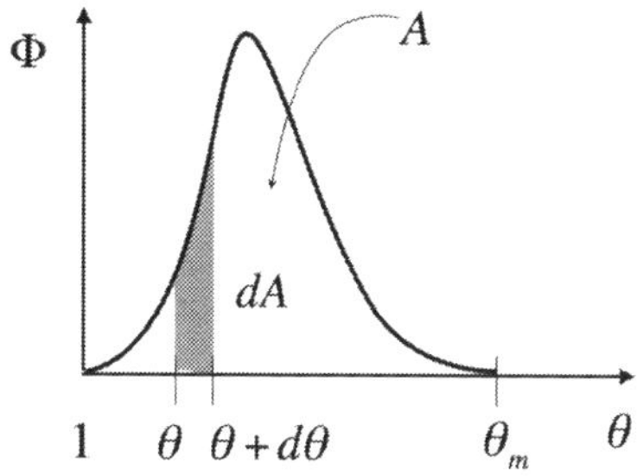

Fig. 8.27 *Distribution function of the fibre cross-sections. A is the total cross-sectional area.*

If we assume that all of the fibres have the same mechanical behaviour, we can establish a relationship between their engineering stress s and elongation λ given by

$$s = \begin{vmatrix} 0 & \text{if} & \lambda < 1 \\ s(\lambda) & \text{if} & 1 \le \lambda < \lambda_R \\ 0 & \text{if} & \lambda \ge \lambda_R \end{vmatrix} \qquad (84)$$

in which we have assumed that there is no resistance to compression ($\lambda<1$), since the fibres would buckle, and also when the elongation to failure λ_R is exceeded.

With these ingredients, we can write the force exerted by the fibres with laxity between θ and $\theta + d\theta$, when the bunch has an elongation Λ, which will be equal to

$$dF = s(\lambda)dA = s\left(\frac{\Lambda}{\theta}\right)\Phi(\theta)d\theta \qquad (85)$$

and the total force on the bunch of fibres corresponding to the elongation Λ is obtained by integrating the above expression

$$F(\Lambda) = \int_{\max\{1,\,\Lambda/\lambda_R\}}^{\min\{\theta_m,\,\Lambda\}} s(\Lambda/\theta)\Phi(\theta)d\theta \qquad (86)$$

since the function $\Phi(\theta)$ is defined between $1 \le \theta \le \theta_m$ and also $s \ge 0$ only if $1 \le \lambda = \Lambda/\theta < \lambda_R$, from which $\Lambda \ge \theta \ge \Lambda/\lambda_R$.

The application of Lanir's model to a uniaxial tensile test to rupture of a human tendon (*extensor digitorum longus*; Schechtman *et al.*, 2002) is shown in Figure 8.28. In order to simplify the situation it has been assumed that the collagen fibres have a linear-elastic behaviour followed by a section of yielding before rupture (Figure 8.28a). The distribution function has been taken to be a constant value.

In Figure 8.28b we see that, despite the simplicity of a model with only four parameters $(E, \lambda_y, \lambda_R, \theta_m)$, it is possible to reproduce the complete tensile test quite well over the whole range; in the initial region of progressively increasing stiffness, in the middle linear zone, in the region of maximum load, and also in the zone where the load falls due to the progressive failure of the fibres. Furthermore, the fitting parameters lie well within the expected range for collagen fibres which we know constitute tendons, whose elastic modulus is about 1GPa (see section 3.4.2).

The results above illustrate the usefulness of models which take into account the length and alignment of the fibres when modelling soft biological materials.

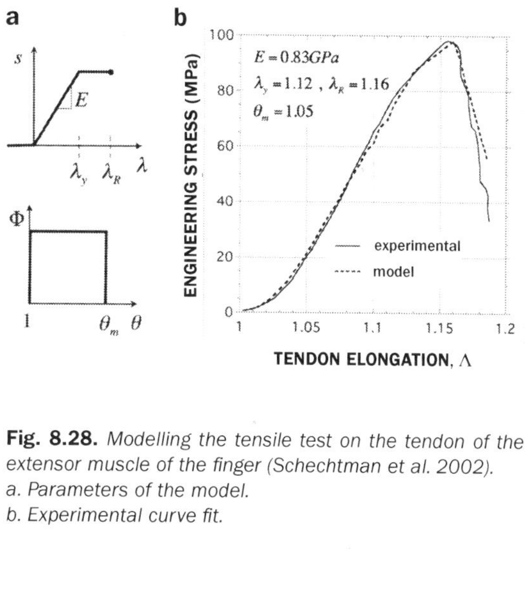

Fig. 8.28. *Modelling the tensile test on the tendon of the extensor muscle of the finger (Schechtman et al. 2002).*
a. Parameters of the model.
b. Experimental curve fit.

Modelling the strength or toughness of a composite material is not a simple task because, unlike homogeneous materials, fracture in composite materials may be due to various mechanisms which depend upon the type of loading and the microstructure of the composite material, geometry and orientation of the reinforcement, the bonding between the reinforcement and the matrix, and changes in the matrix as the fracture progresses. These diverse failure mechanisms, in each case, lead to numerous failure criteria. In the third part of this chapter detailed comment is provided on the modelling of some failure mechanisms and the means of estimating the strength of these biological materials –for wood, shells or bones.

EXERCISE 67

The figure shows three-point bending tests performed by Wang and co-workers (2001) on prismatic abalone and mother-of-pearl shells. As will be noted later, these materials are formed of micrometric ceramic laminas (calcium carbonate in an aragonite structure) of thickness and average diameter of 0.3 and 7 µm, respectively, combined in a polymeric cement which constitutes approximately 5% of the total volume.

The experiments were conducted in seawater, in two different directions, with the aim of determining the anisotropy of the material. The specimens had dimensions (depth x width x length) of 0.5 x 2.5 x 12 mm. The values shown correspond the mean value and standard error.

Discuss the applicability of the Voigt and Reuss models to these materials.

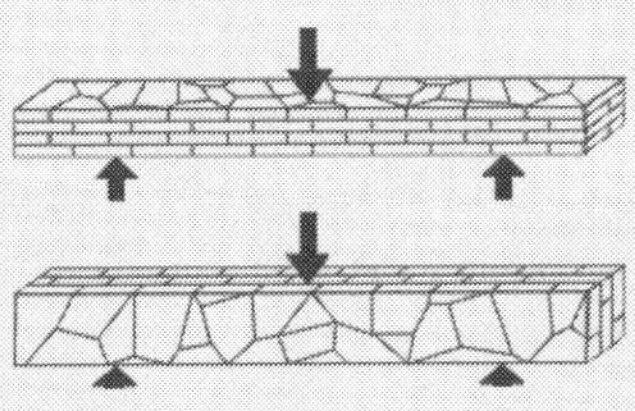

$E(GPa)$	Abalone	Mother-of-pearl
(upper arrangement)	69 ± 7	77 ± 12
(lower arrangement)	66 ± 2	81 ± 4

SOLUTION:

For the purpose of the analysis we shall assume that we are dealing with a composite material formed of a polymeric matrix ($v_m = 0.05$) reinforced with ceramic laminas ($v_f = 0.95$). Observing the experimental setup, we might suppose initially that the Voight model is more appropriate, since in neither arrangement is the interface with largest contact surface between the ceramic laminas subjected to tension/compression. In both cases the same types of interface are subjected to tension/compression and this is reflected in the results, which show no statistically significant differences.

However, the macroscopic dimensions of the specimens invalidate the hypothesis that the reinforcing laminas are essentially continuous, required for a good fit to the Voigt model. In this case, the problem can be approached by observing that there are a large number of interfaces along the tension/compression directions, which we can estimate at $12\,mm / 7\ \mu m \approx 1700$. When the loading direction is essentially perpendicular to the interfaces the Reuss model would seem to be more relevant.

In order to evaluate both models, let us take elastic modulus values for the polymeric matrix (m) and the ceramic reinforcement (f) proposed by Jackson (1988), $E_m = 4\,GPa$ and $E_f = 100\,GPa$.

With these values, Equations (77) and (78) lead to

$$E_{Voigt} = v_m E_m + v_f E_f = 0.05\ 4 + 0.95\ 100 = 95.2\,GPa$$

$$E_{Reuss} = (v_m / E_m + v_f / E_f)^{-1} = (0.05/4 + 0.95/100)^{-1} = 45.4\,GPa$$

The values obtained are far removed from the experimental measurements and, therefore, we conclude that it is necessary to employ more sophisticated models which take into account the geometry of the reinforcing laminas. In any case, it is interesting to point out that the Voigt and Reuss models provide upper and lower bounds to the stiffness of the material.

8.3.4. Cellular materials

1. Introduction

Cellular structures are present in many biological materials. Honeycomb, a classical paradigm of these structures, has caught the attention of mathematicians, physicists and biologists since antiquity; Euclid admired them for their regularity and Pliny cited the case of a man who dedicated his life to the study of these structures. Wood also shows us its cellular structure. One of the first materials examined by Robert Hooke in his new microscope (around 1660) was cork, which allowed him to see and identify the basic unit of biological structures; it was Hooke who baptised the structures he saw in cork cells.

Cellular materials can be **classified** into two major groups: two-dimensional cellular solids commonly called *honeycomb* structure– formed of prismatic cells, such as bee honeycomb, and *three-dimensional cellular solids* –commonly called *foams* or *sponges*– formed of polyhedral cells, such as spongy bone or actual sponges. Biological materials with a foam structure come in many different forms; woods, for example, exhibit highly elongated regular cells very similar to bee honeycomb. Spongy bone looks more like a three-dimensional framework. In other materials, such as corals or cuttlefish shells, the cells are orientated and this confers a marked mechanical anisotropy.

The **characteristic parameters** of cellular materials are the *relative density,* the shape of the cells, and the material properties of the walls or of the framework.

The *relative density* is the ratio of between the density of the cellular material, taken together (apparent density) and the density of the solid that forms the cell walls or trabeculae. Its value is equal to the proportion, by volume, of solid material.

In terms of the *shape of the cells*, it is convenient to distinguish between open cells –connected– in which the solid material is only distributed along the edges of the polyhedral which they form, and closed cells, in which the faces are formed of solid walls.

The *basic material properties* are obtained from the solid that forms the edges or faces of the cellular material polyhedral. Some of these basic materials are, in turn, composite materials, such as in the case of the trabeculae of spongy bone.

The **mechanical response** of cellular materials loaded in compression or tension is illustrated in Figure 8.29, in which some differences are apparent depending upon whether the cellular materials are elastomers (Fig. 8.29a and 8.29c) or elastoplastics (Fig.8.29b and 8.29d).

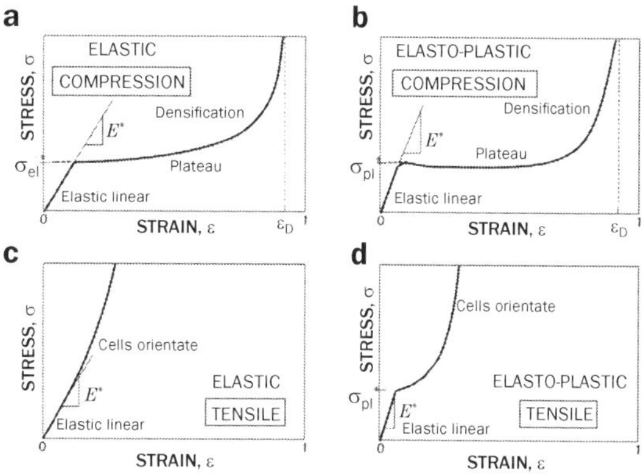

Fig. 8.29 *Schematic stress-strain curves in uniaxial tests performed on cellular materials.*

The stress-strain curves obtained in **compression** tests (Figs. 8.29a and 8.29b) have three characteristic stages. The first stage, an elastic linear region, corresponds to bending of the beams and columns which form the edges, or to the deformation of the walls which form the faces, of the polyhedra. In the second stage, where the stress increases slowly or remains constant, the cells deteriorate through elastic buckling, or by rupture, depending upon the nature of the solid from which they are made. And, finally, there is a stage of densification, due to collapse and direct contact of the solid material. Cellular materials which have low relative densities –between 0.10 and 0.20– allow large strains to be achieved –between 70 and 80%– before densification, as shown in Figures 8.29a and 8.29b.

The stress-strain curves obtained in **tensile** tests are shown in Figures 8.29c and 8.29d. For small strains the response is elastic and linear, as was the case in compression tests. As the strain increases, the cells orientate themselves in the direction of the applied force and the stiffness of the cellular material increases until rupture occurs.

The mechanical behaviour of cellular materials can be **modelled** in several ways: analytically, numerically or, by simpler means, using dimensional analysis.

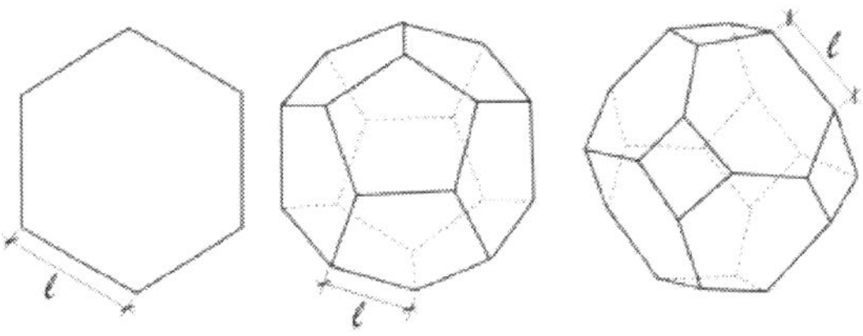

Fig. 8.30. *Regular unit cells*
a. Hexagon
b. Dodecahedron
c. Tetradecahedron

Initially, the models used deduced the mechanical response analytically from the unit cell deformation (see Fig. 8.30); in two dimensions, a hexagon and in three dimensions a dodecahedron, a twelve-faced polyhedron, or the tetradecahedron, a polyhedron with 14 faces (Zhu *et al.*, 1997, Warren and Kraynik, 1997). The nature of these unit cells is related to the need for a polyhedron that fills the space and whose area per unit volume is a minimum. Until 1992 it was thought that the tetradecahedron (proposed by Kelvin in 1887) was the optimum polyhedron, but it has been determined, using numerical methods, the other forms are better, although only by 0.3% (Weaire and Phelan 1994). These models are not very realistic if the material is formed of cells of different shapes and sizes, as happens in many biological materials.

Currently it is common to tackle the problem using numerical methods based on finite elements (Silva *et al.*, 1995, van der Burg *et al.*, 1997). These techniques allow the modelling of structures with cells of different shapes and arranged randomly, rather than periodically. Furthermore, they admit the introduction of defects which, in some cases, are essential for capturing, for example, rupture of the material. Numerical procedures can be used in conjunction with imaging techniques, such as computerised tomography, which make the study of more realistic cellular structures possible (Chen *et al.*, 2001).

Far simpler methods, based on dimensional analysis, permit the modelling of deformation and failure mechanisms which are observed in cellular structures without the need to specify the exact geometry of the cell (Gibson and Ashby, 1982a, b, 1997). This technique assumes that the geometry of the cells is similar for foams with different relative densities. It provides results in terms of the relative density and the properties of the solid which form the cell, but it needs some experiments, as will be seen later, in order to obtain the constants related to the cell geometry.

2. Two-dimensional cellular structures. Honeycomb structures

Gibson and Ashby 1997 studied the response of a beehoneycomb subjected to mechanical loading, particularly structures with equal, ordered, hexagonal cells. With the advent of powerful computers the problem has been generalised to structures with cells of different sizes and shapes, arranged randomly, and allowing, moreover, the introduction of defects.

The aim of this section is to emphasise the most pertinent results of Ashby and Gibson's studies. Strictly speaking, only those results applicable to structures with regularly repeating the unit cells are considered, since, as already indicated, they are able to reproduce some cellular biological material parameters, as will be seen later when wood and spongy bone are discussed (Sections 9.2 and 9.3).

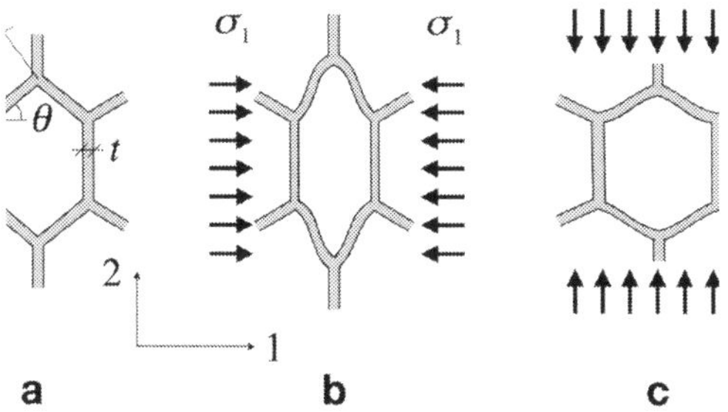

Fig. 8.31. *Deformations of a two-dimensional unit cell.*
a. Initial, undeformed cell.
b and c. Flexural deformations of the cell walls.

Figure 8.31a illustrates the geometrical characteristics of the unit cell, an irregular hexagon in which the sides l and h do not have to be equal. The wall thickness is constant and equal to t, which is taken to be much smaller than l or h. The cell wall material is assumed to be linearelastic and isotropic, characterised by an elastic modulus E_s and Poisson's ratio.

The relative density of the honeycomb, d^*/d_s, is given by

$$\frac{d^*}{d_s} = \frac{h/l + 2}{2\cos\theta\left(h/l + \sin\theta\right)}\frac{t}{l} \qquad (87)$$

d^* is the density of the honeycomb, d_s the density of the solid material and the angle, θ, is defined in Figure 8.31a. (Note that this is also the quotient of the area of the cell and the corresponding area of its walls).

The above result indicates that the relative density is proportional to t/l and that the constant of proportionality is a function of the unit cell geometry. For example, for a regular hexagon ($h = l$, $\theta = 30°$) the constant is $2/\sqrt{3}$, or approximately 1.15.

In the following treatment it will be assumed that the wall thickness t, is much less than the length of the walls, l or h.

Response to loading parallel to the plane of the honeycomb

Compression of the cells

When the honeycomb is subjected to compressive forces located in the plane of the hexagons, the response in the form of the stress-strain diagram is of the type shown in Figures 8.29a and 8.29b, in which it is possible to identify three stages.

In the **initial stage**, under small strains, the behaviour is linearelastic and the walls of the cell are deformed as indicated in Figures 8.31b and 8.31c (depending upon whether the stress act in the 1-or 2-direction). Applying Strength of Materials concepts it is possible to deduce (see Exercise 68) that the ratio of the honeycomb elastic modulus, E^*, and that of the solid material E_s is given by

$$\frac{E_1^*}{E_s} = \frac{\cos\theta}{\left(h/l + \sin\theta\right)\sin^2\theta}\left(\frac{t}{l}\right)^3 \qquad (88)$$

when the compressive forces act in the 1-direction, and

$$\frac{E_2^*}{E_s} = \frac{h/l + \sin\theta}{\cos^3\theta}\left(\frac{t}{l}\right)^3 \qquad (89)$$

when they act in the 2-direction.

For regular hexagons ($h = l$, $\theta = 30°$) both results coincide, and

$$\frac{E_1^*}{E_s} = \frac{E_2^*}{E_s} = \frac{4}{\sqrt{3}}\left(\frac{t}{l}\right)^3 \approx 2.31\left(\frac{t}{l}\right)^3 \qquad (90)$$

These results indicate that the relative elastic moduli of the honeycomb are proportional to $(t/l)^3$ and that the constant of proportionality depends upon the geometry of the cell, that is to say upon h/l and θ. With this information it is possible to draw the initial slope in Figures 8.29a and 8.29b.

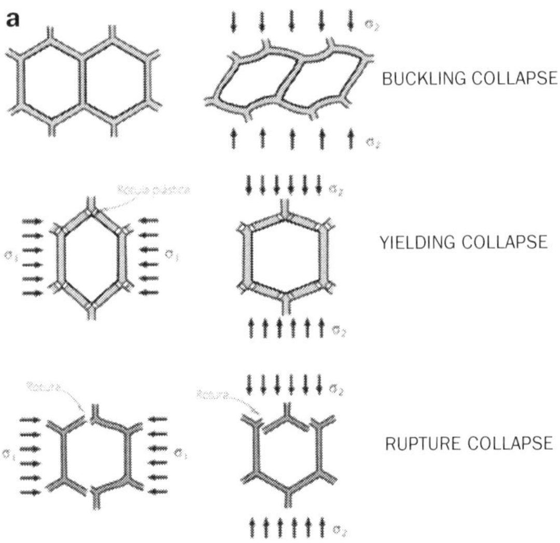

Fig. 8.32. *Collapse modes in a two-dimensional unit cell.*
a. By elastic buckling.
b. By formation of plastic hinges.
c. By rupture of the walls.

As the load increases, cells begin to collapse and the **second stage** (in Figures 8.29a and 8.29b) begins. Collapse may be due to several causes depending upon the cell wall material; by *buckling*, if elastic, by *yielding* (the formation of plastic hinges) if it is elasto-plastic, or by *rupture*, if fragile. These three modes are shown schematically in Figure 8.32. Ashby and Gibson have calculated the respective stresses which determine the end of the first stage and the beginning of the plateau.

- When the cells are subjected to compression in the 2-direction, the *elastic buckling* stress is calculated from the force required to cause buckling of the columns aligned with the loading direction (see Figure 8.32a) and is given by

$$\frac{\sigma^*_{el,\,2}}{E_s} = \frac{n^2\pi^2}{24\cos\theta}\,\frac{t^3}{lh^2} \tag{91}$$

where n is a factor which depends upon the rotational restraint imposed at nodes (whether very flexible or otherwise) and also upon h/l, typically varying between 0.686 ($h=l$) and 0.806 ($h=2l$). Only the collapse stress in the 2-direction is quoted, since it is the most unfavourable. In the case of regular hexagons the 1- and 2-directions are the same. The result for this situation is:

$$\frac{\sigma^*_{el}}{E_s} = 0.22\left(\frac{t}{l}\right)^3 \tag{92}$$

- The stress for collapse by *plastic deformation* is calculated from the stresses that create plastic hinges in the cell walls (as illustrated in Figure 8.32b) and is given, depending upon the loading direction, by:

$$\frac{\sigma^*_{pl,\,1}}{\sigma_{ys}} = \frac{1}{2(h/l+\sin\theta)\sin\theta}\left(\frac{t}{l}\right)^2 \tag{93}$$

for compression in the 1-direction, where σ_{ys} is the yield stress of the cell wall material. For compression in the 2-direction,

$$\frac{\sigma^*_{pl,\,2}}{\sigma_{ys}} = \frac{1}{2\cos^2\theta}\left(\frac{t}{l}\right)^2 \tag{94}$$

In regular hexagons, once again, both values coincide and

$$\frac{\sigma^*_{pl}}{\sigma_{ys}} = \frac{2}{3}\left(\frac{t}{l}\right)^2 \tag{95}$$

- The stress for collapse by *cell wall rupture* (Figure 8.32c) is calculated from the moment required to break the walls, obtained from the flexural strength of the material σ_f and is given by:

$$\frac{\sigma^*_{r,\,1}}{\sigma_f} = \frac{1}{3(h/l+\sin\theta)\sin\theta}\left(\frac{t}{l}\right)^2 \tag{96}$$

for compression in the 1-direction, and also

$$\frac{\sigma^*_{r,\,2}}{\sigma_f} = \frac{1}{3\cos^2\theta}\left(\frac{t}{l}\right)^2 \tag{97}$$

in the 2-direction. For regular hexagons both values coincide, giving

$$\frac{\sigma^*_{r}}{\sigma_f} = \frac{4}{9}\left(\frac{t}{l}\right)^2 \tag{98}$$

In the **final stage**, the stress increases rapidly with the strain due to the *densification* of the cells. The densification strain ε_D (Figure 8.29) for the ideal case must be equal to the porosity of the material, that is to say, $1-d^*/d_S$. In practice, the cells are actually kinked, with small deformations, before exhausting all of the available pore space. Gibson and Ashby then propose the following expression for ε_D (see Figures 8.13a and 8.13b)

$$\varepsilon_D = 1 - 1.4\,d^*/d_s \tag{99}$$

EXERCISE 68
Derive Equation (88).

SOLUTION:
The figure shows the deformation of the cell when compressed along the 1-axis. We shall assume that all of the displacements are small.

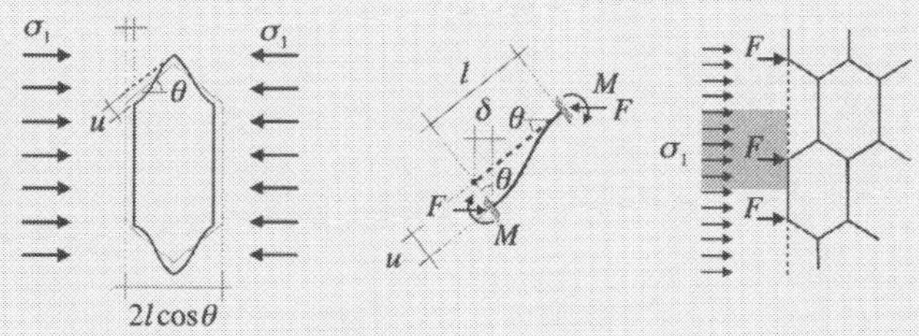

Due to the bending of the upper and lower sides the cell shortens in the 1-direction by an amount 2δ. The strain in this direction will be

$$\varepsilon_{11} = \frac{2\delta}{2l\cos\theta} = \frac{\delta}{l\cos\theta}$$

The value of δ is related to the relative displacement of the ends of the sides, u, through the expression

$$\delta = u \sin\theta$$

In order to quantify the strain we model the sides as beams, built-in at the ends, as indicated in the figure. At each end of the beam there is a bending moment M, and a force F acts in the 1-direction, as shown, since there are no applied forces in the 2-direction. Then, for moment equilibrium,

$$F = \frac{2M}{l \sin\theta}$$

In order to obtain F, we consider equilibrium of the forces on the cell along the 1-axis. The force acting at each node, F, due to the stress σ_1 will be that corresponding to its area of influence

$$F = \sigma_1 (h + l \sin\theta) b$$

and, therefore

$$M = \frac{\sigma_1}{2}(h + l\sin\theta) bl \sin\theta$$

Strength of Materials theory provides the value of u, for a beam of length l, second moment of area I and elastic modulus E_s, subjected to equal moments M

$$u = \frac{Ml^2}{6E_s I}$$

since, in our case, $I = bt^3/12$, where b is the depth of the cell, we arrive at

$$\delta = \frac{2Ml^2}{E_s bt^3}\sin\theta = \frac{\sigma_1 (h + l\sin\theta) l^3 \sin^2\theta}{E_s t^3}$$

and the strain ε_1 will be

$$\varepsilon_{11} = \frac{\sigma_1 (h/l + \sin\theta) l^3 \sin^2\theta}{E_s t^3 \cos\theta}$$

Finally, we can calculate the apparent elastic modulus E^* of the honeycomb as

$$E_1^* = \frac{\sigma_1}{\varepsilon_{11}} = \frac{E_s t^3 \cos\theta}{(h/l + \sin\theta) l^3 \sin^2\theta}$$

from which we obtain Equation 88).

EXERCISE 69

Investigate the type of anisotropy shown by two-dimensional cellular materials with the honeycomb structure in Figure 8.31.

SOLUTION:

Figure (a) depicts the arrangement of the cells and their dimensions. In the direction of the 3-axis, perpendicular to the 1-2 plane, it is assumed that the cell walls do not have openings. The wall material is considered to be homogeneous and isotropic.

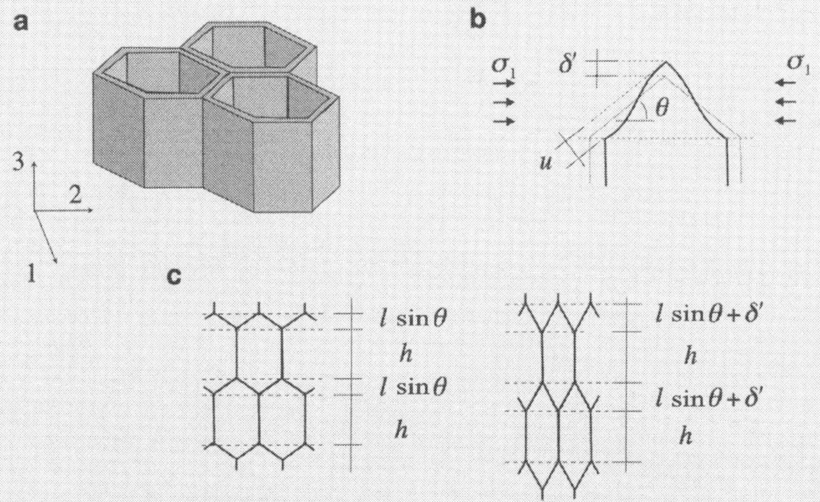

The equivalent homogeneous material must have the symmetry of the cell material, which, as we have seen, has a symmetry with respect to the planar coordinates $x_1=0$, $x_2=0$ and $x_3=0$. It will, therefore, be an **orthotropic** material which can be represented by a linear-elastic model with nine independent parameters, as was seen in Section 8.3.2.

Two of these parameters are the moduli of elasticity E^*_1 and E^*_2 given by Equation (88) and (89).

Furthermore, from the results of Exercise 68 it is easy to determine the Poisson ratio v^*_{12}, (following the nomenclature of Exercise 65). In order to calculate it, we observe that when a compressive load σ_1 is applied in the 1-direcion each vertex of the cell is displaced in the 2-direction by an amount $\delta'=u\cos\theta$ *(Figure b). Therefore, the strain (elongation) in that direction (Figure c) is*

$$\varepsilon_{22}=\frac{\delta'}{h+l\,\sin\theta}=\frac{u\cos\theta}{h+l\,\sin\theta}$$

Since the strain in the 1-direction was a shortening, of value

$$\varepsilon_{11}=\frac{u\,\sin\theta}{l\,\cos\theta}$$

the final result is

$$v^*_{12}=-\frac{\varepsilon_{22}}{\varepsilon_{11}}=\frac{\cos^2\theta}{(h/l+\sin\theta)\sin\theta}$$

The Poisson ratio v^*_{21}, can be calculated directly from the strain produced in the 1-direction when apply a load in the 2-direction, or by making use of the symmetry relationship

$$\frac{\nu^*_{21}}{E^*_2} = \frac{\nu^*_{12}}{E^*_1}$$

which we saw in Exercise 20, together with the values of the elastic moduli. In both cases the result obtained is

$$\nu^*_{21} = \frac{(h/l + \sin\theta)\sin\theta}{\cos^2\theta}$$

> It is worth noting that the two Poisson ratios can have very different values. For a regular hexagonal network, it is found that $v^*_{12} = v^*_{21} = 1$. If $\theta < 0$ negative values are obtained, which implies the under a compressive (or tensile) load, the network contracts (or expands) as much in the direction of the load as perpendicular to it.

The other six independent parameters of the orthotropic model are: the elastic modulus E^*_3, Poisson ratios v^*_{31} and v^*_{32}, and shear moduli G^*_{12}, G^*_{23} and G^*_{31}.

The value of E^*_3 is obtained directly from proportion of solid material, which will be equal to the ratio of the cross-sectional area of the cell walls to their total area

$$\frac{E^*_3}{E_s} = \frac{h/l + 2}{2(h/l + \sin\theta)\cos\theta} \frac{t}{l}$$

The Poisson ratios v^*_{31} and v^*_{32} are, directly, $v^*_{31} = v^*_{32} = v_s$, where v_s is the Poisson ratio of the solid material, since a compression (or traction) in the 3-direction does not induce bending in the cell walls, and a uniform expansion (or contraction) will result in the 1-2 plane. The complementary Poisson ratios $v^*_{13} = v^*_{23}$ are readily obtained from the symmetry relationships (Exercise 65)

$$\frac{\nu^*_{31}}{E^*_3} = \frac{\nu^*_{13}}{E^*_1} \quad \text{and} \quad \frac{\nu^*_{32}}{E^*_3} = \frac{\nu^*_{23}}{E^*_2}$$

which yields

$$\nu^*_{13} = \frac{2\cos^2\theta}{(h/l + 2)\sin^2\theta}\left(\frac{t}{l}\right)^2 \nu_s \quad \text{and} \quad \nu^*_{23} = \frac{2(h/l + \sin\theta)^2}{(h/l + 2)\cos^2\theta}\left(\frac{t}{l}\right)^2 \nu_s$$

which are practically zero, since t/l is small.

Calculation of the **shear modulus** is somewhat more complicated, and Gibson and Ashby's book may be consulted.

Let us finally observe that when the honeycomb has regular hexagonal cells, symmetry makes the cell material **transversely isotropic.**

Traction of the cells
When the cells are subjected to **tensile** forces lying in the plane of the hexagon the associated stress-strain diagram is shown in Figures 8.29c and 8.29d.

The initial elastic modulus, E^*, can be calculated using the approach outlined above. The second stage, in which the curve increases more rapidly compared to that compression tests, an alignment of the cells is produced. Rupture occurs, almost always, by propagation of a crack originating from a defect in the honeycomb.

If the honeycomb material is brittle, Gibson and Ashby suggest that the fracture toughness, K^*_{IC}, of the honeycomb can be estimated from the following expression

$$\frac{K^*_{IC}}{\sigma_f\sqrt{\pi l}} = 0.3\left(\frac{t}{l}\right)^2 \qquad (100)$$

where σ_f is –as before– the flexural strength (or rupture modulus) of the solid material. The assumptions and limitations of this expression are discussed in Gibson and Ashby 1997).

So far, only uniaxial compressive and tensile loadings have been considered. When honeycombs are subjected to biaxial loading their mechanical response in the elastic regime is different because the deformation mechanisms are different; under uniaxial loading the dominant mechanism is flexure of the cell walls, a mechanism that may be almost completely eliminated by the judicious choice of biaxial loads. In these circumstances the cell walls may end up failing by plastic yielding, creep or rupture, and all of these will result in a greater honeycomb strength.

Response to loading normal to the plane of the honeycomb
In some biological materials cells are found in the middle of sandwich-type structures, with the aim of optimising their mechanical behaviour, for example, in the leaves of some plants, in the skull of birds, or in cuttlefish shells.

In these circumstances, if the loading is normal to the plane of the honeycomb, the mechanical function is to transmit the forces through the cell walls parallel to the axis of the hexagonal prisms. Now the walls are stretched or compressed, but not subjected to bending as before, and for this reason, the elastic modulus is greater and the loads which initiate collapse are also greater.

Gibson and Ashby (1997) have provided expressions for the initial elastic modulus and values of the initial collapse loads when the cells are compressed in the direction normal to the plane of the honeycomb (3-axis). For structures with regular hexagonal cells ($l = h$) the following results are obtained:

– for the elastic modulus E^*_3,

$$\frac{E^*_3}{E_s} = \frac{d^*}{d_s} = 1.15\frac{t}{l} \qquad (101)$$

– for collapse by elastic buckling

$$\frac{\sigma^*_{el,3}}{E_s} = 5.2\left(\frac{t}{l}\right)^3 \qquad (102)$$

in which it has been assumed the Poisson ratio of the wall material is 0.3.

– for collapse by plastic buckling

$$\frac{\sigma^*_{pl,3}}{\sigma_{ys}} = 5.6\left(\frac{t}{l}\right)^{5/3} \qquad (103)$$

where σ_{ys} is the yield stress of the cell wall material.

If buckling is avoided, whether elastic or plastic, the yield stress and the rupture strength of the honeycomb can be calculated by multiplying the relative density d^*/d_S by the corresponding properties (yield stress or failure load) of the cell wall material.

Final remarks

The brief analysis performed above (based on Gibson and Ashley book) indicates the following:

The mechanical properties of honeycombs (if the Poisson ratio is ignored) depend upon three parameters: the ratio t/l, the cell geometry (through h/l and θ) and the properties of the cell walls (E_s, σ_{ys}, σ_f).

The ratio t/l is proportional to the relative density d^*/d_s and the percentage by volume of solid material.

For loading in the **plane of the honeycomb**, where the dominant mechanisms are bending or buckling of the walls, the elastic modulus varies with the cube of the relative density, while the collapse stresses (elastic buckling, yielding or rupture) vary with the square or cube of the relative density.

For loading normal to the **plane of the honeycomb**, where the dominant mechanism is stretching or compressing the cell walls, the elastic modulus and the strength (if buckling is eliminated) are proportional to the relative density.

3. Three-dimensional cellular structures: foams and sponges

Numerous biological materials have a sponge structure. The reason, in many cases, is mechanical; for example, in spongy bone this is to produce a lighter stronger structure. In other cases the aim is to try to optimise fluid transport or thermal insulation, such as in the stems of plants or in sponges.

The geometry of spongy materials is more complicated than that of honeycombs. In some materials the cells are open and, in reality, their structure is a framework of beams and columns (called trabeculae). In others, the cells are closed and often filled with fluids.

As already observed, the analysis of these structures may follow one of several approaches. The simplest is based on dimensional analysis, in a manner analogous to that described when dealing with honeycombs. Since the geometry of the cells is more complicated, what is obtained are proportionality relationships between the required parameters and the relative density of the material. The constant of proportionality is obtained from empirical data, as will be seen below. This process is based upon the assumption that the material cells, with different relative densities, are geometrically similar. Furthermore, the results obtained from an analysis of cubic cells can be generalised to other geometries, provided that the deformation and fracture mechanisms are the same. This conclusion has been validated in several experimental studies.

When testing these spongy materials in compression, or in tension, curves similar to those in Figure 8.29 are obtained, where, in the compression tests, it is possible to identify three stages, as will be discussed below.

Compression of foams and sponges

In **compression** tests, in a first stage, with small strains, the behaviour is linear elastic. The elastic modulus for open cells can be estimated using a cubic model, as shown in Figure 8.33a, in which it is assumed that the principal contribution to the strain is due to beam bending, shown in Figure 8.33b. The model yields the following result (see Exercise 70) for the elastic modulus of the cellular material, E^*,

$$\frac{E^*}{E_s} = A\left(\frac{t}{l}\right)^4 = C_1\left(\frac{d^*}{d_s}\right)^2 \tag{104}$$

in which A and C_1 are constants, and it has been noted that the relative density is proportional to $(t/l)^2$, t being the average diameter of the trabeculae.

The subindices s indicate, as before, reference to the solid material. This analysis provides the elastic modulus of the cellular material in terms of the relative density and the modulus of the solid material. The constant C_1 encompasses the geometric characteristics of the unit cell.

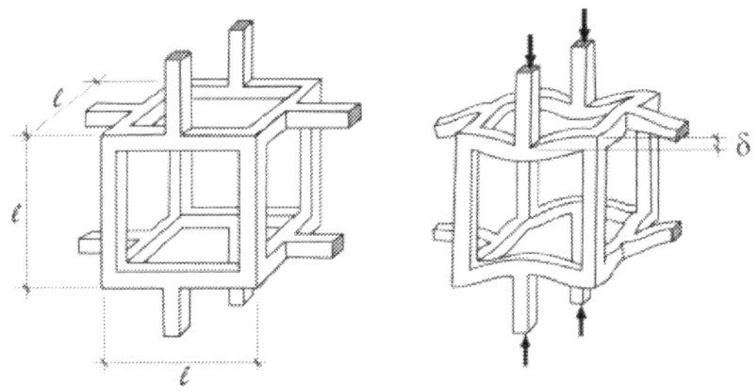

Fig. 8.33. *Deformation of a three-dimensional unit cell.*
a. Initial cell configuration, before deformation.
b. Bending deformation of the spans perpendicular to the loading direction.

Fitting Equation (104) to numerous experimental results, Ashby and Gibson obtained a value of C_1 close to unity. Warren and Kraynik (1997) performed an analysis of a tetradecahedron, rather than a cube, and yielding a value of C_1 = 0.98. Another estimation of C_1 was derived from a three-dimensional numerical analysis with Voronoi unit cells (randomly arranged and of different sizes). The calculation produced a value of C_1 = 0.8 (Vajjhala *et al.*, 2000).

In an analogous treatment, Ashby and Gibson, proposed the following expression for the shear modulus G^* of the cellular material for the special case of isotropic open *cells* with a Poisson ratio equal to 1/3:

$$\frac{G^*}{E_s} = \frac{3}{8}\left(\frac{d^*}{d_s}\right)^2 \tag{105}$$

When closed liquid-filled cells were tested in compression, the fluid opposed the bending of the walls, and this affects the elastic modulus. Gibson and Ashby proposed the following expression for the new elastic modulus, E^*:

$$\frac{E^*}{E_s} = B_1\left(\frac{d^*}{d_s}\right)^2 + B_2\frac{d^*}{d_s} \tag{106}$$

in which the second term accounts for the increase in stiffness due to the fluid. The coefficients B_i are usually obtained experimentally.

EXERCISE 70
Using a Strength of Materials beam theory analysis, prove Equation (104).

SOLUTION:

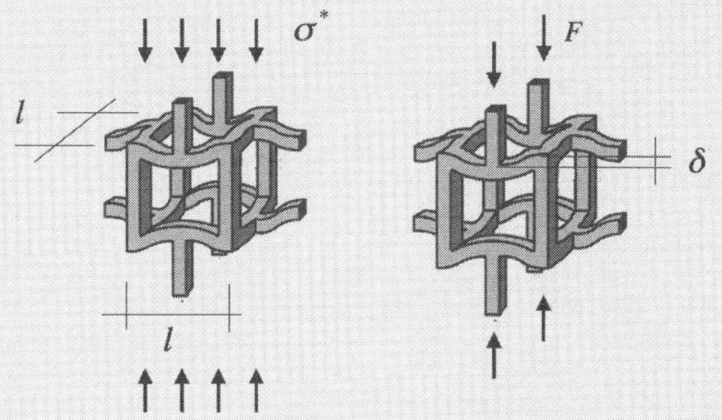

If we consider the cubic cell in Figure 8.33 and model its trabeculae as elastic beams, Strength of Materials shows us that the displacement δ under an applied load F must be proportional to the cube of its length l, according to

$$\delta \sim \frac{Fl^3}{E_s I}$$

where I is the second moment of area of the beam section and E_S the elastic modulus of the solid material from which it is made.

The strains in the cell are proportional to δ/l so we will have

$$\varepsilon \sim \frac{\delta}{l} \sim \frac{\sigma^* l^4}{E_s t^4}$$

since the force acting must be $F \sim \sigma^* l^2$, proportional to the applied stress. We have also taken $I \sim t^4$, t t being the diameter of the trabeculae.

On the other hand, the ratio of the densities will be equal to that of the cell volumes. The volume of solid material will be proportional to $t^2 l$ therefore

$$\frac{d^*}{d_s} \sim \frac{t^2 l}{l^3} \sim \left(\frac{t}{l}\right)^2$$

Finally, the elastic modulus of the cellular material will be proportional to the ratio of the stresses and strains which, with the previous expressions, results in

$$E^* \sim \frac{\sigma^*}{\varepsilon} \sim E_s \left(\frac{t}{l}\right)^4 \sim E_s \left(\frac{d^*}{d_s}\right)^2$$

which is the required expression.

The influence of **defects** on the elastic modulus of cellular materials with open cells has been investigated by several authors, particularly the effect of removing trabeculae in the cells. Gibson and Ashby summarised results in Figure 8.34, in which the elastic modulus of the damaged material, normalised with respect to the intact modulus (E^*/E^*_0) is plotted as a function of the percentage reduction in the relative density. These results were determined numerically; for three-dimensional cellular materials ordered tetradecahedra and random cells (Voronoi polygons) were used, for honeycombs these were ordered hexagons and random polygons (Voronoi polygons). The reduction in density was produced by eliminating bars in the models. The results indicate that the influence of the loss in density is less acute in three-dimensional structures than in two-dimensional ones. These findings may be important in the process of osteoporosis of spongy bone where a gradual loss of bone density occurs.

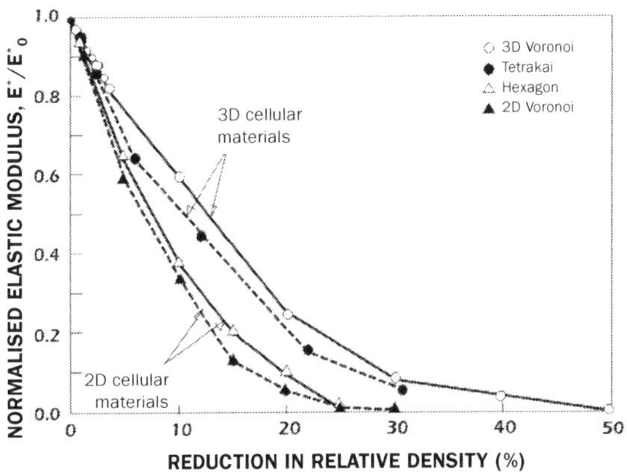

Fig. 8.34. *Influence on the elastic modulus of the (random) reduction of trabeculae for two- and three-dimensional cellular materials.*

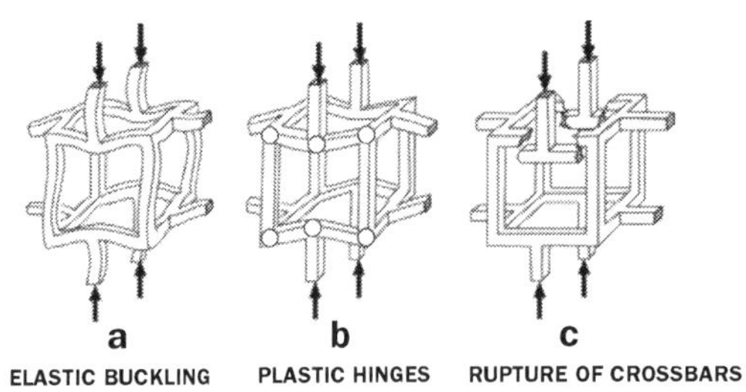

Fig. 8.35. *Collapse modes in open cells.*
a. By elastic buckling.
b. By the formation of plastic hinges in the corners of the trabeculae.
c. By rupture of crossbars.

The final linear-elastic stage occurs when, for various reasons, the stress reaches a critical value in one of the cell components; by elastic buckling of the elements which work in compression, through the formation of plastic hinges at the corners, or by the rupture of a crossbar.

These collapse mechanisms in open cells are shown, in simplified forms, in Figure 8.35 for a cubic cell.

Gibson and Ashby (1997) proposed the following expressions for **open cells:**

- for collapse due to elastic buckling:

$$\frac{\sigma^*_{el}}{E_s} = C_2\left(\frac{d^*}{d_s}\right)^2 \tag{107}$$

  the value of C_2 obtained empirically is $C_2 \approx 0.05$. An analysis with tetradecahedrical cells gives values between 0.10 and 0.18.

- for plastic collapse resulting from plastic hinges at the corners

$$\frac{\sigma^*_{pl}}{\sigma_{ys}} = C_3\left(\frac{d^*}{d_s}\right)^{3/2} \tag{108}$$

  where σ_{ys} is the yield stress of the solid material and $C_3 \approx 0.3$, also obtained empirically.

- for collapse by rupture

$$\frac{\sigma^*_r}{\sigma_f} = C_4\left(\frac{d^*}{d_s}\right)^{3/2} \tag{109}$$

  where σ_f, rupture modulus of the solid material. The value of C_4, also obtained empirically, is approximately 0.2.

When the cells are **closed**, the walls also contribute to the mechanical response of the cellular material. If the cells are formed from liquids –as is often the case– surface tension pulls the material to the edges and leaves a very thin membrane on the faces. In these cases, even though the cells are closed, the membranes contribute little to the stiffness of the material when loaded, and usually break soon. The values found in open cells may be a good approximation in these cases.

If the cell walls are thick, as in the leaves of many plants, the values presented above are not valid, and the dependency on relative density is not so simple. The interested reader can find expressions for the stiffness of these cellular materials in Gibson and Ashby (1997).

As the compressive forces increase a point is reached when the cells collapse and the stress-strain curve increases rapidly, as indicated in Figure 8.29, its slope, as it reaches the maximum strain, ε_D, approaches the value the elastic modulus of the cell material, E_s. The value of ε_D would be expected to be that of the porosity, since, a that instant, it has been eliminated, that is to say $\varepsilon_D = 1-d^*/d_s$. In reality, as already noted in the discussion of honeycombs, the cells stack up on top of one and other and this ideal value is not achieved. Gibson and Ashby proposed the following expression for the maximum strain

$$\varepsilon_D = 1 - 1.4\frac{d^*}{d_s} \tag{110}$$

Traction of foams and sponges

When the loading is **tensile** the material breaks due to the initiation and unstable propagation of a crack. For brittle cellular materials the principles of Linear-Elastic Fracture Mechanics can be applied. In these cases, for open cells, the fracture toughness of the cellular material K^*_{IC}, can be expressed in the following form:

$$\frac{K^*_{IC}}{\sigma_f\sqrt{\pi l}} = C_5\left(\frac{d^*}{d_s}\right)^{3/2} \tag{111}$$

where σ_f is the rupture modulus of the cell material, l the average length of a trabecula and C_5, is an experimentally-determined constant, approximately equal to 0.65.

With the availability of powerful computers it is becoming easier to perform realistic studies of cellular materials; it is no longer necessary to start with equal periodic structures, it is possible to introduce defects, more elaborate constitutive equations, and to simulate complex loadings, more in keeping with reality.

This short Section ends with two observations upon the influence of variable loads and the presence of fluids in the cells.

Fatigue of cellular materials, under repeated compressive loading, is of interest in the study of spongy bones, since it is thought that half of the fractures in patients with osteoporosis are caused as a result of this loading. Fatigue test on materials usually consists of obtaining classical S-N curves, in which the stress amplitude is plotted as a against the number of fatigue cycles to failure. In general these curves have a great deal of scatter. In cellular materials it is observed that when fatigued in compression the strain decreases slowly initially but to increase more rapidly later. The number of cycles in which this sudden transition occurs is usually taken to be the corresponding fatigue value for the load being used.

We must note that, generally, a cellular material always contains **fluid** in its interior, so its effect on the macroscopic mechanical behaviour will depend upon whether it is gaseous or liquid, and also if the cells are open or closed.

If the cellular material is **closed**, the fluid (liquid or gas) always increases the rigidity and the value of the yield stress, the more, the lower its compressibility. Furthermore, in the case of gases, there will be a significant dependence upon temperature, as their compressibility is very temperature-dependent.

When it comes to **open** cellular materials, a very important stiffening effect may occur, and both the elastic modulus and the yield stress will usually depend upon the fluid viscosity, the size of the openings between cells and the strain rate.

In the case of *gaseous fluids* the effect is usually negligible, since due to their low viscosity the gases can circulate without difficulty through the interior of the material when deformed, offering little or no resistance. For the opposite to happen, the size of the cell (or better, the size of the openings between cells) must be very small, below ten microns, and –at the same time– the strain rate very high, above $10s^{-1}$.

In the case of a *liquids*, the situation changes because the viscosity is much higher. This is often the case in biological materials, in which the fluid is similar to water in its mechanical properties. However, when the openings are large, millimetre-sized, such as, for example, in trabecular bone, the effect is negligible since the liquid, moves without difficulty even at impact velocities ($>1s^{-1}$). Nonetheless, with small pores, such as are found in the interior of cortical bone (~40μm), there is a strong dependency of the mechanical properties upon the strain rate, as will be shown in Section 9.3.2.

EXERCISE 71

Determine how the value of the yield strain ε^*_{ys} varies with the apparent density d* in open isotropic cellular materials such as those represented in Figure 8.33, when they are loaded in a compression test.

Apply the results to the analysis of the compression tests on the human subchondral trabecular bone of different apparent densities, performed by Hayes and Carter (1976) and shown in the Figure.

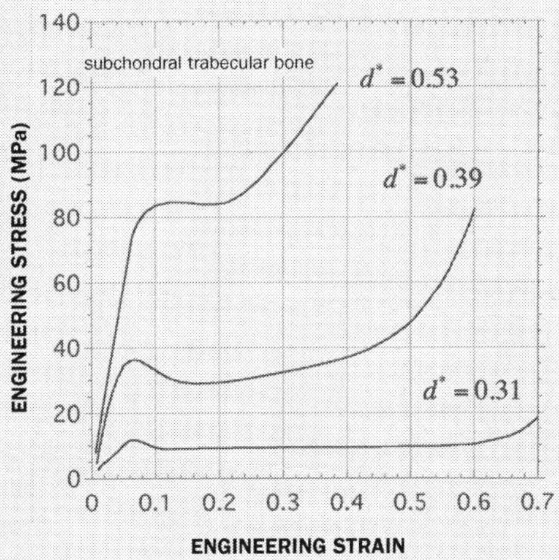

SOLUTION:

Equations (107) and (108) provide the stresses corresponding to the initial collapse of cells, corresponding to the two damage mechanisms operating in compression tests. Their value marks the end of the linear-elastic behaviour of the material and the beginning of the densification phase. To locate the boundary between the two behaviours, linear and nonlinear, we will estimate the values of the strains at this point by dividing the stresses by the apparent elastic modulus E^*.

Using the cited equations and Equation (104) which establishes the relationship between E^* and the apparent density we obtain:

– for the elastic buckling collapse mechanism $\varepsilon_{ys} = \dfrac{\sigma^*_{el}}{E^*} = \dfrac{C_2 E_S \left(d^*/d_s\right)^2}{C_1 E_S \left(d^*/d_s\right)^2} = constant$

– for the plastic hinges formation collapse mechanism $\varepsilon_{ys} = \dfrac{\sigma^*_{pl}}{E^*} = \dfrac{C_3 E_S \left(d^*/d_s\right)^{3/2}}{C_1 E_S \left(d^*/d_s\right)^2} = \dfrac{constant}{\sqrt{d^*}}$

We observe that in the first case, the dependency of the collapse stress and the elastic modulus upon $(d^*)^2$ cancel, giving rise to a value of the strain ε^*_{ys} which is independent of the apparent density. This phenomenon has been observed by several authors and will be discussed later in more detail. In the second mechanism, a hyperbolic dependency is obtained and the value of ε^*_{ys} decreases with the apparent density. These relationships are valid as long as the values for the solid material are identical, which, in practice, and in relation to biological materials, requires comparison of materials from the same source, which can be difficult to achieve.

Hayes and Carter's curves confirm that the compressive yield stress has little dependence upon the apparent density, and we can see (Figure a) that yield is initiated, in the three cases tested, at a value of ε^*_{ys} close to 0.065.

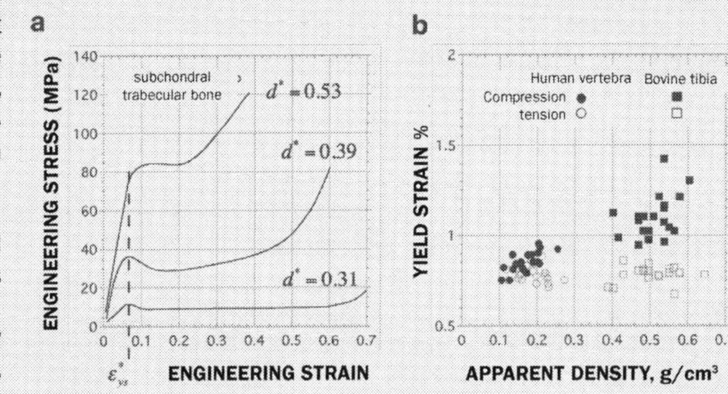

The lack of correlation between the yield strain and the apparent density has also between observed in tensile tests. The results of Kopperdahl and Keaveny (1998) are summarised in Figure (b), indicating a lack of correlation between ε^*_{ys} and d^* in tension and, if anything, a very weak correlation in compression. Note that the yield strain in tension is always less than that in compression, probably because the damage and rupture mechanisms occur much sooner.

9 CHAPTER Examples of Biological Materials

The aim of this chapter is to illustrate the application of some concepts, developed in the previous chapter, related to the mechanical properties of several biological materials of interest: mollusc shells, wood, bone and cartilage.

The format of each example is the same: The structure of the material is described –at the macro-, meso- or microscopic level– as appropriate, and taking into account the, more or less, hierarchical nature of these materials. The next step consists of reviewing their *mechanical properties* and attempting to model their behaviour –with the tools discussed in Chapter 8– with the aim of understanding their properties and to make predictions. The examples investigated relate to, more or less, rigid materials, which hardly deform, and in which it is possible to use the linear-elastic approximation, which simplifies the modelling. Finally, some comments are made on possible applications: as **biological material**, as **biomaterial** (to correct or replace tissue, organs or functions in the human body) and in the field of **biomimetics**.

In the first example, **mollusc shells** are examined. These are composite materials, in which a weak organic matrix (between 1 and 5%) is reinforced with calcium carbonate particles.

The structure of this composite is complex and organised in such a way that its strength and toughness are far greater than the ceramic material from which it is formed. In this example, only two simple structures are analysed: the parallel lamellar and crossed lamellar.

Modelling is usually based on a representative volume formed of ceramic tablets embedded in a cohesive matrix. With much simpler models, based on the rule of mixtures, it is possible to estimate maximum values of the initial elastic modulus and the tensile strength.

As a *biological material,* shells have little or no practical use but great value as decorative items or jewellery. As a *biomaterial*, they have been used since ancient times for dental implants and currently, on occasion, to regenerate bone. The impact strength and large toughness of shells serve as an inspiration, in the field of *biomimetics*, for the design of armours.

The second example is dedicated to **wood**. This is also a composite material, at the microstructural level, of cellulose fibres in a hemicellulose or lignin matrix.

The structure of wood, at the macroscopic level, and in certain circumstances, can be considered as if it were a continuous homogeneous orthotropic material. At the mesoscopic level, it may be thought of as an aggregate of two types of cell walls. The study of cell walls belongs in the microscopic domain. However, it is possible to go even deeper by examining the structure at the nanoscopic level, if the cellulose micofibrils are considered.

The mechanical properties depend significantly upon temperature and humidity –the behaviour is viscoleastic– but, even so, it is possible to make reliable predictions of the mechanical behaviour.

The material microstructure can be modelled as a composition of cellulose fibres embedded in a hemicellulose or lignin matrix and the mesostructure modelled as a cellular honeycomb material (see Section 8.3.4), with properties calculated at the microstructural level. With this information it is possible to obtain satisfactory results for the elastic modulus, rupture load and the fracture toughness, at the macroscopic level.

As a *biological material*, wood has always played an important role in construction since its specific mechanical properties are very competitive. As a *biomaterial*, it has long been of interest –there are examples of wooden prostheses from about 2,600 years ago– and tibia prostheses survive to this day. In the field of *biomimetics* so-called biomorph ceramics are being investigated, in which the cellular structure of wood is replicated using ceramic materials.

The third example concerns **bones**, once again, a composite material. The matrix of the bone tissue is formed of collagen fibres and the reinforcement is calcium phosphate crystals. Bone does not fulfil the same functions throughout its life, and it is remodelled according to the loading it receives; it is a living material.

The structure is complex, although it may be idealised at distinct levels. At the microscopic level, the mineralised collagen fibres play the leading role. At the mesoscopic level these fibres are arranged in different ways giving rise to different types of bone. At the macroscopic level two types of structure can be identified, spongy (cancellous) bone and compact bone.

The mechanical properties depend, to a large extent, upon the relative density (quotient of the density of the bone and that of the walls). Various strategies have been applied to modelling; the analysis of a representative volume, using the techniques developed for cellular materials or modelling the bone numerically using finite elements.

Bone as a *biological material* has no relevance. As a *biomaterial*, it has. It is used for transplants and serves as a model for the fabrication of artificial prostheses, since it is essential to simulate, as closely as possible, the properties of the bone that it is to replace. The distribution of bone tissue in some bones –when the spongy bone is surrounded by compact bone– has provided inspiration in the area of *biomimetics*, since this bone architecture optimises the strength with a minimum weight possible.

Cartilage is investigated in the fourth example, in particular articular or hyaline cartilage. This is also a composite material, although between 70 and 80% of its weight is water. The cells that create and maintain it –called chondrocytes– are surrounded by a matrix formed of collagen and proteoglycans (hyaluronic acid and aggrecans).

The structure is hierarchical, but it may be considered, somewhat arbitrarily, as having two levels: At the microscopic level cartilage can be regarded as the composite material just mentioned, the collagen of the extracellular matrix forms a fine strong mesh which traps the proteoglycan molecules and carries the stresses due to the internal pressure. At the mesoscopic level an anisotropic heterogeneous structure is apparent, since the collagen fibres are distributed in different positions according to their distance from the contact surface.

As far as their mechanical properties are concerned, there is a strong time-dependency, and the pressure and internal fluid movement are also very important. For this reason emphasis is placed upon the role played by the osmotic pressure due to the difference in ionic concentration between the cartilage and the external fluid; cartilage covers the articulating ends of the bones and allows pressure transmission at the lowest possible values –the osmotic pressure reduces the external pressure– and with minimum friction. Two viscoelastic models for the study of the behaviour of cartilage have been proposed: the linear viscoelastic and poroelastic models.

Cartilage as a *biological material* has no relevance. As a *biomaterial*, it has been of interest in eastern pharmacopoeia, as a remedy for a multitude of diseases. In the field of *biomimetics* it may be very interesting to reproduce and improve the mechanisms responsible for ultra-low friction.

9.1. MOLLUSC SHELLS

9.1.1. Introduction

Mollusc shells have piqued the interest of mankind since the earliest times; pierced shells –which were possibly parts of necklaces– have been found in Neolithic sites. It is known that shells –cowry shells– were also used as bargaining "money", and pearls –a byproduct of shells– have always been coveted gems. The iridescent light reflected by pearls and the nacre of some mollusc shells form part of the attraction of these biological materials. However, it is the extraordinary mechanical properties of these materials which make them worthy of the special attention granted to them in this book.

Shells are organic matrix *composite materials*; between 95 and 99% is calcium carbonate (in the form of aragonite or calcite) and the rest, between 1 and 5%, is a protein matrix. Surprisingly, it is this tiny proportion of organic cement and a complex microstructure which confer on nacre, or mother-of-pearl, a resistance to fracture of a magnitude two or three times greater than that of aragonite or calcite.

The excellent mechanical properties of shells invite a study of their composition and microstructure in order to understand their strength and toughness. Moreover, mollusc shells can suggest means whereby brittle materials, such as aragonite, might be transformed into tough materials, if we know how to intelligently introduce the right matrices and to give them structure.

The word *mollusc* was introduced in the seventeenth century from the Latin *molluscus*, which means soft, and borrowed because of its similarity to the name of a thin-shelled nut called *mollusca*. However, its inventor also applied the name to barnacles, which are crustaceans, suggesting that it has never been easy to decide what is and what is not a mollusc, since they are extraordinarily varied. At the most humble extreme there are molluscs similar to worms and at the other we find fascinating creatures like octopuses. Currently, the mollusc phylum is divided into eight living classes, of which the three most important are *bivalves, gastropods* and *cephalopods*.

In *bivalves* the shell is composed of two valves, which are joined dorsally by an elastic ligament and articulated through a hinge. The valves are closed by means of powerful adductor muscles.

The most prominent characteristic of *gastropods* is torsion; at some point in their development, the part of the body containing internal organs rotates relative to the foot. (Some groups have reversed what appeared to be a strong evolutionary trend and undone the process) The shell of these molluscs also grows in a spiral, but through a different process, to gain more space for the visceral mass in a basically tubular shell. (Some groups have also reduced the shell or even lost it altogether).

In present day *cephalopods* the shell is wound in the form of a spiral –like *nautiloids*; however, in the majority of cephalopods the shell is reduced or absent.

The architecture of the humble mollusc shell or the precious nacre of pearls has inspired new tough high-performance ceramic materials, but the information contained in mother-of-pearl does not end here. Mollusc shells have to perform as authentic protective barriers for the invertebrates; they must resist attacks from predators, withstand enormous pressures in the deep hydrothermal vents in which molluscs proliferate, or have the toughness required to stand the onslaught of waves in intertidal zones. Furthermore, they must combine their resistance with other functions; to facilitate the access of nutrients, eliminate waste, and allow growth and reproduction. To any engineer, the design of a material with all of these characteristics would create huge problems. Molluscs achieve these with common materials and water, at ambient temperature and under normal pressures (if we exclude those that live in the depths of the ocean, in hydrothermal vents), using the fundamental techniques of biomolecular engineering.

EXERCISE 72

In many snail shells and other gastropods it has been observed that the growth lines can be approximated to logarithmic spirals of the type $r = A\ exp\ (B\theta)$, where A and B are constants and the variables r and θ are defined in the figure (Thompson, 1961).

Justify this assertion by assuming that, during the growth of the shell, in a time interval dt, the ratio of the supply of material in the radial direction and perpendicular directions remains constant.

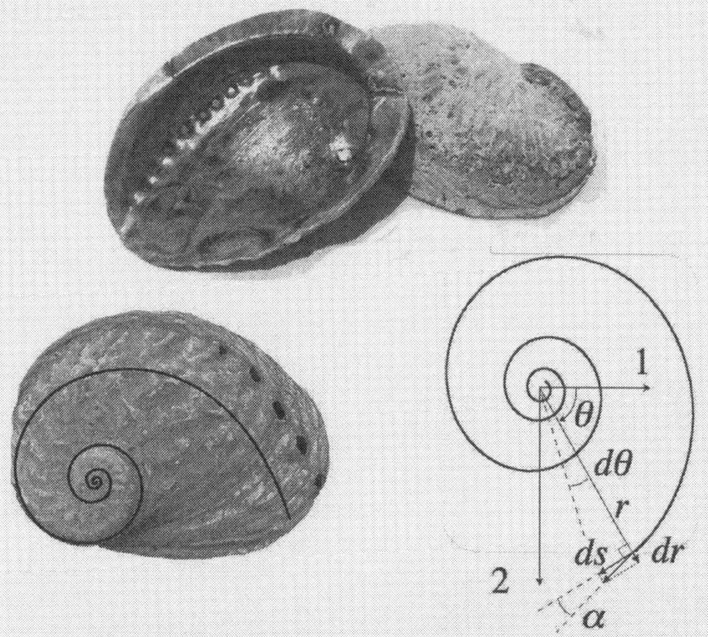

SOLUTION:

From the assumption, $dr/ds = constant = \tan\alpha$, where α is the angle between ds and the spiral.

Since $ds = rd\theta$, then:

$$\frac{dr}{r} = \tan\alpha\ d\theta$$

from which, by integration, we obtain

$$\ln r = \theta\tan\alpha + constant$$

or

$$r = constant\ e^{(\tan\alpha)\theta}$$

which is the expression given in the problem statement

A reader interested in the shapes of shells is encouraged to read, in addition to Thompson's classic book, the monograph by Meinhardt (2003).

9.1.2. Structure

Mollusc shells are formed of densely packed microscopic ceramic bricks, bonded with a fine biopolymer cement. The hierarchical structure of these biological materials is the result of millions of years of evolution, and these shells are so well arranged that their strength and toughness are far superior to those of the ceramic from which it is made.

Curry (1974) has described the various types of microstructure from which the structure of many bivalve, gastropod and cephalopod shells may be fashioned. The casuistry is large and complex, and for that reason this chapter will focus only on the most representative structures which have been extensively studied: *parallel-lamellar structures*, typical pearly shells, and *crossed-lamellar structures*, very complex but common among molluscs with shells.

Parallel-lamellar structure

The two shells with this structure which have attracted the most attention are (Figure 9.1): mother-of-pearl (*Pinctada maxima*), of the bivalve class, and abalone (*Haliotis rufescens*), of the gastropod class.

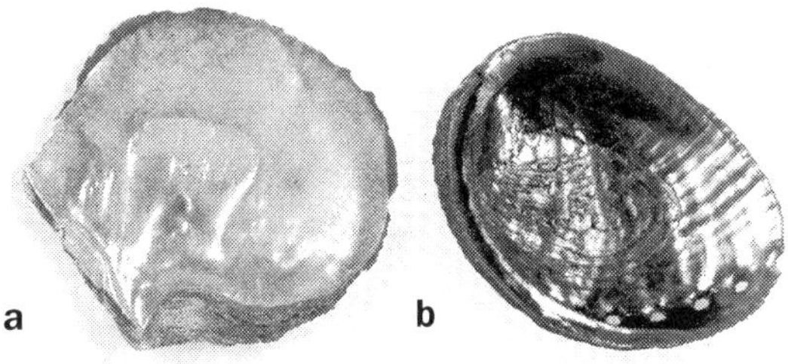

Fig. 9.1. *a. Mother-of-pearl* (Pinctada maxima).
b. Abalone (Haliotis rufescens).

The hierarchical structure of these shells, shown in a simplified form in Figure 9.2, spans six orders of magnitude; from millimetres to nanometres.

Figure 9.2a is a macroscopic section (note the *millimetre* scale) of a shell in which the following layers can be seen: A superficial layer, or periostrack. An intermediate layer formed of hard and brittle prismatic calcite grains. And an internal layer –mother-of-pearl or nacre– formed of a lamellar aragonite structure, not as hard but tougher than the calcite layer.

Figure 9.2b is a close-up, at the mesoscopic level (note the *micron* scale), of the previous figure. Growth bands (possibly annual) of about 200 μm width, separated by denser zones of 20 μm, can be seen in the figure.

A further enlargement, at the microscopic level (note the *nanometric* scale), of the previous figure is presented in Figure 9.2c, in which 200 nm thick aragonite ceramic laminas can be seen (in other samples the thickness may be greater, up to 900 nm), and the average diameter oscillates between 5 and 8 μm separated by 20 or 30 nm thick cement.

This structure is similar in the two cited shells. The principal difference lies in the stacking of the aragonite crystals: In the Haliotis shell the structure is laminar-columnar, as shown in the sketch in Figure 9.2c, in which the plates are almost superimposed (the common area is about 2/3 of the total area of the plate; Barthelat *et al.* 2007). In the *Pinctada* shells, the structure is laminar-irregular, in which the plates are stacked randomly, as has been attempted to indicate in Figure 9.2c.

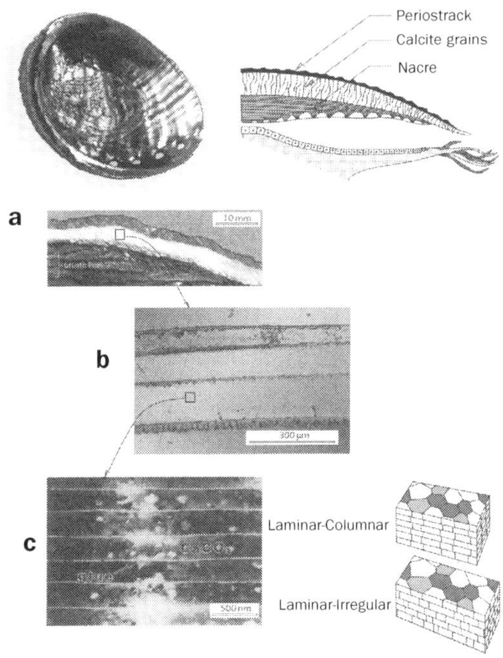

Fig. 9.2. *Hierarchical structure of the shell of the* Haliotis rufescens.
a. Macroscopic level (mm). Three layers.
b. Mesoscopic level (μm). Growth bands.
c. Microscopic level (nm). Aragonite cemented with an organic polymer.

Crossed-lamellar structure
Shells with a crossed-lamellar structure and which have been studied in most detail belong to the class of gastropods, specifically from the Strombus and Conus families, both represented in Figure 9.3.

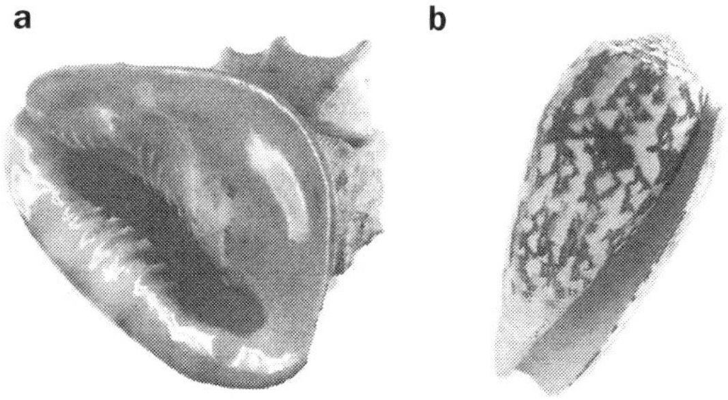

Fig. 9.3. *a. Conch* (Strombus gigas).
b. Cone (Conus striatus).

The shell which has attracted most attention, due to its size and beauty, is the Caribbean conch, *Strombus gigas*. Heuer and Li, among others, (Su *et al.* 2004, Ji *et al.* 2020), have performed a detailed study of its structure, whose most prominent features are described below.

The hierarchical structure of this shell is presented, in a simplified form, in Figure 9.4, spanning six orders of magnitude, from millimetres to nanometres.

The external layer is the *periostrack*, an organic membrane which protects the shell and provides a base upon which calcium carbonate is formed in a thin columnar sheet, some 2 μm thick, on which the crossed-lamellar structure will be built.

At the macroscopic level it is possible to distinguish three *layers*, indicated schematically in Figures 9.4a and 9.4b, known as internal, middle and external, and whose width can vary between 0.5 and 2 mm.

Within each layer, at the meso and microscopic levels, it is possible to observe three types of *lamellae*: those of the first order are shown in Figure 9.4b and are arranged in parallel in each sheet and perpendicularly between adjacent layers. Inside each first order lamella a second family can be found, whose transverse dimensions range from 5 to 60 μm width and 5 to 30 μm thickness. Between the adjacent first-order lamellae, the directions of the second-order lamellae are perpendicular to each other (see Figure 9.4c). Next, second-order lamellae are composed of another smaller family, those of the third order, which are tiny aragonite crystals, whose cross-sections oscillate between 140 and 300 nm wide by 130 nm thick (Figure 3.1.4d) and their length can reach several microns. Finally, small twins have been observed in the crystals.

The microstructure of the *Conus* family is similar to that described above; the shell wall is formed of three layers and in each one of them the distribution of the lamellae is also crossed-lamellar (Currey and Kohn, 1976).

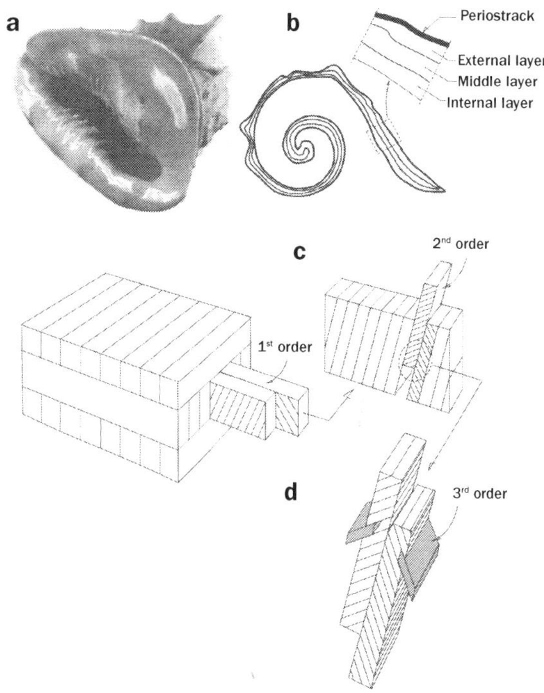

Fig. 9.4. *Hierarchical structure of* Strombus gigas.
a and b. Macroscopic level (mm). Three layers.
c. Second-order lamellae (μm).
d.Third-order lamellae (nm).

Bonding cement

The architecture of shells is a complex lattice of ceramic bricks bonded with a small amount of protein “mortar”, ranging from 0.1 to 5% of the total.

Studies performed on shell of the Haliotis family (parallel-lamellar structure) indicate that the organic mortar is not a homogeneous material, but one which possesses a sandwich structure (Sarikaya *et al.* 1994, Meyers *et al.* 2008). The stiffer, middle part, is a polymer sheet rich in chitin. The lamina is covered in proteins, whose secondary structure are β leaves, and these are covered, in turn, with other proteins which are in contact with the ceramic bricks (the proteins contain aspartate which preferentially binds the calcium atoms of the aragonite).

In the case of shells of the *Strombus* family (crossed-lamellar structure) no such detailed study of the protein mortar that bonds the ceramic bricks has been made. The analysis performed by Osuna-Mascaró (Osuna *et al.* 2014) shows that the amount of protein material is less than that found in pearly structures (0.05% against 1% by dry weight). Another difference encountered is the near absence of chitin (at least in the part soluble in acids). The novelty was the discovery of mannose, the principal monosaccharide which occurs in the acid-soluble part.

The crossed-lamellar structure of shells is often considered to be the most successful from an evolutionary point of view and is characteristic of the biggest and heaviest shells. Nonetheless, as will be seen later, the resistance to fracture in bending tests is lower than that in shell with a parallel-lamellar structure, but its toughness is greater. One possible reason for this fact may be based upon the cost of shell calcification which is due, mainly, to the synthesis of the protein matrix rather than to the deposition of calcium carbonate. That is to say, molluscs which synthesise greater quantities of matrix use up more energy in producing and repairing their shells. From this perspective, it may be deduced that the inferior mechanical properties of crossed-lamellar structures when compared to shells with parallel-lamellar structures is offset by the energetic savings in the fabrication and repair of their shells.

Another issue associated with the protein matrix is related to biomimetics. If materials which imitate structures offered by nature are to be fabricated, it is essential to understand how the protein matrix manages to control the nucleation of inorganic crystals, their shape, size, and growth. Only in this way, when we have a better understanding of the role of the protein matrix, will it be possible to design materials which can effectively self-assemble the building blocks of materials of the future.

9.1.3. Mechanical properties

The high strength and toughness of some shells is well known and, surprisingly, far greater than aragonite, the material from which they are made, even more so, considering that the composition of these shells is 95% aragonite and, in some cases, even 99%. Understanding this phenomenon is an interesting challenge for the scientist and may provide a clue to manufacturing new high performance materials. For this, the first step is precision measurement of their mechanical properties and then modelling the material, not only to reproduce the experimental results but also to design materials with improved properties.

Experiments on shells are fraught with difficulty. The first is the fabrication of suitable specimens because the thicknesses of the shells are usually small (several millimetres) and the material is neither homogeneous nor isotropic. This results in problems in machining a manageable specimen which is representative, so a degree of dispersion in the results is only to be expected. Other issues arise from the test conditions; the values obtained depend upon the relative humidity and the rate of loading. The scatter in the results and a lack of information, in some cases, make it difficult to compare the values reported by different investigators.

Modelling is also complicated because, as has already been mentioned, these biological materials have a very hierarchical structure spanning six orders of magnitude. For this reason, a piece of shell is more like a structure than a homogeneous material. The models developed in Section 8.3.3 for heterogeneous composite materials are of little use, as will be seen later. The most promising approach is to consider the material as a structure, and to estimate the mechanical performance from the mechanical data of its components. This procedure is feasible thanks to the calculating power of current computers. This approach was unthinkable in the 60s and 70s when interests in the mechanical properties of shells began to awaken.

To facilitate presentation, the order of presentation in the previous section will be maintained: first the properties of *mother-of-pearl,* with the *parallel-lamellar* structure which is the simplest, and then the properties of conches with a *crossed-lamellar* structure will be addressed.

The mechanical properties most representative of shells with a *parallel-lamellar* structure are summarised in Table 9.1, measured on abalone (Haliotis) and mother-of-pearl (Pinctada), both in the natural state (wet) and dry. For reference, the corresponding average values of aragonite from several authors have also been added. (It should be remembered that it is the major component – between 95 and 99%– of the shell)

It can be seen that the stiffness (measured from the initial elastic modulus) is similar to that of aragonite. The flexural strength of the shells is an order of magnitude greater than that of aragonite, and the specific fracture strength (measured by the area under the force-displacement curve) is, at least, two orders of magnitude superior to that of aragonite. These results are worth analysing in some detail because of their possible implications in the design of materials.

TABLE 9.1 MECHANICAL PROPERTIES OF THE PARALLEL-LAMINAR STRUCTURE

SHELL	INITIAL MODULUS	FLEXURAL STRENGTH E (GPa)	COMPRESSIVE STRENGTH σ_{FT} (MPa)	FRACTURE TOUGHNESS σ_C (MPa)	FRACTURE TOUGHNESS K_C (MPa $m^{1/2}$)	REFERENCE
Haliotis		(180 ± 20)*		(7 ± 3)*	(700)*	Sarikaya *et al.* (1995)
Haliotis		177 – 197	235 – 540 (548 – 735)**			Menig *et al.* (2000)
Haliotis	(70)*	(194 – 223)*	(340)*			Wang *et al.* (2001)
Haliotis	90 (70)*		250 (120)*			Barthelat *et al.* (2007)
Pinctada					200 – 280 (210 – 300)*	Currey and Taylor (1979)
Pinctada	70 – 73 (60– 64)*			3 – 6	350 – 900 (500 – 1800)*	Jackson *et al.* (1988)
Pinctada	86	200 – 300		3 – 5	260 – 500 (480 – 600)*	Pastor and Elices (2001)
Pinctada	(77 – 81)*	(227 – 248)*				Wang *et al.* (2001)
Aragonite	50 – 100	20 – 30		< 1	(0.6 – 10)	***

* Tests with hydrated specimens.
** Dynamic tests.
*** Average values from various authors.

The types of specimens most commonly used to measure the quoted properties are shown in Figure 9.5. The tensile strength is usually measured directly on specimens with a reduced central section ("dog bone"-type specimen) in order to avoid breaking at the grips. The shear strength is measured using special fixtures. Both types of specimen are shown in Figure 9.5a.

Measurement of the flexural strength involves specimens in the form a beam subjected, almost always, to three-point loading. The *parallel-lamellar* structure of shells suggested that the orientation of the laminas must be taken into consideration, as indicated in Figure 9.5b.

Measurement of the fracture toughness is usually made on specimens in the form of a beam with a central crack, attempting to imitate the type of specimen recommended in the Standards (ASTM, for example). It is also useful, in these cases, to consider the orientation of the laminas, as shown in Figure 9.5c.

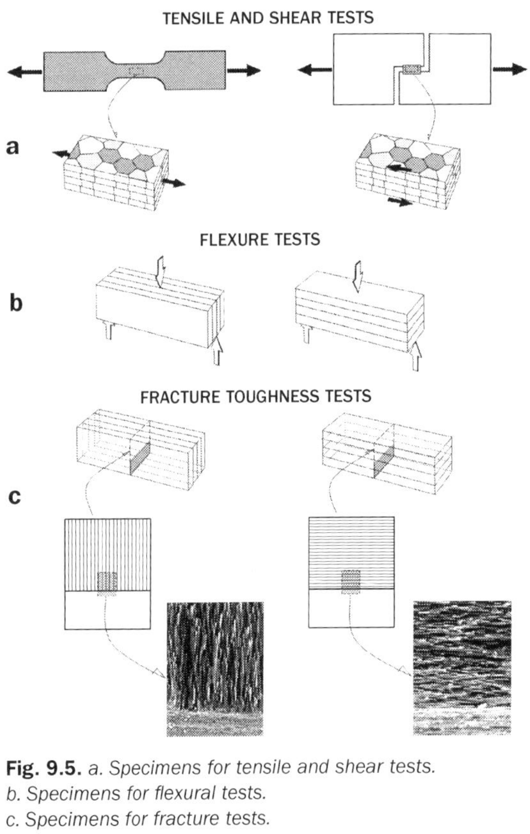

Fig. 9.5. *a. Specimens for tensile and shear tests.*
b. Specimens for flexural tests.
c. Specimens for fracture tests.

Several engineering stress-strain and load-deflection curves, representative of the tests mentioned above, are presented in Figure 9.6. Comparison between curves obtained by different authors is not straightforward for reasons already mentioned; the dispersion of the results due to specimen machining difficulties and different test conditions.

The **tensile** tests on Haliotis shells presented in Figure 9.6a (Barthelat *et al.* 2007) indicate a clear difference in the behaviour of dry and hydrated specimens. The tests were performed in the direction of the laminas (see Figure 9.5a) and represented two dry and two hydrated specimens. In the latter cases it can be seen that after unloading there is a permanent deformation. Also, corresponding curves of pairs of **shear strength** tests are shown in Figure 9.6a. Once again, the difference between dry and hydrated specimens is notable. Also, a significant residual deformation is apparent after unloading.

The effect of anisotropy in the **flexure** tests on Haliotis and Pinctada specimens (data from Wang *et al.* 2001) is apparent in Figure 9.6b. In both materials the behaviour is similar, both in the parallel and perpendicular directions. Tests were carried out with hydrated specimens.

A curve obtained from a **fracture** test on a wet Haliotis specimen (Barthelat *et al.* 2007) is shown in Figure 9.6c, in which it may be observed that there is some degree of linear-elastic behaviour before the initiation of failure. However, the increasing amounts of energy consumed during crack propagation– which has to proceed in a zigzag manner between the ceramic lamellae– make the behaviour of the material deviate quickly from the linear-elastic fracture model (in which the energy consumed is constant during crack propagation). The results of numerous experiments on (dry and hydrated) *Pinctada* specimens performed by Jacksonet *et al.* (1988) are also included in the figure. The figure is illustrative because it supports the hypothesis mentioned above: experimental values of the specific fracture energy G_C (J/m^2) are shown in the vertical axis and on the horizontal axis the values of G_C deduced from the expression $K_C^2 = E\,G_C$, which is valid only for the model of linear-elastic fracture behaviour. We see that, for the most part, the experimental data are far greater than those from the linear-elastic model. It must be borne in mind at both the values of G_C as well as K_C, for hydrated specimens have been corrected, as if they were specimens with some degree of plasticity (see the quoted reference for more details).

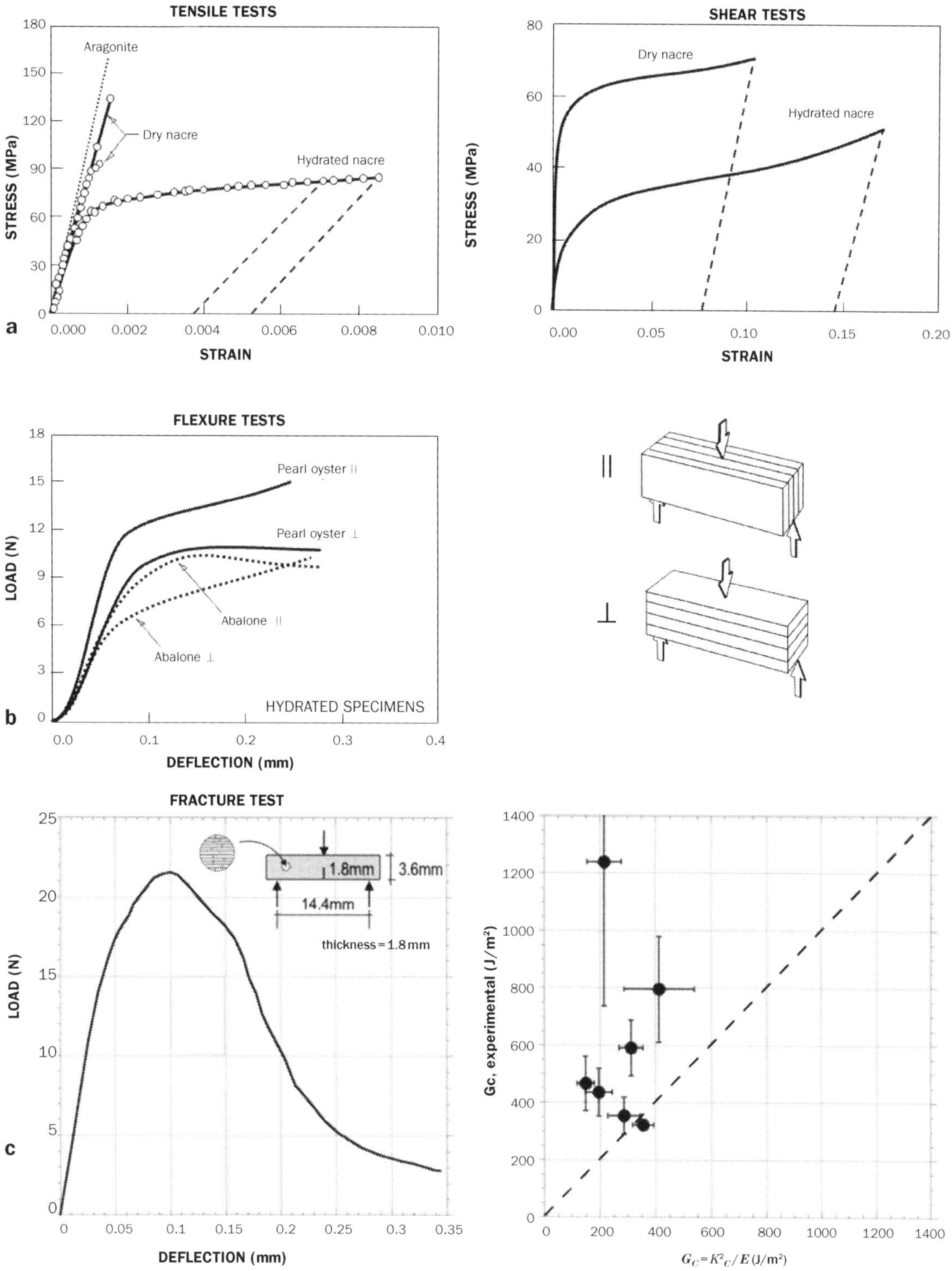

Fig. 9.6. *a. Stress-strain curves obtained in tensile and shear tests; note the influence of the hydration of the specimen.*
b. Load-deflection curves, for nacre, obtained in flexure tests; note the influence of anisotropy.
c. Load-deflection curves obtained in a stable fracture test. Correlation between the experimental and theoretical values of G_C.

In the quoted article of Jackson *et al.* (1988), an attempt was made to estimate the values of the elastic modulus, tensile strength and fracture toughness using highly simplified calculations based upon composite materials models developed in the 1970s. In contrast, more sophisticated models are also mentioned. These were more like structural calculations with finite elements, which allowed investigation of other aspects of the complex structure of mother-of-pearl, like those used by Barthelat *et al.* (2007).

Very simple models such as those of Voigt and Reuss, previously introduced in Section 8.3.3, may be used to predict the elastic modulus. For this, it is assumed that mother-of-pearl is a two-phase composite material: *particles* (aragonite plates) with an elastic modulus E_p and *matrix* (organic polymer) with an elastic modulus E_m. The relative proportions of both phases are V_p and $(1 - V_p)$. The Voigt model, which implies that both phases are in parallel and, therefore, undergo the same strain, yields an expression for the elastic modulus of the composite material E_c.

$$E_c = V_p\, E_p + \left(1 - V_p\right) E_m \qquad (1)$$

The Reuss model, which implies that both phases are in series and, therefore, carry the same stress, produces the following expression for E_c

$$1/E_c = V_p/E_p + \left(1 - V_p\right) E_m \qquad (2)$$

Jackson *et al.* (1988) used the following values to calculate E_c: $E_p = 100$ GPa, $E_m = 4$ GPa and $V_p = 0.95$, values which are quite different from those measured experimentally; 62 GPa (dry conch) or 72 GPa (hydrated conch). These discrepancies are due to a number of causes, among these that neither model accounts for the distribution of stresses along the length of the plates (which are assumed to be infinitely long), nor the transverse shear modulus, G, of the matrix. For plates of finite length Padawer and Beecher (1970) proposed the following corrections to the Voigt model

$$E_c = V_p\, E_p\,[1 - (tan\,h(u)/u)] + \left(1 - V_p\right) E_m \qquad (3)$$

where $u = s\{G\, V_p/[E_p(1 - V_p)]\}^{0.5}$, in which s is the ratio of the length to width of the ceramic plates. As can be seen, when the values of s and G are very large, Equation (3) tends to Equation (1). Even so, the model can be further refined if account is taken of the interactions between adjacent plates, according to Riley's model (Lusis *et al.* 1973). This new correction provides the following result

$$E_c = V_p\, E_p\,\{[1 - \ln(u + 1)]/u\} + \left(1 - V_p\right) E_m \qquad (4)$$

Values of the elastic modulus of mother-of-pearl E_c according to Equations (1), (2), (3) and (4) are shown in Figure 9.7, as functions of the transverse shear modulus G of the matrix, together with the average experimental values for dry (G = 4.6 GPa) and hydrated specimens (G = 1.4 GPa), according to Jackson *et al.* In the curves it was assumed that s = 8 and the contribution of the second term $(1 - V_p)\, E_m$ compared to the first has been neglected. It may be observed that the experimental values for hydrated specimens, 62 GPa (see Table 9.1) fall within the interval 58-72 GPa produced by the Riley and Padawer models. The same is true of the dry specimens, 72 GPa against 68-82 GPa.

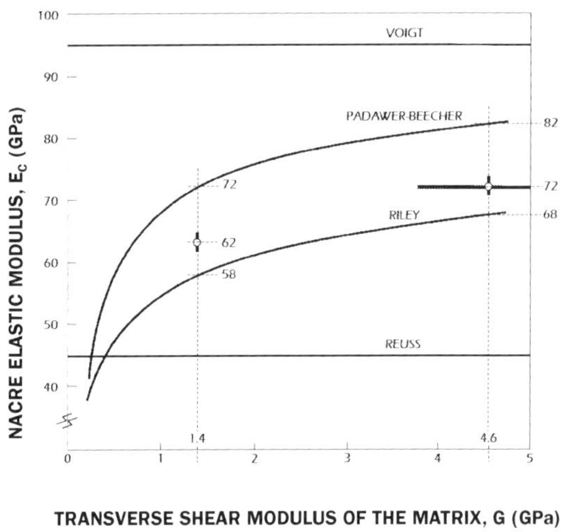

Fig. 9.7. *Elastic modulus of mother-of-pearl as a function of the transverse shear modulus of the organic matrix, for various models.*

EXERCISE 73

From the three-point bending test data presented in Figure 9.6c, determine if the assumption of linear-elastic behaviour is justified during specimen cracking, and estimate the values of the toughness and fracture energy of the material. Assume that, according to the authors, the elastic limit of the material is equal to its flexural strength, estimated at 70MPa, and that the generalised elastic modulus is $E'=E/(1-\nu^2)=50$ GPa.

SOLUTION:

Standard test methods for measuring fracture parameters, and their validity conditions were discussed in Section 8.2.2. The specimen illustrated in Figure 9.6c is beam type, one of the standard geometries used in the ASTM Standard E1820. Equation (27) in Section 8.2.2 is used to calculate the value of the fracture toughness K_I:

$$K_Q = \frac{6P_Q}{BW^{1/2}}\sqrt{\alpha}\frac{1.99-\alpha(1-\alpha)(2.15-3.93\alpha+2.7\alpha^2)}{(1+2\alpha)(1-\alpha)^{3/2}}$$

where P_Q is the crack initiation load, B the thickness, W the depth and $\alpha=a/W$ the relative depth of the crack (see Figure 8.14a).

To obtain the load P_Q at fracture initiation, the procedure shown in Figure 8.14c, and previously used in exercise 54, is followed, with which the values $P_{max}=21.6$N and $P_Q=15$ N are obtained. The value corresponding to the toughness (for $\alpha=0.5$, $B=1.8$mm and $W=3.6$mm) is $K_Q=1.48$ MPa m$^{1/2}$.

To verify that K_Q can be taken to be a proper value of K_{IC} we perform the checks set out in the standard:

– first, we see that $P_{max}/P_Q=1.4>1.1$ does not satisfy the linearity requirement before the peak load

– on the other hand

$$B=b=W-a=1.8\,mm>2.5\left(\frac{K_Q}{\sigma_{ys}}\right)^2=1.1\,mm$$

which does meet the requirements of the standard.

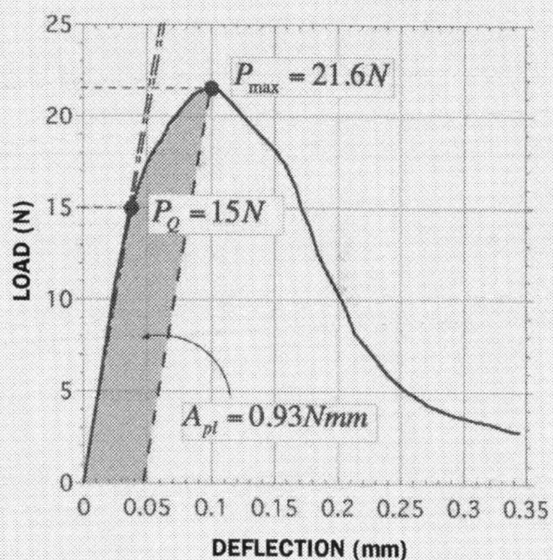

Since we do not have a valid value of K_{IC}, we calculate the fracture energy G_C using Equation (31) in Section 8.2.2. The value of J_{pl} is obtained using Equation (32) from the area A_{pl} of the force-deflection curve up to the maximum load (see the figure)

$$J_{pl}=\frac{1.9A_{pl}}{B(W-a)}=545\ \text{J/m}^2$$

with which we obtain, from Equation (31) in Section 8.2.2

$$G_C=\frac{K_Q^2}{E'}+J_{pl}=44+545=589\ \text{J/m}^2$$

Finally, the checks of Equation (37) give

$$B=b=1.8\,mm>\frac{25G_Q}{\sigma_{ys}}=0.2\,mm$$

which indicates that the calculated value of G_C is valid, according to the standard.

The calculated value of the fracture energy for this material (Haliotis shell in the hydrated state, Table 9.1) is in good agreement with data from other authors and is almost two orders of magnitude greater than that of the ceramic base material (aragonite). We also see that the most important contribution to the value of the fracture energy is that due to the plastic part, in line with Figure 9.6c.

Currently, thanks to increased computing power, better detailed models are available for simulating the microstructure more accurately, allowing reproduction of the stress-strain curves rather than just maximum, or rupture values, such as in the quoted article by Jackson *et al.* (1988). The curves, obtained from the tensile and shear tests, and shown in Figure 9.6a, were reasonably reproduced with the finite element model proposed by Barthelat *et al.* (2007).

Various studies of the behaviour of nacre in **tensile** tests indicate that the principal mode of deformation is sliding of the ceramic bricks, which can be considered linear elastic, and that there is an inelastic component which is due to the mortar which bonds them. The tensile stresses are transmitted between the bricks through shear stresses in the regions of overlap between the bricks. When these reach a maximum stress, the bricks stop sliding over each other and gaps are created (Barthelat *et al.* 2007). The hardening, which is observed when the strain in the hydrated specimen is increased, has been attributed to several causes: the nature of the interface, the presence of nanometric asperities on the faces of the bricks (Evans *et al.* 2001) and waviness, at the micron scale, of the bricks (Barthelat *et al.* 2007). Judging from numerical simulations, the last mechanism seems to be the dominant one.

The model proposed by Barthelat *et al.* consisted of studying a representative volume element of the material, formed by ceramic tablets embedded in a cohesive matrix. The size, shape and distribution of the tablets was obtained from optical images (an example of which is shown in Figure 9.8a). A three-dimensional finite element model of two adjacent layers, which reproduced an area of nacre of 72 x 78 mm, was based on these contours. Two types of boundary between the tablets were considered: horizontal *interfaces* and vertical *joints* (Figure 9.8b). The bricks (tablets) were modelled as transversely isotropic linear-elastic solids, with property data corresponding to aragonite crystals (with five elastic constants). The cohesive elements were located in both the interfaces and the joints. The chosen parameters were based on the mechanisms associated with the behaviour of biopolymers, the nano-asperities and the small mineral bridges between the tablets. A representation of the cohesive function used is provided in Figure 9.8b.

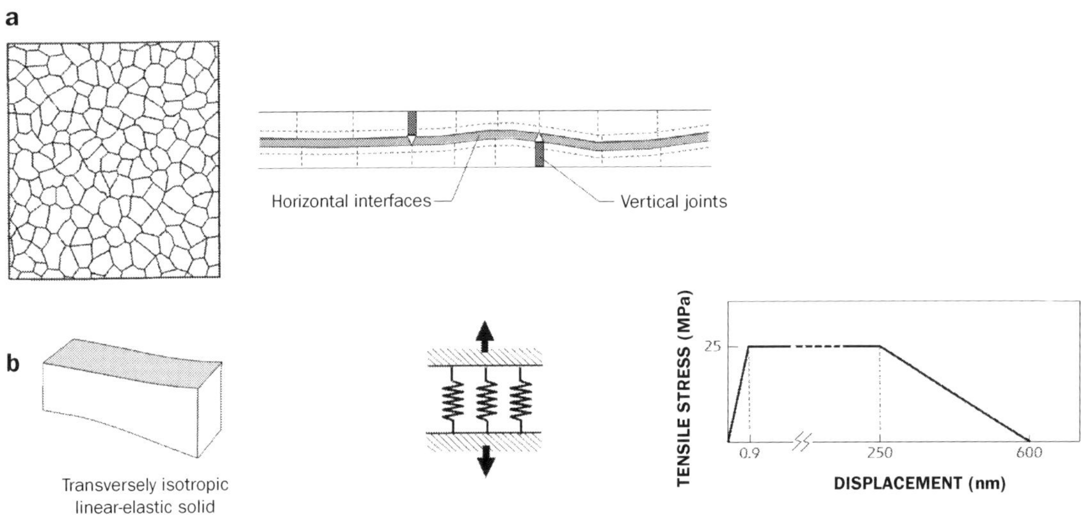

Fig. 9.8. *Model of mother-of-pearl (Barthelat et al. 2007).*
a. Plan and transverse section of the model.
b. Details of the cohesive model for the joint material.

The most significant results obtained with this model are illustrated in Figure 9.9a; the presence of waviness provides good agreement with the experimental results. However, when the tablets are flat, the calculated strain is much lower. The influence of the *joints* (see Figure 9.8b) is reflected in Figure 9.9b. In the initial part of the curve, it is observed that the joints increase the stiffness of the mother-of-pearl (when removing them, the initial modulus falls from 70 to 40 GPa) and the strength is also reduced when they are not considered. Note that, even though the joints add stiffness and strength, they do not appear to influence the strain hardening (since both curves are practically parallel). Another interesting aspect is that this model allows investigation of the *distribution* of the platelets; whether regular or random. To this end an ordered distribution of hexagonal platelets of the same area was studied. The results are shown in Figure 9.9c for hexagonal platelets with, and without, surface undulations. In both cases the results are significantly different from those of the experiments, and they demonstrate the importance of incorporating randomly distributed platelets in the microstructural design of materials aimed at imitating shells.

Perhaps, the most interesting result of this study is the importance of surface undulations in the platelets. Previously, various explanations were given to justify the strain hardening in hydrated mother-of-pearl, nanometric asperities (Evans *et al.* 2001, Wang *et al.* 2001) and the interfacial strength. The asperities can account for hardening by sliding up to 15 or 20 nm, but not for the 100 or 200 nm observed. The strength provided by the biopolymer interfaces, certainly, contributes to strain hardening, but it alone is not sufficient (Barthelat *et al.* 2007). The undulations in the platelets appear to be able to explain this effect.

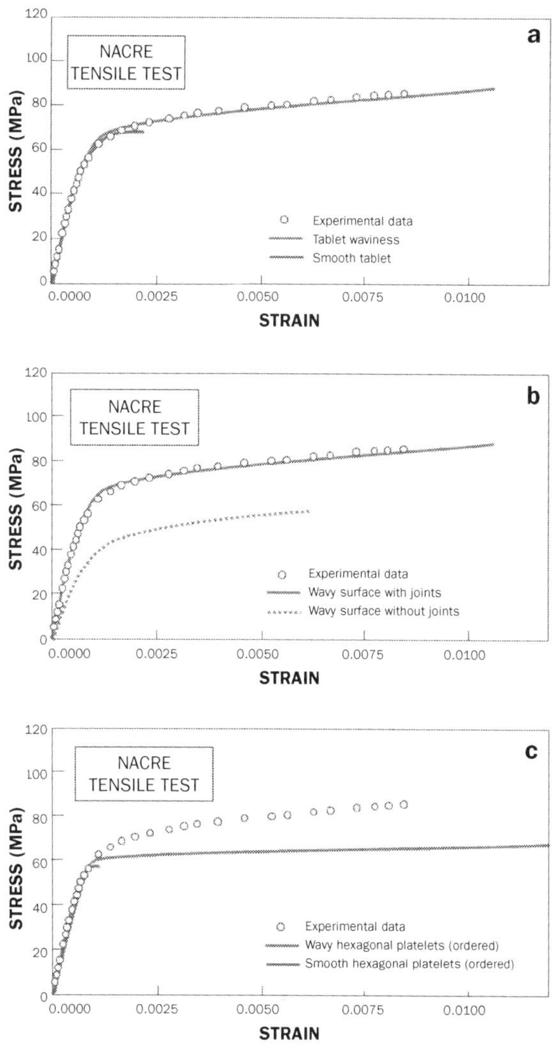

Fig. 9.9. *Stress-strain curves. Circles; experimental data. Curves; numerical data from the model of Barthelat et al.*
a. Influence of waviness of the tablets.
b. Influences of the vertical joints.
c. Influence of the distribution of the ordered tablets.

Values of the mechanical properties of gastropod shells, specifically of *Conus striatus* and *Strombus gigas*, are presented in Table 9.2. If the values are compared to those in Table 9.1, corresponding to bivalves (Haliotis and Pinctada), it may be observed that the specific fracture energies are much greater, while the flexural strengths are less; approximately half. All of this seems to suggest that the *crossed-lamellar* structure provides greater toughness than the *parallel-lamellar*, but with lower flexural strength.

The data in Table 9.2 also illustrate the influence of humidity and loading rate upon the values obtained. These aspects will be discussed briefly, below, using experimental data obtained from *Strombus gigas* shells.

TABLE 9.2. MECHANICAL PROPERTIES OF THE CROSSED-LAMELLAR STRUCTURE

SHELL	INITIAL MODULUS E (GPa)	FLEXURAL STRENGTH σ_{FT} (MPa)	COMPRESSIVE STRENGTH σ_C (MPa)	FRACTURE ENERGY G_C (J/m^2)	REFERENCE
Conus		70 – 200 (3P)			Currey and Kohn (1976)
Strombus		100 (4P)			Laraia and Heuer (1989)
Strombus	60	56 ± 22 (4P) (84 ± 49)*		4000 ± 2000 (13000 ± 7000)*	Kuhn *et al.* (1996)
Strombus		29 – 74	166 – 218 (249 – 361)**		Menig *et al.* (2001)

3P, 4P Tests on beams under three- or four-point loading.
* Tests on dry specimens.
** Dynamic tests (10 – 25)·10^3 GPa/s.

Figure 9.10a contains stress-strain curves from compression tests on specimens loaded in different directions; with the outer face of the shell (shaded in the figure) perpendicular and parallel to the loading direction (Mening *et al.* 2001). The results indicate that the strength is greater when the load is parallel to the external surface. In all of the tests there was quite a lot of scatter, but the average values obtained (shown in Table 9.2) were 166 MPa in the perpendicular and 218 MPa in the parallel directions. These tests were performed slowly; between 3 and 300 MPa per second. When they were repeated at higher velocities, between 10,000 and 25,000 GPa per second, the values obtained were 50% greater; 249 MPa in the perpendicular and 361 MPa in the parallel directions. All of these results are for dry specimens.

Force-deflection curves for dry and hydrated specimens in four-point bending tests (Kuhn-Spearing *et al.* 1996) are shown in Figure 9.10b. The relative position of the external load is shown in the adjoining figure. It can be seen from the curves that the hydrated specimens are more flexible. The average values of the maximum stress (calculated assuming linear-elastic behaviour) were 56 ± 22 MPa (from 13 dry specimens) and 84 ± 49 MPa (from 9 hydrated specimens). These curves were used to estimate the specific fracture energy from the procedure indicated in Section 8.2.2; measuring the area under the curve and dividing it by the specimen cross-sectional area (a fairly approximate process because of the difficulty in determining the fracture area bearing in mind the profusion of cracks created in the test). The values obtained were about $4 \pm 2\,kJ/m^2$ (from 11 dry specimens) and $13 \pm 7\ kJ/m^2$ (from 8 hydrated specimens). A schematic representation of the appearance of cracks in the inner layer is provided in the adjoining figure, together with subsequent bifurcations as they penetrate the middle lamellae which have a different orientation. The large toughness of shells with a crossed-lamellar structure seems to be due to several factors, among these are: crack bifurcation, formation of bridges between the cracks and sliding between the lamellae. All of the elements mentioned when describing the microstructure come into play in these processes; the three layers and the first- and second-order lamellae, at least. This analysis may serve as the inspiration for the design of high-performance ceramics if the advantages of structuring the material at different scales can be included. For example, overlapping alternating laminas (stacking sequence 0/90/0 degrees) and within each lamina, incorporating sub-laminas with a stacking sequence of ±45 degrees.

Finally, Figure 9.10c contains force-deflection curves from a four-point bending test (Kuhn-Spearing *et al.* 1996). It may be observed that the behaviour is far from being linearelastic and that, moreover, it exhibits hysteresis, even when the loads are not high (compared with those of Figure 9.10b). All of this must be borne in mind when interpreting data from tests with these types of biological materials in which certain simplifying assumptions must be made when attempting to interpret the results.

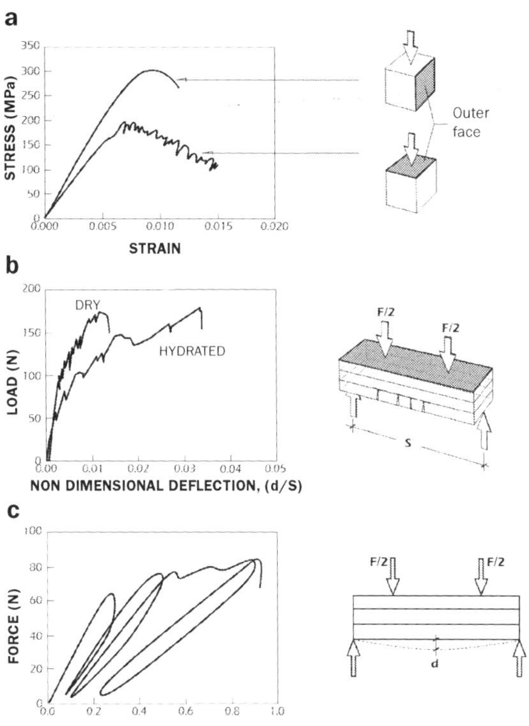

Fig. 9.10. Strombus gigas *shells.*
a. Stress-strain curves obtained in compression tests. Note the orientation information.
b. Force-deflection curves obtained in four-point bending tests. Note the influence of hydration.
c. Force-deflection curves obtained in four-point bending tests. Note the hysteresis.

EXERCISE 74

For a parallel-laminar structure, it has been proposed that the tensile strength in a direction perpendicular to the laminas is due, to a large extent, to the presence of mineral bridges between the laminas. An acceptable value for this strength is 3 MPa (Meyers *et al.* 2008). Song *et al.* (2003) have measured the size and number of these bridges, whose image is shown in the figure on the right.

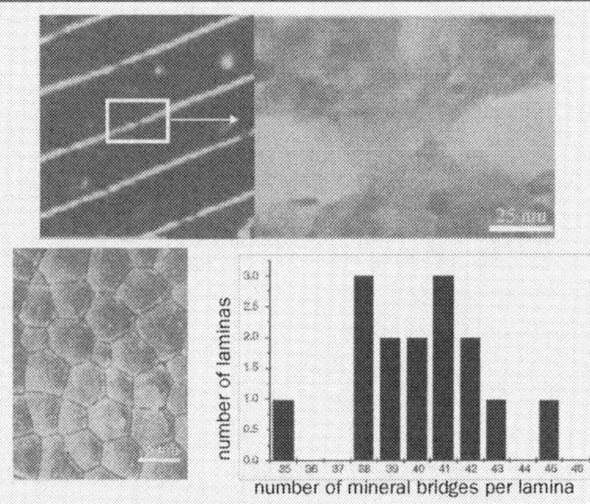

Assuming, that the bridges are approximately cylindrical, calculate their fracture stress and discuss the possible presence of defects within them.

SOLUTION:

From the figure in the problem statement, we see that the average number of bridges per lamina is n = 40, and that the laminas have an average diameter of approximately l = 4.5 µm. With these values we can estimate that the average force required to break the bridges will be

$$F = \frac{\sigma_c}{n}\frac{\pi l^2}{4} = 1.19\mu N$$

where σ_c = 3 MPa is fracture stress of the interface as measured by Meyers and co-workers.

Estimating, also from the figure in the problem statement, an approximate bridge diameter of D = 50 nm, the breaking stress of these will be

$$\sigma_R = \frac{F}{\pi D^2/4} = 0.5GPa$$ which is the required fracture stress.

In order to consider the presence of defects which might have been the cause of failure of the mineral bridges, we compare the value obtained in the calculation with the theoretical strength of the crystal, whose value is estimated at $\sigma_{theor} \approx E/30$, where E is the elastic modulus of the material (Hertzberg, 1996).

From Table 9.1 we see that $E_{aragonite}$ = 50 to 100 GPa, which yields a theoretical strength lying between 1.7 and 3.3 GPa, values far beyond the strength.

We conclude, therefore, that the rupture stress of the mineral bridges must be affected by the presence of defects which reduce their strength.

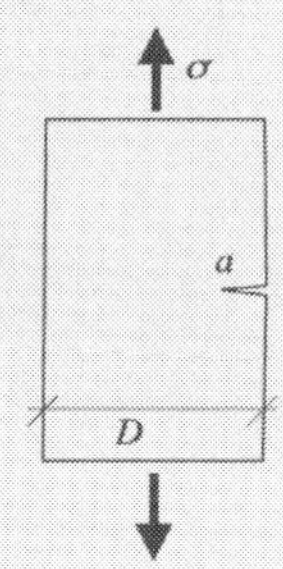

If the rupture of the mineral bridges occurs due to brittle fracture, it is possible to estimate the size of the defects present, with the aid of the fracture mechanics model shown in the figure on the right. In it, the defect has been idealised as a small crack in the surface of the bridge since, if the crack were deep, it would have probably been detected in the electron microscope images.

For this configuration, the value of the stress intensity factor is $K(\sigma,a) = 1.12\sigma\sqrt{\pi a}$

Using the linear-elastic fracture criterion $K = K_C$ (Equation 25a in Section 8.2.2) and a value of $K_C \approx 1\,\text{MPam}^{1/2}$, we find that the size of the crack which produces failure at a stress σ_R

$$a = \frac{1}{\pi}\left(\frac{K_C}{1.12\sigma_R}\right)^2 = 1015nm$$ which we see is several orders of magnitude greater than the size of the mineral bridge mineral itself.

Even if we were to refine the model, it would be difficult to justify a mechanism of failure by brittle fracture, so that it seems reasonable to assume that, either there will be some other damage mechanism, or perhaps that the material properties of the mineral bridges do not correspond exactly to those of crystalline aragonite.

9.1.4. Final remarks

Mollusc shells offer us the possibility of considering their wider application from three points of view: as *biological materials*, as *biomaterials* and as models for *biomimetics*.

As **biological materials**, shells have enjoyed little practical application but are highly valued as decorative items and jewellery. One of the few references to the practical use of shells Moir (1990), in which he comments on the inhabitants of the Takuu Atoll, in New Guinea, using the giant shells of the *Tridacna gigas* (which can measure up to 1 metre and weigh 300 kilos) to fabricate knives and hoes, due to a lack of suitable minerals in the Atoll. With this exception, the mother-of-pearl of shells, for decorative uses, and the pearls, in jewellery, are the most familiar applications of some shells.

Pearl has been recognised as a gem since the earliest times because, as well as its rarity and mysterious origins, it has been credited with healing powers; virtue symbols and mystical properties. It appears that the custom of adorning oneself with pearls has its origin in India, from which the Phoenicians introduced it to Europe. The Greeks called pearls *margarites*, and there are those who think that this comes from the Persian *murwari*, which means daughter of light. The harsh reality is that the pearl is the response of the mollusc to the irritation produced when a material enters from outside its shell. To protect its delicate tissue, the mollusc coats the foreign body with fine layers of calcium carbonate. Currently, pearls do not have as much value as in the past because the majority of them are now cultivated and produced in large quantities. The most prized are those cultivated in seawater. Those cultivated in freshwater are smaller (between 3, 5 and 7 mm), often asymmetrical, and less brilliant, and their production is widespread (more than 1500 tonnes per year). The iridescent colours of the light reflected by the mother-of-pearl is another of the attractions of the shells used for decorative purposes. This special glow –also called *orient*– is due to the process of optical interference when light rays pass through and are reflected in the fine layers, whose thickness is of the order of the wavelength of visible light.

Shells, as a **biomaterial**, have also been used since ancient times. In 1931 a fragment of jaw, from the Mayan civilisation, was found with dental implants made of mother-of-pearl. X-rays showed that the roots of the implant were well integrated into the bone. The discovery was made by the archaeologist couple Dorothy and Wilson Popenoe during an excavation at the *Playa de los Muertos*, in the Ulua Valley in Honduras. The jaw, dated to around AD 600, belonged to a young woman who was about 20 years old. She had three implanted mother-of-pearl incisors, which were initially believed to have been inserted after death, but a detailed x-ray analysis showed that compacted bone had formed around the implants (Bobbio, 1972).

Recently, Evelyne López and co-workers (Atlan *et al.* 1997) used the *mother-of-pearl* of the mollusc *Pinctada maxima* as a biomaterial; they ground it finely and mixed it with the blood of eight patients, women between 45 and 55 years old, who had suffered bone loss in the upper jaw. This solution was injected into the tissue affected by the bone loss, and these were biopsied after 6 months. The results showed that there was no inflammation, and that the mother-of-pearl had been accepted by the tissue, as if it were the bone of the patients themselves. Furthermore, the osteoblasts (specialised cells which secrete the bone matrix) were activated and formed healthy bone tissue in the implant. The new bone tissue remained welded to the mother-of-pearl particles without any intermediate soft or fibrous tissue. The osteoclasts (cells which participate in the process of bone tissue absorption) remained active around the bone, but did not interact with the mother-of-pearl.

As **models for biomimetics**, mollusc shells suggest procedures by which brittle materials, such as aragonite, can be transformed into tough materials, if they can be suitably structured and bonded. Nature has done this, improving and refining these structures over millions of years. The Science and Engineering of Materials attempts to take advantage of these findings through a new branch, the interdisciplinary area of *Biomimetics* (Crossman *et al.* 2019).

The first attempts at the biomimetics of shells date to the 1980s and 1990s (see, for example, Meyers *et al.* 2008). Imitating the structure of mother-of-pearl, they fabricated pieces of silicon carbide (SiC) which were much tougher than conventional forms. Working on SiC small plates coated in graphite, they managed to increase the fracture toughness from 4 up to 18 MPa $m^{1/2}$, and the fracture energy went from 30 to 6000 Jm^{-2}. In this new material the SiC took on the role of the aragonite, and the graphite that of the protein glue.

Monolithic fibrous ceramics are a second generation of tough ceramics, developed at the beginning of this century. The excellent properties of these materials result from their design (at the micron scale), their structure (small strong hexagonal bricks), and the fine but weak adhesive which bonds the bricks. Silicon nitride Si_3N_4 cells –in the form of 250 μm diameter fibres– are cemented in a 10 to 20 μm thick boron nitride BN mortar. The rupture modulus of this material is about 510 MPa at ambient temperature and about 290 MPa at 1300°C. The laminates formed of Si_3N_4 bonded with BN, also exhibit excellent fracture behaviour, almost two orders of magnitude greater than monolithic Si_3N_4.

A third, even more promising generation, developed by Deville *et al.* (2007), is based upon *freeze casting*. On freezing an aqueous solution of ceramic material, the ice crystals expel the ceramic particles and form ice columns and laminas several microns thick. By sublimating the ice, a ceramic lattice is left, which can be infiltrated with a polymer or metal. Using this technique, Ritchie and co-workers (Munch *et al.* 2008) fabricated materials which imitate the structure of mother-of-pearl; manufacturing an alumina scaffold (by freeze casting), sintering it and then infiltrating it with an alloy of aluminium and silicon. The mechanical properties achieved are impressive; flexural strength between 120 and 210 MPa, fracture toughness values of 5 MPa $m^{1/2}$ (for crack initiation) and G_C values between 5 and 8 kJ/m^2 (obtained from the area under the force-deflection curve). These properties have been improved upon by increasing the contact between the aluminium alloys and the alumina, and adding small amounts of titanium to the alloy, they managed to reduce the surface tension and improve impregnation. With 0.5% of titanium, the fracture toughness was increased 10 MPa $m^{1/2}$ and the strength approached 600 MPa. Possibly, by reducing the surface tension, the aluminium alloy infiltrated more effectively. In the article by Gao *et al.* (2020) there is a summary of this *freeze casting* technique.

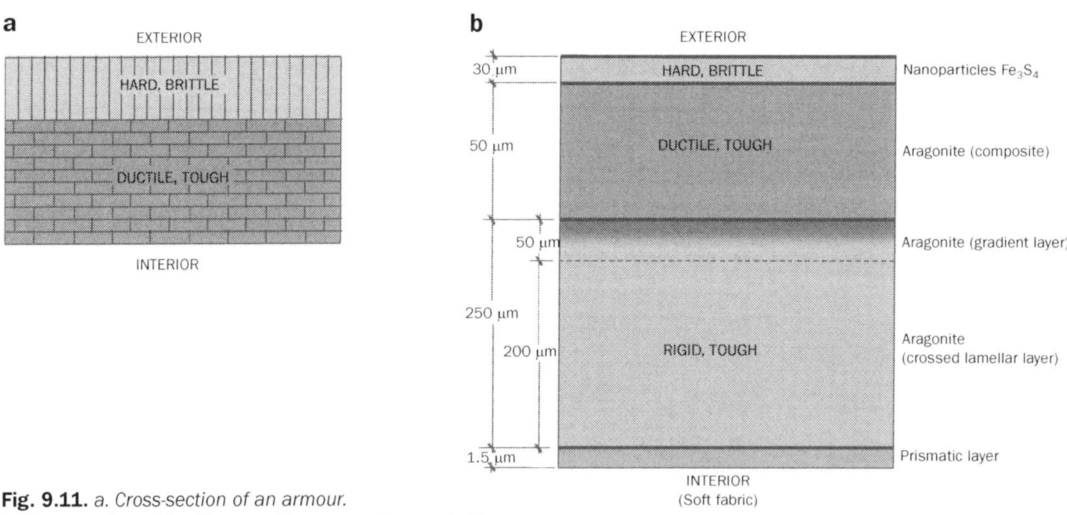

Fig. 9.11. *a. Cross-section of an armour.*
b. Cross-section of the Chrysomallon squamiferum *shell.*

It is worth noting that, although the microstructure of these artificial materials is similar to that of the shell biological materials, the fracture mechanisms are different. It has been observed that one of the most effective mechanisms in the fracture of mother-of-pearl is *pull-out* of the rather rigid, wavey, bricks, a mechanism which is more effective when the shell is hydrated because the mortar is more ductile and facilitates this process. However, in the artificial materials, it is believed that the dominant mechanism is the plastic deformation of the aluminium bricks.

Molluscs have developed a shell structure which is highly effective in resisting the impact of predators, very similar to military armour. A simplified form of the cross-section of armour (Sánchez-Gálvez, 2012) is shown in Figure 9.11a; the external layer is a hard, brittle material (almost always a ceramic) in order to splinter the projectile, and the internal layer is a tough, deformable, ductile material (steel or aluminium) to absorb the excess kinetic energy of the projectile. Note the similarity of Figure 9.11a to the cross-sections of the shells shown in Figures 9.2a and 9.4a, in which the external layer is formed of calcite (a brittle material) columns and the interior is a lamellar structure composed of bonded aragonite (a more ductile tough material) plates.

The external layer does not have to be monolithic; it may also imitate the structure of mother-of-pearl, such as in the LAST® armour, manufactured by Foster-Miller, and which consists of hexagonal SiC or B_4C tiles wrapped in Kevlar® and joined tightly with a Velcro® type adhesive, which facilitates the repair of the armour when a plate is broken.

Mollusc shells which survive at great depths in the presence of hot springs may possess special structures to withstand such aggressive conditions, structures that are worth investigating. With this in mind, Ortiz and co-workers studied the shell of the gastropod *Chrysomallon squamiferum*, which lives in hydrothermal chimneys more than 3000 m deep (Yao *et al.* 2010). The cross-section of a representative fragment of the shell is shown in Figure 9.11b, where three layers, unlike other known shells, can be distinguished: An intermediate layer of organic material, relatively thick and flexible, is paired with two rigid, highly mineralised layers; the exterior rich in iron sulphide and the interior formed of aragonite.

The authors of the quoted article modelled the structure using finite elements, considering an elasto-plastic behaviour in each layer, and obtaining values of the elastic modulus and elastic limit for each layer from nano-indentation tests. They simulated the response of the shell subjected to an external indenter and demonstrated that cracks may initiate in the internal aragonite layer. When these cracks progress towards the exterior, they may be blocked by the, more ductile, intermediate layer. The thin external layer –a material composed of iron sulphide nanoparticles (about 20 nm in diameter) embedded in an organic matrix– seems to take on a role as the first line of defence against impacts.

Mollusc shells provide a good example of how nature designs both a material and a structure. Materials engineers should keep this in mind when designing new materials; thinking not only of the material in isolation, but rather as in integral element of the structure.

9.2. WOOD

9.2.1. Introduction

Wood is one of the oldest biological materials. Ancient uses, as a *biological material*, were in structural applications and also as *biomaterial* as evidenced by the wooden prosthesis for the big toe of an Egyptian woman who lived more than 2,600 years ago.

It should be noted that the word *material*, derived fom the Latin *materia*, was used to designate the wood from a tree; specifically, the trunk of trees that were used in construction.

Wood possesses very interesting mechanical properties, as will be seen later, especially when specific properties (that is, per unit mass) are considered, because under this criterion they can approach values comparable to steel. The mechanical properties exhibited by wood are due to the hierarchical structure of its components. Optimised by nature over millions of years, trees have had to withstand adverse loads (wind, weather, predators, etc.).

Currently, the annual production of wood is more than 1,500 million tonnes, a similar quantity to that of steel. About 20% of this production is for structural purposes: beams, posts and panels. Other uses range from paper to furniture, boats, and musical instruments (García Esteban *et al.* 2002).

Trees are the tallest and longest-living lifeforms populating our planet.

Among the oldest trees are the sequoias (*Sequoia sempervirens*), which have survived in the natural state only in North America. One of them (the Founder's tree) is 112 m high (in October 1996) and a little over 500 years old, another (the General Sherman) is 85 m high with a base circumference of 24 m.

A Rocky Mountain pine (*Pinus aristata,* called Methuselah) is one of the oldest trees; it has lived for 4,700 years, at an altitude of 3,050 m, but is only 12 m tall. A dragon tree (*Dracaena draco*), in the Canary Islands, could be more than 6,000 years old. The baobab tree (*Adansonia digitata*), in Kenya, could be 5,000 years old, with a 40 m base circumference. The oldest giant sequoias could be over 4,000 years old. In Tasmania, there is a royal holly (*Lomatia tasmanica*) whose shoots come from a mother stock which is still alive after 40,000 years. This tree was a contemporary of Upper Paleolithic humans (Fischesser, 2009). In Europe, it is possible that the oldest tree is a pine, in Bosnia (*Pinus heldreichii*), which could be 1,075 years old (New Scientist, 2016).

The structure of wood, its mechanical properties and principal applications are briefly described below.

9.2.2. Structure

Trees are classified into two families: *coniferous* plants (gymnosperms, or softwoods) and *leafy* plants (angiosperms, or hardwoods).

Coniferous wood contains cells, called *tracheids*, which serve three functions: transport of sap, provision of strength and water storage in the sapwood. Hardwood contains vessels, tracheid, various types of sclerenchyma fibres and parenchyma cells: the vessels, which are responsible for the conduction of sap and the *fibres*, which underpin the strength.

Another tree classification is based upon whether the leaf is perennial or falls. Most conifers have perennial leaves, while most hardwoods have leaves that fall.

In general, the average density (specific gravity) of coniferous wood (0.41) is less than that of hardwoods (0.50), but there is some degree of overlap between the two families; the density of conifers can vary between 0.29 and 0.60, while that of hardwood ranges from 0.32 to 0.81.

The wood of conifers, being a very ancient species (they appeared during the Triassic Period, between 248 and 206 million years ago), is quite homogeneous due to the limited diversity of its woody cells. Cells in hardwoods are more heterogeneous due to their great specialisation, characteristic of species that originated later (during the Cretacious period, between 144 and 65 million years ago). For more details see Gil (2013) or Fischesser (2000).

The structure of wood can be studied at various levels: macroscopic, mesoscopic and microscopic.

Figure 9.12 shows the most characteristic details at the **macroscopic** level. Three distinct regions can be identified in the cross-section of a tree trunk: the *xylem*, the *cambium* and the *bark*. In the xylem, which takes up practically the entire section, two regions can be identified; the central, darker one, also called the *heartwood*, is formed of dead cells and the outer, lighter zone, known as the *sapwood*, is formed of living cells. All cells in the wood are produced by the *cambium*, a thin ring which surrounds the xylem. The *bark* is situated on the exterior of the cambium.

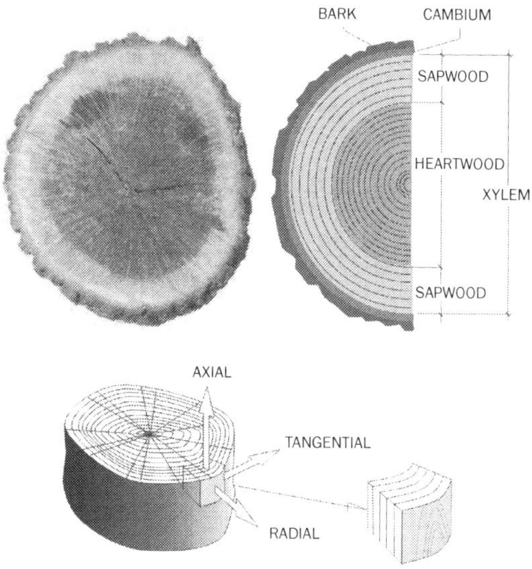

Fig. 9.12. *Macrostructure of the section of a trunk, and the notation used.*

A series of concentric rings can also be observed in the cross-section of the trunk. In each ring there is a lighter zone which is formed at the start of the growing season and another, narrower and darker, produced at the end of the season. With the exception of trees in tropical regions, the *cambium* produces only one ring each year and, for this reason, these are known as annual growth rings, and are a great help in dating past events. The science on which this phenomenon is based is dendrochronology (from the Greek δενδρον, tree).

Generically, wood can be considered as a homogeneous, orthotropic continuous material with three planes of symmetry: radial, tangential and axial (Figure 9.12). The stiffness and strength are greatest in the axial direction, parallel to the trunk. In the radial and tangential directions, these values are less and go from 1/2 to 1/20 of the axial properties, depending upon the tree under consideration.

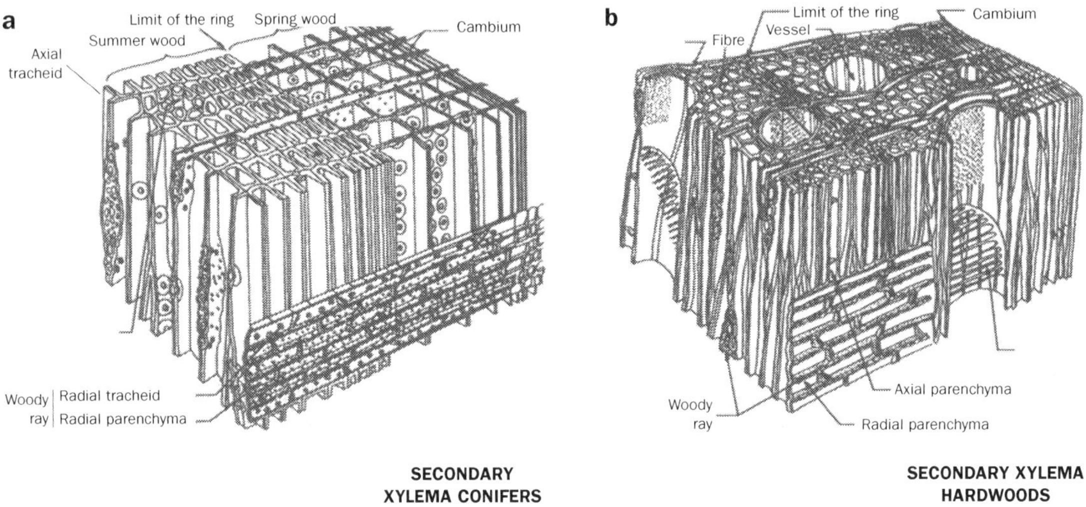

Fig. 9.13. *Mesoscopic structure of wood (Gil, 2003).*

At the **mesoscopic** scale (Figure 9.13) the structure of wood can be considered as an aggregate of cellular walls, of two types; some in the form of fibres and others more parallelepipedic, as shown in Figure 9.13.

The **fibrous cells**, which are aligned with the longitudinal (or axial) direction of the trunk, have relatively thick walls and are quite wide (the ratio of the length and diameter may be about 20). Their principal function is mechanical, that of support. In conifers these cells are the *tracheids* (Figure 9.13a) and, in hardwoods, the fibres (Figure 9.13b). In conifers the tracheids also perform transport and water storage functions.

The **nonfibrous cells** are grouped in tissues, oriented in the axial and radial directions, and perform transport and storage functions (starch, sugars, lipids, proteins, etc.). They form vessels in hardwoods and *radial* and *axial parenchyma* in both conifers and hardwoods.

The different types of cells and their principal functions are summarised in Table 9.3, for both conifers and hardwoods.

TABLE 9.3 FUNCTION OF CELLS IN WOOD

FUNCTION	CONIFERS	HARDWOODS
Mechanical	Tracheids	Fibres
Transport	Radial Parenchyma Tracheids	Radial Parenchyma Vessels
Storage	Axial Parenchyma	Axial Parenchyma

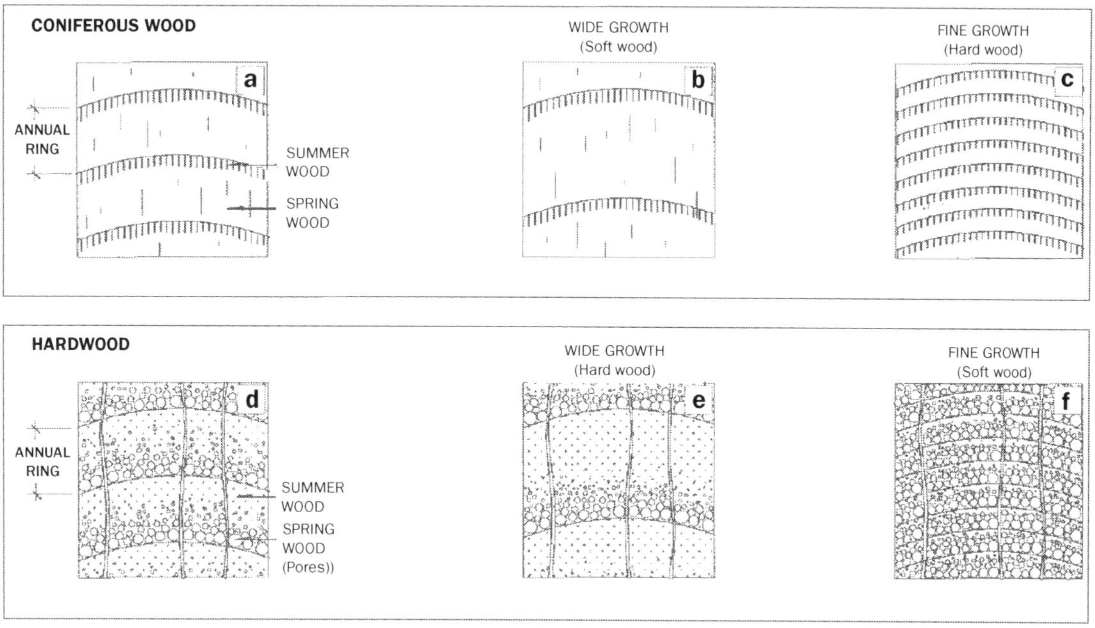

Fig. 9.14. *Structure of the growth rings in conifers and hardwoods.*

The **mesostructure** of conifers is quite uniform. The tracheids which take up 90 to 95% of the volume, have lengths between 3 and 5 mm and diameters typically ranging from 25 to 45 m. The mechanical properties of this wood are principally determined by the tracheids.

A detailed representation of the cross-section of a conifer is provided in Figure 9.14a, in which the growth rings are highlighted. The spring wood is lighter than the summer wood, which is denser and more coloured. In spring wood, the cell width is large and wall thickness thin. In summer wood, width is small and thickness thick. When the rings are wide, the texture is weak and the wood soft. On the contrary, if the growth rings are thin, the texture is strong and the wood hard (see Figures 9.14b and 9.14c).

The mesostructure of hardwood is more complex because there are, at least, four types of cells involved. The *fibres* occupy, a minimum of, 50% of the volume and have a great influence on the mechanical properties of these woods.

The *vessels* may occupy about 20% of the volume and have the role of transporting water and nutrients. *Pores* may be observed in the cross-section of the vessels in Figure 9.14d. The mechanical properties depend, to a large extent, upon number of pores per unit volume. The *radial parenchyma* may represent 17%, or more, of the volume, while the *longitudinal (axial) parenchyma* can range between 1 and 24% of the volume.

A detail of the cross-section of a hardwood is shown in Figure 9.14d, in which the appearance of the growth rings and pores present in them are highlighted; the spring wood is very porous and the summer wood has the finest pores. When the rings are very thin, the density of the pores is large and the wood is soft; when the rings are wide, the density of the pores is less and the wood is harder (see Figures 9.14e and 9.14f). The presence of pores allows classification of the woods; if vessels are not visible, the wood is a conifer.

A study of the cell wall lies on the **microscopic** domain. The cell wall structure is shown schematically in Figure 9.15. This is formed of a series of layers and all of these contain cellulose microfibrils –except the intermediate lamina– with different orientations with respect to the wall axis.

The *primary wall* (P) is very thin and associated with the *intermediate lamina* (M) which has the role of bonding the walls of adjacent cells. The microfibrils of the primary wall have no preferential orientation. In the *secondary wall* (S) the microfibrils are aligned and densely packed. This wall is subdivided into three layers: the exterior (S_1), intermediate (S_2) and interior (S_3). The intermediate layer S_2, the thickest, is the one with the greatest influence on the mechanical properties. Table 9.4 contains data on the relative thicknesses of the layers, the angle of the microfibrils with respect to the vertical axis, and the numer of laminas which constitute each layer.

TABLE 9.4 CELL WALL STRUCTURE OF SPRUCE

LAYER TYPE	RELATIVE THICKNESS (%)	MICROFIBRIL ANGLE*	NUMBER OF LAMINAS
M + P	5	Random	–
S_1	10	50° – 70°	4 – 6
S_2	75	10° – 30°	30 – 150
S_3	10	60° – 90°	0 – 6

* The angle is measured relative to the vertical axis.

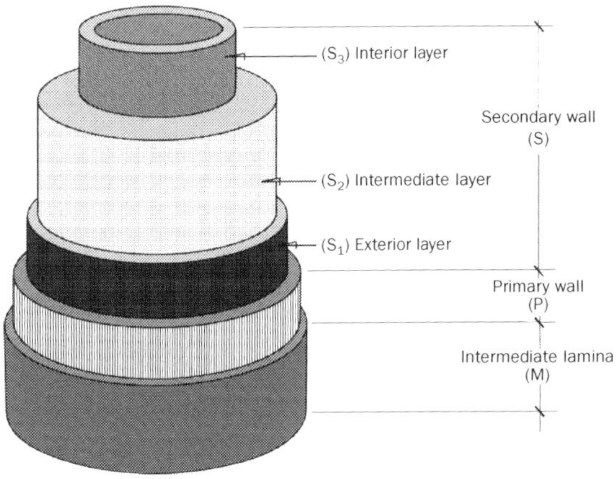

Fig. 9.15. *Microscopic structure of the cell wall.*

It is possible to go to an even deeper level in the mesostructure if the details of the *cellulose microfibrils* are considered. In this, the approach is **nanoscopic**, since the cross-section of these fibres is 3.5 x 10 nm.

Cellulose is a polymer formed of long glucose chains (between 5000 and 10000 units) without branches (see Section 7.2.1). These chains are arranged to form microfibrils in which there are highly ordered (crystalline) and less-ordered regions (amorphous zones). The (more compact) crystalline regions hardly absorb water, while the amorphous regions may absorb more water, changing the mechanical properties of the microfibrils. This may explain the variation in the properties of wood with changes in humidity and temperature, since between 40 and 50% of the composition of wood is cellulose, whose density is about 1550 kg/m^3.

Other components of the cell wall, at the nanoscopic level, are *hemicellulose* and *lignin* which have the function of binding the cellulose fibres; to some extent the cell walls may be modelled as a composite material constituted of cellulose fibres embedded in a matrix of hemicellulose and lignin.

Hemicellulose is a polysaccharide but, unlike cellulose, its chains are short (between 500 and 3000 monomer units); they are branched and, largely, amorphous. Due to its structure, it is easily hydrolysed by acids and enzymes. In conifers it represents 25% by weight and in hardwoods 35%. Its density is about 1500 kg/ m^3. Lignin is a complex phenolic (a diverse group of aromatic compounds), and its structure is amorphous.

Lignin is the main adhesive which bonds both the components of the cell wall and other cells. In hardwoods it represents between 20 and 30% by weight. In conifers it is somewhat less. Its density is about 1400 kg/m^3. It begins to soften at about 170 °C, a charactieristic which is exploited in the fabrication of wood products.

9.2.3. Mechanical properties

Modelling the mechanical properties of wood is a complex task, as might be imagined from its intrinsic structure. In addition, the microstructure varies, not only between different species but also within species, because it changes during different growth stages. Wood is an orthotropic material, and its properties depend upon humidity and temperature. Time also comes into play since it exhibits viscoelastic behaviour. Even so, it is possible to make reliable predictions of wood's mechanical behaviour.

Ideally, starting from the *microstructure*, it should be possible to estimate the mechanical properties of the cell walls, modelling them as a material composed of cellulose fibres embedded in a hemicellulose or lignin matrix. And, in a second stage, starting from the *mesostructure*, wood may be modelled as a cellular material with a honeycomb structure whose walls have the previously calculated properties. Despite these simplifications, the results obtained are quite satisfactory, in particular if they are limited to the linear-elastic modulus, the compressive strength, and the fracture toughness.

Easterling *et al.* (1982) studied the structure and mechanical properties of balsa wood by means of experiments performed in a scanning electron microscope and observing how the cell walls behaved. Initially, they examined how the cell walls deformed elastically and, then, how they underwent plastic collapse or rupture. The results showed that the elastic modulus, rupture load and anisotropy of the wood were controlled, in part, by the mechanical properties of the cell walls and, also, by their dimensions and shape. All of this indicated that the mechanical properties of the wood depended strongly upon its density. Data from other woods suggest that these findings can be generalised.

The model used by Easterling *et al.* for deducing the mechanical behaviour is illustrated in Figure 9.16a. It consists of a cellular honeycomb structure and is intended to be a simplification of the mesoscopic structure shown in Figure 9.13. The structure is composed of elongated cell blocks, separated by partitions (spokes or radii) formed of smaller cells, and of other shapes. The model was inspired by photos of balsa wood, as in Figure 9.16b.

The *growth rings* form cylindrical surfaces normal to the radii and are visible to the naked eye, but they become diffuse when viewed in the electron microscope. For this reason, these differences have not been included in the model, which only considers the two materials mentioned: the elongated cell bundles and the spoke cell panels. (This may create some confusion when simplified models are developed, in which the role of the growth rings is ignored, as in Easterling's model.)

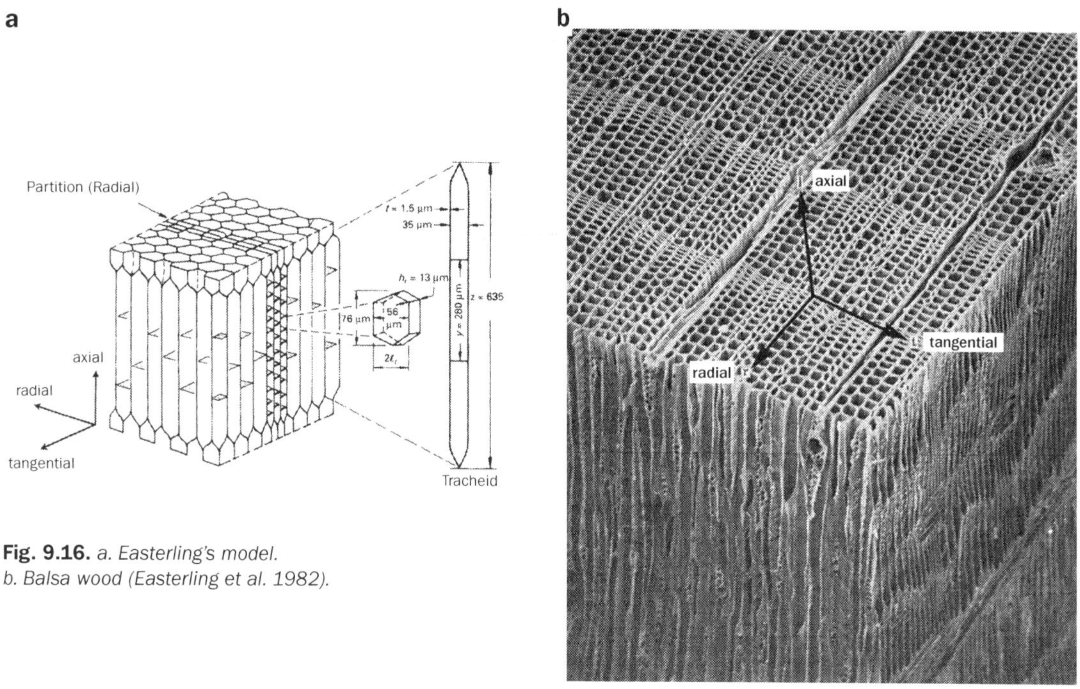

Fig. 9.16. *a. Easterling's model. b. Balsa wood (Easterling et al. 1982).*

Elastic modulus

Figure 9.17 shows average values of the stress-strain curves for various woods of a range of densities (beech, 750 kg/m^3; willow, 400 kg/m^3; balsa, 110 kg/m^3), obtained from axial (in the direction parallel to the fibres) and radial (in the direction perpendicular to the fibres) compression tests. Results from tests in the tangential direction are very similar to those in the radial direction.

For small strains (less than 0.02%), the behaviour is linear elastic in both directions. The elastic modulus in the axial direction is always greater than that in the radial (this effect cannot be observed in the figure). Beyond the linear-elastic region the curves exhibit a plateau, which can extend to strains of the order of 0.2% (for beech wood) and 0.8% (for balsa). Then the stress increases rapidly.

The deformation mechanisms in the cell walls during the compression test are shown in Figure 9.18. These can be, more or less, confirmed by the scanning electron microscope studies performed by Easterling *et al.* (1982).

In the *axial compression* tests, the cell walls are initially compressed; after, with increasing strain, the approximately prismatic cells collapse by folding and breaking of the cell bases (Figure 9.18a). This mechanism is more apparent in low-density woods, such as balsa; the cells are coupled together, like the teeth of two opposing combs, until the bases collapse and, a load drop is detected. The process is repeated in the next row of cells, and the resulting curve has a sawtooth profile with multiple breaks until the wood is practically dense at which point the load increases.

In compression tests in the *tangential direction* (and also in the *radial*), the cells bend and the walls end up collapsing (Figures 9.18b and c); in some woods, a small load drop is detected after the elastic limit, which is usually associated with this process.

As mentioned in Section 8.3.4, these tests can be modelled with a cell or honeycomb structure.

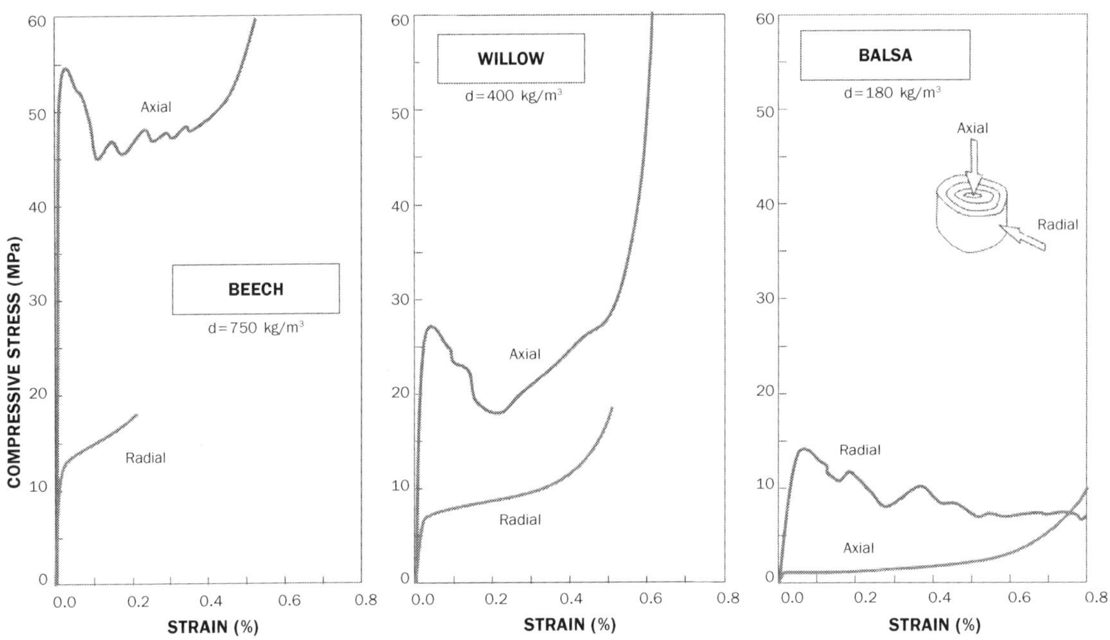

Fig. 9.17. *Engineering stress-strain curves obtained from compression tests in the axial and radial directions, for woods of very different densities.*

A model for balsa wood, according to Easterling *et al.* (1982), is shown in Figure 9.16. The tracheids and fibres can be idealised as hexagonal prisms in which, occasionally, there are some transverse membranes. The radii are transverse bundles with rectangular sections, as shown in the figure. The model could be refined, if necessary, by adding growth rings and channels.

Assuming an approximately hexagonal cellular structure which deforms by bending of the walls (Figure 9.18b) we can estimate the elastic modulus in the tangential direction, E_T, by means of Equations (87) and (88) in Section 8.3.4:

$$\frac{E_T}{E_c} = C_1 \left(\frac{d}{d_c}\right)^3 \quad (5)$$

where E_c and d_c are the modulus and density of the cell wall material, respectively. C_1 is a constant which depends upon the geometry of the honeycomb. To avoid confusion due to the anisotropy of the cell wall E_c, it is usual to take it as equal to the wall modulus in the axial direction (~35 GPa). The value of d_c is –as we have seen– about 1500 kg/m^3. d is density of the wood.

The elastic modulus in the radial direction, E_R, might be expected to be similar to E_T, but the presence of the radii, which have a greater density and modulus, make the structure appear stiffer in this direction.

$$E_R \simeq 1.5\, E_T \quad (6)$$

The elastic modulus of the wood in the axial direction, E_A, is greater than E_T because, in this case, the cell walls are loaded in compression, rather than bending. The result is the same as loading the honeycomb in the direction perpendicular to the plane. The result which was obtained in Section 8.3.4, Equation (101), is simply

$$\frac{E_A}{E_c} = \frac{d}{d_c} \quad (7)$$

The theoretical values of E_A and E_T have been represented in Figure 9.19 as a function of the relative densities, d/d_c, together with numerous experimental results for E_A, E_T and E_R, following Gibson and Ashby (1997). The predictions for the axial modulus, E_A, are quite good. The predictions for E_T and E_R are also acceptable for high densities. For relatively low densities –of the order of 0.1– the values of E_R are somewhat greater, perhaps for the reasons mentioned previously.

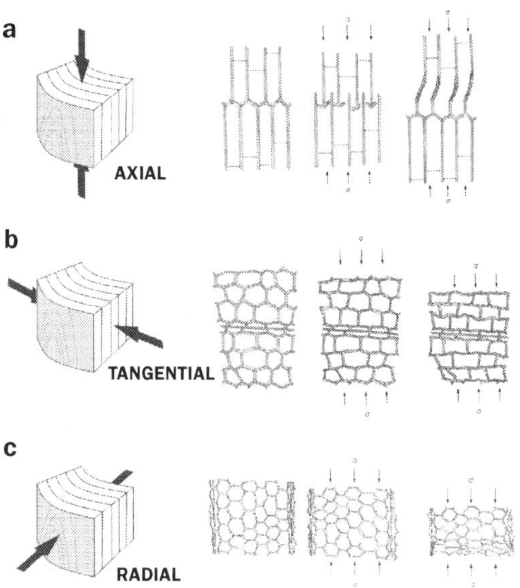

Fig. 9.18. *Deformation mechanisms in cell walls subjected to compression.*
a. Axial direction. b. Tangential direction. c. Radial direction.

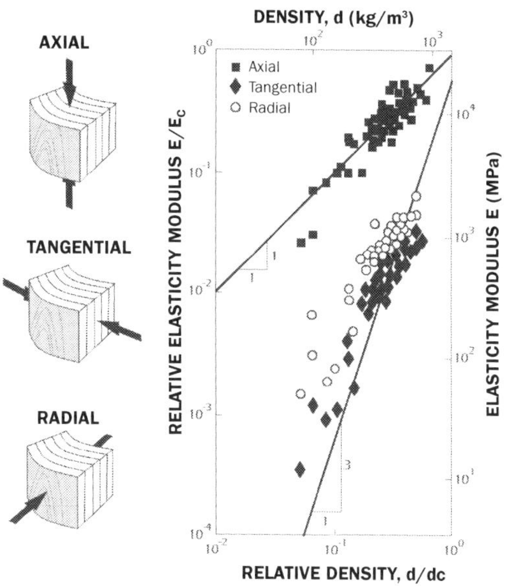

Fig. 9.19. *Values of the elastic moduli (axial, tangential and radial) relative to those of the cell walls, as a function of the relative density.*

EXERCISE 75
Show that Equation (2) $-E_R \approx 1.5\ E_T-$ gives **elastic modulus** in the **radial** direction as a function of the elastic modulus in the **tangential** direction.

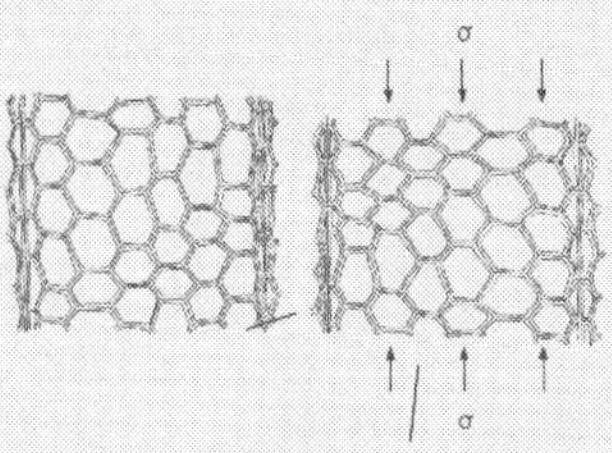

SOLUTION:
To evaluate E_R, we assume that the wood is a material composed of two components: the **tracheid cell** zone as one part, and the **ray** zone as the other (Figure 9.16).

On applying a radial compression (Figure 9.18c) both components act in parallel, according to the Voigt model shown Figure 8.24. As a result, the radial modulus E_R is given by Equation (107), and it will be a combination of the modulus of the region formed by the tracheid cells, which we can approximate by E_T, and the modulus E_{CR} of the zone occupied by the rays:

$$E_R = V_R E_{CR} + (1 - V_R) E_T$$

where V_R is **volume fraction of the rays;** about 10% in hardwoods and 20% in conifers.

On the other hand, if d_R is the density of the wood in the ray zone, the modulus E_{CR} will be

$$\frac{E_{CR}}{E_c} = C_1 \left(\frac{d_R}{d_c}\right)^3$$

since the deformation of the ray cells is also due to bending of their walls, which we assume is similar to that of the tracheid cells.

Combining the previous expression with Equation (1), we obtain

$$\frac{E_{CR}}{E_T} = \left(\frac{d_R}{d}\right)^3 = R^3$$

where R is the ratio of the ray and wood densities. R usually lies between 1.1 and 2.

With this result we can calculate the value of E_R

$$E_R = V_R R^3 E_T + (1 - V_R) E_T = \left[1 + V_R (R^3 - 1)\right] E_T$$

For **balsa** $V_R = 0.14$ and $R = 2$, which leads to $E_R \approx 2\ E_T$.

For **woods of greater densities,** R is practically 1 so that $E_R \approx E_T$.

Generally, Gibson and Ashby (1997) propose an intermediate value of $E_R \approx 1.5\ E_T$.

Compressive strength

The compressive strength can also be modelled using the results obtained in Chapter 8, employing cellular honeycomb models.

For loading in the *tangential* direction when the structures bend yield or buckle, the collapse stress due to plastic deformation, $\sigma_{R,T}$, is given by (Section 8.3.4, Equations 93 and 94):

$$\frac{\sigma_{R,T}}{\sigma_{y,c}} = C_3 \left(\frac{d}{d_c}\right)^2 \quad (8)$$

where $\sigma_{y,c}$ is the value of the yield stress of the cell wall material (~350 MPa), and d_c is its density.

For loading in the *radial* direction, the rays act as reinforcement and, once again, (according to Gibson and Ashby 1997) applying the rule of mixtures the strength in the radial direction, $\sigma_{R,R}$, for balsa wood can be estimated as

$$\sigma_{R,R} = 1.4\, \sigma_{R,T} \quad (9)$$

This result is more approximate for low-density woods (Easterling, 1982), but the difference between $\sigma_{R,R}$ and $\sigma_{R,T}$ decreases as the density increases.

Finally, for loading in the *axial* direction, failure is due to the plastic collapse of the bases of the prisms in low-density woods (such as balsa) or yielding and buckling of the prisms in higher-density woods (such as spruce). In any case, the failure is initiated by axial plastic strain, but the mechanism is not quite as simple as that described in Section 8.3.4. Easterling *et al.* (1982) have proposed the following expression:

$$\frac{\sigma_{R,A}}{\sigma_{yc}} = C_4 \left(\frac{d}{d_c}\right) \quad (10)$$

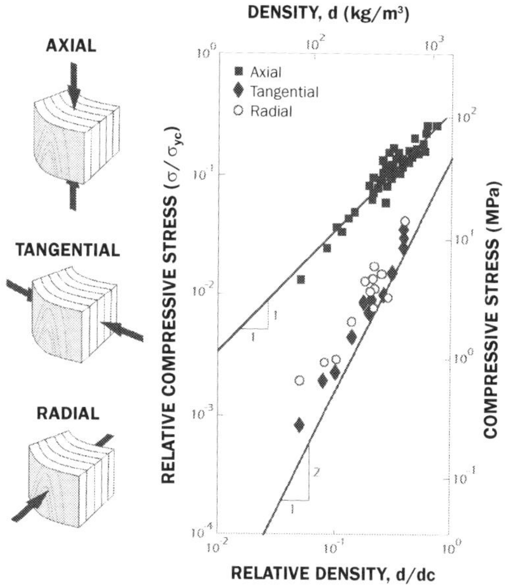

Fig. 9.20. *Values of compressive strength (axial, tangential and radial) relative to the compressive strength of the cell walls, as a function of the relative density.*

Theoretical values of $\sigma_{R,T}$ and $\sigma_{R,A}$ are shown in Figure 9.20 as a function of the relative density, together with numerous experimental results for $\sigma_{R,A}$, $\sigma_{R,T}$ and $\sigma_{R,R}$, according to Gibson and Ashby (1997). As may be observed, the predictions are quite good and illustrate a marked anisotropy which, moreover, is a function of the density; the ratio of the axial to transverse strength is about 2 for dense woods but may reach 10 for woods such as balsa.

Fracture and toughness

The study of the fracture of wood is tackled using the tools of Fracture Mechanics for linear-elastic homogeneous isotropic materials, even though it is well known that wood is orthotropic, nor is its behaviour linear-elastic for large deformations.

However, despite wood having a variable and heterogeneous microstructure, the proposed approximation yields reproducible results and data which, in most cases, are useful for design purposes. In what follows, account has been taken of the works of Ashby (Ashby *et al.*, 1985, and Gibson and Ashby 1997). A more refined study of the fracture of wood can be found in Stanzl-Tschegg (2006) or in Reiterer *et al.* (2002).

Values of the fracture toughness of ten woods are shown in Figure 9.21, as a function of their relative densities. The toughness was measured using standard procedures. Their values are referred to as K_{Ic}^{*n}, when the initial crack is normal to the fibre direction and K_{Ic}^{*P} when parallel to the fibre direction (Figure 9.21). Since, strictly, these are not true toughnesses –because of the simplifications discussed above– an asterisk is added to reflect this approximation. Note that the values of K_{Ic}^{*n} are, in general, ten times more than those of K_{Ic}^{*P}. The experiments were performed at ambient temperature, with a relative humidity of about 10% and with a loading rate of 0.01 mm/s, since it has been demonstrated that the fracture toughness values –like the elastic modulus and strength– depend upon the temperature, humidity and loading rate (Jeronimidis, 1980). A notable aspect of Figure 9.21 is that, under normal circumstances, the fracture toughness depends primarily upon the relative density of the wood.

Microscopic observation of the crack propagation in mode I allows identification of some failure mechanisms at the cell level. The most characteristic are shown schematically in Figure 9.22 (Ashby *et al.* 1985). For tests in the *direction parallel* to the fibres, in general, it was observed that in low-density woods ($d / d_c < 0.2$), the cracks advance breaking ofthe cell walls (Figure 9.22a). In dense woods ($d / d_c > 0.2$) two mechanisms occur; failure of the cell walls (Figure 9.22a) and debonding of the cell walls (Figure 9.22b). For tests in the *direction perpendicular* to the fibres, this second mechanism dominates (Figure 9.22b); the crack progresses seeking out the easiest route and when energetically favourable the walls disbond rather than break. This phenomenon produces a tortuous path such as that represented in Figure 9.21. It is also observed that when the crack encounters a large pore, or a channel, it often stops because the stress concentration at the crack tip is reduced (Figure 9.22c).

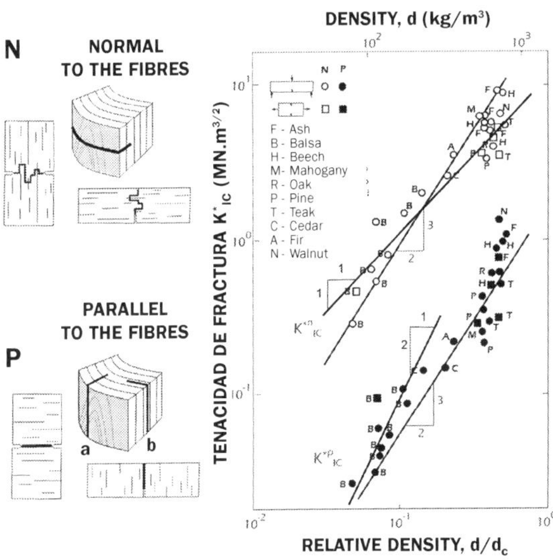

Fig. 9.21. *Values of the fracture toughness depending upon the crack location (normal or parallel to the fibres) for ten types of wood as a function of their relative densities.*

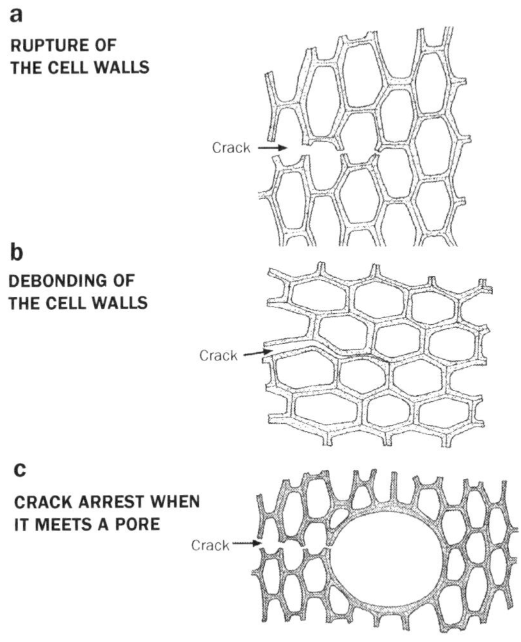

Fig. 9.22. *Schematic representation of various rupture mechanisms at the cell level.*

A macroscopic view of the cracks is shown in simplified form in Figure 9.21; when the initial notch is *parallel to the fibre direction,* the crack grows in the same plane and when the notch is *normal* to the fibre direction, the trajectory is more complex. In this case, the crack tends to propagate a certain distance in the direction perpendicular to the plane of the notch. Moving away from the plane of the notch, the stress is increased and the cell walls rupture following propagation in a plane parallel to the initial notch. Then, the above process is repeated. This is known as the Cook and Gordon (1964) fracture mechanism.

The trend of the results depicted in Figure 9.21 can be justified, following the reasoning of Ashby (1985), from the basic concepts of linear-elastic Fracture Mechanics, and from the simplifications introduced above to model the structure as a honeycomb. Following Ashby, we introduce the values of the specific fracture energy of the wood for rupture of the cell wall G_r, and for disbonding between the walls G_d. We can assume that the values G_r and G_d are essentially constant (although they depend upon the ambient temperature and humidity) for diverse woods, since the composition and structure of the cell wall vary little between species.

When crack propagation **parallel to the fibres** is produced, the primary mechanism is –as we have seen– rupture in the case of low densities and rupture combined with cell wall debonding in the case of high densities.

Consider the propagation of a crack whose plane is perpendicular to the tangential direction and which grows in the radial direction (Figure 9.21-P a). The following approximate relationship may be established for the value of the stress intensity factor at the crack tip K_I and the energy available for crack propagation G given by Equation (26) in Section 8.2.2.2.

$$K_I^2 = E_T G \tag{11}$$

In the previous equation, we have taken the elastic modulus in the tangential direction to be representative of the elastic modulus of the material, that is to say, in the direction perpendicular to the plane of the crack since, when it propagates, it produces a release of elastic energy principally due to strains perpendicular to the plane of the crack.

When propagation is associated with **debonding** of the walls, we have

$$\left(K_{IC}^{*p}\right)^2 = E_T G_d \tag{12}$$

Taking into account the result previously obtained for E_T, (Equation 1) we see that:

$$K_{IC}^{*p} = \sqrt{E_T G_d} = \sqrt{C_1 E_c G_d}\left(\frac{d}{d_c}\right)^{3/2} = C_5\left(\frac{d}{d_c}\right)^{3/2} \tag{13}$$

A similar dependency of K_{IC}^{*p} with density is obtained if we consider a crack *perpendicular* to the radial direction (Figure 9.21-Pb) which grows in the tangential direction (in this case the modulus E_R is used). The result, represented in Figure 9.21, shows good agreement –on average– with recorded experimental data.

When the propagation is produced by rupture of the cell wall, the energy consumed will be G_r. In this case, the fraction of area occupied by the cell walls in the plane of the break is, for regular panels, approximately $t/(2l\cos\theta) = t/(l\sqrt{3})$ (Figure 8.31a). Using Equation (87) in Section 8.3.4 we see that $t/(l\sqrt{3}) = (1/2)(d/d_c)$ and, for unit crack area, a fracture energy of 0.5 (d/d_c) G_r is consumed.

For propagation *parallel* to the fibres which go through the cell walls, we can write

$$K_{IC}^{*p} = \sqrt{0.5 E_T G_r\left(\frac{d}{d_c}\right)} = \sqrt{0.5 C_1 E_c G_r\left(\frac{d}{d_c}\right)^4} = C_6\left(\frac{d}{d_c}\right)^2 \tag{14}$$

When the crack propagates in a direction **normal to the wood fibres,** progressing through the rupture of the cell walls, the energy consumed per unit area of crack propagation will be equal to (d/d_c) G_r, since the proportion of ruptured surface is d/d_c.

Repeating the reasoning of the previous paragraphs, but now using the axial elastic modulus E_A whose dependency on density is given by Equation (7), it is found that:

$$K_{IC}^{*n} = \sqrt{E_A G_r\left(\frac{d}{d_c}\right)} = \sqrt{E_c G_r\left(\frac{d}{d_c}\right)^2} = C_7\left(\frac{d}{d_c}\right) \tag{15}$$

Representations of Equations (13), (14) and (15) are shown in Figure 9.21. It may be observed that the experimental results indicate that the behaviour lies approximately between the two extreme cases studied. Equation (14) has been adjusted only to the experimental density values of less than 0.2.

EXERCISE 76

Derive Equation (5) $-\sigma_{R,R} \approx 1.4\ \sigma_{T,R}-$ which expresses **the compressive strength** in the **radial** direction in terms of the compressive strength in the **tangential** direction for balsa wood.

SOLUTION:

The hypothesis that was made to simplify the problem –for balsa wood– is outlined in the following sketch (Easterling *et al.* 1982).

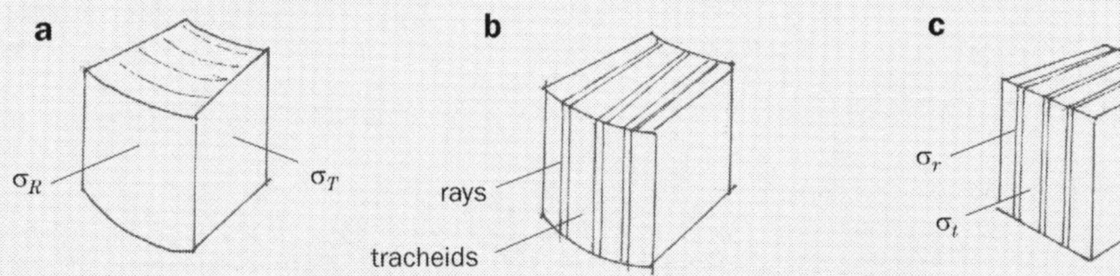

The notation employed is shown in figure *a*; the essential elements that need to be considered are highlighted in figure *b*: rays and tracheids (note that the effect of the growth rings is ignored), and figure *c* illustrates the most appropriate material model (Voigt), along with the notation.

The radial stress σ_R, according to this model, is given by $\sigma_R = V_r\, \sigma_r + (1-V_r)\ \sigma_t$ where, V_r is the volume fraction of the radial component, and σ_r and σ_t are shown in figure *c*.

To calculate the strength, it is assumed that the two materials considered (rays and tracheids) are elastic, perfectly plastic, as indicated in figure *d*. Structural collapse at $\sigma_{R,R}$ will occur when the yield strain of the most flexible material (tracheids) is reached. Therefore, $\sigma_{R,R} = V_r\, \sigma^*_r + (1-V_r)\ \sigma^*_t$ where σ^*_r and σ^*_t are indicated in figure *d*.

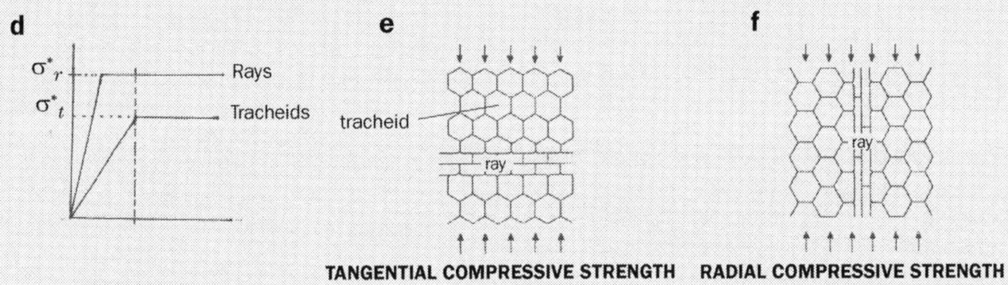

Figures *e* and *f* contain sketches of the structures of the elements of the rays and tracheids which suggest the hypothesis that the compressive strength of the **wood** in the tangential direction $\sigma_{T,R}$ must be similar to that of the **tracheids** in the radial direction σ^*_t, a hypothesis which leads to $\sigma_{R,R} = V_r\, \sigma^*_r + (1-V_r)\, \sigma_{T,R}$.

To calculate σ^*_r the honeycomb cell model may be used, from which $\sigma^*_r / \sigma_{yc} = C_3\ (d_r/d_s)^2$ where d_r is the density of the radial cells and σ_{yc} and d_s the values for balsa wood, and account has been taken of the fact that t/l is proportional to d_r/d_s.

On the other hand, according to Equation 8, $\sigma_{T,R} / \sigma_{yc} = C_3\ (d_t/d_s)^2$ where d_t is, now, the density of the trabecular region.

The ratio between σ^*_r and $\sigma_{T,R}$ will be $\sigma^* / \sigma_{T,R} = (d_r/d_t)^2 = R^2$.

For balsa wood, the following approximate values may be taken (Easterling *et al.* 1982), d_r is twice d_t, so that R=2 and the volume fraction of the radial component, V_r = 0,4.

Finally, from these data, $\sigma_{R,R} = V_r\, \sigma^*_r + (1-V_r)\, \sigma_{T,R} = [1-(R^2-1)\, V_r]\, \sigma_{T,R} = (1+3\, V_r)\, \sigma_{T,R} \simeq 1{,}4\, \sigma_{T,R}$.

9.2.4. Final remarks

Wood, as a **biological material,** has recently generated an interest in studying the relationship between structure and mechanical properties. One reason for this is the growth in the use of wood as a construction material, contrary to the trend of the last few years when it was thought that new materials were going to replace it, its low density (which facilitates transport and makes its specifc properties competitive), low thermal conductivity and very low electrical conductivity, together with low production costs, have made this material so attractive in both Civil Engineering and in Architecture. Another reason which has given impetus to the study of the relationship between the microstructure and properties is the growing interest in biomimetics. Many biological materials exhibit mechanical properties –such as strength, stiffness, or toughness– far superior to the those of their basic constituents, as is the case in wood. These properties are not accounted for using only composite materials models, something more is needed, it is necessary to take into account the hierarchical structure and other peculiarities. The aim is to unravel these aspects in order to design more competitive materials.

To design wooden structural elements, account has to be taken of the orthotropic nature of wood whose axes are longitudinal (in the direction of the trunk), radial, and tangential (as indicated in Figure 9.12). Furthermore, wood is a heterogeneous and hygroscopic material. Nonetheless, it is possible to design wooden structural components, assuming that the material is homogeneous and that its characteristics remain constant within a range of relative humidity, temperature and loading rate. The interested reader can refer to the book by Smith, Laudis and Gong (2003).

Wood, a biological material par excellence, enjoyed a certain role –years ago– as as **biomaterial,** in the sense of "material to treat, correct or *replace* tissues, organs or functions in the human body" (European Society for Biomaterials, 9th European Conference on Biomaterials, Chester, UK, 9-11 September 1991, ed. P.J. Docherty), since it was used to fabricate prosthetic devices.

A wooden prosthetic toe was discovered in an Egyptian mummy of a 50-year-old woman who lived 2,600 years ago. The prosthesis was articulated and must have been functional, as was demonstrated by replicas made at Salford University. Herodotus tells that in the year 550 B.C., the soldier Hegesistratus underwent amputation of a foot and was given a wooden implant to allow him to walk. There are numerous references to wooden prostheses in more recent times, with the best known being those involving pirates and, amongst these, the famous Cornelius Jol (1597-1641), nicknamed *Peg Leg* (*Houtebeen*, in Dutch). He was an admiral in the Dutch East India Company during the Eighty Years' War that the Low Countries waged against Imperial Spain. Jol, more than a soldier was a privateer who plundered Spanish and Portuguese vessels in the Atlantic Ocean. His nickname came from the use of a wooden prosthesis which replaced a leg lost in battle.

In the field of **biomimetics,** attempts have been made to replicate the cell structure of wood with ceramic materials. These are called biomorphic ceramics. Honeycomb structures have been fabricated with ceramic cells of the order of 50 μm much less than conventional ceramic cells (see for example Zollfranc and Sieber, 2005). These materials can be competitive for use at high temperatures, heat exhangers, filters or catalyst supports.

Another strategy based on densifying the wood by chemical treatment plus compression and high temperature (Song *et al.* 2018) produces a high-performance material for structural uses – increasing the failure load, toughness and impact resistance by an order of magnitude– and making it competitive with many metals on cost, weight and mechanical properties.

EXERCISE 77

A construction will employ single-piece pine beams whose dimensions are: length $L = 5$ m, depth $D = 20$ cm, width $B = 10$ cm. These beams are supported at their ends and carry a working load of $F = 1{,}000$ N acting at the middle.

It is required to calculate:

a) The central deflection of the beams.
b) By how much the load may be increased before rupture.
   What will be the safety factor, defined as the ratio rupture load/working load?
c) If one beam has a defect located in the centre of the beam –in the form of a through-the-thickness crack, such as that in Figure 8.14a– whose size is $a = 1$cm, what is the new safety factor?
d) The longest defect, such as that in the previous paragraph, that can be tolerated by the beam under the working load.

NOTE. Assume linear-elastic behaviour. The value of the Stress Intensity Factor for this example may be found in any manual or in the book of Elices and Guinea (2021).

SOLUTION:

a) The geometry of the problem is illustrated in Figure 8.9a. Equation (23) in Section 8.2.1 gives the central deflection of *an uncracked beam* subjected to three-point bending

$$\delta_{3P} = \frac{FL^3}{48EI} = \frac{4FL^3}{EBD^3}$$

in which we have substituted the second moment of area of the rectangular section B x D

$$I = \frac{1}{12}BD^3$$

The elastic modulus, E, is taken to be the modulus in the axial direction E_A, (Section 9.2.3) bearing in mind that –from the dimensions of the beam– the wood fibres will be parallel to the length of the beam, and the bending strains will occur in the axial direction. Equation (7) gives the value

$$E_A = E_c \frac{d}{d_c} = 13\text{GPa}$$

in which we have taken the ratio $d/dc = 0.37$, from Figure 9.21 and $E_c \approx 35$ GPa, the modulus of the cell wall in the axial direction. The value obtained is slightly higher than the usual range for Scots pine for structural use (8 to 11 GPa). With E_A and the beam dimensions, the resulting deflection is

$$\delta = 0.048\,\text{m}$$

b) To estimate the failure load for the *uncracked beam*, we calculate the maximum compressive stress that can be supported in the direction parallel to the fibres, from Figure 9.20. For $d/dc = 0.37$ we obtain $\sigma_{R,A} \approx 40$ MPa. This value is quite close to the experimental results obtained in compression tests in the fibre direction.

In Exercise 53 the maximum stress in a rectangular beam was calculated as

$$\sigma_{max} = \frac{3FL}{2BD^2}$$

This value is equal to the *maximum compressive stress* on the upper face of the beam and, also, the *maximum tensile stress* on the lower face. If we specify that the maximum compression in the fibre must not exceed $\sigma_{R,A}$, then

$$F_R = \frac{2BD^2}{3L}\sigma_{R,A} = 21.3\text{kN}$$

the maximum load that the beam can support without collapse.

In practice, the value of F_R is greater, since collapse usually occurs by tensile rupture, which in the case of pine is at a stress of about 100 MPa. In this case, the linear-elastic model that we are using ceases to be a good approximation. It induces –among other phenomena– a compressive yielding in the upper fibres, and a stress distribution which is no longer symmetrical about the central fibre of the beam. Nonetheless, supposing that the above formula can be extrapolated to the point of tensile rupture, a value $F_R = 53.3$ kN is obtained.

The safety factor of the structure will then be 21.3/5 = 4.3, or 53.3/5 = 10.7, for one failure load or another.

c) In this case we consider a beam with a central crack. The equations used in the previous paragraph are no longer valid.

Specifically, calculation of the deflection produced in the presence of the crack is complex and beyond the scope of this text. Suffice it to say that the deflection of a cracked beam –in the linear-elastic regime– is the sum of the beam deflection without the crack (calculated previously) plus a contribution which depends upon the ratio of the crack length to the depth of the beam a/D, which diverges when a tends to D.

To calculate the rupture load *when there are cracks*, we must resort to fracture mechanics. For a linear-elastic material, as we saw in Section 8.2.2, structural failure occurs when $K_I = K_{IC}$ (the local approach to fracture). The value of K_{IC} for our pine wood can be estimated from the data in Figure 9.21, from which we obtain $K_{IC} \approx 3.3$ MPam$^{1/2}$.

Equation (27) in Section 8.2.2 provides the value of K_I for a rectangular cross-section beam in bending with a length to depth ratio $(L/D)=4$. In our case $L/D=25$ so the formula is not strictly applicable, and we must refer to manuals which tabulate the value of K_I (for example, Elices and Guinea, 2021). To simplify the calculation, we take the problem to be equivalent to that of a cracked beam of rectangular cross-section, subjected to a bending moment $M=FL/4$, whose solution is

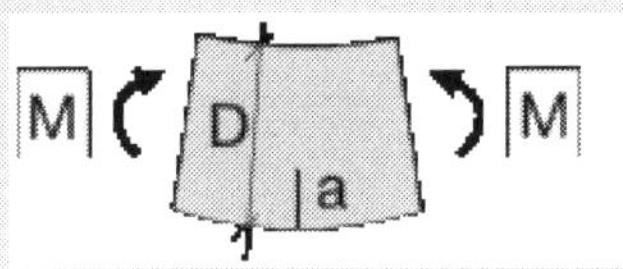

$$K_I = \frac{6M}{BD^2}\sqrt{\pi a}\left[1.122-1.40(a/D)+7.33(a/D)^2-13.08(a/D)^3+14(a/D)^4\right]$$
$$= \frac{3FL}{2BD^2}\sqrt{\pi a}\left[1.122-1.40(a/D)+7.33(a/D)^2-13.08(a/D)^3+14(a/D)^4\right]$$

For $a=1$ cm, the fracture load that causes $K_I=K_{IC}$ is $F_R=0.845$ kN, which is below the working load F_T = 5 kN. The safety factor is then about 0.16.

d) To calculate the maximum crack size compatible with the working load, it is necessary to solve the equation $K_I\ (F_T, a_c) = K_{IC}$ taking the value of a_c as an unknown, that is

$$K_{IC} = \frac{3F_TL}{2BD^2}\sqrt{\pi a_c}\left[1.122-1.40(a_c/D)+7.33(a_c/D)^2-13.08(a_c/D)^3+14(a_c/D)^4\right]$$

whose solution gives $a_c = 0.26$ mm.

We see that the beams are extremely sensitive to the presence of cracks. This fact is true for any cracked structure –within the hypothesis of linear-elastic behaviour– the presence of these defects drastically reduces the maximum loads that may be supported.

9.3. BONE

9.3.1. Introduction

Bone tissue contains an abundant extracellular matrix, the osteoid matrix, which surrounds widely spaced cells, together this is called **bone**, and in humans it is composed of approximately 30% collagen fibres, 55% crystal salt minerals, and 15% water.

Bone is formed in embryos from collagen. It has already been observed that collagen has good properties when it is subjected to tension; but since it is very flexible, it cannot support compression or bending which are the loads that bones have to resist. For this reason, during birth, animals add small amounts of hydroxyapatite to the collagen fibres, which provide the bone with rigidity. In adult animals, some bone may contain as much as 67% hydroxyapatite.

Bones, like other biological materials, possess enviable properties for an engineer since they are capable of self-repair and, also remodel themselves in response to mechanical loading. When the stresses carried by the bone decrease, the masses of the bones also decrease; a phenomenon recognised by astronauts after long periods in space. On the other hand, osteocytes (mature bone cells) detect the deformations induced when the stresses increase, and they release chemical substances which stimulate the osteoblasts (bone-forming cells) in order to generate new bone tissue.

The bones of mammals and birds can be considered, at the microscopic level, as a **material composed** of a collagen fibre matrix reinforced with calcium phosphate (hydroxyapatite) crystals. The structure of bone is not simple, it is hierarchical, as will be seen later. Bone tissue, furthermore, contains water (which plays an important role in the mechanical properties), living cells –*osteocytes, osteoblasts* (previously mentioned), and *osteoclasts* (cells which reabsorb the bone)– and blood vessels. Bone is covered with a compact conjunctive tissue called the *periosteum*. Tendons and ligaments are inserted into bones, and their extremes present a uniform transition from pure collagen fibres through to a *fibrocartilage* zone which is progressively calcified until it fuses with the bone.

Other tissues, such as antlers and dentin, may also be considered as bone tissue; the antlers present in the deer family are totally bone when fully developed. On the other hand, the *horns* of bovids are not bones but are hollow sheaths of the keratinised epidermis which are arranged around a bone core which emerges from the skull.

Bone is also a **multifunctional material**. Bones have evolved to be able to fulfil many functions, both within the same animal species and between different species. Three examples, taken from Currey (2002) and Cotterell (2010), serve to illustrate this point. The femur of mammals performs two classic functions: to support and facilitate movement. Femurs must, therefore, be strong and stiff. Antlers, which have to withstand the blows of the confrontations of the male deers in the mating season, must be tough (more resistant to fracture) than normal bone, and they need not be very stiff. The third example is provided by the middle-ear bone of whales, which need not be so stiff and strong as those already cited but which must be very dense in order to transmit and pick up sound. All of these properties are summarised in Table 9.5, in which density, stiffness (in terms of the elastic modulus) and the toughness (specific fracture energy) are shown.

TABLE 9.5 PROPERTIES OF VARIOUS TYPES OF BONE

PROPERTY	DEER ANTLER	HUMAN FEMUR	WHALE TYMPANUM
Density (kg/m^3)	1860	2060	2470
Stiffness Elastic modulus (GPa)	7.4	13.5	33.0
Toughness Fracture energy (J/m^2)	6190	1710	200

Bone is a **living material,** dynamic tissue, in which growth and restructuring are complex processes which implies *destruction* –achieved by cells which reabsorb the bone (osteoclasts)– and *construction* –due to bone-forming cells (osteoblasts). For example, in a long bone such as the femur, the internal medulary cavity becomes larger as a result of reabsorption of the internal surface of the surrounding bone and at the same time it adds new bone by deposition. The growth of the bone is regulated by various hormones. These, together with a derivative of vitamin D, are responsible for keeping the level of calcium in the blood constant. After menopause, women lose about 5% of their bone mass annually, which may lead to a condition known as osteoporosis and an increased risk of bone fractures.

Bone density is an important parameter which allows characterisation of several mechanical properties, as will be seen later. **Trabecular, cancellous** or **spongy** bone consists of a framework of bone tissue with a structure of interconnected voids. The direction of the tiny beams or trabeculae is orientated so as to optimise the resistance to the stresses that the bone must withstand under normal conditions. All bone is initially formed as spongy bone, but some become compact as a result of later remodelling. **Compact** bone (also known as **cortical** bone) is dense and apparently solid to the naked eye. In many bones, the external part is formed of compact bone and the internal, spongy bone. In some cases, such as the extremities of long bones and in vertebrae, this configuration minimises the weight of the bones and improves the distribution of load between them. In other cases, such as in the cranial vault or in the iliac crest, the bone structure adopts the form of a sandwich –with compact bone on the outside and spongy bone on the inside. This arrangement is much more mechanically efficient than homogeneous bone with the same mass.

The greater part of this chapter is dedicated to the knowledge of the mechanical behaviour of spongy bone, which is very important in biomedicine. In patients with *osteoporosis*, bone mass reduces until eventually bones may fracture under loads which might be regarded as normal in healthy individuals. These breaks are common in the femoral head, in vertebrae, and in the wrist, and are due, in part, to a reduction of spongy bone. Spongy bone is also involved in *osteoarthritis*, a degenerative disease of the joints in which the lubricating cartilage is eroded and may even disappear; in this condition, the distribution of forces which act on the joints depends upon the mechanical properties of the underlying spongy bone. Consequently, if the bone structure is altered, and with it its properties, changes in the distribution of forces on the joint may damage its lubrication system.

In order to quantify the mechanical behaviour of bone, it is necessary to know, in some detail, its microstructure (which is addressed in Section 9.3.2) and, from that information, to model it. To this end, previously described models for composite and cellular materials are used (which are applied in Section 9.3.3).

9.3.2. Structure

The structure of bone is complex, and in order to model the mechanical behaviour it is necessary to consider it at three distinct levels and to try to simplify it –idealising it– in each one of these levels. Going from bottom to top, the first level is that of the mineralised collagen fibres (microscopic level). These fibres can be arranged in various ways giving rise to different types of bone (mesoscopic level). The third level, which can be seen with the naked eye (macroscopic level) consists of structures formed by combinations of the different types of bone mentioned above.

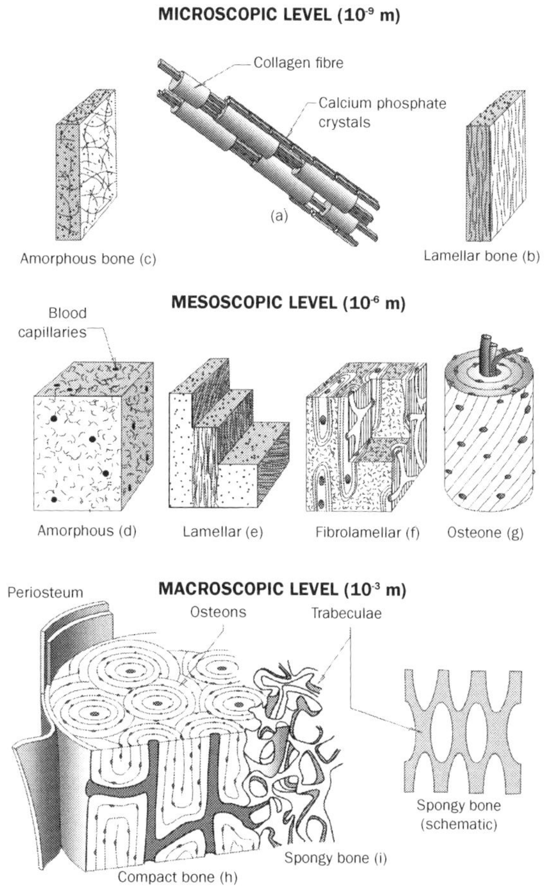

Fig. 9.23. *Bone structures that can be discerned at different levels of observation.*

Microscopic level

The basic component of bone is collagen fibre mineralised with calcium phosphate crystals as shown schematically in Figure 9.23a. The crystalline structure of the mineral phase is not easy to identify, due to the nanometric size of its crystals, but it is known to belong to the apatite group, specifically hydroxyapatite. The presence of small amounts of carbonate suggests that there could also be dahllite.

The mineralised collagen fibres can be arranged in two ways: ordered, giving rise to *lamellar* bone, or random, creating a tissue called *amorphous* (or *woven*) bone. Both types are shown, in simplified form, in Figures 9.23b and 9.23c, respectively. The fibres of amorphous bone are between 0.1 and 0.3 microns in diameter. The fibres in lamellar bone are thicker, between 2 and 3 microns, and arranged regularly to form plates whose thickness varies form 2 to 6 microns.

Mesoscopic level

The different combinations adopted by lamellar and amorphous bone are shown in Figures 9.23d-g.

Amorphous bone regions are found in young animals, in the foetus, and in the callouses which are produced when bone is repaired after fracture. This is a fast-growing bone. Some blood channels which traverse the bone are included in Figure 9.23d.

Figure 9.23e is similar to Figure 9.23d, with orientated fibres. This type of bone is usually called *primary lamellar* or *laminar*. It is common in reptiles, and also in birds and mammals in the form of circular layers; particularly in the periosteum. Its growth is slower than that of amorphous bone.

Figure 9.23f shows a mixture of the two previous types of bone: lamellar and amorphous. This type of bone is usually called *fibrolamellar*.

Figure 9.23g contains a representation of an *osteone*, the basic histological unit of bones. These are essentially lamellar bone cylinders, whose diameter measures about 200 m, with a central canal (also called the Havers or Haversian canal) through which arteries, veins and nerves may pass. This type of bone is also called Haversian. As indicated in the figure, there are also osteocytes connected through canals in the osteones.

Bone tissue is formed, in essence, by osteones cemented together and interconnected by blood vessels and nerves, where the cellular matrix dominates.

Macroscopic level

To the naked eye, bone appears to be a solid homogeneous material, but when sectioned it is far from homogeneous and exhibits a complicated structure; the outer part is dense compact bone while on the inside a spongy structure is observed. Both structures are shown in Figures 9.23h and 9.23i. The four previously described mesoscopic structures are involved in the formation of **compact bone.** On the other hand, **spongy** or *trabecular* **bone** is formed principally of laminar bone and osteone fragments. The structure of spongy bone is cellular; when the density is low, the structure resembles that of sponge, while when it is high, it is more like a foam with perforated bubble walls.

Compact bone is dense, with a **porosity** –percentage of the total volume not occupied by bone– almost always less than 15%. When the porosity is greater than 50%, bone is then considered *spongy*. The transition between compact and spongy bone is usually clear and occurs within a narrow range; it is rare to find porosities between 15 and 50% (Currey 2002, p. 23). The **relative density** –the ratio between the total or *apparent* density of the bone and the density of the wall material (1800 kg/m^3)– is also used for classification; in practice, the relative density of spongy bone ranges from 0.05 to 0.5 (technically any bone with a relative density less than 0.7 is classified as spongy).

Bone grows in response to mechanical loads, as already mentioned, everything seems to indicate that the trabeculae tend to orient themselves in the direction of the principal stresses following the isostatics shown in Figure 9.24. It is understood that the density of spongy bone depends upon the magnitude of the stresses that it supports, while the mesostructural anisotropy is a function of the direction and values of the principal stresses.

For a more detailed explanation of the different structures cited, the reader is directed to Currey (2002), chapter 1.

a **b**

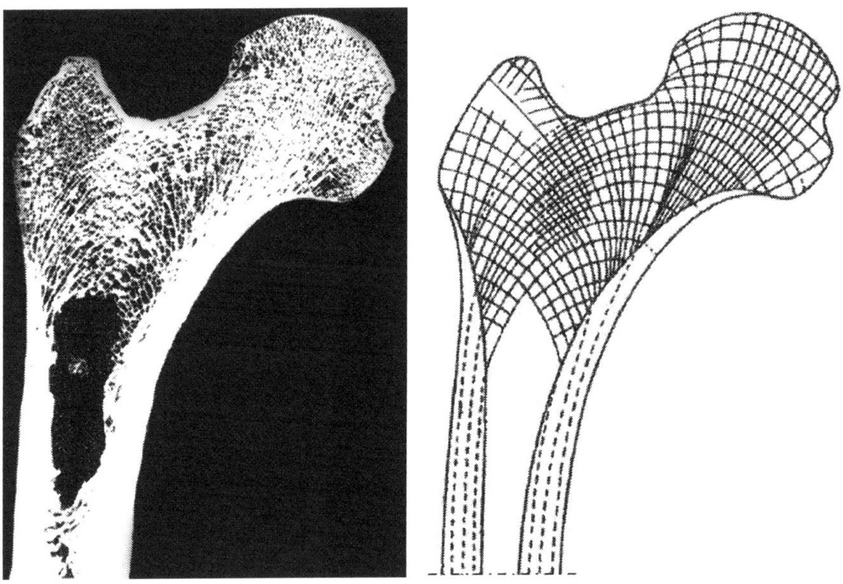

Fig. 9.24. *a. Longitudinal section of the femoral head.*
b. Isostatics (zones of pure compression or tension) obtained for an elastic model when the head is subjected to physiological loads.

9.3.3. Mechanical properties

A knowledge of the mechanical properties of bone is required for an understanding of their behaviour and function, whether they are healthy or damaged; consider, for example, the risk of fractures in osteoporosis or the advantage of knowing the stress distribution in the design of implants. For these purposes, models are needed for the prediction of the mechanical behaviour of bones.

The simplest models are based upon homogenising the mechanical analysis, considering the average properties of a representative volume of the material for compact bone, or a representative cell, or set of cells, in the case of spongy bone. In all of them, it is essential to have information about the basic components; data which are obtained experimentally for each case.

Another, even simpler, way of tackling the global problem consists of using dimensional analysis to model the deformation and failure mechanisms observed in cellular materials without specifying exactly the geometry of the cells. This is the approach adopted by Gibson and Ashby (1997, 2010). For spongy bone, these models assume that the cell geometries are similar, independent of their density. The mechanical properties are obtained as functions of the relative density and of the values of compact bone. Even so, experimental data are needed to provide the constants related to the cell geometry, as will be seen later.

A third approach is based upon detailed modelling of the bone using finite element analysis, a procedure that is gaining increased favour as computing capacity increases and, with it, the possibility of observing and recording, in each case, the mesoscopic and microscopic bone structure, such as with computerised tomography. This method allows consideration of local geometrical effects, such as notches, or imperfections.

From now on, and in order to make the section more succinct, the treatment is basically restricted to dimensional models.

Stiffness and strength

Figures 9.25 and 9.26 contain the stress-strain curves obtained from compression (Figure 9.25) and tensile (Figure 9.26) tests on various types of bone. One of the purposes of these figures is to show the diversity which is obtained depending upon the type of bone, the anisotropy and the type of test.

The relative density of the bone has a great effect in compression tests (Figure 9.25); compact bone is very stiff and not very flexible, while spongy bone suffers reduced stiffness and increased flexibility as its porosity increases.

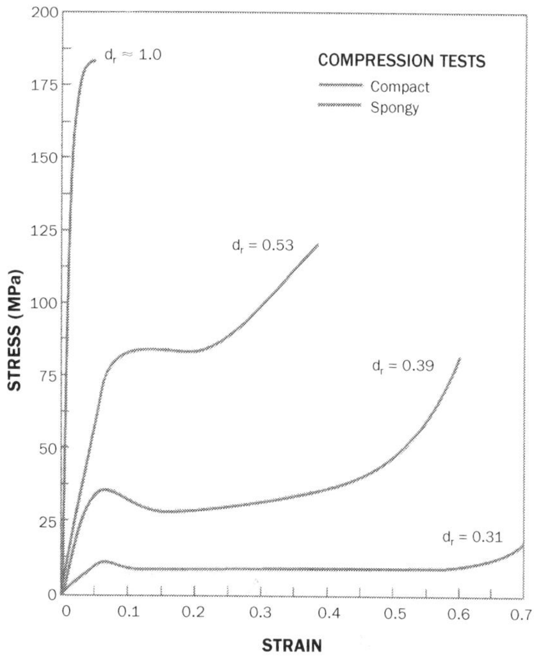

Fig. 9.25. *Influence of the relative density* (d_r) *in compression tests on compact and spongy bones (data from Hayes and Carter (1976) and Hayes and Gerhart (1985)).*

The influence of the test direction, whether longitudinal or transverse, is clear in the tensile tests (Figure 9.26) performed on compact bone. Other tensile test results for spongy bone have been included in the figure to illustrate the different values of strength and strain to failure.

Bone, as has been mentioned several times, is a complex material in which, as well as the density and anisotropy, time also influences the mechanical properties, as will be discussed later. For these reasons, modelling the mechanical behaviour of bone is difficult. Figure 9.27 shows, in schematic form, the mechanism responsible for the deformation of spongy bone subjected to compression (Figure 9.27a) and tension (Figure 9.27b).

A great deal of data (strength, strain to failure and intitial elastic modulus) can be found in Currey (2002) for compact human and bovine bones, tested in the longitudinal, radial and circumferential directions.

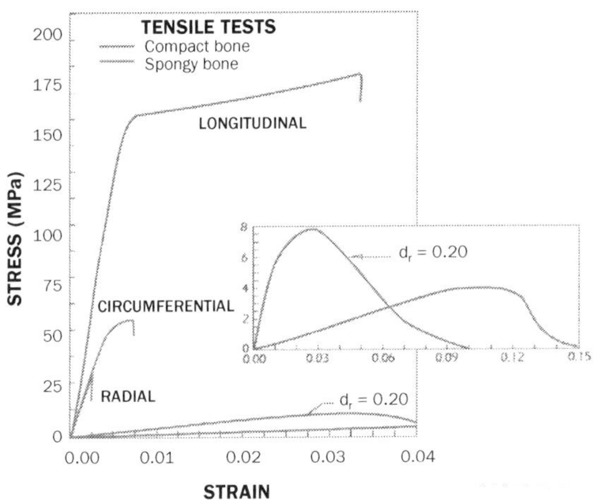

Fig. 9.26. *Influence of anisotropy and relative density (d_r) in* ***tensile*** *tests on compact and spongy bones (data from Carter et al. (1980) and Kefalas and Eftaviopoulos, 2012).*

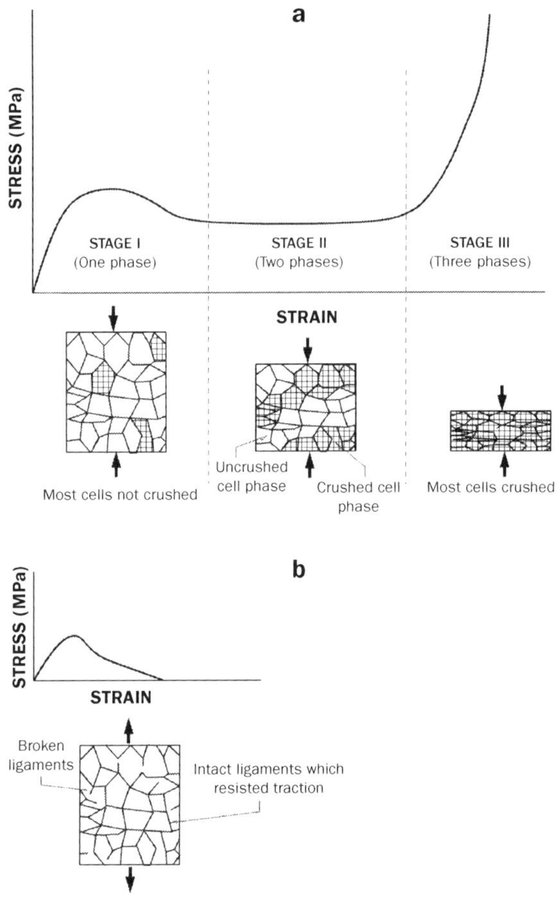

Fig. 9.27. *Mechansims responsible for the deformation of spongy bone when tested in compression (a) and tension (b).*

Next, two essential aspects are analysed in some detail in order to understand the mechanical behaviour of bone: the *stiffness* and *strength*.

The **stiffness** of bone, measured in terms of the *elastic modulus*, may vary from 7 GPa (for example, for the compact femoral bone of a young, 3-month-old polar bear) to 30 GPa (for the femur of a deer). The value for femoral compact bone in a human adult is around 17 GPa. These values are much lower than the, approximately, 120 GPa of hydroxyapatite. The average modulus of collagen –the other component of bone– is of the order of 1 GPa.

Several attempts have been made to model bone. The first, and simplest, for compact bone were based on **composite material** models. Katz (1971) considered two extreme cases: When the load is in the direction of the reinforcement (Voigt model) and when it is perpendicular (Reuss model). The corresponding elastic moduli are given by (see Section 8.3.3).

$$E\ (Voigt) = V_m\, E_m + V_c\, E_c \tag{16}$$

and

$$1/E\ (Reuss) = V_m/E_m + V_c\,/E_c \tag{17}$$

where the subindices m and c refer to the mineral and collagen, respectively, and V is the volume fraction of each constituent. Both extreme cases are represented in Figure 9.28, together with the experimental values taken from Currey (2002). This model has little predictive capability and was later refined (Katz 1981, Wagner and Weiner 1992, Pidaparti *et al.* 1996, and Jager and Fratzl 2000, among others). Upper and lower bounds of a more refined model, that of Hashin-Shtrikman, are also shown in Figure 9.28, also considered by Katz. The dentine region has been included because in that type of bone the fibres are highly orientated in the loading direction and for that reason the Reuss model gives a reasonable prediction.

The elastic modulus in bone –which, as has already been noted, is essentially a material composed of hydroxyapatite, collagen and water– is basically a function of two variables: the percentage of mineral content and the porosity. The relationship between the elastic modulus and mineral content can be seen in Figure 9.28. The trend is clear, but if the contribution of porosity is added, as will be seen later, the correspondence of these two variables improves markedly.

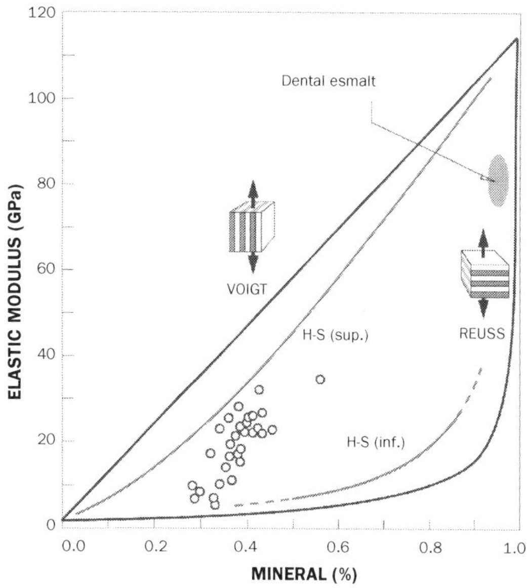

Fig. 9.28. *Values of the elastic modulus (for different compact bones) as a function of their mineral content. The predictions of various models (see text) are also shown in the figure. The experimental data are from Currey (2002).*

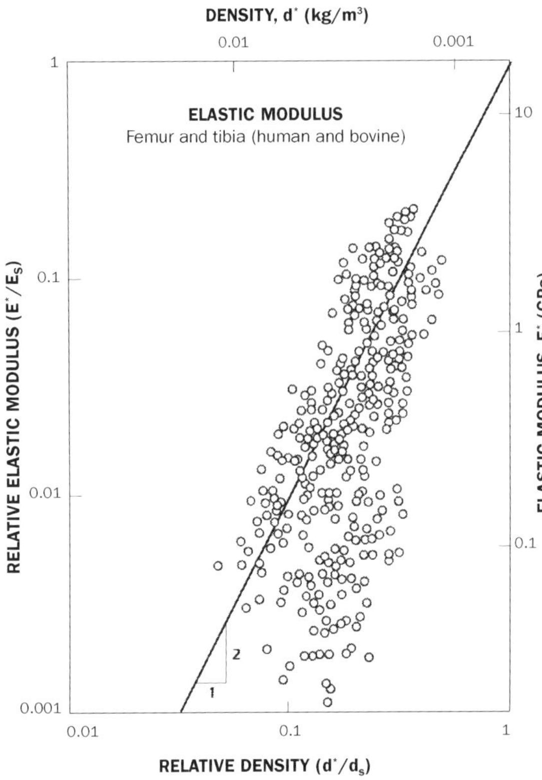

Fig. 9.29. *Values of the elastic modulus (for different bones) as a function of their relative density (based on Gibson et al., 2010).*

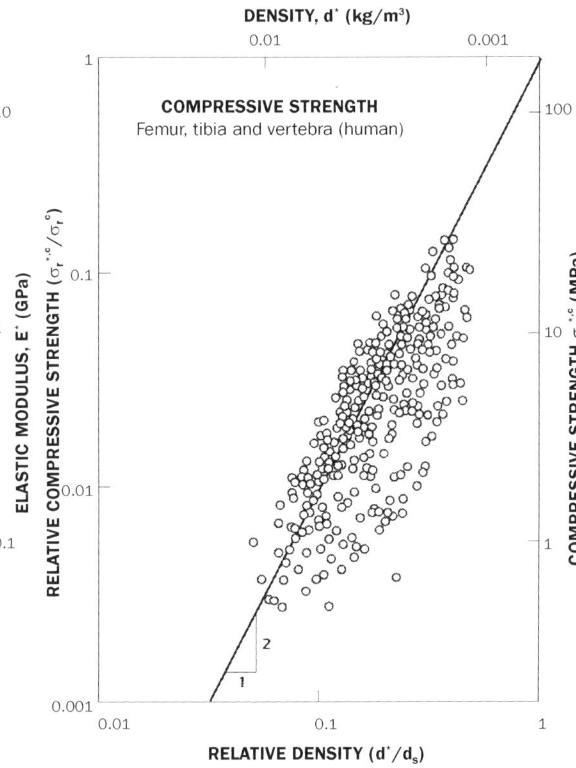

Fig. 9.30. *Values of the compressive strength (for human bone) as a function of their relative density (based on Gibson et al., 2010).*

The contribution of porosity, more specifically for spongy bone, has been analysed in detail by Ashby and Gibson using the **cellular model** described in Section 8.3.4.

In order to estimate the initial elastic modulus, E^*, it may be assumed that the deformation of the cellular structure is linear elastic and that the trabeculae are tiny beams which deform in bending. In these circumstances, it is found –Section 8.3, Equation (104)– that the relative elastic modulus (E^*/E_s) varies as the square of the relative density (d^*/d_s), that is to say:

$$\frac{E^*}{E_S} = C_1 \left(\frac{d^*}{d_S}\right)^2 \tag{18}$$

The experimental results from numerous measurements on human and bovine bones (principally on tibias and femurs) are presented in Figure 9.29, and reasonable agreement with Equation (18) is apparent. The elastic modulus E^* and the density d^* of the spongy bones have been normalised with respect to the trabecular wall values of the bone (indicated in Table 9.6); 18 GPa and 1800 kg/m^3, respectively. The value of C_1 is, approximately, unity (Gibson *et al.* 2010).

TABLE 9.6 AVERAGE VALUES OF THE TRABECULAE OF HUMAN SPONGY BONE

DENSITY KG/M³	ELASTIC MODULUS (GPa)	σ_r^c (COMPRESSION) (MPa)	σ_r^t (TENSION) (MPa)
1800	18	182	115

The dispersion in the experimental results is due to several causes: the marked *anisotropy* of the spongy bone and the fact that measured values depend upon the direction in which the load is applied (see, for example, the data obtained by Williams and Lewis 1982). The *duration of the tests* also influences the results, as will be seen later; the strain rate may vary by five orders of magnitude. Finally, the *water content,* or the relative humidity during the test, can also affect the values obtained.

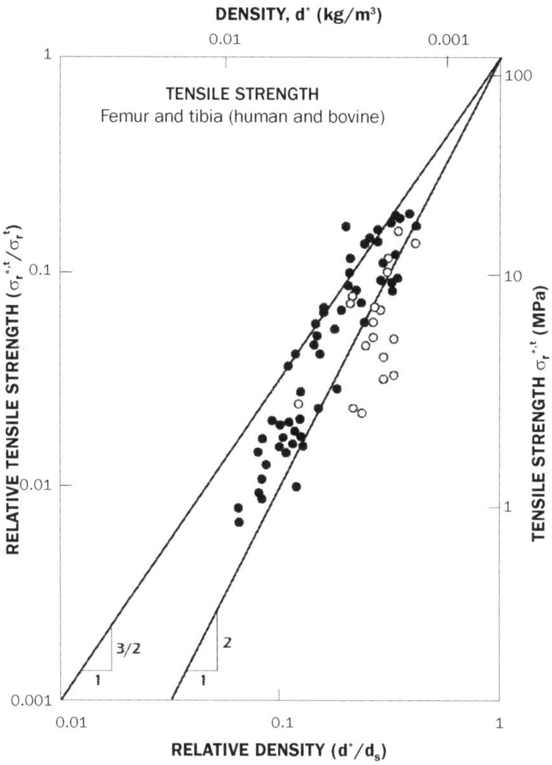

Fig. 9.31. *Tensile strength (for human and bovine bones) as a function of the relative density. The filled symbols correspond to specimens in which longitudinal direction coincides with the bone axis (based on Gibson et al., 2010).*

The values of the **strength**, in tension as well as compression, do not show such a clear dependence upon the mineral phase content as the elastic modulus and the classical models for composite materials yield poor predictions. The cellular models provide more acceptable results for *spongy bone.*

In order to estimate the *compressive strength* of spongy bone, it may be assumed that the structure collapses by buckling of the trabeculae, according to the analysis in Section 8.3.4, Equation (107), in which it was found that:

$$\frac{\sigma_{rot}^{*,c}}{E_s} = C_2 \left(\frac{d^*}{d_s}\right)^2 \tag{19}$$

The experimental measurements on human bone are presented in Figure 9.30, in which good agreement with Equation (19) is apparent. The rupture load $\sigma_{rot}^{*,c}$ must be normalised by the corresponding elastic modulus, since this datum is difficult to obtain the rupture stress in compression in compact bone has been used (= 182 MPa). The density is normalised, as before, with that of the cell wall of the bone (1800 kg/ m³). The dispersion in the results is due to the same causes as mentioned before.

The *tensile strength* of spongy bone may also be estimated with the cellular model assuming that rupture is produced by the formation of plastic hinges in the trabeculae as was analysed in Section 8.3.4, Equation (108), from which:

$$\frac{\sigma_{rot}^{*,t}}{\sigma_r^t} = C_3 \left(\frac{d^*}{d_s}\right)^{3/2} \tag{20}$$

Figure 9.31 contains Equation (20) together with experimental results obtained with human and bovine bones. The rupture stress has been normalised by the tensile strength of compact bone (σ_r^t = 115 MPa) instead of σ_{ys} and the density by that of the bone wall (1800 kg/m³). The tests in which the longitudinal direction of the specimen coincides with the bone axis (femur or tibia) are represented by filled symbols and fall within lines of slopes 3/2 and 2. The tests represented by open symbols correspond to specimens tested transversely.

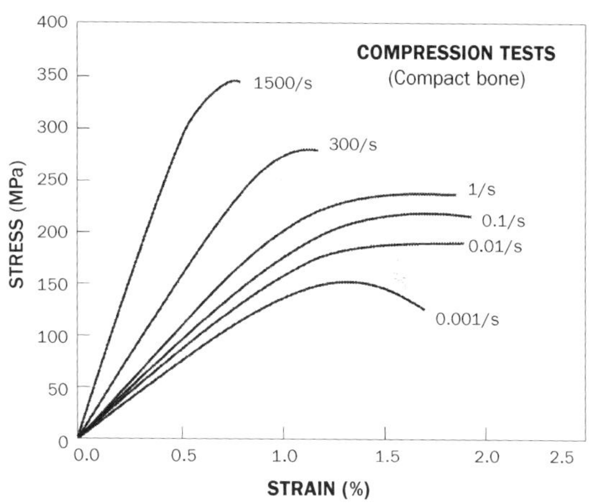

Fig. 9.32. *Compression tests on compact bone (human and bovine) performed at different strain rates (McElhaney and Byars 1966).*

EXERCISE 78

Compression tests have been carried out in the three directions of the edges of a cube removed from cortical bone, as shown in the figure below.

In all of the tests, the bone was compressed uniaxially up to the same value of strain, and the strain produced in the other directions measured. The table below presents the values obtained.

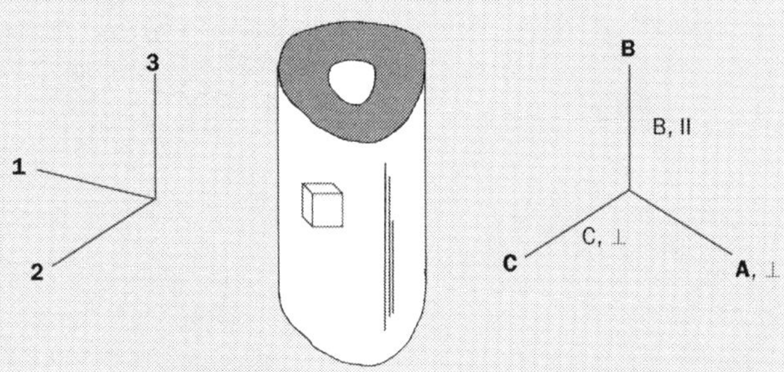

COMPRESSION SENSE	ε_{AA} (%)	ε_{BB} (%)	ε_{CC} (%)	ε_{AB}, ε_{AC}, ε_{BC}
axis A	**–0.067**	0.022	0.036	0
axis B	0.022	**–0,067**	0.022	0
axis C	0.037	0.023	**–0.067**	0

Using the results of Section 8.3.2 and Exercise 65 gives reasoned responses to the following questions:

a) Which mechanical behaviour –orthotropic or transversely isotropic– better fits the results obtained?
b) What is the direction of the osteons? We will call this the parallel axis (||) and the axes at 90º, perpendicular axis (⊥).
c) What is the value of the Poisson ratio $\nu_{\perp\perp}$?
d) What is the value of the Poisson ratio $\nu_{\perp||}$?

SOLUTION:

a) and b) As has been seen, cortical bone is an anisotropic material which responds in different ways according to the direction of the loading. In the experiments performed, we observe that the shear stresses ε_{ij} are zero, since the A, B, C directions coincide with the axes of othotropy, as deduced from the structure of the flexibility tensor presented in Exercise 65:

$$\begin{bmatrix} \varepsilon_{11} \\ \varepsilon_{22} \\ \varepsilon_{33} \\ \varepsilon_{23} \\ \varepsilon_{31} \\ \varepsilon_{12} \end{bmatrix} = \begin{bmatrix} 1/E_1 & -\nu_{21}/E_2 & -\nu_{31}/E_3 & & & \\ -\nu_{12}/E_1 & 1/E_2 & -\nu_{32}/E_3 & & & \\ -\nu_{13}/E_1 & -\nu_{23}/E_2 & 1/E_3 & & & \\ & & & 1/(2G_{23}) & & \\ & & & & 1/(2G_{31}) & \\ & & & & & 1/(2G_{12}) \end{bmatrix} \begin{bmatrix} \sigma_{11} \\ \sigma_{22} \\ \sigma_{33} \\ \sigma_{23} \\ \sigma_{31} \\ \sigma_{12} \end{bmatrix}$$

in which 1,2, 3 are the directions of orthotropy. This tensor is symmetrical so that

$$-\frac{\nu_{21}}{E_2} = -\frac{\nu_{12}}{E_1} \quad -\frac{\nu_{31}}{E_3} = -\frac{\nu_{13}}{E_1} \quad \text{and} \quad -\frac{\nu_{32}}{E_3} = -\frac{\nu_{23}}{E_2}$$

In the case of a transversely isotropic material, where the 3-axis is that of isotropy, then

$$\frac{1}{E_2} = \frac{1}{E_1} \quad \frac{\nu_{31}}{E_3} = \frac{\nu_{13}}{E_1} = \frac{\nu_{32}}{E_3} = \frac{\nu_{23}}{E_2} \quad \text{and} \quad \frac{1}{2G_{23}} = \frac{1}{2G_{31}} \quad \frac{1}{2G_{12}} = \frac{1+\nu_{12}}{E_1}$$

When a load is applied along the axis of transverse isotropy, the strain along the other two axes is the same, a situation that does not occur if the material is orthotropic. We see that this is our case when the compression is applied along the B axis, which we will identify as the axis of transverse isotropy, axis 3. This axis will also be the axis of the osteons, the parallel axis (||).

For small numerical differences (0.037 vs. 0.036 and 0.022 vs. 0.023), we see that the strains fit the **transversely isotropic** model, axes A and C being axies 1 and 2, or perpendicular axes ($\perp$).

c) In order to obtain the Poisson ratio $\nu_{\perp\perp}$ which we identify with ν_{12}, we consider the load along the axis A (or axis 1) and the strain along C (or axis 2). In this case it will be

$$\varepsilon_{AA} = -0.067\% \quad \varepsilon_{CC} = +0.036\% \quad \text{and} \quad \nu_{12} = -\frac{\varepsilon_{CC}}{\varepsilon_{AA}} = 0.54$$

If we consider compression along the other perpendicular axis, C (or axis 2) , and the strain produced in the axis A (or axis 1) we will have

$$\varepsilon_{AA} = +0.037\% \quad \varepsilon_{CC} = -0.067\% \quad \text{and} \quad \nu_{21} = -\frac{\varepsilon_{AA}}{\varepsilon_{CC}} = 0.55$$

then, approximately, $\nu_{12} \approx \nu_{21}$.

d) The Poisson ratio $\nu_{\perp||}$, which we identify with ν_{13}, is obtained in a similar manner, considering now the compressive load along the axis A and the strain produced along B:

$$\varepsilon_{AA} = -0.067\% \quad \varepsilon_{BB} = +0.022\% \quad \text{and} \quad \nu_{13} = -\frac{\varepsilon_{BB}}{\varepsilon_{AA}} = 0.33$$

or also, loading along C and considering the strain along B:

$$\varepsilon_{BB} = 0.023\% \quad \varepsilon_{CC} = -0.067\% \quad \text{and} \quad \nu_{23} = -\frac{\varepsilon_{BB}}{\varepsilon_{CC}} = 0.34$$

leading again to $\nu_{13} \approx \nu_{23}$.

Creep and relaxation

The experimental results of mechanical tests indicate that the values obtained may depend upon time, since they are a function of the loading rate and the duration of the test. Figure 9.32 shows several results from compression tests on small compact bone, human and bovine. Tests were performed at various strain rates –from very slow to very rapid, = 1500/s– (McElhaney 1966). The rupture stress ranges from 150 MPa, for slow tests, to 350 MPa for very fast ones. The elastic modulus increases and the elongation under maximum load decreases with increasing test speed. In another type of experiment it is observed that the residual deformation after unloading depends upon the maximum load and the duration of the cycle. All of this indicates that bone is a visco-elasto-plastic material.

These aspects are important for an understanding of the mechanical behaviour of bone. For example, during vigorous activity, long bones may sustain strain rates of 0.03/s in the human body, and 0.08/s in galloping horses. Impact fractures could be produced with higher strain rates, between 0.3/s and 1/s (Currey 2002). Note that for a race horse the values are even higher, and this may be why frequent fractures are found. If the strain in a bone is maintained for a while, it is observed that the stress relaxes. On the other hand, if the stress is held constant, the strain increases and the bone *creeps*.

Figures 9.33a and b show the results of two *creep* tests, in which it is observed that the strain increases with time when the bone specimen supports a constant stress σ_0. The data in Figure 9.33a is for spongy bone from the tibia of a bovine under a constant stress of σ_0=19 MPa. Figure 9.33b is for compact bone from the femur of a cow, subjected to a constant stress of σ_0=86 MPa.

It may be observed that in both figures from the start of the test, the strain rate (tangent to the curve) decreases rapidly with time until it reaches a practically stable value. This first stage is known as primary creep. The stage when the strain rate is constant is called secondary or stationary creep. Finally, a later stage, when the strain rate increases rapidly, is called tertiary creep, and rupture of the specimen quickly follows. Currey (2002) has suggested that it is possible to find a threshold stress below which the strain rate in the stationary phase is negligible; for human bones this threshold is about 65 MPa. Data from Rimnac and co-workers (1993) for bovine compact bone (Figure 9.33b) suggest that the threshold –if it exists in this case– must be below 86 MPa.

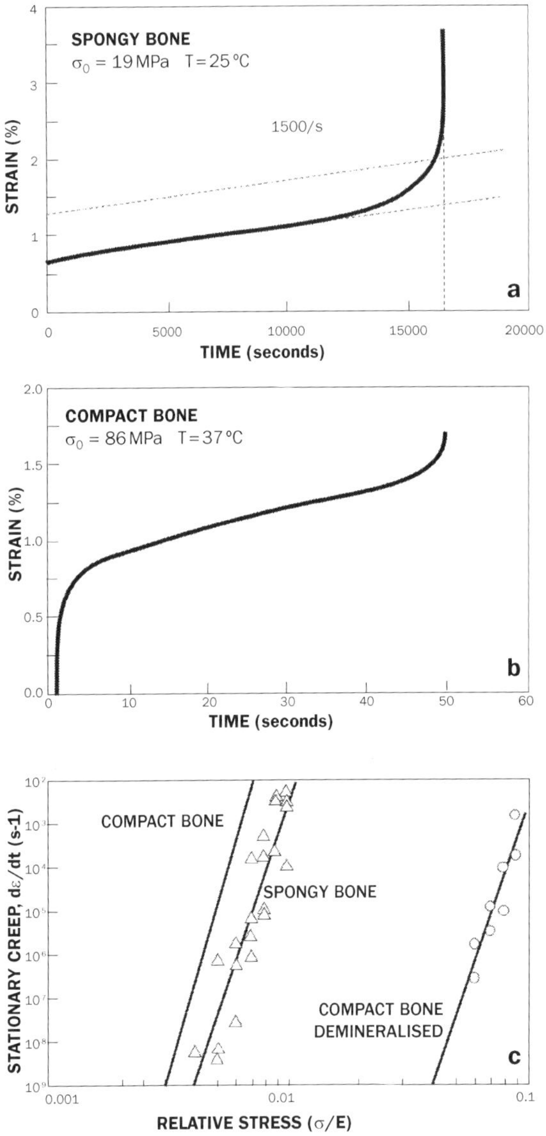

Fig. 9.33. *Creep tests.*
a. Spongy bone. Bovine tibia (Bowman et al., 1998).
b. Compact bone. Cow femur (Rimnac et al., 1993).
c. Compact, spongy and demineralised bone (Bowman et al., 1999).

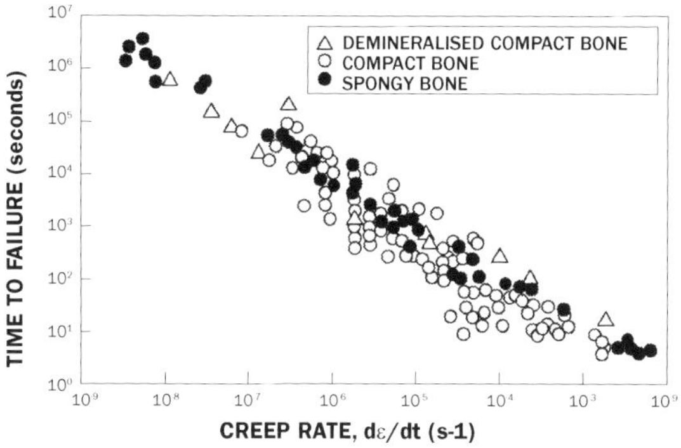

Fig. 9.34. *Time to failure as a function of the stationary creep rate for different types of bone: compact, spongy, and demineralised (after Bowman et al., 1999).*

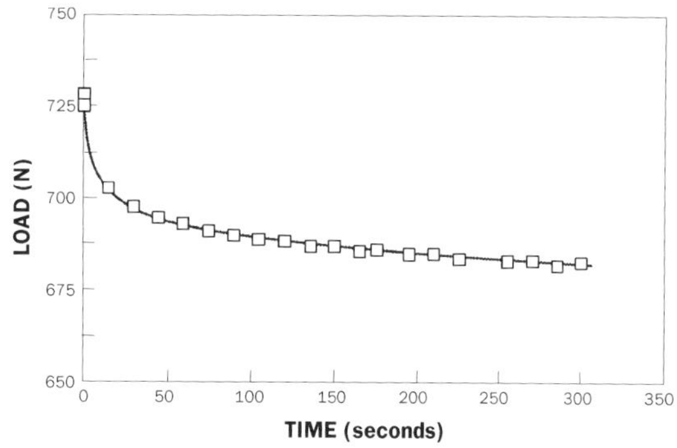

Fig. 9.35. *Relaxation test on spongy bone. Bovine femur (Quaglini et al., 2009).*

If the values of the strain rate in the stationary regime (secondary creep) are plotted as a function of the stress normalised with respect to the elastic modulus (σ_0/E), on a logarithmic scale, the results are quite well aligned, as shown in Figure 9.33c, in which data for spongy, demineralised compact bones are presented (taken from Bowman *et al.* 1999), together with a line corresponding to compact bone (according to Fondrk *et al.* 1992. The individual data have not been drawn). Two fairly obvious conclusions emerge from this figure:

The first is that the experimental results can be represented as a power function of (σ/E), that is to say:

Compact bone	$d\varepsilon/dt = 3.38 \cdot 10^{38} \cdot (\sigma/E)^{18.9}$
Spongy bone	$d\varepsilon/dt = 1.18 \cdot 10^{28} \cdot (\sigma/E)^{15.6}$
Demineralised compact bone	$d\varepsilon/dt = 6.74 \cdot 10^{12} \cdot (\sigma/E)^{15.5}$

where the constants which multiply (σ/E) have dimensions of s^{-1}. Gibson *et al.* (2010) have suggested that an exponential fit is more realistic than the power law indicated above.

The second conclusion lies in the exponents of (σ/E); the similarity of the exponents for compact, spongy, and dimineralised bones suggest that the demineralised component of the bone (which reflects the role of collagen) is what is responsible for the creep, as much for compact bone as for spongy. In both, the mineral component plays the part of a hard elastic inclusion.

Figure 9.34 shows the time taken to break a bone specimen in a creep test as a function of the strain rate in the stationary region (both logarithmic scales). The data come from Bowman *et al.* (1999). The results for the previous three types of bone are a close fit to a straight line given by:

$$t\,(failure) = 0.03\,(d\varepsilon/dt)^{-0{,}91}$$

where t is given in seconds and $(d\varepsilon/dt)$ in $(\text{seconds})^{-1}$. The result obtained is characteristic of creep phenomena and known as the Monkman-Grant formula.

Figure 9.35 contains results of a *relaxation* test in which it is apparent that the stress decreases with time when the bone specimen supports a constant strain. The tests on spongy bone indicate that the behaviour does not conform to a linear viscoelastic model (Quaglini *et al.* 2009).

Anisotropy is an other issue which complicates the analysis of these tests (Iyo *et al.* 2004). In general, the procedure used to understand this phenomenon is based on separately considering the contribution of the collagen matrix (an integral part of the isotropic component) and the contribution of the bone structure (responsible for the anisotropy). It is unusual to observe rupture with time in relaxation tests (as opposed to creep tests), what is observed –when the initial strain is above that corresponding to the yield stress– is the generation of damage in the bone, a process which is reflected in rupture tests (Nagaraja *et al.* 2007).

Visco-elasto-plasticity also manifests itself in bone subjected to *cyclic loading*. In the first part of this book –under linear viscoelasticity in Section 2.5.3– the storage (G_1) and loss moduli (G_2), as well as the loss tangent were defined. Both G_2 and *tan* δ are indicators of the energy dissipated in each cycle, while G_1 is a measure of the elastic modulus.

Lakes *et al.* (1979) studied the viscoelastic response of human and bovine compact bone as a function of temperature, using cyclic torsion tests and came to the conclusion that temperature was important and that experiments must be performed at body temperature if tests are to be relevant to in vivo situations. More recently, Yamashita *et al.* (2001) studied the behaviour of human femur compact bone by means of cyclic flexure tests at a frequency of 1 Hz and at 37 °C; they measured the loss tangent, *tan* δ, and the storage modulus G_1. A typical set of results from these tests is represented in Figure 9.36. The intention of this study was to keep the loads low to use the formalism of linear viscoelasticity, but they found that for very low loads the dispersion in the results was large. It was necessary to find a threshold load that resulted in little scatter and which was sufficiently low to stay within the bounds of the linear hypothesis (these values depend upon the size and type of specimen). The moisture content in the specimens has a strong effect on the value of *tan* δ; dry specimens produce lower values than wet ones.

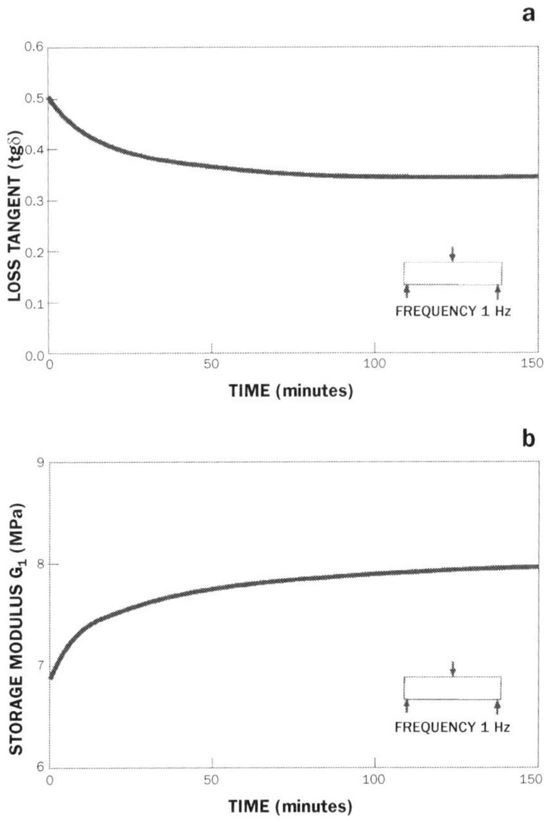

Fig. 9.36. *Cyclic load tests. Application of the linear viscoelastic model to human femur compact bone (Yamashita et al., 2001).*
a. Loss tangent.
b. Storage modulus.

EXERCISE 79

The figure below shows the compressive relaxation experiments of Quaglini *et al.* (2009), on 10 mm diameter cylindrical trabecular bone specimens. Using the results of Section 2.5.3 shows that the linear viscoelastic model is inapplicable to this material.

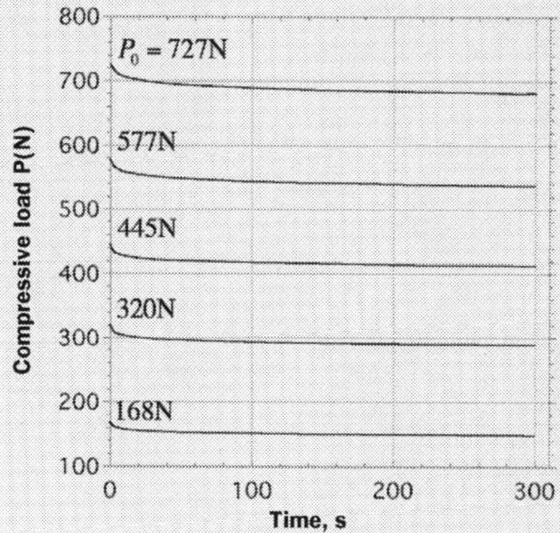

SOLUTION:

In Section 2.5.3, on linear viscoelasticity, we see that in a relaxation experiment the stress σ on the specimen must satisfy Equation (40)

$$\sigma(t) = \varepsilon_0 G(t)$$

where ε_0 is the initial applied strain, which is held constant throughout the test, and $G(t)$ is the relaxation modulus, a function of time which depends only upon the material. If we divide the above expression by the values at time $t = 0$ we obtain

$$\sigma / \sigma_0 = P / P_0 = G(t) / G(0)$$

so that the stress will be proportional to the applied load P.

Plotting the values of the relative load P/P_0 against time, we see that a **single** function $G(t)/G(0)$ is **not obtained,** as the linear viscoelastic model demands.

As discussed in Section 2.5.3, strict applicability of the linear viscoelastic model is quite limited and no biological material –bone included– satisfies the requirements for its application.

Even so, the linear viscoleastic model can be used to model trabecular bone –and other materials– provided the load and time intervals considered are limited.

As an example, in our case we see that the curves corresponding to P in the range 445 to 727 N, and up to 300 seconds are reasonably concurrent, so that a linear viscoelastic model with an average function $G(t)$ obtained from this relaxation can be sufficiently accurate for loads and times within the specified ranges.

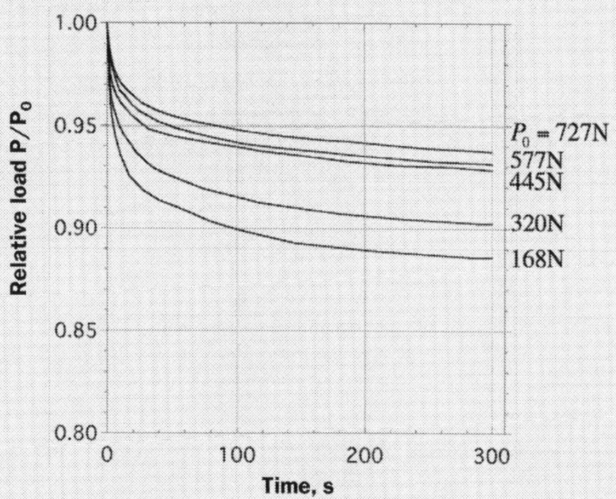

Fracture

Fractures in healthy bone are due to overloading (loads higher than normal) or suddenly applied loads. In both cases the crack responsible for the fracture is initiated in a place where a stress concentration or excessive strain is produced: a notch or a cavity. Other types of fracture are due to the bone being weakened and fracturing under normal loads. The weakening, or damage, originates from various causes; *osteoporosis* is the most frequent, but repeated loads, *fatigue*, can also weaken bone tissue, *creep*, as already noted, can be another cause but it is rare under normal circumstances. Below, both processess, overloading and fatigue, are briefly analysed.

Fractures due to static loading are initiated by the generation and unstable propagation of cracks, and for their study it is possible, in principle, to apply the fracture criteria described in Section 8.2.2. To this end, fracture tests have been performed on cracked specimens, in the form of beams (SEN) or compact tension specimens (C), and values of the fracture toughness (K_c) and the specific fracture energy (G_c) measured. These results must be treated with caution since bone does not always behave as a linar-elastic material (and therefore the concept of K may lose its validity), is anisotropic, and the results depend upon how the specimen was machined. Humidity, temperature and time also affect the results. Therefore, when values are taken from the literature, note must be taken as to how they were obtained.

In Figure 9.37 the common geometries used to measure fracture toughness are shown. The specimens are machined so that the crack propagates in the direction of the osteons (longitudinal, L) or transverse to them (T). The results depend upon the orientation of the specimens as illustrated in the adjoining graphic, the values of K_c double when the crack propagates transversely to the direction of the osteons. (The data correspond to human compact humerus bone; Ritchie *et al.* 2005.)

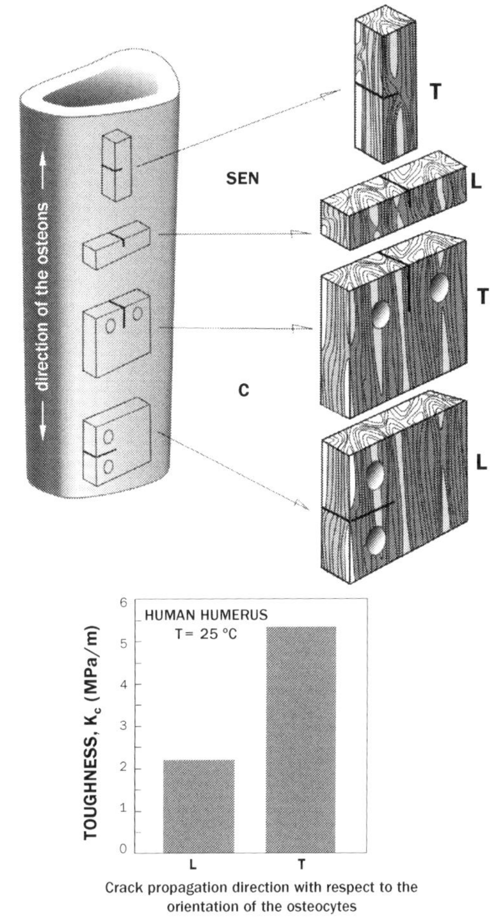

Fig. 9.37. *Common geometries and specimen arrangement for fracture toughness measurement. The data correspond to human humerus compact bone (Ritchie et al., 2005).*

In Table 9.7 values of K_c and G_c are included for different species, bone types and specimens machined in different orientations (Ritchie *et al.* 2005).

TABLE 9.7 FRACTURE TOUGHNESS OF COMPACT BONE (RITCHIE ET AL. 2005)

SPECIES	BONE	ORIENTATION	K_c (MPa $m^{1/2}$)	WHALE TYMPANUM	SPECIMEN GEOMETRY
Human	Femur	L	6.4 ± 0.3	–	SEN
Human	Humerus	T	2.2 ± 0.2	–	SEN
Human	Humerus	L	5.3 ± 0.4	–	SEN
Bovine	Femur	L	2.4 – 5.2*	920 – 2780	C
Bovine	Tibia	L	4.5 – 5.4*	760 – 2130	C
Bovine	Tibia	L	2.8 – 6.3*	630 – 2880	C
Baboon	Femur	T	1.8 ± 0.5	–	C
Baboon	Femur	L	6.2 ± 0.7	–	SEN

* Specimens performed at different loading rates. SEN: cracked beam specimens. C: compact tension specimens. L and T: orientations shown in Figure 9.37.

It is shown in Figure 9.38 that the values of the toughness can be a function of the crack propagation before reaching an unstable size; the R-curve behaviour discussed when studying fracture in Section 8.2.2. These experimental results illustrate the existence of intrinsic hardening mechanisms which hinder crack propagation, a topic which will be discussed next. Figure 9.38a shows, once more, that bone is an anisotropic material and that it is more difficult to cross the osteons than to propagate parallel to them (data from An *et al.* 2011). In Figure 9.38b the cracks propagate parallel to the direction of the osteons, but it indicates that the mechanisms which impede their propagation decrease with age, not only the resistance to propagation (slope of the line) but also the initial toughness (ordinate at the origin) (data from Koester *et al.* 2011). Therefore, once a crack is induced, it is more likely to be arrested in a young person than in an old individual.

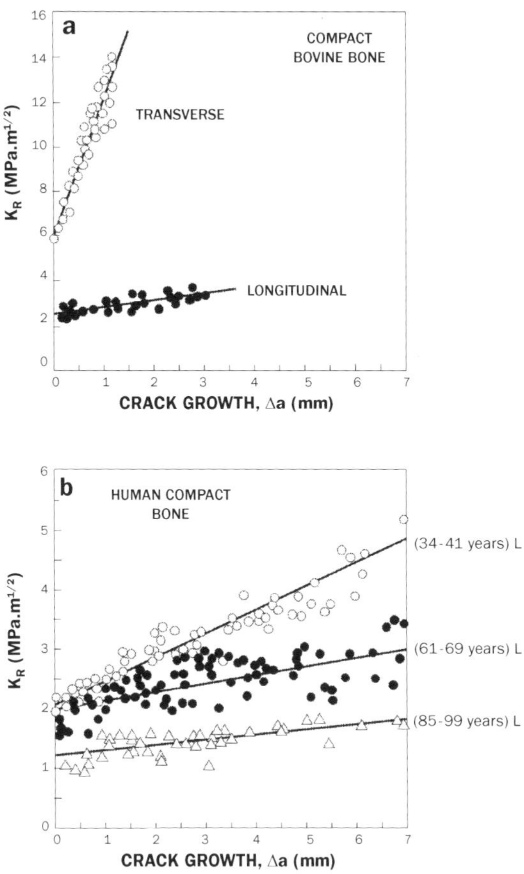

Fig. 9.38. *Values of the fracture toughness, K_R, as a function of the crack growth.*
a. Effect of anisotropy (An et al., 2011).
b. Effect of age (Koester et al., 2011).

In order to explain the increase in toughness with crack progression, several mechanisms have been proposed, and the most accepted are illustrated in Figure 9.39. The most important is due to ligaments, or *bridges*, between the two crack faces, see Figure 9.39a. If the number of bridges increases as the crack grows, so will the toughness. In some cases, when the number of ligaments created is equal to the number destroyed when the crack advances, the R-curve (indicated in Figure 9.38) stops increasing and remains horizontal until unstable fracture. Another mechanism which impedes crack advance, because the stress intensity factor decreases, caused by crack deviation in the vicinity of osteons (Figure 9.39b), when the crack plane is transverse (see Figure 9.37, direction T). The interfaces between the osteons are highly mineralised (usually called *cement lines*) and are microstructurally weak and, therefore, are the preferred paths for cracking. Finallly, crazing around the crack (Figure 9.39c) can also increase the toughness if the microcracking is accompanied by dilatation. All of these mechanisms are analysed in detail in Nalla *et al.* (2005) and Ritchie *et al.* (2005).

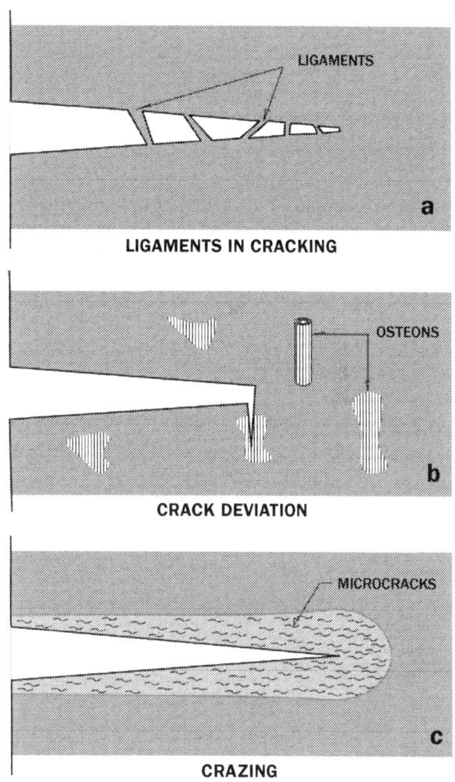

Fig. 9.39. *Mechanisms explaining the increase in toughness with crack growth. a. Formation of bridges between crack faces. b. Promotion of crack deviation. c. Crazing (diffuse cracking) accompanied by dilatation.*

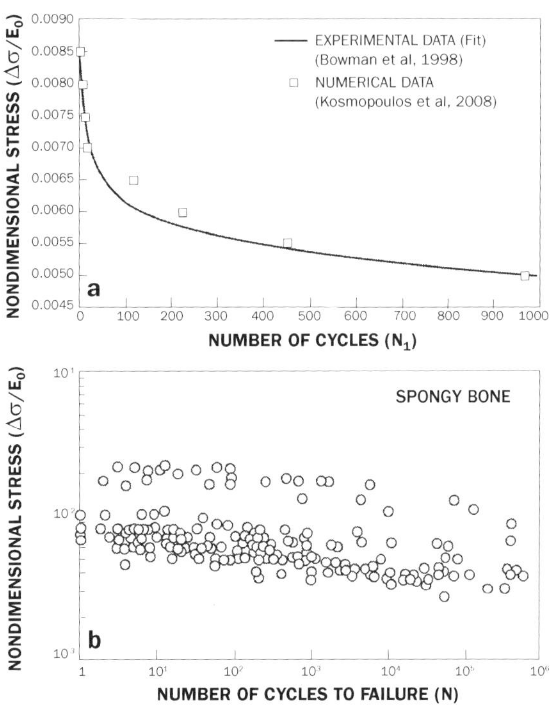

Fig. 9.40. *Compressive fatigue tests.*
a. Bovine spongy bone. Curve fitted to experimental data (Bowman et al., 1998) and data from a numerical model (Kosmopoulos et al. 2008).
b. Log-log plot of $(\Delta\sigma/E)$ *against number of cycles to failure* N_f*. Various experiments (Gibson et al., 2010).*

Fracture due to variable loads, *fatigue fracture*, is due to the generation of stable cracks over time until a critical crack size, when unstable fracture is produced. It is estimated that in humans half of vertebra fractures and 10% of hip fractures are due to the variable loads which we support in our daily lives and not due to a traumatic event (Meyers and Wilson 1997). In both cases, most of the loads are borne by spongy bone working in compression and for this reason almost all of the investigations into the fatigue fracture of bones have been carried out with cyclic compression tests on spongy bone.

The simplest tests consist of subjecting bone specimens to cyclic compressive loads and recording the number of cycles to failure. S-N curves, previously discussed in Section 8.2.2, are produced by this procedure. Various authors have suggested highly simplified models to explain these results.

Figure 9.40a shows the results from a compressive fatigue test on bovine spongy bone. The values obtained conform quite well with the following expression (Bowman *et al.* 1998):

$$N_f = 1.77 \cdot 10^{-23}\ (\Delta\sigma/E_0)$$

in which N_f is the number of cycles to failure, $\Delta\sigma$ is the stress range, and E_0 is the initial elastic modulus: 2,690 MPa.

The previous results lie on a straight line when the scales of the axes are logarithmic. However, in Figure 9.40b, we see that the data from the fatigue tests of five authors (according to Gibson *et al.* 2010) do not provide such a satisfactory fit. This may be due to the inclusion of different species of bone (human and bovine), different types of specimen tested in different orientations, to the contribution of creep and to the specific damage produced by the cyclic loads. In any case, a similar trend is observed in all of the results, the line which represents them has a slope of about -12.

Attempts at modelling the fatigue behaviour of spongy bone have used the finite element method with various architectures: two-dimensional hexagonal honeycombs or three-dimensional Voronoi polyhedra. Cracks were introduced into these low-density idealised spongy bone structures, and Paris' formula was used to calculate their propagation (Guo *et al.* 1994, Schaffner *et al.* 2000 or Makiyama *et al.* 2002). The results provided a potential relationship between N_f and $(\Delta\sigma/E_0)$, but with a somewhat different exponent; -4.5 rather than -12. It is possible that these differences are due to the fact that the models ignore the crack initiation phase and the damage produced by microcracking. In any case, the exponent -12 is more appropriate for dense spongy bone, while the models discussed refer to low-density spongy bone.

Kosmopoulos *et al.* (2008) have proposed a model for dense spongy bone which gives an exponent of -12.2, very close to the experimental results. The model assumes that the repeating loads which are produced during everyday life produce microdamage in the bone, which initiates a gradual reduction in the elastic modulus and strength before the appearance of microcracks. The numerical results agree well with those of the experiments, as may be seen in Figure 9.40a.

For a more detailed account of the different methods of testing bone, the reader is advised to consult the book of An and Draughn (2000). More information on the mechanical properties can be found in Currey (2002).

9.3.4. Final remarks

Bone as a **biological material.**
The material of bone, as previously mentioned, is a network of collagen intermixed with calcium phosphate, and water. Collagen is widespread throughout the animal kingdom, and it is perhaps surprising that other realms have not used the combination of collagen and minerals to create strong tough structures. Combinations of organic material and minerals are rarely found; in mollusc shells where calcium carbonate is associated with an organic material, the constitution of the latter is hardly 5%. As for silica, another material used by living organisms, it does not form composites with organic materials, even though it is the skeletal material of various protozoas and the spicules of some sponges.

Stiffness in the plant kingdom is achieved through the formation of highly hierarchical structures with organic materials, as discussed in the analysis of wood. In invertebrates stiffness is produced by minerals in the form of small crystals, also with a very hierarchical structure, bonded by a low protein matrix (between 1 and 5%) or with highly structured organic materials, as happens in arthropods, whose cuticles are formed of chitin fibres held together with protein cross-links. The drawback of these cuticles is that they are not reabsorbed as the animal grows and must be shed periodically. Animals with bones do not have this inconvenience; the bone is remodelled.

Bone does not perform the same functions during its lifetime and changes as the individual develops and in response to the loadings it receives. Currey (2002) has investigated the mechanical properties of dry human femur from the age of 3.5 to 40 years. The elastic modulus and strength (in flexural tests) increase rapidly up to 10 years of age and then grow more slowly or stabilise. The impact strength, on the other hand, decreases rapidly up to 10 years and then decreases more slowly. It is possible to correlate these results with human behaviour if it is borne in mind that children are subjected to different selective pressures than adults; children are quite curious and have a tendency to fall, but don't have to run long distances in search of prey. For the young, it is very useful not to have bones break easily, while in grown-ups, strength is more advantageous than toughness. As has been mentioned repeatedly, it is very difficult to simultaneously achieve high strength and toughness.

The changes produced in bone after 40 years of age, to take an approximate date, are not usually of the adaptive type but rather of the senile nature. The structure of spongy bone changes with age because its mass is reduced; the already thin trabeculae go on reducing their thickness and many are reabsorbed, which results in a marked reduction in the elastic modulus and strength of the bone. Guo and Kim (2002) have developed a microstructural model of spongy bone which allows simulation of the trabecular thinning, including their progressive disappearance. The results indicate that the loss of trabeculae is more damaging than the thinning, as much for the elastic modulus as the strength. For example, a random reduction by *disappearance* of 10% in the number of trabeculae induces a reduction of 40% in the elastic modulus and the strength, while the same reduction in the *thickness* of the trabeculae only produces a loss of 20%. These results indicate that the usual medical treatments for osteoporosis –which, basically, provoke an increase in the thickness of the trabeculae, but not their number– are incapable of totally recovering the initial properties of the bone.

Bone, as a structural material, had a pivotal role during the earliest human settlements. In regions where caves or sufficient wood were not available for construction, dwellings were made with mammoth bones. Archeologists have discovered the remains of homes in Siberia, up to 8 metres in diameter, whose structure –now collapsed– was made of mammoth bones. Of course, these structures would have been covered with skins.

It should not surprise us, taking into account the above, that bone has played an important role in the arts, particularly in music. In 2006, a flute, made from a griffon vulture bone, was discovered in a cave in Germany. The instrument was 34 cm long and 8 mm in diameter with five finger holes and a mouthpiece adequate for its function, it appears to have been made 35,000 years ago (https://www.nationalgeographic.com/culture/article/bone-flute-is-oldest-instrument-study-says). Also, in a nearby cave, two flutes made from swan bones were found, dated between 42,000 and 43,000 years ago. Another impressive find, in a cave in Slovenia in 1995, is a piece of femur from a cave bear with four artificial holes, which could well have been a flute. Interestingly, it has been dated at about 60,000 years ago and fabricated by Neanderthals (https://en.wikipedia.org/wiki/Divje_Babe_flute).

Bone as a **biomaterial.**
In the sense of a material being used "to treat, correct or replace tissue or organs in the human body" (European Society for Biomaterials, 1991), bone is a *biomaterial*, because it has been used for grafts.

Bone grafts are used in surgery of the locomotor system, vertebral infusions or post-traumatic bone defects, to name a few examples. Autologous bone graft or autograft (obtained from the patients themselves) is the material of choice of surgeons, but it has disadvantages related to the morbidity of the donor site (usually the iliac crest) which restricts its application. Available alternatives are a bone graft obtained from another patient (allograft) or other synthetic materials (biomaterials).

The first successful allograft operation appears to be by McEwen, in 1881, to repair a humerus (McEwen 1881). Since then, its use has spread in centres that treat complex pathologies. In the USA more than 150,000 bone allografts are performed each year (Tomford *et al.*, 1999). The advantages of these grafts over autografts are: greater availability, absence of problems in the donor zone, and shorter surgical time. Among the inconveniences are longer incorporation times and lower osteoconductive and osteoinductive capacities than autografts. Compatibility and the risk of transmission of infections improve with proper donor selection (Calvo *et al.*, 2011).

A knowledge of the relationships between structure and mechanical properties of bone not only allows understanding of its behaviour, but it also provides a valuable aid to the design of implants because it is essential to simulate, as well as possible, the properties of the bone which it is going to substitute. In femoral head implants, for example, the differences between the properties of the implant and the surrounding bone are believed to be one of the reasons why the prostheses decouple after 10 or 15 years. In these cases, the replacement of the prosthesis is then not so straightforward or safe as before. The design of prostheses that are more like bone would have considerable economic repercussions since the number of femoral implants in the world is extremely high; in the USA alone, more than 200,000 were performed in 2010.

To conclude, it is worth pointing out, as has been done in the previous examples, some ideas which this extraordinary biological material may offer to **biomimetics**. It has been observed that spongy bone is surrounded by compact bone, and this bone architecture has the effect of optimising the strength for the minimum possible mass.

Long bones can be considerd as if they were thick-walled cylinders of compact bone with enlarged ends and in whose interior is spongy bone. The wall thickness is optimised so that the stiffness of the bone is maximised for the function it performs (Alexander *et al.* 1990).

Flat bones have a sandwich structure: two compact bone laminas with spongy bone in between. In this type of bone the transition between the compact and spongy bone is quite abrupt, contrary to what happens in long bones where the transition is gradual. This structure is also optimised, for the functions it has to perform, so that the mass is minimal (Gibson *et al.* 2002).

In some circumstances engineers also design structures to optimise the weight according to purely mechanical criteria and at other times they are inspired by nature. The design of the structure of the fuselage and wings of aircraft are based on maximum strength with minimum weight. During the Second World War, the wings of the English multirole "Mosquito" aircraft were fabricated with a sandwich structure using birch for the surfaces with a balsa core. This was made possible, among other things, by the development of suitable adhesives. Nature already used these principles millions of years ago to design the skeletons of birds.

9.4. CARTILAGE

9.4.1 Introduction

There are three types of cartilage tissue in human beings, *hyaline* or *articular cartilage* –the one that will concern us most in this section– *fibrous cartilage*, a special case of articular cartilage with a higher degree of collagen fibre reinforcement and which is situated primarily in the intervertebral discs and menisci, and *elastic cartilage*, which is yellowish and more elastic than articular cartilage due to the presence of elastin fibres. Elastic cartilage is present in the epiglottis, the larynx, the external auditory canal and the Eustachian tube.

Articular cartilage, also called hyaline because of its whitish/bluish appearance, is more abundant in human organisms and contains type II collagen fibres. Its presence strengthens the nasal skeleton, the larynx, trachea and the bronchi and forms the costal arches of the ribs (Figure 9.41). Its principal function is, however, to cover the articular extremes of bones so that mechanical loads are transmitted with minimum friction (Figure 9.42). Like other cartilage, it is avascular, alymphatic and aneural.

Articular cartilage is composed of 70-80% by weight of water. The rest is made up of cells that create and sustain it, called chondrocytes, and an extracellular matrix formed fundamentally of type II collagen (~80% of dry weight) and proteoglycans (~20% ofdry weight) (Mow *et al.* 1990).

The cellular volume is about 10%, its contribution as dry weight being insignificant. Cartilage is one of the tissues with the least cell density in the body, possessing an average of a million cells per cm^3 compared to 100 million for most other tissues. This fact makes tissue regeneration difficult, once damaged. The low cellular density is related a low capacity to receive nutrients which can only reach the cells through diffusion, since they lack a vascular network.

The extracellular matrix is formed of collagen and complex molecules called proteoglycans, whose structure is shown in Figure 9.43. This is a combination of proteins and mucopolysaccharides (also known as glycosaminoglycans, or GAGs for short) which produce a fabric capable of retaining large quantities of water.

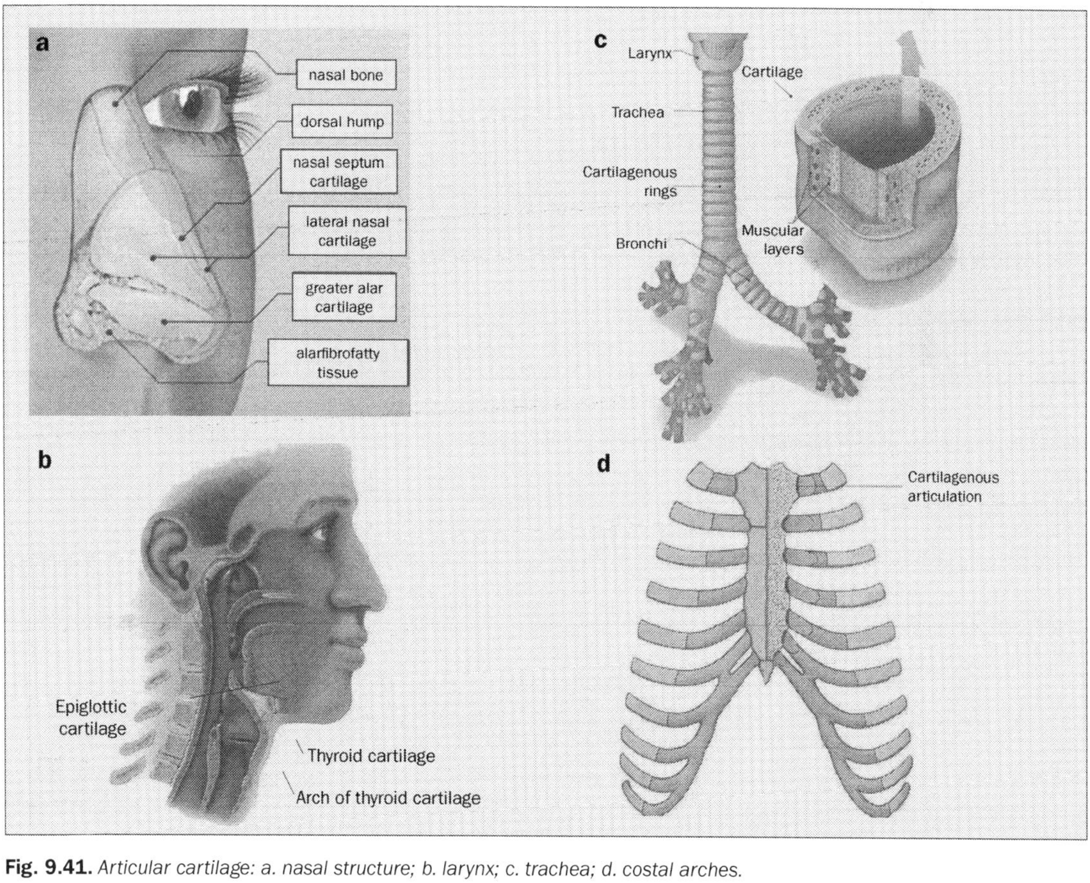

Fig. 9.41. *Articular cartilage: a. nasal structure; b. larynx; c. trachea; d. costal arches.*

Articular cartilage covers joints in thicknesses less than a few millimetres, and it is capable of supporting large loads with the lowest coefficients of friction, as shown in Table 9.8. To this end, the joints are encapsulated in approximately 1 ml of lubricating fluid called synovium (from the Latin *syn* –with– and *ovum* –egg) since its colour and consistency are reminiscent of egg white (Figure 9.42). Synovial liquid is a thixotropic fluid which reduces its viscosity with motion, in a way that further reduces the friction. Its composition includes hyaluronic acid secreted by the cells of the synovial membrane which surrounds the joint, a proteoglycan called lubricin which is produced by the chondrocytes of the cartilage, and interstitial fluid from blood plasma.

Due to the effects of movement, articular cartilage is almost perpetually subjected to alternating loads which, in the case of the hip or knee, may reach ten times the body weight, acting on surface areas of the order of cm^2. These loads induce motion of the internal fluid of the cartilage and facilitate transport of nutrients and waste products. The structure of cartilage is capable of supporting these loads without loss of hydration, maintaining its low coefficient of friction.

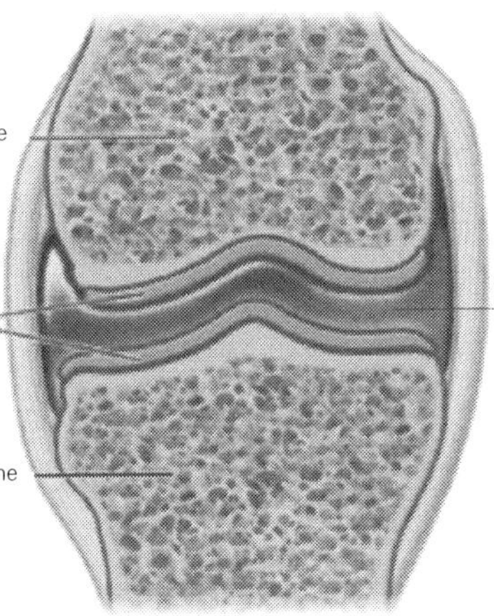

Fig. 9.42. *Articular end.*

TABLE 9.8 COEFFICIENT OF FRICTION OF ARTICULAR CARTILAGE AND OTHER MATERIALS

MATERIAL	COEFFICIENT OF FRICTION
Articular cartilage**	
Human knee	0.005 - 0.02
Porcine shoulder	0.02 - 0.35
Canine ankle	0.005 - 0.01
Other materials	
Rubber on concrete*	1.0
Brake material on iron*	0.1 - 0.4
Graphite on steel*	0.1
Hickory on dry snow**	0.08 - 0.18
Ice on ice**	0.02

* Static

** Dynamic

9.4.2. Structure

Articular cartilage, like other biological materials, has a highly hierarchical structure from the molecular to the anatomical level. However, to facilitate description and analysis, we will distinguish –with some arbitrariness– the microscopic (molecular) and mesoscopic levels in its structure.

Microscopic level

At this level, cartilage can be considered as a material composed of water, cells (chondrocytes) and an extracellular matrix formed of collagen and proteoglycans (hyaluronic acid and aggrecans).

Hyaluronic acid is a fundamental component of cartilage and its presence is essential for it to correctly fulfil its biological purpose.

Hyaluronic acid is a glycosaminoglycan formed by the repetition of disaccharide glucuronic acid/N-acetyl-glucosamine, which constitutes the backbone of the large molecules of the cartilage. In an average- weight man (70 kg), there are about 15 g of hyaluronic acid, and about a third of this is degraded and synthesised every day (Stern 2004). Its name comes from *hyalos* –in Greek, "transparent glass"– since it was first observed in the vitrius humor, and from the uronic acid which forms it. Hyaluronic acid is used widely in the treatment ofdiseases of the cartilage such as arthritis or osteoarthritis and is obtained principally from natural sources such as the crests of roosters and the umbilical cord (Boeriu *et al.* 2013).

Protein chains are covalently bonded to the hyaluronic acid chains which, in turn, have cross-links to other GAG chains, called **aggrecans,** as shown in Figure 9.43. The principal aggrecans present in cartilage are chondroitin sulphate and keratan sulphate, both –like the rest of the GAGs– constructed of a repeated disaccharide unit. In the case of chondroitin sulphate it is glucuronic acid and N-acetyl-galactosamine and in keratan sulphate it is galactose and N-acetyl-glucosamine. The average molecular weight of the long proteoglycan chains is about 200-300MDa.

The fact that the sugars in the chondroitin sulphate and keratan sulphate chains are heavily sulphated is the key for the correct functioning of the cartilage, since the sulphate groups ($-SO_3H$), as well as the acid groups (–COOH) tend to easily lose their proton (H^+) and are active osmotically in attracting cations such as Na^+, Ca^{2+} or K^+. On average, there are about 50 g of GAG saccharides per litre of cartilage and each disaccharide, again on average, is ionised with a charge–2e. The average molecular weight is about 502 Da from which we can estimate that in 0.2 M (M=moles/litre) the negative electric charge isassociated with the cartilage. Given that this charge cannot be displaced and is anchored to the GAGs, it is given the name *fixed charge density* (FCD), and its role is crucial in the development of the osmotic pressures which sustain the structure and functioning of the articular cartilage.

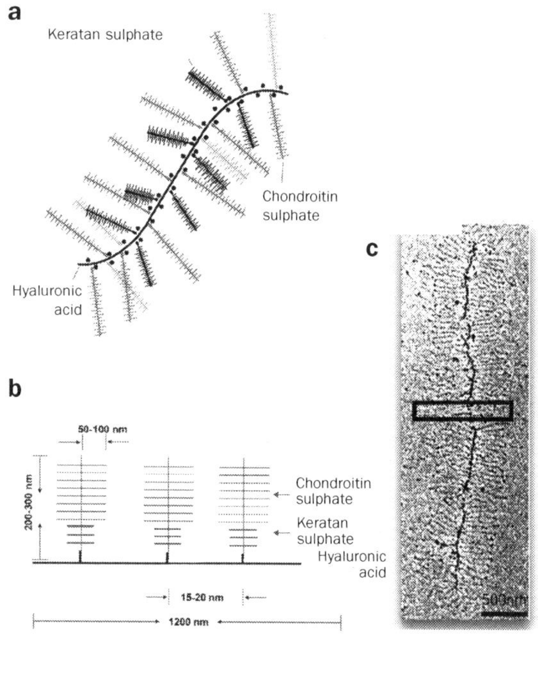

Fig. 9.43. *Proteoglycans of articular cartilage:*
a. Structure of the proteoglycans formed by hyaluronic acid.
b. Molecular distances.
c. Transmission micrograph of bovine articular cartilage showing the principal hyaluronic acid chain and side chains (inset).

The **collagen** present in the extracellular matrix forms a fine strong mesh that traps the proteoglycan molecules and supports the stresses due to the internal pressure. For the most part, they are type II collagen fibrils (between 80 and 95%), finer than those of type I and with a higher degree of cross-linking. There are also even finer types IX and XI collagen fibrils interlacing the type II collagen fibres with each other and with proteoglycans, especially in immature cartilage.

Chondrocytes are the only cells present in cartilage, and they produce *collagen*, proteoglycans and hyaluronic acid. Their concentration is very low since their survival is subject to the diffusion of nutrients and metabolites from the articular surface, cartilage being an avascular tissue. Small pores, called lagoons (lacunae), are commonly found isolated in the matrix.

Mesoscopic level

Collagen fibres in the articular collagen take up different positions according to their distance from the contact surface, as shown in Figure 9.44a, and they give rise to three layers known as *superficial or tangential, middle and deep* (Glenister 1976).

The *superficial* layer, which can be as much as one-fifth of the thickness, has collagen fibres arranged in leaves parallel to the surface. This layer is responsible for carrying the frictional forces produced during articulation and has the greatest concentration of collagen fibres and water content. Split lines can also be seen in the superficial layer, which are produced when the surface is punctured, and they are marked with ink. The fibre distribution results in a local crack in the surface tissue which indicates the direction of the collagen (Figure 9.44b).

The *middle* layer contains the highest concentration of proteoglycans as well as a lot of water and collagen fibres. The last of these lose their orientation parallel to the surface to become, more or less, perpendicular to the contact with the underlying bone. This layer can be as much as half of the total cartilage thickness and carries the combined compression and shear forces.

The *deep* layer includes the gradual junction of the articular cartilage and the bone, since the cartilage is progressively calcified. This is the zone of lowest water content, collagen and proteoglycans and, typically, occupies one-third of the thickness. A *tidemark,* which separates the calcified collagen from the rest, is located in this layer (Figure 9.44a).

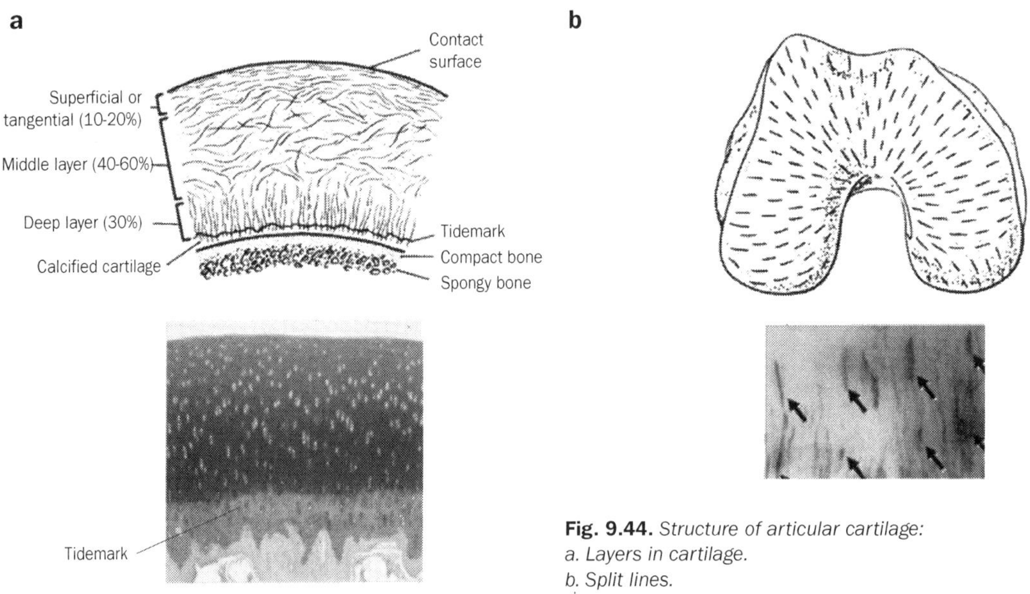

Fig. 9.44. *Structure of articular cartilage:*
a. Layers in cartilage.
b. Split lines.

9.4.3. Mechanical properties

Turgidity

Turgidity (swelling) in cartilage is due to the **osmotic pressure** caused by the ionic balance between the synovial liquid and the liquid in the interior of the cartilage, whose walls act like a semi-permeable membrane (see Figure 9.45).

The presence of the negative charge associated with the collagen molecular structure –the fixed charge density or FCD mentioned in the previous section– allows cartilage to retain water molecules even in the presence of significant external pressure. Thanks to the swelling, cartilage can bear large external pressures, counteracted by the osmotic pressure.

To better understand this phenomenon, it is helpful to examine in detail, first how the osmotic pressure, Π, can be quantified as a function of the ion concentration inside and outside the cartilage (Equation 23) and, secondly, how to calculate the effective pressure, p-Π, which the cartilage supports, and turns out to be much smaller than the external applied pressure p (see Figure 9.45c).

Figure 9.45 illustrates this situation in a schematic form. For greater simplicity we will only consider equilibrium of the Na^+ and Cl^- ions, which are the most abundant in the interstitial fluid.

Under equilibrium conditions, the Na^+ ions displace the H^+ protons and tend to occupy positions close to the negative charges of the GAGs ($-SO_3^-$, $-COO^-$), increasing their concentration in the interior of the cartilage. This increase is counteracted by the entrance of water molecules, which induce an increase in the osmotic pressure.

If $[Na^+]_e$ and $[Cl^-]_e$, and $[Na^+]_i$ and $[Cl^-]_i$ are the ion concentrations in the external and interior fluids, respectively, the equations governing the problem are:

Neutrality of the exterior fluid

$$[Na^+]_e = [Cl^-]_e \tag{21}$$

Neutrality of the fluid in the cartilage

$$[Na^+]_i = [Cl^-]_i + FCD \tag{22}$$

and the Gibbs-Donnan equation, obtained by equating the Gibbs functions and ignoring the coefficients of activity, which relate the concentrations of the electrically charged species on either side of the membrane.

$$\frac{[Cl^-]_e}{[Cl^-]_i} = \frac{[Na^+]_i}{[Na^+]_e} \tag{23}$$

Observe that to establish electro-neutrality of the cartilage, it is necessary to consider the *FCD* coming from the *GAGs*.

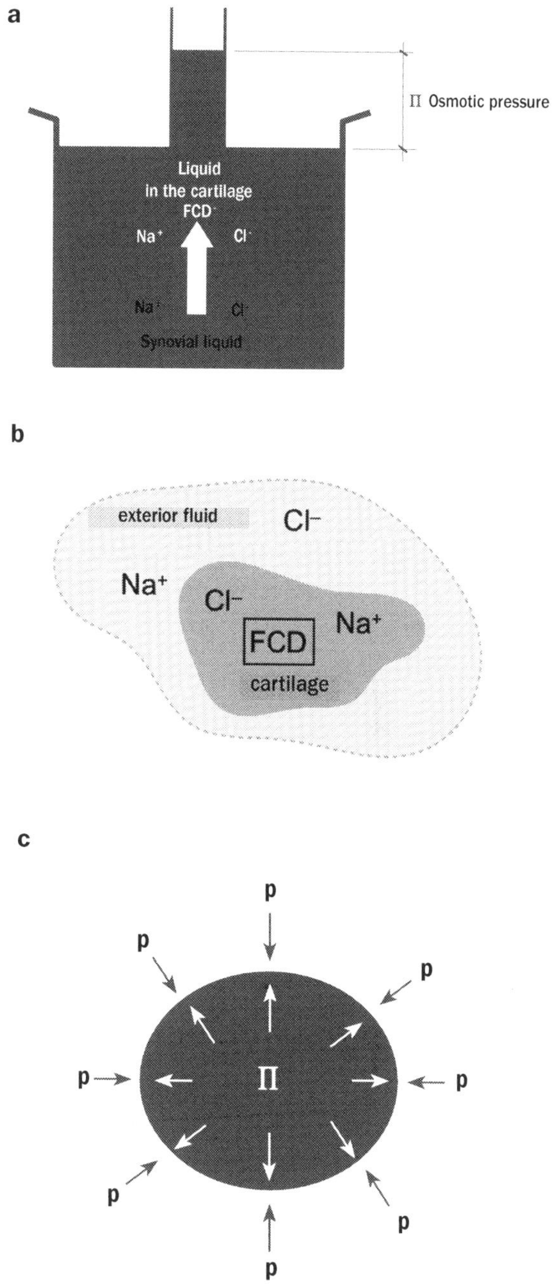

Fig. 9.45. *Ionic equilibrium model in cartilage.*
a. Osmotic pressure Π.
b. Cartilage and synovial liquid (schematic).
c. Effective pressure: p - Π

If the external equilibrium concentration of NaCl is C, it is easy to solve the three previous equations to obtain the concentrations:

$$[Na^+]_e = [Cl^-]_e = C \quad (24)$$

$$[Na^+]_i = \frac{FCD}{2} + \sqrt{\left(\frac{FCD}{2}\right)^2 + C^2} \quad (25)$$

$$[Cl^-]_i = -\frac{FCD}{2} + \sqrt{\left(\frac{FCD}{2}\right)^2 + C^2} \quad (26)$$

whose values are shown in Figure 9.46a as a function of FCD/C, the ratio of the fixed charge density and the external saline concentration.

The difference in ionic concentration between the cartilage and the external fluid provokes displacement of the water molecules and the appearance of an osmotic pressure which can be calculated by means of van't Hoof's equation:

$$\Pi = RT\{[Na^+]_i + [Cl^-]_i - [Na^+]_e - [Cl^-]_e\} = $$
$$= RT(\sqrt{FCD^2 + 4C^2} - 2C) \quad (27)$$

where R is the Gas Constant (R=8.31 J/mol/K) and T the absolute temperature. Figure 9.46b depicts its variation with saline concentration.

For an isotonic external concentration ([NaCl]=0.15M), at physiological temperature (37 °C) and with fixed charge density of 0.2 M the osmotic pressure works out at 0.16 MPa, a value approximately 60% higher than the atmospheric pressure of 0.1 MPa (recall that the internal osmotic pressure of the cartilage is a gauge pressure, above atmospheric pressure).

If the fixed charge density is reduced by a factor of ten, FCD=0.02M (through loss of $GAGs$, for example), the value of the osmotic pressure falls drastically by two orders of magnitude to 0.0017 MPa. This fact highlights the importance of keeping the content of the cartilage GAGs as high as possible.

Since the osmotic pressure is intimately linked to the FCD and that to the GAG content, it is interesting to estimate the concentration of the latter. The GAG content in cartilage can be determined from a measure of the salt in the external fluid and the $[Na^+]_i$ content in the internal tissue. To obtain this last value the ^{23}Na isotope is used, which is detected by nuclear magnetic resonance. Knowing the concentrations of $[Na^+]_e$ and $[Na^+]_i$ the FCD is calculated as:

$$FCD = \frac{[Na^+]_i^2}{[Na^+]_e} - [Na^+]_e \quad (28)$$

obtained directly from Equations (25), (26) and (27).

With this value of the FCD, the average electric charge of –2 electrons per disaccharide, and the average molecular weight of the GAG disaccharides (502.5 Da) we obtain

$$GAGs = \frac{FCD \text{ x } (502.5 \text{ g/mol})}{2} \quad (29)$$

Typical values lie between 15 and 65 g/l depending upon the age and location of the cartilage (~50 g/l for FCD=0.2 M).

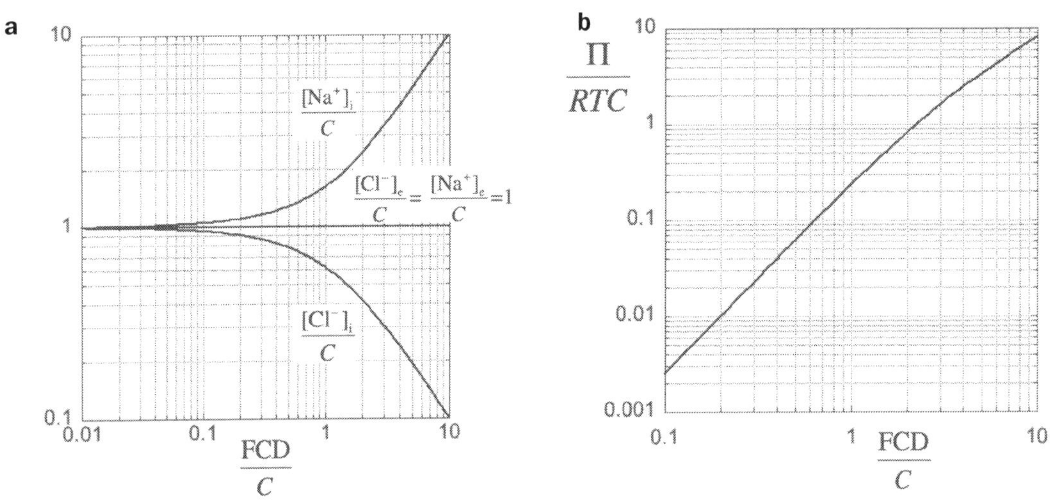

Fig. 9.46. *Ionic equilibrium in cartilage: a. equilibrium concentrations; b. osmotic pressures.*

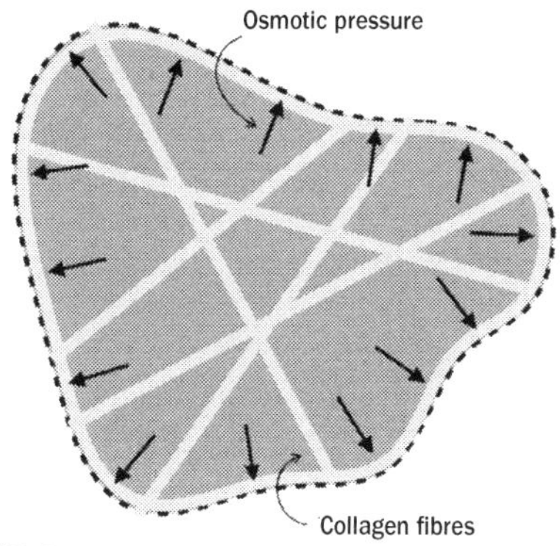

Fig. 9.47. *Structural network of the articular cartilage.*

Finally, with the data obtained, it is possible to estimate the net pressure supported by the cartilage as the difference between the external applied pressure and the osmotic pressure. The osmotic pressure in the cartilage is supported by the structure of the collagen fibres in its interior, in such a way that a network with internal pre-tensions is generated, which not only holds its shape and retains water, improving lubrication, but it is also stable under applied tensile or compressive loads (Figure 9.47).

It is interesting to observe that when the cartilage is compressed, the concentration of the fixed charge increases, as the GAG structures are approached. This phenomenon produces an increase in the osmotic pressure which –in turn– reduces the net pressure which supports the cartilage structure. A simple calculation reveals the importance of this mechanism.

The net pressure will be the difference between the applied pressure and the osmotic pressure,

$$p_{net} = p - \Pi = p - RT(\sqrt{FCD^2 + 4C^2} - 2C) \qquad (30)$$

where FCD will be the fixed charge density in the compressed volume, which we will call V. If we consider the bulk modulus K of the cartilage, defined as the change in pressure required to cause a unit reduction in volume:

$$K = \frac{\Delta p_{net}}{-(\Delta V/V)} \qquad (31)$$

we can integrate the previous expression with the initial condition $V = V_0$, $p_{net} = 0 - \Pi_0$ and obtain

$$V = V_0 \exp\left(-\frac{p_{net} + \Pi_0}{K}\right) \qquad (32)$$

In which Π_0 is given by

$$\Pi_0 = RT(\sqrt{FCD_0^2 + 4C^2} - 2C) \qquad (33)$$

The FCD in the compressed volume can be easily calculated by means of the relationship

$$FCD\ V = FCD_0\ V_0 \qquad (34)$$

since the total charge in the $GAGs$ must be conserved.

The combination of Equations (30), (32), (33) and (34) allows calculation of p_{net} as a function of the applied external pressure p knowing C, FCD_0, T and K.

Figure 9.48 shows p_{net} as a function of p for various values of the *GAG* content in the cartilage, within the physiological range. The following values were used in the calculation: T=37 °C, physiological saline concentration (0.15M) and the cartilage bulk modulus 0.2MPa (Rotter *et al.*, 2002).

In Figure 9.48 we see that the pressure supported by the collagen structure and GAGs p_{net} is much lower than the applied external pressure p, and that the osmotic pressure supports the greater part of the load. This is a very effective mechanism in avoiding overload of the weak cartilage structure.

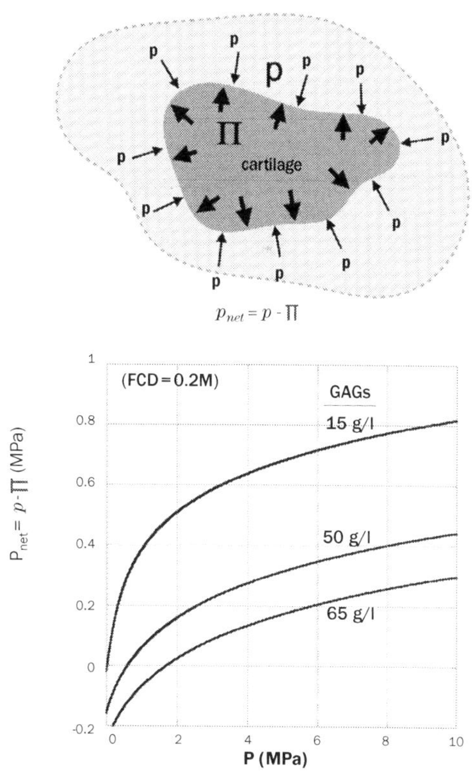

Fig. 9.48. *Net pressure supported by the cartilage as a function of the applied external pressure [NaCl]=0.15 M, T=37 °C, K = 0.2 MPa.*

Stiffness and strength

Articular cartilage is a porous material with a high water content, and it is very sensitive to the applied loading rate. Consequently, values determined experimentally are usually equilibrium values, allowing a steady state to be achieved after application of the external load. Otherwise, the loading rate should be specified, since measurements may differ by an order of magnitude.

Figure 9.49 shows the results from tensile tests performed by Woo *et al.* (1976) on specimens taken from bovine articular cartilage. The test pieces, standard dog bone specimens 250 μm thick, were taken from different depths in the cartilage, as indicated in the figure (numbered 1, 2 and 3). Tests were carried out in the directions parallel and perpendicular to the lines of collagen alignment (*split lines*).

The curves have a characteristic J-shape corresponding to tissue in which the reinforcing fibres group together and progressively align as the deformation increases. The highest stress values are obtained in the tests parallel to the collagen fibres (that is, parallel to the *split lines*). As the sample is taken from deeper inside the cartilage this effect is attenuated, since as we have noted, the orientation of the fibres changes and is no longer parallel to the surface and the *split lines*.

Table 9.9 summarises the data obtained by Woo *et al.* (1976), of the stiffness in the linear zone of the tensile curve, once the fibres are aligned. Their values are obtained from the stresses produced by letting the specimen relax between two points in the linear zone, typically those from strains of 10 and 20% (ε_1=0.1, ε_2=0.2):

$$E_{equilibrium} = \frac{\sigma(\varepsilon_2,\infty) - \sigma(\varepsilon_1,\infty)}{\varepsilon_2 - \varepsilon_1} \quad (35)$$

The values in Table 9.9 clearly show the anisotropy of this material. Similar values are obtained for human patellar cartilage, as illustrated in Figure 9.50 which is taken from Bae *et al.* (2008). As we see, the values are also dependent on the position and orientation of the specimen.

TABLE 9.9 VALUE OF THE EQUILIBRIUM MODULUS OF THE BOVINE ARTICULAR CARTILAGE AS A FUNCTION OF POSITION AND ORIENTATION (WOO ET AL., 1976)

DIRECTION OF THE SPLIT LINES	LAYER/DEPTH	TENSILE STRENGTH (MPA)
Parallel	1	23.0
	2	11.2
	3	5.0
Perpendicular	1	12.2
	2	6.6
	3	3.6

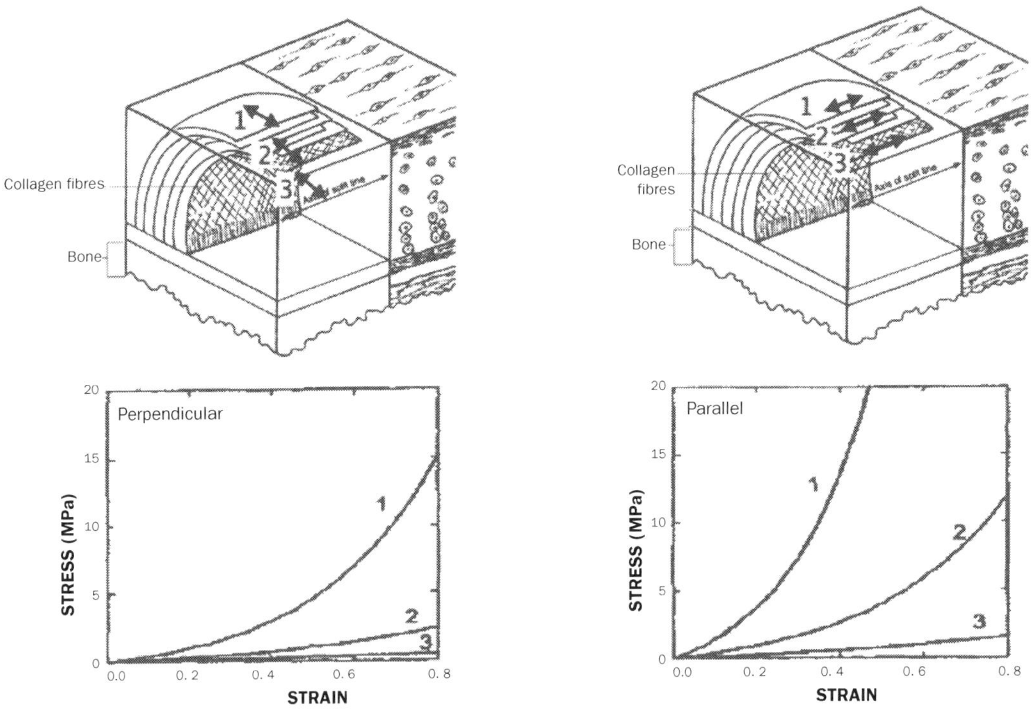

Fig. 9.49. *Tensile tests of bovine cartilage specimens (Woo et al. 1976).*

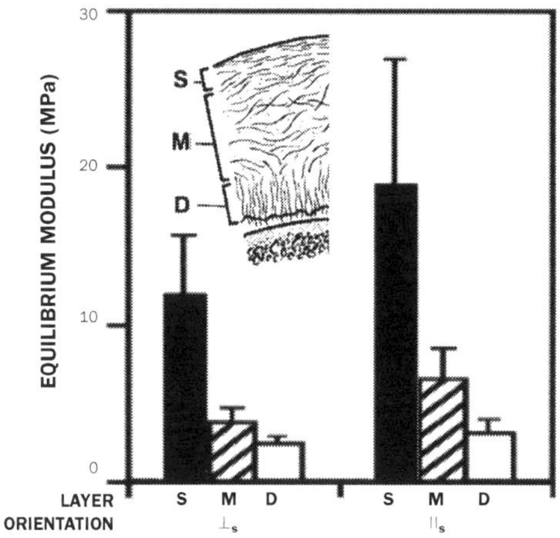

Fig. 9.50. *Values of the equilibrium modulus for human patellar cartilage specimens (Bae et al. 2008).*
S: Superficial.
M: Middle.
D: Deep.

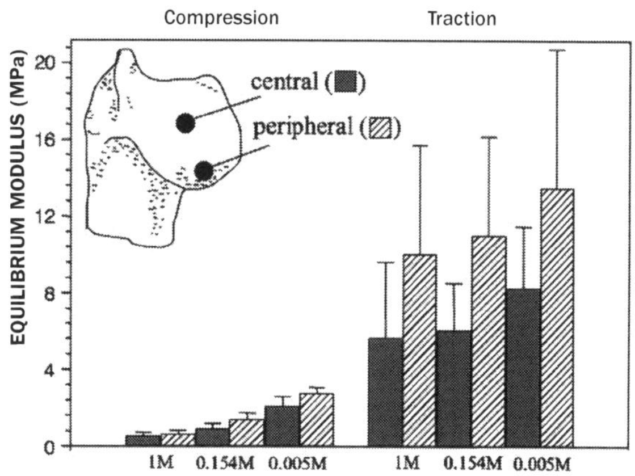

Fig. 9.51. *Values of the equilibrium modulus for bovine articular cartilage as a function of the external saline concentration (Korhonen et al. 2010).*

The stiffness dependence (measured by the equilibrium modulus) of the cartilage with the external saline concentration is shown in Figure 9.51. The mechanical properties of this tissue are very sensitive to the electrochemical equilibrium, as we have shown in previous paragraphs. This fact means that the variations in ion concentration (through changes in saline concentration or in the fixed charge density, for example) can bring about important biomechanical changes.

Like the elastic modulus, the tensile strength of articular cartilage is dependent upon the rate of loading in the test, the zone and orientation of the specimen, so that published data show large variations. Williamson *et al.* (2003) measured the tensile strength of articular bovine cartilage, obtaining values between 3 and 10 MPa, with strains to failure between 40 and 70%.

Figure 9.52 shows Kempson's results for specimens from the surface layer of human femoral cartilage, for men and women between 8 and 90 years old. The specimens were orientated in the direction of the collagen fibres. The strengths obtained were almost all over 10 MPa, reaching average values of 40 MPa for ages between 20 and 30 years.

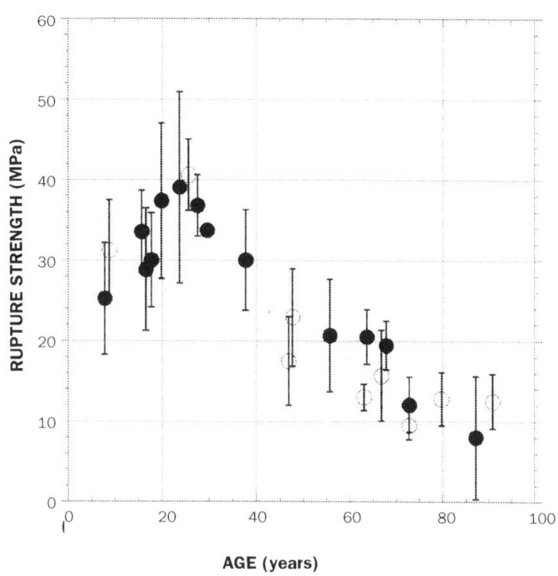

Fig. 9.52. *Rupture strength of femoral cartilage. Filled points correspond to male, and open symbols to women (Kempson 1982).*

Time-dependent behaviour

As has been discussed, the response of articular cartilage during the application of a load is not constant with time. Fluid motion in its interior causes a delayed response which –if the load is constant– may take some time to reach equilibrium.

Figure 9.53a shows the values obtained by Salak *et al.* (1987) for tensile tests at three different strain rates and Figure 9.53b contains the results of Radin *et al.* (1970) for the variation in elastic modulus of bovine cartilage with strain rate. The results of Oloyede *et al.* (1992), summarised in Figure 9.53c, were obtained from compression tests at different loading rates. Bearing in mind that in running or jumping strain rates of over 0.05/s can be encountered, it is evident that in order to analyse the biomechanical behaviour of cartilage under physiological loads, it is essential to consider its dependency on time.

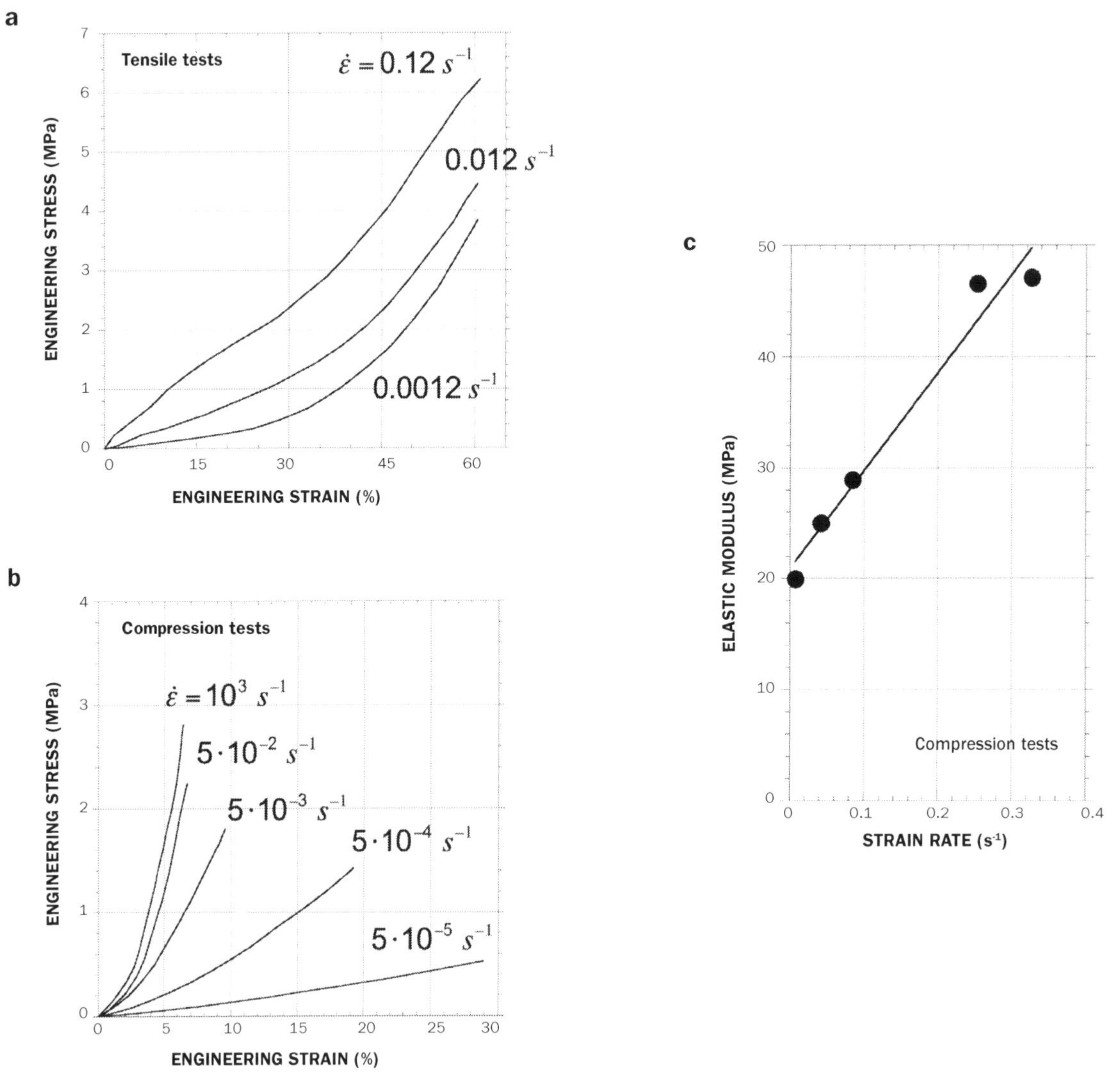

Fig. 9.53. *Dependency of the mechanical properties of cartilage with time.*
a. Tensile tests (Salak et al. 1987).
b. Compression tests (Oloyede et al. 1992).
c. Elastic modulus at a strain of 10% (Radin et al. 1970).

Models

Two different, but complementary, approximations have been proposed in order to model cartilage. The easiest is to consider cartilage as a linear *viscoelastic* material and, in its simplest form, to represent its behaviour as a Prony series, obtained by an empirical fit to the parameters (G_∞, G_i, τ_i). The model, depicted in Figure 9.54, shows the viscoelastic parameters measured in dynamic experiment with a sinusoidal cyclic strain.

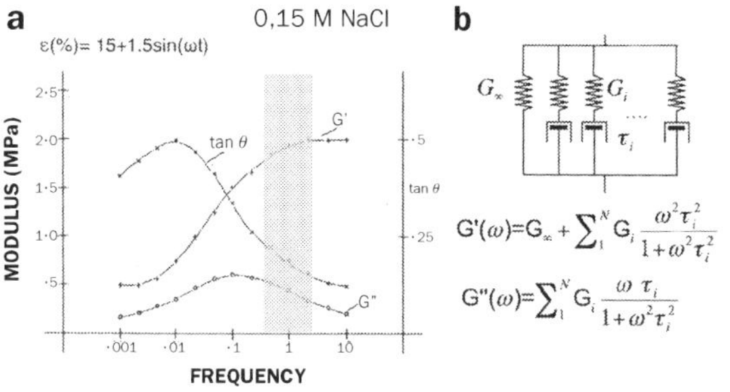

Fig. 9.54. *a) Dynamic experiments with cyclic strains applied to bovine cartilage. Storage modulus G', loss modulus G'', tangent of the phase angle between strain and stress waves $\tan\theta = G''/G'$. The shaded zone corresponds to the walk-run frequencies (Frank 1980).*
b. Prony series approximation with discrete relaxation time steps.

Parameters obtained from the viscoelastic model depend upon the load level or applied strain. A quasi-linear viscoelastic approach proposed by Fung (1993) provides some improvement in modelling.

Poroelastic models assume that the cartilage is a connected porous material through which fluid similar to synovial liquid and various ions carrying charge can move. These models are an alternative method of representing cartilage which, due to their complexity, are not widely used since they require numerical calculation and need parameters that are difficult to estimate directly, such as the viscosity of the fluid moving inside the cartilage, its porosity or the mechanical behaviour of its collagen matrix and GAGs.

A brief description of the more important poroelastic models is provided below. The interested reader may find more detail in Fung (1993).

The biphase model (Mow *et al.* 1980) assumes that cartilage is composed of a *hyperelastic* matrix and a viscous fluid, both of which are incompressible, and the effect of the electric charge is neglected. A solution to the problem is based upon conservation of mass (continuity), equilibrium (without movement) and the constitutive equations of the binary composite. Figure 9.55 shows the application of the model in a confined compression test, in which the lateral strain of the cartilage is constrained.

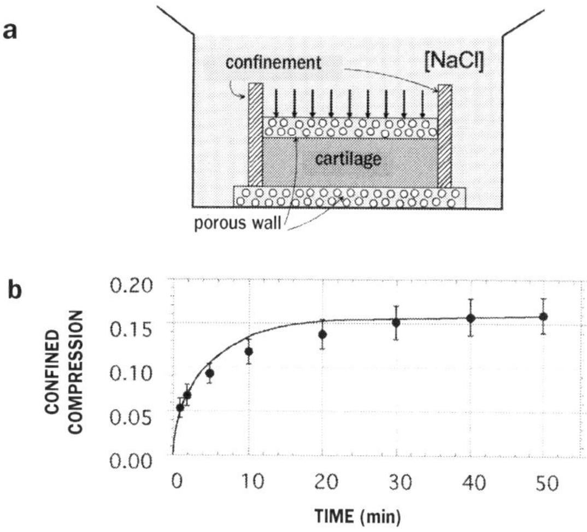

Fig. 9.55. *Confined compression test on bovine cartilage and curve fit from the biphase poroelastic model.*
a. Experimental configuration.
b. Creep curve under compression. Applied pressure 0.1 MPa (Mow et al. 1980).

A development of the biphase poroelastic model includes the (+) and (–) ions as additional phases taking into account their electrochemical equilibrium (Lai *et al.* 1991). In contrast to the biphase model (and to the viscoelastic models), the inclusion of the ionic interaction allows account to be taken of the effect of the saline concentration of the external fluid, as illustrated in Figure 9.56. This shows the axial contraction tests on bovine cartilage performed by Mow *et al.* (1984) and the fit of the previously cited triphase model of Lai *et al.*

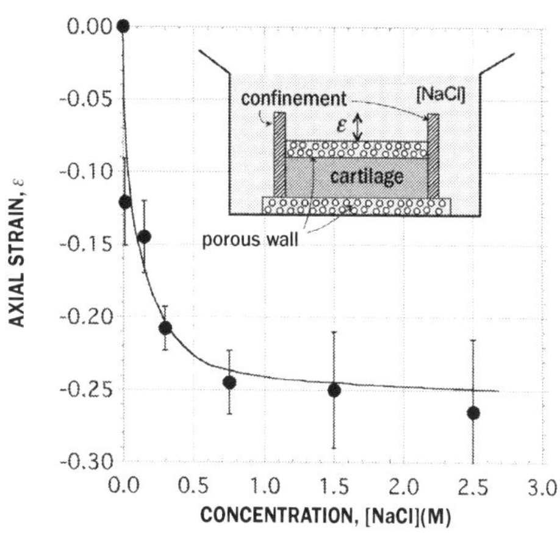

Fig. 9.56. *Fit of the triphase poroelastic model to the experimental axial contraction of bovine cartilage in NaCl (Lai et al. 1991). The cylindrically shaped specimen is confined in all directions except the axial. The value [NaCl] = 0 corresponds to tests in distilled water.*

EXERCISE 80

Analyse the cartilage axial contraction data of Mow *et al.* (1984) shown in Figure 9.56 using the biphase model which assumes that the collagen fibre matrix and proteoglycans of the cartilage have a constant aggregate modulus, H^*, and the interstitial fluid behaves according to the electrochemical equilibrium Equations (17) - (19).

SOLUTION:

The sketch below shows a simplified analysis of the axial contraction test. The cartilage is subjected to an osmotic pressure Π and a matrix stress σ^*. Mechanical equilibrium requires that $\Pi = \sigma^*$.

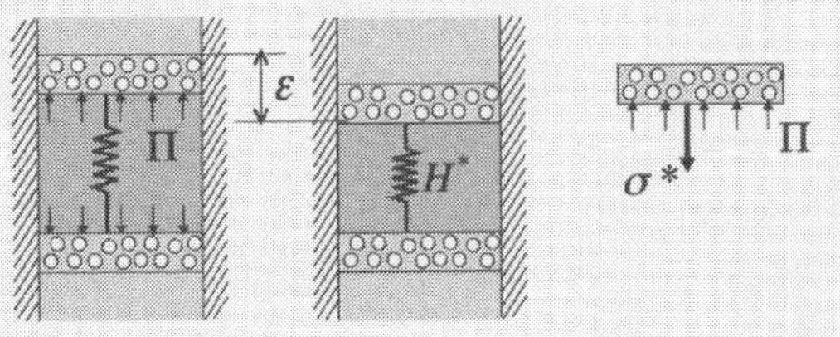

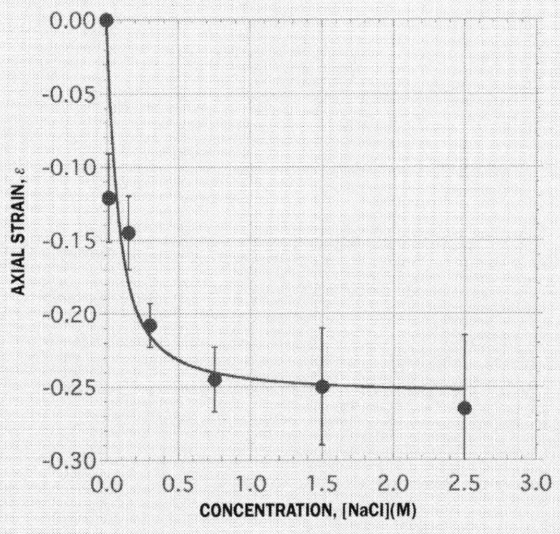

The axial strain, ε, can be calculated from the aggregate modulus as

$$\varepsilon = \frac{\sigma^*}{H^*} = \frac{\Pi}{H^*} = \frac{RT}{H^*}\left[\sqrt{FCD^2 + 4C^2} - 2C\right]$$

in which Equation (23) has been used for the osmotic pressure Π.

Using the data in Figure 9.56, take as a reference the strain that is produced when [NaCl] = 0, and fit the function

$$\tilde{\varepsilon} = \varepsilon - \varepsilon(0) = \frac{RT}{H^*}\left[\sqrt{FCD^2 + 4C^2} - 2C\right] - \frac{RT}{H^*} FCD$$

taking reference values $T = 300\,°\mathrm{K}$ and $FCD = 0.2\,\mathrm{M}$.

The resulting fit is shown in the figure, which represents a good approximation to the experimental data. Note that the fit is very similar to that obtained with the triphase model (Figure 9.56).

Observe also that in this case the only adjustable parameter is the aggregate modulus which turns out to be $H^* = 1.94$ MPa. This value is consistent with that of 2 MPa measured by Mow *et al.* (1984) for this material under hypertonic conditions ([NaCl] = 0).

Fracture

Articular cartilage can crack due to the applied load, wear or friction on the surface, and the differences in stiffness with the subchondral bone especially when the loads are dynamic and the cartilage is very stiff, as we have seen. The causes of the appearance of cracking are not well established, and it is not well understood how a microcrack progresses under the action of physiological loads.

The small thickness of cartilage makes it difficult to fabricate and test specimens with the desired orientation. As a first step, some authors have measured the fracture strength of cartilage by adapting procedures developed for other materials, in particular polymers. Chin-Purcell and Lewis (1996) measured the fracture strength of canine patella cartilage using mixed bone cartilage specimens with a deep crack, so as to minimise the effect of the constrained zone. Figure 9.57a shows the general arrangement of the test. The initial crack passes through the bone and enters the cartilage, penetrating it as it propagates. The specimens have a thickness of about 0.2 mm.

The breaking strengths measured by Chin-Purcell and Lewis were between 0.14 and 1.2 kJ/m^2, in line with those obtained for other tissues containing collagen such as skin.

These authors also performed tear tests on trouser-shaped specimens (Figure 9.57b), obtaining values consistently lower, between 0.24 and 0.8 kJ/m^2.

Xiao *et al* (2015), using the same configuration as Chin-Purcell and Lewis, measured the toughness of porcine articular cartilage and obtained a value of 11 MPam$^{1/2}$.

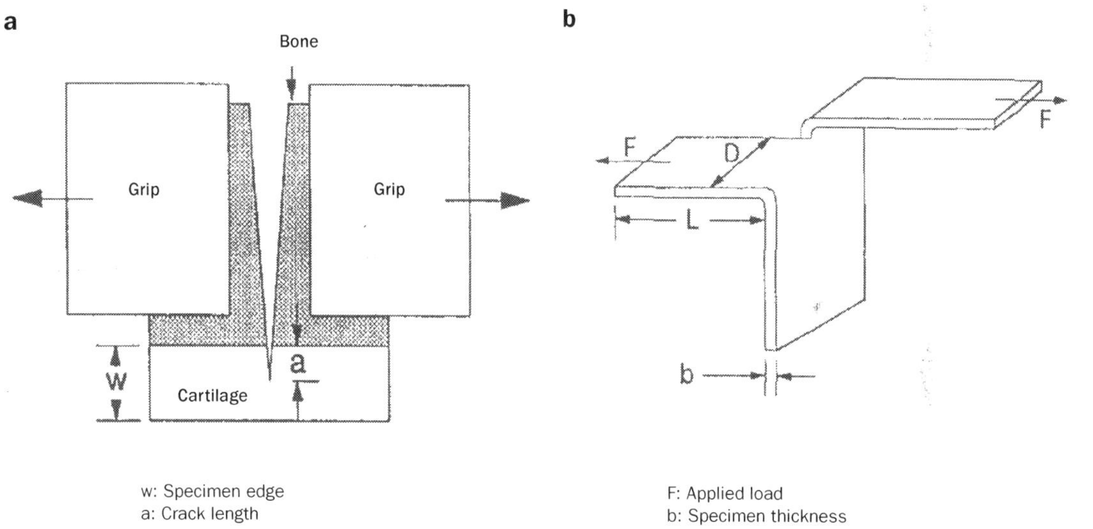

Fig. 9.57. *Fracture tests on canine patellar cartilage (Chin-Purcell et al. 1996).*
a. Notched specimen.
b. Tear specimen.

Friction

Thanks to the capacity of cartilage to retain large amounts of water and the presence of synovial fluid, articular contact is one with the lowest coefficients of friction (Table 9.8), which is vitally important in avoiding tissue wear and tear and in facilitating its regeneration.

The mechanisms thought to be involved in this phenomenon are illustrated in Stribeck's curve (Figure 9.58), which is widely used in the study of friction between surfaces. In this representation, the coefficient of friction is plotted as a function of the Sommerfeld Number ($\eta va/N$), the product of the fluid viscosity η, the sliding velocity v, the contact length a and the inverse of the force applied normal to the surfaces of contact N. In this figure, two basic regimes are identified: *limiting friction* and *hydrodynamic lubrication*, separated by a mixed transition region.

Limiting friction occurs through direct contact of the surfaces. In the case of cartilage, this situation compresses the tissue producing an increase in the osmotic pressure, as we have seen, which in turn impedes the loss of fluid. In addition, the proteoglycan lubricin present in the synovial liquid interacts with the proteoglycan chains of the articular surface and helps maintain hydration and the viscosity of the contact.

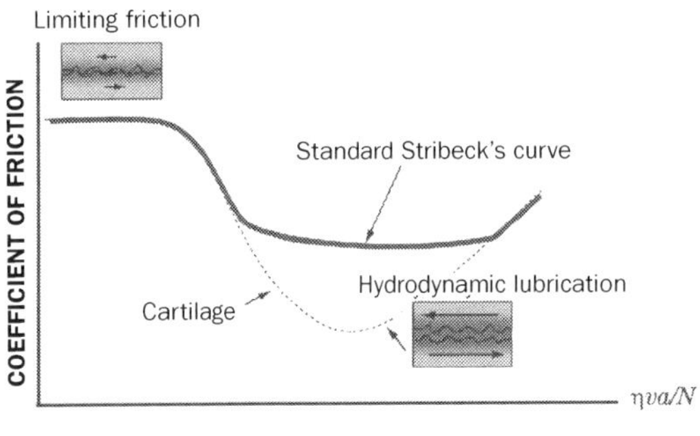

Fig. 9.58. *Stribeck's curve.*

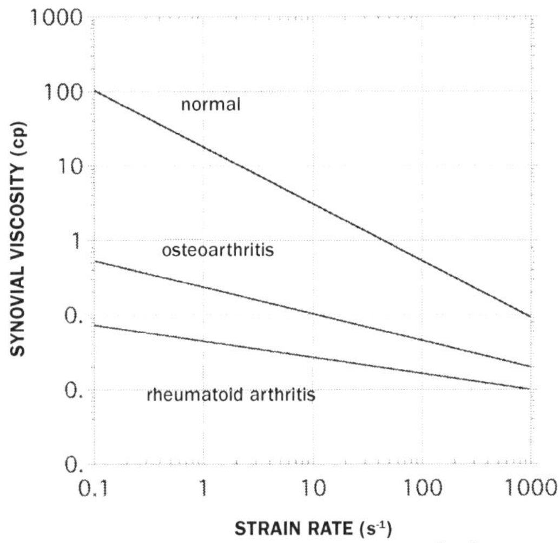

Fig. 9.59. *Variation of the synovial viscosity with strain rate.*

During *hydrodynamic lubrication,* the articular surfaces do not come into direct contact, and load transfer is made through a layer of synovial fluid. This mechanism is more efficient and provides a lower coefficient of friction, although in order to come into play it is necessary to exceed a certain sliding velocity or, alternatively, to reduce the load on the joint.

The fact that the synovial fluid is thixotropic (Figure 9.59) improves both lubrication mechanisms since, at low velocities, the increase in viscosity keeps the fluid between the contact surfaces for a longer time, protecting the proteoglycan structures and impeding the loss of water. In the hydrodynamic regime, the reduction in viscosity when the sliding velocity increases reduces the shear stresses that give rise to friction. When a degenerative process occurs, as we saw in Figure 9.59, the synovial fluid loses viscosity and its thixotropy is also diminished.

9.4.4. Final remarks

Hyaline or articular cartilage is preferentially located in the joints below and plays a fundamental role in the transmission: of forces and movement, maintaining a very low coefficient of friction.

It is a highly hydrated tissue, with neither blood vessels nor nerve ends, and very low cellular density (up to 100 times less than other body tissues) which makes recovery difficult once damaged. Its matrix is made up of a network of negatively charged collagen fibres and proteoglycans which, thanks to osmotic equilibrium, is effective in retaining water molecules.

Its mechanical behaviour is very dependent upon time and is governed by the pressure in internal fluid movement, and in turn influenced by the electrochemical equilibrium. This makes the models which have been proposed for its study very complex and difficult to implement in practice, principally due to tae difficulty in reliably determining their parameters.

Articular cartilage, because of its location and physiological function, is subjected to dynamic and impact loads which in the long term may produce fatigue cracks and cause rupture. Biomechanical factors, together with others of a biological nature, are the basis of the appearance of severe degenerative conditions such as osteoarthritis below. For this reason, it is important to be able to evaluate and understand the biomechanical properties of this tissue in order to act appropriately to prevent, diagnose and treat these diseases.

EXERCISE 81

Calculate the mechanical moment required to mobilise the articulation of the femur with the knee in the configuration of 0° flexure, indicated in the diagram.

Apply the result to a person of mass: a) 50 kg, b) 75 kg and c) 90 kg. Consider the situations that the person would encounter standing on two feet, walking and running. Comment on the results obtained.

DATA:

- Cartilage contact area: A=10 cm^2. Assume a spherical contact of radius R=2.5 cm.
- Maximum force on the knee: W/2 for standing, W when walking and 3W when running. W is the body weight.
- The coefficient of articular friction is shown in the figure below, obtained by LL. Malcom (An experimental investigation of the frictional and deformational responses of articular cartilage interfaces to static and dynamic loading. PhD Thesis, University of California, San Diego, 1976).

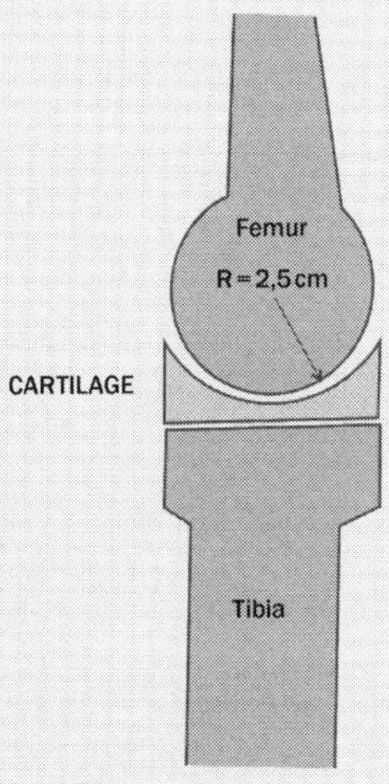

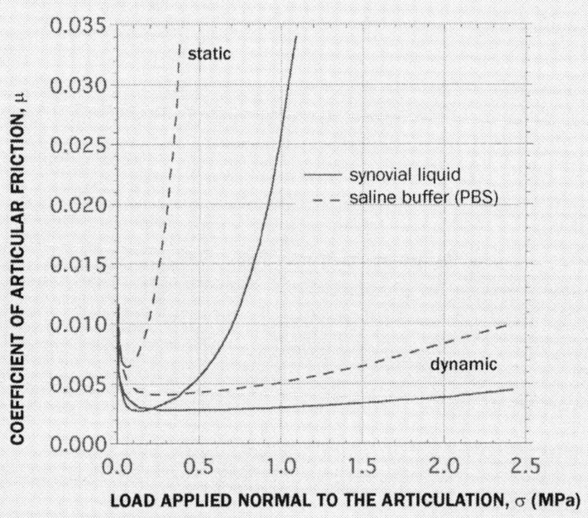

SOLUTION:

A sketch of the forces acting on an infinitesimal element of the contact is provided in the figure below.

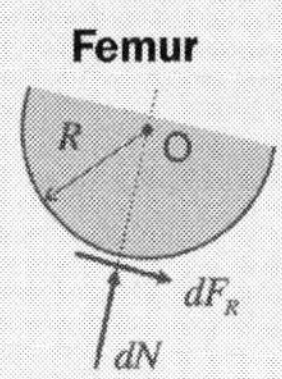

If dN is the normal force acting, the value of the frictional force produced in movement will be: $dF_R = \mu dN$ where μ is the coefficient of friction.

The moment dM about the centre of rotation of the articulation, point O, will be $dM = RdF_R = \mu RdN$

The total value is obtained by integration $M = \int \mu R dN = \mu R \int dN = \mu RN$

In order to estimate the value of the coefficient of friction from Malcom's data, we need to know both the stress normal to the contact, σ, and the nature of the motion (static or dynamic) in each case.

The normal stress is calculated by distributing the total load N across the contact area A

$$\sigma = N/A$$

The table below gives the values corresponding to each of the required situations.

			m (kg)		
SITUATION	**PARAME-TERS**	**UNITS**	**50**	**75**	**90**
standing	N = mg/2	N	245	368	441
walking	mg		491	736	883
running	3 mg		1472	2207	2649
standing	σ = N/A	MPa	0.245	0.368	0.441
walking			0.491	0.736	0.883
running			1.472	2.207	2.649
standing	μ (synovia)		0.0030	0.0039	0.0048
walking			0.0028	0.0028	0.0029
running			0.0033	0.0042	0.0048
standing	μ (PBS)		0.0140	0.0290	>~0.035
walking			0.0042	0.0046	0.0049
running			0.0064	0.0092	0.0110
standing	**M (synovia)**	**Nm**	**0.0184**	**0.0359**	**0.0530**
walking			**0.0343**	**0.0515**	**0.0640**
running			**0.1214**	**0.2318**	**0.3178**
standing	**M (PBS)**	**Nm**	**0.0858**	**0.2667**	**>0.3900**
walking			**0.0515**	**0.0846**	**0.1082**
running			**0.2354**	**0.5077**	**0.7284**

From the data obtained, we can see that the values of the dynamic coefficient of friction are always less than the static ones and, furthermore, their increment when the load applied normal to the articulation increases is much less.

Similarly, the values of the coefficient of friction for articulation lubricated with synovial liquid are significantly less than those corresponding to an articulation lubricated with a saline buffer (PBS), whose behaviour and viscosity are similar to water. Although this situation is unrealistic, it is useful to put into perspective the importance, as we have seen, of synovial fluid which is thixotropic. This fact, together with the hydrodynamic lubrication mechanism, leads to a significant reduction in the value of microns in dynamic situations (mobile articulation). Hence the importance of having the synovial fluid in a healthy state.

We observe also that the values of the frictional force and the mechanical moment grow more rapidly than the weight of the individual, accentuating this phenomenon in static situations. This fact points to the danger of obesity in relation to the deterioration of articular cartilage.

REFERENCES

Abrahams, M. (1967).
Mechanical behaviour of tendon In vitro.
Medical and biological engineering 5 (5); 433-443.

Agianniotis A, Rachev A, Stergiopulos N. (2012).
Active Axial Stress in Mouse Aorta.
J of Biomechanics 45 1924-1927.

Alexander R. McN. (1962).
Visco-Elastic Properties of the Body-Wall of Sea Anemones.
J Exp Biol 39, 373-386.

Alexander R. McN. (1964).
Viscoelastic properties of the mesogloea of jellyfish.
J Exp Biol 41, 363-369.

Alexander R. McN. (1966).
Rubber-like properties of the inner hinge-ligament of pectinidae.
J Exp Biol 44, 119-130.

Alexander R.McN., Ker R.F., Bennett M.B.I (1990).
Optimum stiffness for leg bones.
J Zool, London 222, 471-478.

An B., Liu Y., Arola D. *et al.* (2011).
Fracture toughening mechanism of cortical bone: An experimental and numerical approach.
JMBBM 4, 983-992.

An Y.H., Draughn R.A. (Eds.) (2000).
Mechanical Testing of bone and the Bone-implants Interface.
CRC Press.

Amiel D, Frank C, Harwood F, *et al.* (1984).
Tendon and ligaments: A morphological comparison.
J Orthop Res 1: 257–265.

Andersons, J., Sparnins, E., Joffe, R. *et al.* (2005)
Strength distribution of elementary flax fibres.
Compos. Sci. Technol. 65, 693–702.

Armentano R L, Levenson, J G, Barra J. *et al.* (1991).
Assessment of elastin and collagen contribution to aortic elasticity in conscious dogs.
Am. J. Physiol., 260 H1870–H1877.

Arruda E.M., Boyce M.C. (1993)
A three-dimensional constitutive model for the large stretch behaviour of rubber elastic materials.
J Mech Phys Solids 41 389-412.

Ashby M.F., Easterling K.E., Harrysson R. *et al.* (1985).
The fracture and toughness of woods.
Proc Roy Soc A 398, 261-280.

ASM Handbook Vol. 8 (2000).
Hardness Testing.
ASM, Materials Park, Ohio.

ASTM C39 (2012).
Standard Test Method for Compressive Strength of Cylindrical Concrete Specimens.
Annual Book of ASTM Standards, Vol. 04.02.

ASTM D695 (2010).
Standard Test Method for Compressive Properties of Rigid Plastics.
Annual Book of ASTM Standards, Vol. 08.01.

ASTM D790 (2003).
Standard Test Methods for Flexural Properties of Unreinforced and Reinforced Plastics and Electrical Insulating Materials.
Annual Book of ASTM Standards, Vol. 08.01.

ASTM D2240 (2015).
Standard Test Method for Rubber Property—Durometer Hardness.
Annual Book of ASTM Standards, Vol. 09.01.

ASTM D6272 (2010).
Standard Test Method for Flexural Properties of Unreinforced and Reinforced Plastics and Electrical Insulating Materials by Four-Point Bending.
Annual Book of ASTM Standards, Vol. 08.03.

ASTM E8 (2013).
Standard Test Methods for Tension Testing of Metallic Materials.
Annual Book of ASTM Standards, Vol. 03.01,

ASTM E9 (1995).
Standard Test Methods of Compression Testing of Metallic Materials at Room Temperature.
Annual Book of ASTM Standards, Vol. 03.01.

ASTM E384 (2017).
Standard Test Method for Microindentation Hardness of Materials.
Annual Book of ASTM Standards, Vol. 03.01.

ASTM E399 (2012).
Standard Test Method for Linear-Elastic Plane-Strain Fracture Toughness K_{Ic} of Metallic Materials.
Annual Book of ASTM Standards, Vol. 03.01.

ASTM E1304 (2002).
Standard Test Method for Plane-Strain (Chevron-Notch) Fracture Toughness of Metallic Materials.
Annual Book of ASTM Standards, Vol. 03.01.

ASTM E1820 (2013).
Standard Test Method for Measurement of Fracture Toughness.
Annual Book of ASTM Standards, Vol. 03.01.

Atlan G., Balmain N., Berland S., *et al.* (1997).
Reconstruction of human maxillary defects with nacre powder: Histological evidence of bone regeneration.
C R Acad Sci Paris / Life Sci 320, 253-258.

Atkins A G, Mai Y W (1985).
Elastic and Plastic Fracture.
John Wiley & Sons.

Bäck, M, Gasser T C, Michel J B *et al.* (2013).
Review. Biomechanical factors in the biology of aortic wall and aortic valve diseases.
Cardiovascular Research 99 232–241.

Bae W.C., Wong V.W., Hwang J., *et al.* (2008).
Wear-lines and split-lines of human patellar cartilage: Relation to tensile biomechanical properties.
Osteoarthritis Cartilage 16(7), 841-845.

Ball R.C., Doi M., Edwards S.F. *et al.* (1981).
Elasticity of entangled networks.
Polymer 22, 1010-1018.

Barber E.J.W. (1992)
Prehistoric Textiles.
Princeton Univ. Press.

Barnett V. (1975)
Probability plotting methods and order statistics.
Applied Statistics 24 (1), 95-108.

Barthelat F., Tang H., Zavattieri P. D. *et al.* (2007).
On the mechanics of mother-of-pearl: A key feature in the material hierarchical structure.
J Mech Phys Solids 55, 306-337.

Bazant, Z.P and Planas, J. (1997).
Fracture and Size Effect in Concrete and Other Quasibrittle Materials.
CRC Press, Boca Raton.

Bendit E.G. (1960).
A quantitative X-ray diffraction study of the alpha-beta transformation in wool keratin.
Text Res J 30, 547-555.

Beste L.F., Hoffman R.M. (1950).
Resilience of fibres and fabrics: Quantitative study.
Text Res J 20, 441-453.

Bhattacharya, A.K., Nix W.D. (1988).
Finite element simulation of indentation experiments.
Int J Solids Structures 24, 1287-1298.

Bledzki A.K., Gassan J. (1999).
Composites reinforced with cellulose based fibres.
Prog Polym Sci 24, 221-274.

Billiar KL, Sacks MS. (2000).
Biaxial mechanical properties of the natural and glutaraldehyde treated aortic valve cusp–Part II: A Structural Constitutive Models.
J Biomech Eng 122 327–335.

Bishop MC, Coulston CN. (1989).
Roman Military Equipment.
Shire Publications.

Bishop J E, Lindahl G (1999).
Regulation of cardiovascular collagen synthesis by mechanical load.
Cardiovasc. Res., 42 27–44.

Bobbio A. (1972).
The first endosseous alloplastic implant in the history of man.
Bull Hist Dent 20, 1-6.

Boeriu C.G., Springer J., Kooy, F.K. *et al.* (2013).
Production methods for hyaluronan.
Int J Carbohydrate Chem 2013, 624967.

Boltzmann L. (1874).
Zur Theorie der Elastischen Nachwirkung.
Sitzungsberichte der Akademie der Wissenschuften II 70, 275-306.

Bouligand Y (1972).
Twisted fibrous arrangements in biological materials and cholesteric mesophases.
Tissue & Cell 4 (2) 189-217.

Bowman S.M., Gibson L.J. *et al.* (1999).
Results from demineralized bone creep tests suggest that collagen is responsible for the creep behavior of bone.
J Biomech Eng 121, 253-258.

Bowman S.M., Guo X.E., Cheng D.W. *et al.* (1998).
Creep contributes to the fatigue behavior of bovine trabecular bone.
J Biomech Eng 120, 647-654.

Brinckmann J. (2005)
Collagens at a glance.
Top. Curr. Chem. 247, 1–6.

Buehler M.J. (2006)
Nature designs tough collagen: Explaining the nanostructure of collagen fibrils.
PNAS 103 (33), 12285-12290.

Buehler, M. J. (2008)
Nanomechanics of collagen fibrils under varying cross-link densities: atomistic and continuum studies.
J. Mech. Behav. Biomed. Mater. 1: 59–67.

Burton AC (1954).
Relation of structure to function of the tissues of the wall of blood vessels.
Physiol Rev 34 619-642.

Bustamante C., Smith S.B. *et al.* (2000)
Single-molecule studies of DNA mechanics.
Current Opinion in Structural Biology 10(3), 279-285.

Calvo F., Sanz-Moreno V., Agudo-Ibáñez L. *et al.* (2011).
RasGRF suppresses Cdc42-mediated tumour cell movement, cytoskeletal dynamics and transformation.
Nature Cell Biology 13(7), 819-826.

Capaldi F.M. (2012).
Continuum Mechanics.
Cambridge Univ. Press.

Carrera A, García Páez *et al.* (2001).
Pretensado y comportamiento mecánico del pericardio de ternera utilizado en la construcción de bioprótesis cardíacas.
Biomecánica 9, 1, 3–11.

Carter D.R., Schwab G.H., Spengler D.M. (1980).
Tensile fracture of cancellous bone.
Acta Orthopaed Scan 51, 733-741.

Chan Y.L., Ngan A.H.W. (2010).
Invariant elastic modulus of viscoelastic materials measured by rate-jump tests.
Polymer Testing 29, 558-564.

Chapman G. (1953).
Studies on the Mesogloea of Coelenterates: II. Physical Properties.
J Exp Biol 30, 440-451.

Chawla, K. K. (2016).
Fibrous Materials.
Cambridge Univ. Press , HRS.

Chawla K.K. (2012).
Composite Materials (3rd. Ed.).
Springer.

Chen C., Lu T.J., Fleck N.A. (2001).
Effect of inclusions and holes on the stiffness and strength of honeycombs.
Int J Mech Sci 43, 487-504.

Chen P *et al.* (2012).
Predation versus protection: Fish teeth and scales evaluated by nanoidentation.
J. Mater. Res. 27 100-112.

Chen, X L,Varner S E *et al.* (2003).
Laminar flow induction of antioxidant response element-mediated genes in endothelial cells. a novel anti-inflammatory mechanism.
J. Bio. Chem. 278 703–711.

Chin-Purcell M.V., Lewis J.L. (1996).
Fracture of articular cartilage.
J Biomech Eng 118(4), 545-556.

Chitapalli R K *et al.* (2014).
Fabrication, testing and modeling of a new flexible armor inspired from natural fish scales and osteoderms.
Bioinspir. Biomim. 9 036005.

Choi H S, Vito R P, (1990).
Two dimensional stress-strain relationship for canine pericardium.
J. Biomech. Eng. 112 153–159.

Choung C J, Fung Y C. (1986).
Residual stress in arteries.
In Frontiers in Biomechanics. G. W. Schmid-Schoenbein, *et al.* (Eds.), (pp. 117–129).
New York: Springer.

Christensen R.M. (1979).
Mechanics of Composite Materials.
John Wiley & Sons.

Claes E. (2010).
Estudio mecánico de las arterias coronarias humanas y sus sustitutos vasculares.
Ph. D. Thesis, Universidad Politécnica de Madrid.

Claramunt R, García Páez J M, Alvarez L. *et al.* (2012).
Fatigue behaviour of young ostrich pericardium.
Material Sciences and Engineering C, 32 6 1415-1420.

Claramunt R, García Páez J M, Alvarez L, Spottorno J, Ros A, Casado M C. (2011).
Short-term fatigue testing can predict medium-term pericardium behaviour.
JMBBM 4, 1929-193.

Clark J M, Glagov S. (1985).
Transmural organization of the arterial media: The lamellar unit revisited.
Arterioscler. Thromb. Vasc. Biol. 5 19-34.

Cook R.B., Zioupos P. (2009).
The fracture toughness of cancellous bone.
J Biomech 42, 2054-2060.

Cook J., Gordon J.E. (1964).
A mechanism for the control of crack propagation in all-brittle systems.
Proc Roy Soc A 282, 508-520.

Cotterell B. (2010).
Fracture and Life.
Imperial College Press.

Cowin, S.C., Doty, S.B. (2007).
Tissue Mechanics.
Springer, New York.

Courtman DW, Pereira CA, Kashef V. *et al.* (1994).
Development of a pericardial acellular matrix biomaterial: biochemical and mechanical effects of cell extraction.
J Biomed Mater Res 28 (6) 655–666.

Cowan PM, McGavin S, North ACT (1955).
The polypeptide chain configuration of collagen.
Nature 176: 1062–64.

Cowin, S.C., Doty, S. B. (2007)
Tissue Mechanics.
Springer, New York, USA.

Cubo N, Garcia M, Del Cañizo J F, *et al.* (2016).
3D bioprinting of functional human skin: production and in vivo analysis.
Biofabrication. Dec 5;9(1):015006.

Cugno A M *et al.* (2012).
On the mechanical behaviour of scales from Cyprinus carpio.
JMBBM 7 17-29.

Currey J.D. (2002).
Bones: Structure and Mechanics.
Princeton Univ Press.

Currey J.D., Kohn A.J. (1976).
Fracture in the crossed-lamellar structure of Conus shells.
J Mater Sci 11, 1615-1623.

Currey J.D., Taylor J.D. (1974).
The mechanical behaviour of some molluscan hard tissues.
J Zool Lond 173, 395-406.

Danilatos G., Feughelman M. (1979).
Dynamic mechanical properties of α-keratin fibers during extensión.
J Macromol Sci – Phys B16, 581-602.

Dannenberg H. (1961).
Measurement of adhesión by a blister method.
J Appl Polym Sci 5, 125-134.

Daza R., Cruces J., Arroyo-Hernández M. *et al.* (2015).
Topographical and mechanical characterization of living eukaryotic cells on opaque substrates: Development of a general procedure and its application to the study of non-adherent lymphocytes.
Phys Biol 12, 026005.

Demiray H. (1981).
Large deformation analysis of some soft biological tissues.
J. Biomech. Eng., 103, 73–78.

Deville S., Sáiz E., Tomsia A. P. (2007).
Ice-templated porous alumina structures.
Acta Mater 55(6), 1965-1974.

Doerner M.F., Nix W.D. (1986).
A method for interpreting the data from depth-sensing indentation instruments.
J Mater Res 1(4), 601-609.

Dorrington K., Grut W., McCrum N.G.
Mechanical state of elastin.
Nature, London 255, 476-478.

Drummy L.F., Farmer B.L., Naik R.R. (2007)
Correlation of the β-sheet crystal size in silk fibers with the protein amino acid sequence.
Soft Matter 3, 877-882.

Drymon, M. M. (2008)
Disguised as the Devil: How Lyme Disease Created Witches and Changed History.
Wythe Avenue Press. p. 114. ISBN 0-615-20061-3.

Easterling K.E., Harrysson R. *et al.* (1982).
On the mechanics of balsa and other woods.
Proc Roy Soc A 383, 31-41.

Ebenstein D.M., Coughlin D., Chapman J. *et al.* (2008).
Nanomechanical properties of calcification, fibrous tissue, and hematoma from atherosclerotic plaques.
J Biomed Mater Res 91A, 1028-1037.

Edwards S.F., Vilgis Th. (1986).
The effect of entanglements in rubber elasticity.
Polymer 2, 483-492.

Eichner R., Rew P., Engel A. *et al.* (1985).
Human epidermal keratin filaments: Studies on their structure and assembly.
Ann N Y Acad Sci 455, 381-402.

Ekeland A, Engesaeter L.B., Langeland N. (1982).
Influence of age on mechanical properties of healing fractures and intact bones in rats.
Acta Orthop Scand 53, 527-534.

Ekeland A, Engesaeter L.B., Langeland N. (1981).
Mechanical properties of fractured and intact rat femora evaluated by bending, torsional and tensile tests.
Acta Orthop Scand 52(6), 605-613.

Elices M. (1998).
Mecánica de la Fractura.
ETSICCP, Universidad Politécnica de Madrid.

Elices M., Llorca, J. (2002)
Fiber Fracture.
Elsevier Science.

Elices M., Pérez-Rigueiro J., Plaza G.R. *et al.* (2005).
Finding inspiration in Argiope trifasciata spider silk fibers.
JOM 57(2), 60-66.

Elices M., Plaza G.R., Pérez-Rigueiro J. *et al.* (2011).
The hidden link between supercontraction and mechanical behavior of spider silks.
Journal of the Mechanical Behavior of Biomedical Materials 4(5), 658-669.

Eppell, S. J., B. N. Smith, H. Kahn *et al.* (2006).
Nano measurement with micro-devices: mechanical properties of hydrated collagen fibrils.
J. Roy. Soc. Interface. 3: 117–121.

Escofflier C *et al.* (1989).
Age-related mechanical properties of human skin: an in vivo study.
J Invest Dermatol 93, 353-357.

Evans A.G., Suo Z., Wang R.Z., Aksay I.A., He M.Y., Hutchinson J.W. (2001).
Model for the robust mechanical behavior of nacre.
J Mater Res 16(9), 2475-2484.

Evans F G. (1973).
Mechanical properties of bone.
Charles C Thomas. Springfield IL.

Farag R. Elmogahzy Y. (2009).
Tensile properties of cotton fibres.
In Handbook of Tensile Properties of Fibers (A.R. Bunsell Ed.), Chap. 3.

Feltwell J. (1990)
The Story of Silk.
St. Martin Press, NY.

Fernandes M.C., Aizenberg J., Weaver J.C. *et al.* (2021).
Mechanically robust lattices inspired by deep-sea glass sponges.
Nature Materials 20, 237-241.

Feughelman M. (1997)
Mechanical Properties and Structure of Alpha-Keratin Fibres.
University of New South Wales Press.

Feughelman M. (1954).
The creep of wool fibres in water.
J Text Inst 45, T630-T641.

Feughelman M., Robinson M.S. (1971).
Some mechanical properties of wool fibers in the "Hookean" región from zero to 100% relative humidity.
Text Res J 41, 469-474.

Fidelis, M. E. A., Pereira, T. V. C., Gomes, *et al.* (2013).
The effect of fiber morphology on the tensile strength of natural fibers.
Journal of Materials Research and Technology, 2, 149-157.

Fischesser B. (2000).
El árbol.
El Drac, S.L.

Flory P. J. (1989).
Statistical Mechanics of Chain Molecules.
Hanser Publishers, Munich

Fondrk M., Bahniuk E., Davy D.T. *et al.* (1988).
Some viscoplastic characteristics of bovine and human cortical bone.
J Biomech 21, 623-630.

Frank E.H. (1980).
Electromechanical Dynamics of a Charged Hydrated Polymer.
Thesis (B.S.), MIT.

Fronek K, Schmid-Schoenbein G, Fung Y C. (1976).
A noncontact method for three-dimensional analysis of vascular elasticity in vivo and in vitro.
J Appl Physiol 40, 634-637.

Fung Y C. (1993).
Biomechanics Mechanical Properties of Living Tissues, 2nd Ed.
Springer.

Fung YC. (1975).
Stress, deformation, and atelectasis of the lung.
Circulation Res 37, 481-496.

Fung YC, Lin S. (1991).
Changes of zero-stress state of rat pulmonary arteries in hypoxic hypertension.
J Appl Physiol 70 2455-2470.

Fung YC, Liu S Q. (1989).
Change of residual strains in arteries due to hypertrophy caused by aortic constriction.
Circulation Res. 65 1340-1349.

Fung YC, Fronek K, Patitucci P. (1979).
Pseudoelasticity of arteries and the choice of its mathematical expression.
Am J Physiol 237 H620-H631.

Fung YC, Tong P, Palitucci P. (1978).
Stress and strain in the lung.
J Eng Mech Div Am Soc Civil Eng 104 201-224.

Funk N *et al.* (2015).
Bioinspired Fabrication and Characterization of a Synthetic Fish Skin for the Protection of Soft Materials.
Appl. Mater. Interfaces 7 5972-5983.

Fung YC. (1972).
Stress–strain history of soft tissues in simple elongation.
In Biomechanics, its Foundations and Objectives (YC Fung, N Perrone, M Anliker, Eds.), 181–208, Englewood Cliffs, NJ: Prentice-Hall.

Gao W., Wang M., Bai H. (2020).
A review of multifunctional nacre-mimetic materials based on bidirectional freeze casting.
JMBBM 109, 103820.

Gao H., Chiu C.H., Lee, J. (1992).
Elastic contact versus indentation modeling of multi-layered materials.
Int J Solids Structures 29, 2471-24.

García Esteban L., *et al.* (2002).
La madera y su tecnología.
Fundación Conde del Valle de Salazar. Madrid.

García-Herrera C. (2008).
Comportamiento Mecánico de la Aorta Ascendente: Caracterización Experimental y Simulación Numérica.
Ph. D. Thesis, Universidad Politécnica de Madrid.

García-Páez J M, Carrera A, Herrero J, *et al.* (2006).
Determination of the force necessary for the propagation of tears in ostrich and calf pericardium.
J Biomedical Materials Research: Part B Applied Biomaterials. 79 (2) 229-235.

García-Páez J M, Carrera A, Herrero J *et al.* (2004).
Resistance to tearing of calf and ostrich pericardium: influence of the type of suture material and the direction of the suture line.
J Biomedical Materials Research: Part B Applied Biomaterials. 69 (2) 125-34.

García-Páez J M, Claramunt R, Herrero J *et al.* (2009).
Energy consumption as a predictor test of the durability of a biological tissue employed in cardiac bioprosthesis.
J Biomedical Materials Research Part A. 89A 336–344.

García-Páez J M, Herrero J, Carrera A *et al.* (2001).
Ostrich pericardium, a biomaterial for the construction of valve leaflets for cardiac bioprostheses: mechanical behaviour, selection and interaction with suture materials.
Biomaterials 22 20 2731–2740.

García-Páez J M, Jorge E, Claramunt R *et al.* (2010).
Propagation of tears in pericardium from young bulls: influence of the suture.
Artificial Organs 34 (3) 65-71.

García-Paéz J M, Herrero J, Carrera *et al.* (2000).
The influence of chemical treatment and suture on the elastic behavior of calf pericardium utilized in the construction of cardiac bioprostheses.
J Mater Sci Mater Med. 11 (5) 273–277.

Garrido M.A., Elices M., Viney C. *et al.* (2002).
Active control of spider silk strength: Comparison of drag line spun on vertical and horizontal surfaces.
Polymer 43(4), 1537-1540.

Gasser T C, Gallinetti S, Xing X *et al.* (2012).
Spatial orientation of collagen fibers in the Abdominal Aortic Aneurysm wall and its relation to wall mechanics.
Acta Biomaterialia, 83 091–3103, 201.

Gatesy J., Hayashi C., Motriuk D. *et al.* (2001)
Extreme diversity, conservation, and convergence of spider silk fibroin sequences.
Science 291, 2603-2605.

Gibson L.J., Ashby M.F., Harley B.A. (2010).
Cellular Materials.
Cambridge Univ. Press.

Gibson L.J., Ashby M.F. (1997).
Cellular Solids: Structure and Properties, 2nd Edn.
Cambridge Univ. Press.

Gibson L.J., Ashby M.F., Schajer G.S. *et al.* (1982).
The mechanics of two-dimensional cellular materials.
Proc Roy Soc Lond A382, 25-42.

Gibson L.J., Ashby M.F. (1982).
The mechanics of three-dimensional cellular materials.
Proc Roy Soc Lond A382, 43-59.

Gil L. (2010).
Anatomía vegetal aplicada a las especies forestales.
Fundación Conde del Valle de Salazar. Madrid.

Gil-Duran S, Arola D, Ossa E A. (2016).
Effect of chemical composition and microstructure on the mechanical behaviour of fish scales from Megalops atlanticus.
JMBBM 56 134-145.

Glenister T.W. (1976).
An embryological view of cartilage.
J Anat 122, 323-330.

Gosline J.M., French C.J. (1979).
Dynamic mechanical properties of elastin.
Biopolymers 18, 2091-2103.

Gosline J M. (1971).
Connective Tissue Mechanics of Metridium Senile: I. Structural and Compositional Aspects.
J Exp Biol 55, 763-774.

Gosline J M. (1971).
Connective Tissue Mechanics of Metridium Senile: II. Visco-Elastic Properties and Macromolecular Model .
J Exp Biol 55, 775-795.

Grossman M, Pivovarov D, Bouville F, Dransfeld C, Masania K, Studart A R. (2019).
Hierarchical toughening of nacre-like composites.
Adv. Funct. Mater. 29 1806800.

Grytsan A, Eriksson T, Watton PN, *et al.* (2017).
Growth Description for Vessel Wall Adaptation: A Thick-Walled Mixture Model of Abdominal Aortic Aneurysm Evolution.
https://doi.org/10.3390/ma10090994.

Guinea G V, Atienza J M, Elices M *et al.* (2005).
Thermomechanical behaviour of human carotid arteries in the passive state.
AJP- Heart and Circulatory Physiology 288, 2940-2945.

Guinea G.V., Rojo F.J., Elices M. (2004)
Brittle failure of dry spaghetti.
Eng. Failure Anal. 11, 705.

Guinea G.V., Pérez-Rigueiro J., Plaza G.R. *et al.* (2006)
Volume constancy during stretching of spider silk.
Biomacromolecules 7(7), 2173-2177.

Guinea G.V., Elices M., Pérez-Rigueiro J., Plaza G.R. *et al.* (2023)
Structure and properties of spider and silkworm silk for tissue engineering in Silk Biomaterials (s.c. Kundu Ed.) Chap. 9.

Guinea G.V., Elices M., Pérez-Rigueiro J. *et al.* (2005a)
Stretching of supercontracted fibers: A link between spinning and the variability of spider silk.
J Exp Biol 208(1), 25-30.

Guinea G.V., Elices M., Real J.I. *et al.* (2005b).
Reproducibility of the tensile properties of spider (Argiope trifasciata) silk obtained by forced silking.
Journal of Experimental Zoology Part A - Compartive Experimental Biology 303A, 37-44.

Guo X.E., Kim C.H. (2002).
Mechanical consequences of trabecular bone loss and its treatment: A three dimensional model simulation.
Bone 30, 404-411.

Guo X.E., McMahon T.A., Keaveny T.M. *et al.* (1994). *Finite element modeling of damage accumulation in trabecular bone under cyclic loading.* J Biomech 27, 145-155.

Hakimi O., Knight D.P., Vollrath F. *et al.* (2007) *Spider and mulberry silkworm silks as compatible biomaterials.* Compos Pt B: Eng 38, 324-337.

Harding C. (2004). *The stratum corneum: Structure and function in health and disease.* Dermatologic Therapy 17, 6-15.

Harley, R., D. James, A. Miller *et al.* (1977). *Phonons and the elastic moduli of collagen and muscle.* Nature. 267: 285–287.

Hay J.L., O'Hern M.E., Oliver W.C. (1998). *The importance of contact radius for substrate-independent property measurement of thin films.* Mat Res Soc Symp Proc 522, 27-32.

Hayes W.C., Gerhart T.N. (1985). *Biomechanics of bone applications for assessment of bone strength.* Bone and Mineral Research 3, 259-294.

Hayes W.C., Carter D.R. (1976). *Post yield behavior of subchondrial trabecular bone.* J Biomed Mat Res Symp 7, 537-544.

Hazen A. (1930). *Flood Flows: A Study of Frequencies and Magnitudes.* John Wiley & Sons, New York.

Hebert J.J., Thibodeaux D.P., Shofner F.M. *et al.* (1995). *A new single fiber tensile tester.* Text Res J 65, 440-444.

Heim, A. J., W. G. Matthews, and T. J. Koob. (2006). *Determination of the elastic modulus of native collagen fibrils via radial indentation.* Appl. Phys. Lett. 89, 181902.

Hingorani RV, Provenzano PP, Lakes RS *et al.* (2004). *Nonlinear viscoelasticity in rabbit medial collateral ligament.* Ann Biomed Eng 32: 306–312.

Holzapfel G. A. (2006). *Nonlinear Solid Mechanics.* John Wiley & Sons.

Holzapfel G A, Gasser T C (2001). *A viscoelastic model for fiber-reinforced composites at finite strains: Continuum basis, computational aspects and applications.* Computer Methods in Applied Mechanics and Engineering. 190, 4379–4403.

Hoppmann W H, Wan L. (1970). *Large deformation of elastic tubes.* J Biomech 3, 593-600.

Horný L, Adamek T, Kulvajtova M. (2014). *Analysis of axial prestretch in the abdominal aorta with reference to post mortem interval and degree of atherosclerosis.* JMBBM. 33:93-98.

Horný L, Gultová E, Chlup H *et al.* (2010). *Mullins Effect in Aorta and Limiting Extensibility Evolution.* Bulletin of Applied Mechanics, 6 21 1-5.

Hudspeth M., Nie X. Chen W. *et al.* (2012). *Effect of loading rate on mechanical properties and fracture morphology of spider silk.* Biomacromolecules 13(8), 2240-2246.

Humphrey J D, Yin F C P. (1988). *Biaxial mechanical behavior of excised epicardium.* J. Biomech. Eng. 110, 349-351.

Humphrey J D. (2002). *Cardiovascular Solid Mechanics: Cells, Tissues, and Organs.* Springer.

Ikai A. (2010). *A review on: Atomic force microscopy applied to nano-mechanics of the cell.* Adv Biochem Engin/Biotechnol 119, 47-61.

Ikoma T *et al.* (2003). *Microstructure, mechanical, and biomimetic properties of fish scale from Pagrus major.* J. Structural Biology 142, 327-333.

ISO 6506-1 (2014). *Metallic materials –Brinell hardness test– Part 1: Test method.* International Organization for Standardization.

ISO 6506-2 (2014).
Metallic materials -Brinell hardness test- Part 2: Verification and calibration of testing machines.
International Organization for Standardization.

ISO 6506-3 (2014).
Metallic materials -Brinell hardness test- Part 3: Calibration of reference blocks.
International Organization for Standardization.

ISO 6506-2 (2014).
Metallic materials -Brinell hardness test- Part 4: Table of hardness values.
International Organization for Standardization.

ISO 14577-4 (2007).
Metallic materials -Instrumented indentation test for hardness and materials parameters- Part 4: Test method for metallic and non-metallic coatings.
International Organization for Standardization.

Iyo T., Maki Y., Sasaki N., Nakata M. (2004).
Anisotropic viscoelastic properties of cortical bone.
J Biomech 37, 1433-1437.

Jackson A.P., Vincent J.F.V., Turner R.M. (1988).
The mechanical design of nacre.
Proc R Soc Lond B234, 415-440.

Jäger L., Fratzl P. (2000).
Mineralized collagen fibrils: A mechanical model with a staggered arrangement of mineral parlticles.
Biophysical Journal 79, 1737-1746.

James H.M., Guth E. (1943).
Theory of the elastic properties of rubber.
J Chem Phys 11, 455-481.

Jefferies R., Jones D. M., Roberts J. G., Selby K., Simrnens S. C., Warwicker J. O. (1969).
Current ideas on the structure of cotton.
Cell Chem Tech 3, 255-274.

Jeronimidis G. (1980).
The fracture behaviour of wood and the relations between morphology.
Proc Roy Soc B208, 447-460.

Ji H.M., Yang W., Chen D.L. *et al.* (2020).
Natural arrangement of fiber-like aragonites and its impact on mechanical behavior of mollusk shells: A review.
JMBBM 110, 103940.

Johnson T.P.M., Socrate S., Boyce M.C. (2010).
A viscoelastic, viscoplastic model of cortical bone valid at low and high strain rates.
Acta Biomaterialia 6, 4073-4080.

Katz J.L. (1971).
Composite material models for cortical bone.
In Mechanical Properties of Bone (Cowin S.C., Ed.), 171-184.

Kadler K. (1994).
Extracellular matrix. 1: Fibril-forming collagens.
Protein Profile 1, 519-638.

Kefalas V., Eftaxiopoulos D.A. (2012).
Experimental study of cancellous bone under large strains and a constitutive probabilistic model.
JMBBM 6, 41-52.

Kempson G.E. (1982).
Relationship between the tensile properties of articular cartilage from the human knee and age.
Annals of the Rheumatic Diseases 41, 508-511.

Kerr T. (1937).
The structure of the growth rings in the secondary wall of the cotton hair.
Protoplasma 27(1), 229-241.

King R.B. (1987).
Elastic analysis of some punch problems for a layered medium.
Int J Solids Structures 23, 1657-1664.

Kita M, Negi A, Kawano S. *et al.* (1990).
Measurement of Retinal Adhesive Force in the in Vivo Rabbit Eye.
Investigative Opthalmology and Visual Science, 31, 4, 624-628.

Knets I. Malmeisters A. (1977).
Deformability and strength of human compact bone tissue.
In Mechanics of Biological Solids, Proceedings Euromech. Colloquim 68, Varna, Bulgaria, 1975, Sofia, 123–141.

Koester K.J., Barth H.D., Ritichie R.O. (2011).
Effect of aging on the transverse toughness of human cortical bone: Evaluation by R-curves.
JMBBMer 4, 1504-1513.

Kolken H. M. A., Lietaert K., Sloten T. *et al.* (2020).
Mechanical perfomance of auxetic meta-biomaterials.
JMBBM 404, 103-658.

Kopperdahl D.L., Keaveny T.M. (1998).
Yield strain behavior of trabecular bone.
J Biomech 31, 601-608.

Korhonen R.K., Jurvelin J.S. (2010).
Compressive and tensile properties of articular cartilage in axial loading are modulated differently by osmotic environment.
Medical Engineering & Physics 32, 155-160.

Kosmopoulos V., Schizas C., Keller T.S. (2008).
Modeling the onset and propagation of trabecular bone microdamage during low-cycle fatigue.
J Biomech 41, 515-522.

Kruzic J.J., Kim D.K., Koester K.J. *et al.* (2009).
Indentation techniques for evaluating the fracture toughness of biomaterials and hard tissues.
JMBBM 2, 384-395.

Kuhn-Spearing L.T., Kessler H., Chateau E. *et al.* (1996).
Fracture mechanisms of the Strombus gigas conch shell: Implications for the design of brittle laminates.
J Mater Sci 31, 6583-6594.

Kwan M K, Woo S LY. (1989).
A structural model to describe the nonlinear stress-strain behavior for parallel-fibered collagenous tissues.
J Biomech Eng 111, 361-363.

Lai W.M., Hou J.S., Mow V.C. (1991).
A triphasic theory for the swelling and deformation behaviors of articular cartilage.
J Biomech Eng 113, 245-258.

Lakes, R. (2009)
Viscoelastic Materials.
Cambridge Univ. Press.

Lakes R.S., Katz J.L., Sternstein S. (1979).
Viscoelastic properties of wet cortical bone – I. Torsional and biaxial studies.
J Biomech 12, 657-678.

Landau L D, Lifshitz E M. (1986).
Theory of Elasticity (3rd Ed.).
Pergamon Press.

Lanir Y. (1979).
A structural theory for the homogeneous biaxial stres strain relationships in flat collagenoust issues.
J Biomech 12, 423-436.

Lanir Y, Fung Y C. (1974).
Two-dimensional mechanical properties of rabbit skin. I. Experimental system.
J Biomech 7 (1), 29-34.

Lanir Y, Fung Y C. (1974).
Two-dimensional mechanical properties of rabbit skin. II. Experimental results.
J Biomech 7 (2), 171-182.

Lanne T, Sonesson B, Bergqvist D, *et al.* (1992).
Diameter and compliance in the male human abdominal aorta: influence of age and aortic aneurysm.
Eur J Vasc Surg 6, 178-184.

Laraia V.J., Heuer A.H. (1989).
Novel composite microstructure and mechanical behavior of mollusk shell.
J Am Ceram Soc 72, 2177-2179.

Lawn B. (1993).
Fracture of Brittle Solids (2nd Ed).
Cambridge Univ. Press.

Le Bourhis E. (2008).
Glass: Mechanics and Technology.
John Wiley & Sons.

Lee, C.H, Singla, A. Lee, Y. (2001)
Biomedical applications of collagen.
International Journal of Pharmaceutics, 221, 1-22.

Leikina E, Mertts MV, Kuznetsova N, *et al.* (2002).
Type I collagen is thermally unstable at body temperature.
Proc. Natl. Acad. Sci. USA 99: 1314–18.

Letai A., Fuchs E. (1995).
The importance of intramolecular ion pairing in intermediate filaments.
Proc Natl Acad Sci USA 92, 92-96.

Libonati F., Vergani L. (2016).
Understanding the structure-property relationship in cortical bone to design a biomimetic composite.
Composite Structures 139, 188-198.

Linde A., Lussi A. Crenshaw M.A. (1989).
Mineral induction by immobilized polyanionic proteins.
Calcif. Tissue Int. 44, 286–95.

Lindley P. B. (1972).
Energy for crack growth in model rubber components.
J. Strain Analysis 7, (2), 132-140.

Lusis J., Woodhams R.T., Xanthos M. (1973).
The effect of flake aspect ratio on the flexural properties of mica reinforced plastics.
Polym Engng Sci 13, 139-145.

Lyman M. (2019).
The remarkable life of the skin.
Bantom Press.

Makiyama A.M., Vajjhala S., Gibson L.J. (2002).
Analysis of crack growth in a 3D Voronoi strcture: A model for fatigue in low density trabecular bone.
J Biomech Eng 124, 512-520.

Malcom L.L. (1976).
An experimental investigation of the frictional and deformational responses of articular cartilage interfaces to static and dynamic loading.
Ph.D. Thesis, University of California, San Diego.

Maher, B. (2013).
Tissue engineering: How to build a heart.
Nature, 499, 20–22,
https://doi.org/10.1038/499020a.

Martel A., Burghammer M., Davies R.J. *et al.* (2007).
Thermal Behavior of Bombyx mori Silk: Evolution of Crystalline Parameters, Molecular Structure, and Mechanical Properties.
Biomacromolecules 8, 3548-3556.

Martin C, Sun W. (2014).
Modeling of long-term fatigue damage of soft tissue with stress softening and permanent set effects.
Biomechanics and modeling in mechanobiology. 12 (4), 645–655.

Matsumoto T, Hayashi K (1994).
Mechanical and dimensional adaptation of rat aorta to hypertension.
J. Biomech Eng 116, 278-283.

Matsumoto T, Hayashi K (1996).
Stress and strain distribution in hypertensive and normotensive rat aorta considering residual strain.
J Biomech Eng 118, 62-73.

McDonald D A (1974).
Blood Flow in Arteries.
Edward Arnold, London.

McElhaney J.H., Byars E.F. (1966).
Dynamic response of bone and muscle tissue.
J Appl Physiol 21, 4.

McEwen W. (1881).
Observation concerning transplantation of bone. Illustrated by a case of inter-human osseous transplantation, where by over two-thirds of the shaft of the humerus was restored.
Proc Roy Soc London 32, 232-247.

Meinhardt H. (2003).
The algorithmic beauty of sea shells.
Springer.

Menig R. Meyers M.H., Meyeers M.A. *et al.* (2001).
Quasi-static and dynamic mechanical response of Strombus gigas (conch) shells.
Mater Sci Eng A 297, 203-211.

Menig R., Meyers M.H., Meyeres M.A. *et al.* (2000).
Quasi-static and dynamic mechanical response of Haliotis rufescens (abalone) shells.
Acta Mater 48, 2383-2398.

Mendoza-Novelo B, Avila E E, Cauich-Rodríguez J V, Guinea G V, *et al.* (2011).
Decellularization of pericardial tissue and its impact on tensile viscoelasticity and glycosaminoglycan content.
Acta Biomater, 7 (3), 1241-1248.

Mendoza-Novelo B, Cauich-Rodriguez J V. (2009).
The effect of surfactants, crosslinking agents and L-cysteine on the stabilization and mechanical properties of bovine pericardium.
J Appl Biomater Biomech. 7 (2), 123–131.

Mendoza-Novelo B, Alvarado-Castro D I, *et al.* (2013).
Stability and mechanical evaluation of bovine pericardium cross-linked with polyurethane prepolymer in aqueous medium.
Materials Science and Engineering C33 2392-2398; DOI:10.1016/j.msec.2013.02.001.

Meredith R. (1945).
A comparison of the tensile elasticity of some textile fibers.
J Textile Inst 36(7), 147-164.

Meyer K.H., Ferri C. (1935).
Sur l'élasticité du caoutchouc.
Helv Chim Acta 18, 570-589.

Meyers MA *et al.* (2012).
Battle in the Amazon: Arapaima versus Piranha.
Adv. Eng. Mater. 14, 279-288.

Meyers M.A., Chen P.Y., Lin A.Y.M., *et al.* (2008).
Biological materials: Structure and mechanical properties.
Progress Mater Sci 53, 1-206.

Meyers M.A., Lin A.Y.M., Chen P. Y., *et al.* (2008).
Mechanical strength of abalone nacre: Role of the soft organic layer.
JMBBMer 1, 76-85.

Mirnajafi, Raymer J, Scott M J, Sacks M S. (2005).
The effects of collagen fiber orientation on the flexural properties of pericardial heterograft biomaterials.
Biomaterials, 26(7), 795–804.

Moir B.G. (1990).
Comparative studies of "fresh" and "aged" Tridacna gigas shell*: Preliminary investigations of a reported technique for pretreatment of tool material.*
J Archaeol Sci 17, 329-345.

Morton W.E., Hearle J.W.S. (1993).
Physical Properties of Testile Fibres, 3rd edition.
The Textile Institute, Manchester.

Motriuk-Smith D., Smith A., Hayashi C.Y., *et al.* (2005).
Analysis of the conserved N-terminal domains in major ampullate spider silk proteins.
Biomacromolecules 6, 3152-3159.

Mow V.C., Woo S.L-Y., Ratcliffe A. (Eds.) (1990).
Biomechanics of Diarthrodial Joints, Vol. 1 & 2.
Springer, New York.

Mow V.C., Schoonbeck J. (1984).
Contribution of Donnan osmotic pressure towards the biphasic modulus of articular cartilage.
Trans Orthop Res Soc, 262.

Mow, V.C., Kuei, S., Lai, W.M., *et al.* (1980).
Biphasic creep and stress relaxation of articular cartilage in compression: Theory and experiments.
Journal of Biomechanical Engineering 102, 73-84.

Munch E., Launey M.E., Alsem D.H., *et al.* (2008).
Tough, bio-inspired hybrid materials.
Science 322, 1516-1520.

Murcia S *et al.* (2015).
Temperature effects on the fracture resistance of scales from Cyprinus carpio.
Acta Biomaterialia 14, 154-163.

Murcia S *et al.* (2017).
The natural armors of fish: A comparison of the lamination pattern and structure scales.
JMBBM 73 17-27.

Murcia S *et al.* (2018).
Contributions of the layer topology and mineral content to the elastic modulus and strength of fish scales.
JMBBM 78 ,56-64.

Murphy A.R., Kaplan D. (2009).
Biomedical applications of chemically-modified silk fibroin.
J Mater Chem 19, 6443-6450.

Myers E.R., Wilson S.R. (1997).
Biomechanics of osteoporosis and vertebral fracture.
Spine 22, 25S-31S.

Nagaraja S., Ball M.D., Guldberg R.E. (2007).
Time-dependent damage accumulation under stress relaxation testing of bovine trabecular bone.
Int J Fatigue 29, 1034-1038.

Nakazawa Y., Asakura T. (2002)
High-Resolution 13C CP/MAS NMR Study on Structure and Structural Transition of Antheraea pernyi Silk Fibroin Containing Poly(l-alanine) and Gly-Rich Regions.
Macromolecules 35, 2393-2400.

Nalla R.K., Kruzic J.J., Kinney J.H., *et al.* (2005).
Mechanistic aspects of fracture and R-curve behavior in human cortical bone.
Biomaterials 26, 217-231.

Nawroth J *et al.* (2012).
A tissue-engineered jellyfish with biomimetic propulsión.
Nature Biotechnology 30 (8), 792-797.

New Scientist (2020).
January 15, page 7, January 29, page 9.

Ng B.H., Chou S.M., Lim B.H., Chong, A. (2004).
Strain rate effect on the failure properties of tendons.
Proc Inst Mech Eng H. 218 (3),203-6.

Nielsen L, Chen P (1968).
Young´s modulus of composites filled with randomly oriented fibers .
J. Mater. 3 352-358.

Nimni, M. E., Cheung, D., Strates, B., *et al.* (1987).
Chemically modified collagen: A natural biomaterial for tissue replacement. Journal of Biomedical Materials Research, 21, (6), 741–771.

Nissenbaum A. (1997).
8000 Years Collagen from Nahal Hemar Cave.
Archaeology and Natural Sciences 5, 5-9.

O'Connell M K, Murthy S, Phan S *et al.* (2008).
The three-dimensional micro- and nanostructure of the aortic medial lamellar unit measured using 3d confocal and electron microscopy imaging.
Matrix Biol. 27, 171–181.

Oesser, S., Proksch, E., Schunck, M. (2008).
Prophylactic treatment with a special Collagen Hydrolysate deceases cartilage tissue degeneration in the knee joints.
Osteoarthritis and Cartilage 16,72.

Oliver W.C., Pharr G.M. (1992).
An improved technique for determining hardness and elastic modulus using load and displacement sensing indentation experiments.
J Mater Res 7(6), 1564-1583.

Oloyede A., Flachsmann R., Broom N.D. (1992).
The dramatic influence of loading velocity on the compressive response of articular cartilage.
Connective Tissue Res 27(4), 211-224.

Onions W.J. (1962).
Wool, an Introduction to its Properties, Varieties, Uses and Production.
London: Ernest Benn Ltd.

Orgel JPRO, Irving TC, Miller A, *et al.* (2006).
Microfibrillar structure of type I collagen in situ.
Proc. Natl. Acad. Sci. USA 103: 9001–5.

Osuna-Mascaró A., Cruz-Bustos T., Behamada S., *et al.* (2014).
The Shell organic matrix of the crossed lamellar Queen conch shell (Strombus gigas).
Comp Biochem Physiol B 168, 76-85.

Oyen M.L., Cook R.F. (2009).
A practical guide for analysis of nanoindentation data.
JMBBM 2, 396-407.

Padawer G.E., Beecher N. (1970).
On the strength and stiffness of planar reinforced plastic resins.
Polym Engng Sci 10, 185-192.

Parenteau-Bareil, R., Gauvin, R. and Berthod F. (2010).
Collagen-based biomaterials for tissue engineering applications.
Materials 3, 1863-1887;
https://doi.org/10.3390/ma3031863.

Pashneh-Tala *et al.* (2016).
The Tissue-Engineered Vascular Graft–Past, Present, and Future.
Tissue Engineering B, 22 (1), 68-100.

Pastor J.Y., Elices M. (2001).
Comportamiento mecánico de la madreperla.
Anales de Mecánica de la Fractura 18, 237-243.

Pauling L., Corey R.B. (1953).
Two rippled-sheet configurations of polypeptide chains, and a note about the pleasted sheets.
Proc Natl Acad Sci USA 39, 253.

Pauling L., Corey R.B., Branson H.R. (1951).
The structure of proteins: Two hydrogen bonded helical configurations of the polypeptide chain.
Proc Natl Acad Sci USA 37, 205-211.

Pérez-Rigueiro J., Elices M., Llorca J., *et al.* (2002).
Effect of degumming on the tensile properties of silkworm (Bombyx mori) silk fiber.
Journal of Applied Polymer Science 84(7), 1431-1437.

Perez-Riguero, J., Elices, M., Llorca, J., *et al.* (2001).
Tensile properties of Argiope trifasciata drag line silk obtained from the spider's web.
J Appl Polym Sci 82, 2245-2251.

Pérez-Rigueiro J., Elices M., Llorca J., *et al.* (2001).
Tensile properties of silkworm silk obtained by forced silking.
Journal of Applied Polymer Science 82(5), 1928-1935.

Pérez-Rigueiro J., Viney C., LLorca J., *et al* (2000).
Mechanical properties of silkworm silk in liquid media.
Polymer 41(23), 8433-8439.

Perez-Riguero, J., Viney, C., Llorca, J, *et al.* (1998).
Silkworm silk as an engineering material.
J Appl Polym Sci 70, 2439-2447.

Pérez-Rigueiro J., Elices M., Plaza G.R., *et al.* (2005).
The effect of spinning forces on spider silk properties.
J Exp Biol 208(14), 2633-2639.

Peters L., Woods H.J. (1956).
Protein fibres
In Mechanical Properties of Textile Fibres (Meredith R., Ed.), North-Holland, Amsterdam.

Pidaparti R.M.V., Chandran A., Takano Y., *et al.* (1996).
Bone mineral lies mainly outside collagen fibrils: Predictions of a composite model for osteonal bone.
J Biomech 29, 909-916.

Planas, J., Elices, M, Guinea G.V. (1994).
Cohesive cracks as a solution of a class of nonlocal models.
In Fracture and Damage in Quasibrittle Structures (Bazant *et al.* Eds.) E & FN Spon, London, 131-144.

Planas, J., Guinea, G.V., Elices M. (1994).
Stiffness associate with quasi-concentrated loads.
Materials and Structures 27, 311-318.

Planas J., Guinea G.V., Elices M. (2007).
Constitutive model for fiber-reinforced materials with deformable matrices.
Phys Rev E 76, 041903.1-9.

Plaza G.R., Guinea G.V., Pérez-Rigueiro J., *et al.* (2006).
Thermo-hygro-mechanical behavior of spider dragline silk: Glassy and rubbery states.
Journal of Polymer Science Part B – Polymer Physics 44(6), 994-999.

Plaza, G. R., Pérez-Rigueiro J., Riekel C., *et al.* (2012).
Relationship between microstructure and mechanical properties in spider silk fibers: two regimes in the microstructural changes.
Soft Matter 8, 6015–6026.

Polzer S, Bursa J, Gasser T C, Staffa R, *et al.* (2013).
Numerical implementation to predict residual strains from the homogeneous stress hypothesis with application to abdominal aortic aneurysms.
Annals of Biomedical Engineering, 41, 1516–1527.

Purslow P P. (1983).
Measurement of the fracture toughness of extensible connective tissues.
J Materials Sci 18, 3591-3598.

Qu Q *et al.* (2015).
New genomic and fossil data illuminate the origin of enamel.
Nature, https://doi.org/10.1038/nature15259.

Quaglini V., La Russa V., Corneo S. (2009).
Nonlinear stress relaxation of trabecular bone.
Mechanics Research Communications 36, 275-283.

Rachev A, Stergiopulos N, Meister J J. (1998).
A model for geometric and mechanical adaptation of arteries to sustained hypertension.
ASME J Biomech Eng 120, 9-17.

Radin E.L., Paul I.L., Lowy M. (1970).
A comparison of the dynamic force transmitting properties of subchondral bone and articular cartilage.
J Bone Joint Surg 52A, 444-456.

Raghavan M L, Vorp D A (2000).
Toward a biomechanical tool to evaluate rupture potential of abdominal aortic aneurysm: Identification of a finite strain constitutive model and evaluation of its applicability.
Journal of Biomechanics 33, 475–482.

Ramshaw JAM, Shah NK, Brodsky B. (1998).
Gly-X-Y tripeptide frequencies in collagen: A context for host-guest triple-helical peptides.
J. Struct. Biol. 122:86–91.

Reiterer A., Sinn G., Stanzl-Tschegg S.E. (2002).
Fracture characteristics of different wood species under mode I loading perpendicular to the grain.
Mat Sci Eng A 332, 29-36.

Rémi E, Khelil N, Di Centa I, *et al.* (2011).
Pericardial Processing: Challenges, Outcomes and Future Prospects; Chapter 22. Biomaterials Science and Engineering. R. Pignatello, ed. DOI: 10.5772/24949.

Rico F., Roca-Cusachs P., Gavara N., *et al.* (2005).
Probing mechanical properties of living cells by atomic force microscopy with blunted pyramidal cantilever tips.
Physical Review E 72, 021914.

Rimnac C.M., Petko A.A., Santner T.J., *et al.* (1993).
The effect of temperature stress and microstructrure on the creep of compact bovine bone.
J Biomech 26, 219-228.

Ritchie R.O., Kinney J.H., Kruzic J.J., *et al.* (2005).
A fracture mechanics and mechanistic approach to the failure of cortical bone.
Fatigue Fract Engng Mater Struct 28, 345-371.

Riveros F, Chandra S, Finol E A, *et al.* (2013).
A pull-back algorithm to determine the unloaded vascular geometry in anisotropic hyperelastic AAA passive mechanics.
Annals of Biomedical Engineering, 41, 694–708.

Rivlin R.S., Thomas A.D. (1953).
Rupture of rubber: I. Characteristic energy for tearing.
J Polym Sci 10, 291-318.

Roach M R, Burton A C (1957).
The reason for the shape of the distensibility curves of arteries.
Can J Biochem Physiol 35 (8), 681-690.

Rojo F J, García Páez J M, Guinea G V, *et al.* (2010).
Optimal selection of biological tissue using the energy dissipated in the first loading cycle.
J. of biomedical materials. Part B, Applied biomaterials, 95B, 2, 414-420.

Rogers G.E. (2000).
Genetic engineering for novel fibers.
J Text Inst 91, 24-31.

Romero de Ávila, M.D., Escudero, R., Ordóñez, J.A., *et al.* (2014).
Weibull analysis characterizes the breaking properties of dry-cured ham slices.
Meat Science 97, 451-458.

Rotter N., Tobias G., Lebl M., *et al.* (2002).
Age-related changes in the composition and mechanical properties of human nasal cartilage.
Archives of Biochemistry and Biophysics 403, 132-140.

Roy C S. (1881).
The elastic properties of the arterial wall.
J. Physiol. 3, 125–159.

Rudykh S, Ortiz C, Boyce M C. (2015).
Flexibility and protection by design: imbricated hybrid microstructures of bio-inspired armor.
Soft Matter 11, 2547-2554.

Sacks M S, Smith D B, Hiester ED. (1997).
A small angle light scattering device for planar connective tissue microstructural analysis.
Annals of Biomedical Engineering, 25, 678-589.

Sacks M S, Chuong C J. (1998).
Orthotropic Mechanical Properties of Chemically Treated Bovine Pericardium.
Annals of Biomedical Engineering, 26, 892–902.

Sacks M S. (1999).
A method for planar biaxial mechanical testing that includes in-plane shear.
J Biomech Eng 121, 551-555.

Sacks M S, Sun W. (2003).
Multiaxial mechanical behavior of biological materials.
Annu Rev Biomed Eng 5, 251-284.

Sakar-Deliormanli A., Güden M. (2006).
Microhardness and fracture toughness of dental materials by indentation method.
Journal of Biomedical Materials Research, Part B Applied Biomaterials 76B(2), 257-264.

Sánchez-Gálvez V. (2012).
Materiales para la Defensa.
ISDEFE-UPM, Cuadernos 10.

Sarikaya M., Liu J., Aksay I.A. (1995).
Nacre: Properties, Crystallography, morphology, and formation.
In Biomimetics: Design and Processing of Materials (Sarikaya, Aksay Eds.) AIP Press.

Sasaki, N. and Odajima S. (1996).
Stress-strain curve and Young's modulus of a collagen molecule as determined by the x-ray diffraction technique.
J. Biomech. 29:655–658.

Sasser P. F., Shofner F. M., Chu Y. T., *et al.* (1991).
Interpretations of single fiber, bundle, and yarn tenacity data.
Text Res J 61, 681-690.

Shadwick R E, Russell A P, Lauff R F. (1992).
The structure and mechanical design of rhinocerus dermal armour.
Phil Trans R Soc Lond B. 337, 419-428.

Schaffner G., Guo, X.E., Silva M.J., *et al.* (2000).
Modelling fatigue damage accumulation in two-dimensional Voronoi honeycombs.
Int J Mech Sci 42, 645-656.

Schechtman H., Bader D.L. (2002).
Fatigue damage of human tendons.
J. Biomech. 35, 347-353.

Schechtman, H., Bader, D. L. (1997).
In vitro fatigue damage of human tendons.
J. Biomech. 30 (8), 829-835.

Schwartz P. (Ed.). (2008).
Structure and Mechanics of Textile Fiber Assemblies.
Woodhead Pub. Cambridge.

Seki Y, Schneider M.S., Meyers M.A. (2005).
Structure and mechanical behavior of a toucan beak.
Acta Mater 53, 5281-5296.

Sellaro T L, Hildebrand D, Lu Q J., *et al.* (2007).
Effects of collagen fiber orientation on the response of biologically derived soft tissue biomaterials to cyclic loading.
J. of Biomedical Materials Part A, 80A 194–205.

Shen, Z.L., Dodge, M.R., Kahn, H., *et al.* (2008).
Stress-Strain Experiments on Individual Collagen Fibrils.
Biophysical Journal 95, 3956-3963.

Sherman V R *et al.* (2018).
A comparative study of piscine defense: The scales of Arapaina gigas, Latimera chalumnae and Atractosteus spatula.
JMBBM 73, 1-16.

Shoemaker P. A. *et al.* (1986).
A constitutive model for two-dimensional soft tissues and its application to experimental data.
J Biomech 19, 695-702.

Shoulders M.D. and Raines R.T. (2009).
Collagen Structure and Stability.
Annu. Rev. Biochem.78, 929–58.

Silva M J, Hayes W C, Gibson LJ. (1995).
The effects of non-periodic microstructure on the elastic properties of two dimensional solids.
Int J Mech Sci 37, 1161-1177.

Simpson W.S., Crawshaw G.H. (2002).
Wool: Science and Technology.
Woodhead Pub. Ltd.

Sire J Y *et al.* (2009).
Origin and evolution of the integumentary skeleton in non-tetrapod vertebrates.
J. Anat. 214, 409-440.

Skalak R., Chien S. (1987).
Handbook of Bioengineering.
McGraw Hill, New York.

Smart J., Williams J.C. (1972).
A comparison of single-integral non-linear viscoelastic theories.
J. Mech. Phys. Solids, 20, 313-324.

Smith, C.A. and Wood, E.J. (1991).
Biological Molecules.
Chapman & Hall, London.

Smith I., Landis E., Gong M. (2003).
Fracture and Fatigue in Wood.
John Wiley & Sons.

Soderhall, C., I. Marenholz, T. Kerscher *et al.* (2007).
Variants in a novel epidermal collagen gene (COL29A1) are associated with atopic dermatitis.
PLoS Biology 5,1952-1961.

Sokolis D P. (2007).
Passive mechanical properties and structure of the aorta: segmental analysis.
Acta Physiol. 190, 277–289.

Song F. Soh A.K., Bai Y.L. (2003).
Structural and mechanical properties of the organic matrix layers of nacre.
Biomaterials 24, 3623-3631.

Song J., Chen C., Zhu S., *et al.* (2018).
Processing bulk natural wood into a high-performance structural material.
Nature 554, 224-228.

Sparrow J.T. (1973).
The Fracture of Cotton.
Ph.D. Thesis, Univ. of Manchester.

Spatz, H.C, O'Leary, E.J., Vincent J.F.V. (1996).
Young's moduli and shear moduli in cortical bone.
Proc R Soc Lond B263, 287-294.

Speakman J.B. (1927).
Intracellular structure of the wool fibre.
J Text Inst 18, T431-T453.

Stanzl-Tschegg S.E. (2006).
Microstructure and fracture mechanical response of wood.
Int J Fracture 139, 495-508.

Stern R. (2004).
Hyaluronan catabolism: A new metabolic pathway.
Eur J Cell Biol 83(7), 317-325.

Sun W, Sacks M, Fulchiero G *et al.*, (2004).
Response of heterograft heart valve biomaterials to moderate cyclic loading.
Journal of Biomedical Materials Part A 69A, 658-669.

Su X.W., Zhang D.M., Heuer A.H. (2004).
Tissue regeneration in the shell of the giant Queen conch, Strombus gigas.
Chem Mater 16, 581-593.

Tang B., Ngan A.H.W. (2011).
Nanoindentation using an atomic force microscope.
Philosophical Magazine 91, 1329-1338.

Tanaka T, Fung Y C. (1974).
Elastic and inelastic properties of the canine aorta and their variation along the aortic tree.
J Biomechanics 7, 357-370.

Termonia Y. (2000).
Molecular modeling of the stress/strain behaviour of spider dragline.
In Structural Biological Materials (M. Elices Ed.) Pergamon Press.

Thibodeaux D.P., Hebert J.J., Abd El-Gawad N.S., *et al.* (1998).
Quality measurements - relating bundle strength to mantis single fiber strength measurements.
Cotton Sci 2, 62-67.

Thomson d'Arcy (1961).
On Growth and Form.
Cambridge Univ. Press.

Timoshenko S. P., Goodier J. N. (1982).
Theory of Elasticity.
McGraw Hill.

Timoshenko S, Woinowsky-Krieger S. (1959).
Theory of plates and shells.
McGrawHill.

Tong P, Fung Y. (1976).
The stress-strain relationship for the skin.
J Biomechanics 9, 649-657.

Tonge T K, Atlan L S, Voo L M, *et al.* (2013a).
Full-field bulge test for planar anisotropic tissues: Part I - Experimental methods applied to human skin tissue.
Acta Biomaterialia 9, 5913–5925.

Tonge T K, Voo L M, Nguyen T D. (2013b).
Full-field bulge test for planar anisotropic tissues: Part II - A thin Shell method for determining material parameters and comparison of two distributed fiber modeling approaches.
Acta Biomaterialia 9, 5926–5942.

Tobaruela A. (2016).
Caracterización mecánica de materiales de base colágeno para aplicaciones cardiovasculares.
Ph.D. Thesis, Universidad Politécnica de Madrid.

Tobaruela A, Elices M, Guinea G V. (2016).
Tear and decohesion of bovine pericardial tissue.
JMBBM, 1–9.

Tomford W.W., Mankin H.J. (1999).
Bone banking update on methods and materials.
Orthopedic Clinics of North America 30(4), 565-570.

Treloar L.R.G. (1975).
The Physics of Rubber Elasticity (3rd Ed.).
Clarendon Press.

Treloar L.R.G. (1944).
Stress-strain data for vulcanised rubber under various types of deformation.
Trans Faraday Soc 40, 59-70.

Trowbridge E. A., Crofts C. E., Lawford P., *et al.* (1989).
The mechanical properties of natural and chemically modified bovine pericardium for use with heart valve substitutes.
In Material properties and stress analysis in biomechanics, Chapter 2. Manchester University Press.

Trowbridge E A, Crofts C.E (1989).
The tearing strength of glutaraldehyde fixed bovine pericardium.
Biomaterials, artificial cells and artificial organs. 17(3), 315-328.

UNE-EN 23878 (2001).
Metal duro. Ensayo de dureza Vickers.
Asociación Española de Normalización.

UNE-EN ISO 3738-1 (2011).
Metales duros. Ensayo de dureza Rockwell (escala A). Parte 1: Método de ensayo.
Asociación Española de Normalización.

UNE-EN ISO 3738-2 (2007).
Metal duro. Ensayo de dureza Rockwell (Escala A). Parte 2: Preparación y calibración de bloques patrón.
Asociación Española de Normalización.

Vajjhala S., Kraynik A.M., Gibson L.J. (2000).
A celular solid model for modulus reduction due to resorption of trabeculae in bone.
J Biomech Eng 122, 511-515.

Van der Burg M, Shulmeinster V, van der Geissen E, *et al.* (1997).
On the linear elastic properties of regular and random open-cell foam models.
J Cellular Plastics 33, 31-54.

Van der Geest J P, Sacks M S, Vorp D A (2006).
The effects of aneurysm on the biaxial mechanical behavior of human abdominal aorta.
J Biomechanics, 39, 1324–1334.

Van der Rijt, J. A. J., K. O. van der Werf, *et al.* (2006).
Micromechanical testing of individual collagen fibrils.
Macromol. Biosci. 6, 697–702.

Vawter D, Fung Y C, West J B. (1979).
Constitutive equation of lung tissue elasticity.
J Biomech Eng Trans ASME 101, 38-45.

Veronda D R, Westmann R A. (1970).
Mechanical characterization of skin-finite deformations.
J Biomechanics 3, 111-124.

Viidik A. (1990).
Structure and function of normal and healing *tendons and ligaments.*
In Biomechanics of Diarthrodial Joints, Mow, Ratcliffe and Woo (eds.) 3-38. Springer. New York.

Viidik A. (1968).
A rheological model for unaclcified parallel-fibered collagenous tissue.
J Biomechanics 1, 3-11.

Vincent J. (1990).
Structural Biomaterials.
Princeton Univ Press.

Vogel H G. (1972).
Influence of age, treatment with corticosteroids and strain rate on mechanical properties of rat skin.
Biochim Biophys Acta 286, 79-83.

Vollrath F., Porter D. (2006).
Spider silk as archetypal protein elastomer.
Soft Matter 2, 377-385.

Wagner H.D., Weiner S. (1992).
On the relationship between the microstructure of bone and its mechanical stiffness.
J Biomech 25, 1311-1320.

Wang R.Z., Suo Z, Evans A.G., *et al.* (2001).
Deformation mechanisms in nacre.
J Mater Res 16(9), 2485-2493.

Ward I. M., Sweeney J. (2004).
An Introduction to the Mechanical Properties of Solid Polymers (2nd Ed.).
Wiley.

Ward, A.G., Courts, A. (1977).
The Science and Technology of Gelatin.
Academic Press.

Wan K T, Mai Y M. (1995).
Fracture mechanics of a new blister test with stable crack growth.
Acta Metall Mater 143, 4109-4115.

Warner, S.B. (1995).
Fiber Science.
Prentice Hall.

Warren W.E., Kraynik A.M. (1997).
Linear elastic behavior of a low-density Kelvin foam with open cells.
J Appl Mech 64, 787-794.

Weaire D., Phelan R. (1994).
A counter-example to Kelvin's conjecture on minimal surfaces.
Phil Mag Lett 69, 107-110.

Weidmann G. (1990).
Structural Materials. Chap. 7.
Open Univ. Press.

Weibull, W. (1951).
A statistical distribution function of wide applicability.
J. Appl. Mech.-T. ASME, 18, 293–297.

Weibull W. (1939).
A statistical theory of the strength of materials.
Ingeniors Ventenskaps Akademien Handlingar, 151.

Weis-Fogh T. (1961).
Thermodynamic properties of resilin, a rubber-like protein.
J Mol Biol 3, 520-531.

Weiss JA, Gardiner JC. (2001).
Computational modeling of ligament mechanics.
Crit Rev Biomed Eng 29, 303-71.

Wenger,M.P. E., Bozec,L., Horton, M.A., *et al.* (2007).
Mechanical Properties of Collagen Fibrils.
Biophysical Journal 93, 1255–1263.

Williams J G. (1997).
Energy release rates for the peeling of flexible membranes and the analysis of blister tests.
Int J of Fracture 87, 265-288.

Williams J.L., Lewis J.L. (1982).
Propeerties and an anisotropic model of cancellous bone from the proximal tibial epiphysis.
J Biomech Eng 104, 50-56.

Williamson A.K., Chen A.C., Sah R.L. (2003).
Tensile mechanical propertis of bovine articular cartilage: Variations with growth and relationships to collagen network components.
J Orthop Res 21, 872-880.

Wineman A.S., Rajagopal K.R. (2000).
Mechanical Response of Polymers.
Cambridge Univ. Press.

Wollnsky H, Glagov S. (1967).
A lamellar unit of aortic medial structure and function in mammals.
Circ Res 20, 99-111.

Woo S.L-Y., Akeson W.H., Jemmott G.F. (1976).
Measurements of nonhomogeneous, directional mechanical properties of articular cartilage in tension.
J Biomech 9(12), 785-791.

Work R.W., Young C.T.
The amino acid compositions of major and minor ampullate silks of certain orb-web-building spiders
J Arachnol 15 (1987), 65-80.

Wu K S, Osdol W, Dauskart R. (2006).
Mechanical properties of human stratum corneum: Effects of temperature, hydration, and chemical treatment.
Biomaterials 27, 785-795.

Wu P. D., Giessen E. van der. (1993).
On improved network models for rubber elasticity and their applications to orientation hardening in glassy polymers.
J Mech Phys Solids 41, 427-456.

Xiao Y., Rennerfeldt D.A., Friis E.A., *et al.* (2017).
Evaluation of apparent fracture toughness of articular cartilage and hydrogels.
J Tissue Eng Regen Med 11(1), 121-128.

Yamashita J., Furman B.R., Rawis H.R. *et al.* (2001).
The use of dynamic mechanical analysis to assess the viscoelastic properties of human cortical bone.
J Biomed Mater Res (Appl Biomater) 58, 47-53.

Yang W *et al.* (2013a).
Natural flexible dermal armor.
Adv. Mater. 25, 31-48.

Yang W *et al.* (2013b).
Structure and fracture resistance of alligator gar (Atractosteus spatula) armored fish scales.
Acta Biomaterialia 9, 5876-5889.

Yang W *et al.* (2014).
Protective role of Arapaima gigas fish scales: Structure and mechanical behaviour.
Acta Biomaterialia 10, 3599-3614.

Yao H., Dao M., Imholt T., *et al.* (2010).
Protection mechanisms of the iron-plated armor of a deep-sea hydrothermal vent gastropod.
PNAS 107, 987-992.

Yeoh O. H. (1993).
Some forms of the strain energy function for rubber.
Rubber Chem.Technol. 66, 754-771.

Zener C. (1948).
Elasticity and Anelasticity of Metals.
Univ. Chicago Press.

Zhou Z., Confalonieri F., Medina N., *et al.* (2000).
Fine organization of Bombyx mori fibroin heavy chain gene.
Nucleic Acids Res 28, 2413-2419.

Zhou Z.L., Ngan A.H.W., Tang B., *et al.* (2012).
Reliable measurement of elastic modulus of cells by nanoindentation in an atomic force microscope.
JMBBM 8, 134-142.

Zhu D *et al.* (2012).
Structure and mechanical performance of a modern fish scale.
Adv. Eng. Mater. 4, B185-B194.

Zhu D *et al.* (2013).
Puncture resistance of the scaled skin from stripped bass: Collective mechanisms and inspiration for new flexible armor designs.
JMBBM 24, 30-40.

Zhu J, Wang X, He C, Wang H. (2012).
Mechanical properties, anisotropic swelling behaviours and structures of jellyfish mesogloea.
JMBBM 6, 63-73.

Zhu H.X., Knott J.F., Mills N.J. (1997).
Analysis of the elastic properties of open-cell foams with tetrakaidecahedral cells.
J Mech Phys Solids 45, 319-343.

Zollfrank C., Sieber, H. (2005).
Microstructure evolution and reaction mechanism of biomorphous SiSiC ceramics.
J Am Cerm Soc 88, 51-58.

SUBJECT INDEX

FSC
www.fsc.org
MIX
Papier | Fördert gute Waldnutzung
FSC® C083411

Kolibri 360 GmbH
Ferdinand-Jühlke-Straße 7
99095 Erfurt, Deutschland
produktsicherheit@kolibri360.de